VOLUME ONE HUNDRED AND ONE

ADVANCES IN BOTANICAL RESEARCH

Lipids in Plants and Algae: From Fundamental Science to Industrial Applications

ADVANCES IN BOTANICAL RESEARCH

Series Editor

Jean-Pierre Jacquot
Membre de l'Institut Universitaire de France, Unité Mixte de Recherche INRA, UHP 1136 "Interaction Arbres Microorganismes", Université de Lorraine, Faculté des Sciences, Vandoeuvre, France

VOLUME ONE HUNDRED AND ONE

ADVANCES IN BOTANICAL RESEARCH

Lipids in Plants and Algae: From Fundamental Science to Industrial Applications

Volume Editors

FABRICE RÉBEILLÉ

Laboratoire de Physiologie Cellulaire et Végétale, CNRS, CEA, INRAE, Univ. Grenoble Alpes, IRIG, CEA Grenoble, Grenoble France

ERIC MARÉCHAL

Laboratoire de Physiologie Cellulaire et Végétale, CNRS, CEA, INRAE, Univ. Grenoble Alpes, IRIG, CEA Grenoble, Grenoble France

ELSEVIER

ACADEMIC PRESS

An imprint of Elsevier

Academic Press is an imprint of Elsevier
125 London Wall, London, EC2Y 5AS, United Kingdom
The Boulevard, Langford Lane, Kidlington, Oxford OX5 1GB, United Kingdom
50 Hampshire Street, 5th Floor, Cambridge, MA 02139, United States
525 B Street, Suite 1650, San Diego, CA 92101, United States

First edition 2022

ISBN: 978-0-12-820803-8
ISSN: 0065-2296

Publisher: Zoe Kruze
Acquisitions Editor: Sam Mahfoudh
Developmental Editor: Jhon Michael Peñano
Production Project Manager: James Selvam
Cover Designer: Victoria Pearson

Typeset by STRAIVE, India

Contents

Contributors

Omar Arias-Gaguancela
Department of Biological Sciences, BioDiscovery Institute, University of North Texas, Denton, TX, United States

Philip D. Bates
Institute of Biological Chemistry, Washington State University, Pullman, WA, United States

Anne Berna
Institut de biologie moléculaire des plantes, Plant Isoprenoid Biology team, CNRS, Université de Strasbourg, Strasbourg, France

Aurélien Blary
Institut de biologie moléculaire des plantes, Plant Isoprenoid Biology team, CNRS, Université de Strasbourg, Strasbourg, France

Demé Bruno
Large Scale Structures group, Institut Laue Langevin, Grenoble, France

Valérie Bultel-Poncé
Institut des Biomolécules Max Mousseron, IBMM, Université de Montpellier, CNRS, ENSCM Batiment Balard, Montpellier, France

Kent D. Chapman
Department of Biological Sciences, BioDiscovery Institute, University of North Texas, Denton, TX, United States

Joy Collombat
Laboratoire de Physiologie Végétale, Université de Neuchâtel, Neuchâtel, Switzerland

Sylvain Darnet
Institut de biologie moléculaire des plantes, Plant Isoprenoid Biology team, CNRS, Université de Strasbourg, Strasbourg, France

Valeria Di Dato
Marine Biotechnology Department, Stazione Zoologica Anton Dohrn Napoli, Villa Comunale, Napoli, Italy

Thierry Durand
Institut des Biomolécules Max Mousseron, IBMM, Université de Montpellier, CNRS, ENSCM Batiment Balard, Montpellier, France

Jean-Denis Faure
Institut Jean-Pierre Bourgin, INRAE, AgroParisTech, Université Paris-Saclay, Versailles; AgroParisTech, Paris, France

Jean-Marie Galano
Institut des Biomolécules Max Mousseron, IBMM, Université de Montpellier, CNRS, ENSCM Batiment Balard, Montpellier, France

Thibault Geneste
Institut Jean-Pierre Bourgin, INRAE, AgroParisTech, Université Paris-Saclay, Versailles, France

Angel Gil-Izquierdo
Research Group on Quality, Safety and Bioactivity of Plant Foods, Department of Food Science and Technology, CEBAS (CSIC), 30100 Campus University, Murcia, Spain

Nolwenn Guéguen
Université Grenoble-Alpes, CEA, INRAE, CNRS, IRIG, LPCV, Grenoble, France

Alexandre Guy
Institut des Biomolécules Max Mousseron, IBMM, Université de Montpellier, CNRS, ENSCM Batiment Balard, Montpellier, France

Jouhet Juliette
Univ. Grenoble Alpes, CNRS, CEA, INRAE, IRIG, LPCV, 38000, Grenoble, France

Felix Kessler
Laboratoire de Physiologie Végétale, Université de Neuchâtel, Neuchâtel, Switzerland

Jetty Chung-Yung Lee
School of Biological Sciences, The University of Hong Kong, Pokfulam, Hong Kong SAR

Damien Le Moigne
Université Grenoble-Alpes, CEA, INRAE, CNRS, IRIG, LPCV, Grenoble, France

Sébastien Leterme
Université Grenoble-Alpes, CNRS, INRAE, CEA, IRIG, LPCV, Grenoble, France

Kin Sum Leung
School of Biological Sciences, The University of Hong Kong, Pokfulam, Hong Kong SAR

Sonia Medina
Research Group on Quality, Safety and Bioactivity of Plant Foods, Department of Food Science and Technology, CEBAS (CSIC), 30100 Campus University, Murcia, Spain

Morgane Michaud
Université Grenoble-Alpes, CNRS, INRAE, CEA, IRIG, LPCV, Grenoble, France

Camille Oger
Institut des Biomolécules Max Mousseron, IBMM, Université de Montpellier, CNRS, ENSCM Batiment Balard, Montpellier, France

Hiroyuki Ohta
School of Life Science and Technology, Tokyo Institute of Technology, Yokohama, Japan

Ida Orefice
Marine Biotechnology Department, Stazione Zoologica Anton Dohrn Napoli, Villa Comunale, Napoli, Italy

Giovanna Romano
Marine Biotechnology Department, Stazione Zoologica Anton Dohrn Napoli, Villa Comunale, Napoli, Italy

Juliette Salvaing
Université Grenoble-Alpes, CEA, INRAE, CNRS, IRIG, LPCV, Grenoble, France

Hubert Schaller
Institut de biologie moléculaire des plantes, Plant Isoprenoid Biology team, CNRS, Université de Strasbourg, Strasbourg, France

Venkatasalam Shanmugabalaji
Laboratoire de Physiologie Végétale, Université de Neuchâtel, Neuchâtel, Switzerland

Mie Shimojima
School of Life Science and Technology, Tokyo Institute of Technology, Yokohama, Japan

Bolik Stéphanie
Univ. Grenoble Alpes, CNRS, CEA, INRAE, IRIG, LPCV, 38000; Large Scale Structures group, Institut Laue Langevin, Grenoble, France

Claire Vigor
Institut des Biomolécules Max Mousseron, IBMM, Université de Montpellier, CNRS, ENSCM Batiment Balard, Montpellier, France

Wayne Zita
Laboratoire de Physiologie Végétale, Université de Neuchâtel, Neuchâtel, Switzerland

Preface

The term "lipid" is a general one that encompasses a large group of molecules. The common thing about these molecules is that they are organic compounds with hydrophobic or amphiphilic properties. They derive from two different types of chemical structures that serve as building units and thus form two distinct groups: (i) ketoacyl and (ii) isoprene groups. Lipids deriving from ketoacyl groups comprise fatty acids, glycerolipids, sphingolipids, and polyketides. They are involved in a wide array of metabolic and structural functions; in particular, they are the main constituents of cell membranes and oil bodies. Lipids deriving from isoprene groups comprise sterol and prenol families. They too are involved in a large number of cellular processes, acting in membrane fluidity, protection, electron transport, or in integrated functions such as plant defense.

There is no life without lipids. The most obvious reason is the emergence of membranes as one of the critical steps in the origin of life. The chemistry of the cell is contingent on the laws of thermodynamics. As predicted by the second law of thermodynamics (entropy), spontaneous processes occur in directions that increase the overall disorder. Diffusion is a spontaneous process, and it is obvious that without boundaries the macromolecular components of a cell would spread and disperse into the environment. Life requires highly compacted and ordered processes, which imply diffusion barriers. In a world of water, amphiphilic molecules have emerged as the best candidates to create those membranes that delimit vesicles and compartments in which the reactions of life can occur efficiently. In eukaryotic and most prokaryotic cells, these molecules are essentially diacylglycerolipids (which are composed of a glycerol molecule esterified with two fatty acids), exemplified by phospholipids and glycoglycerolipids. Cell membranes are not just barriers. They also control exchanges with the external environment and participate in the transformation/production of the energy required to maintain cell activity (formation of chemical or electrical gradients; electron transport chains). Indeed, living systems are not closed systems, as defined by thermodynamics. According to the law of entropy, closed systems are in equilibrium. Living systems are open systems that must permanently maintain a state of nonequilibrium by recovering free energy from their environment. In such a nonequilibrium state, high-enthalpy substrates from the external environment are transformed into low-enthalpy products inside the cell, providing energy and metabolic fluxes needed to synthesize the components of life and to maintain cellular activity. To fulfill these functions of exchange and import/conversion of energy, membranes interact with many specific proteins and integrate within their structure a number of other hydrophobic molecules such as sterols, carotenoids, or quinones. Membranes are therefore very complex systems that require many metabolic processes to adjust their composition and plasticity continuously.

Lipids are not only involved in the formation and composition of membranes, but they are also major actors in many other metabolic functions. Fatty acids store carbons and energy, allowing the production of more ATP during their oxidation than sugars with the same number of carbons. This is because carbon atoms in fatty acids are in a more reduced state than those of carbohydrates. From this point of view, oxidation of triacylglycerol (a glycerol molecule esterified with three fatty acids) produces twice more energy than carbohydrates and proteins on a dry weight basis. For this reason, triacylglycerols stored in oil bodies represent the main form of energy storage in animals and in many plants and algae.

Lipids are also the precursors of many compounds that play essential roles in mediating and controlling a large number of cellular processes. A few examples of compounds are as follows. Fatty acids are precursors of oxylipins, such as jasmonic acid in plants, which is involved in the control of ontogenesis and reproductive processes and in the resistance to various pathogens. In the major phylum of photosynthetic stramenopiles, prostaglandins, prostacyclins, thromboxanes, and nonenzymatically produced phytoprostanes are also major oxylipins; their study in terms of diversity of structures, evolution, and role is still in its infancy. Sterols are involved in the synthesis of steroid hormones. For example, brassinosteroids are plant steroids essential for normal growth and development. Cross talks exist between metabolic and signaling roles. Lipid compounds such as phosphatidic acid and diacylglycerol are metabolic hubs controlling various metabolic pathways. In particular, phosphatidic acid interacts with several signaling pathways, including those of the TOR signaling cascade involved in cell stress response, adipogenesis, autophagy, or insulin signaling in animals. Other examples include sphingosine-1-phosphate and phosphatidylinositol phosphates that are, respectively, involved in calcium mobilization and calcium-mediated activation of protein kinase C. These few examples illustrate how lipids encompass an immense molecular diversity; they have structural, metabolic, bioenergetic, and signaling roles, pointing to a sophisticated network of reactions and control mechanisms at the very heart of cellular functions. Today, there is still much to discover and understand.

Why is it important to study lipid metabolism in plants and algae from a societal perspective? There are several reasons for that: health, sustainable production of valuable compounds, and green chemistry. From a nutritional point of view, several fatty acids are considered essential because they are either not or poorly synthesized by human beings. This is the case for linoleic acid (LA, $18{:}2^{\Delta 9,12}$) of the ω-6 series, α-linolenic acid (ALA, $18{:}3^{\Delta 9,12,15}$) of the ω-3 series, and the ω-3 very-long-chain polyunsaturated fatty acids (VLC-PUFAs) such as eicosapentaenoic acid (EPA, $20{:}5^{\Delta 5,8,11,14,17}$) or docosahexaenoic acid (DHA, $22{:}6^{\Delta 4,7,10,13,16,19}$). A deficiency of the first two is rare in humans due to their easy availability in diet (plant oils), but this is not the case for VLC-PUFAs that are not produced at a high level in land plants. The actual diet source for these fatty acids is fish oil, which is not a sustainable source considering the overexploitation of fish stocks and their contamination by toxic substances such as heavy metals. Likewise today, squalene is obtained from shark liver oil. Squalene, the precursor of sterols, is extensively used in the cosmetic industry for its antioxidant properties. Alternative and more sustainable sources have to be found in order to fulfill the high societal demand for these compounds. Marine algae and other marine protists are likely candidates. Today, they are emerging as interesting alternatives for the production of a number of valuable compounds, including VLC-PUFAs (human health), squalene (cosmetic industry), carotenoids, and pigments (food industry). Plants and algae are also promising models for the development of nonpetroleum chemistry. In this context, chemistry based on renewable lipids from plants or algae could replace the controversial use of fossil carbon and indirectly contribute to the reduction of CO_2 emissions. For example, unsaturated oils could serve for manufacturing biosourced polyurethane foams, useful for thermal insulation. Likewise, squalene from algal cultures could serve to produce isoprene units as raw material to synthesize a diverse range of industrial products, including synthetic rubber and a wide variety of elastomers. Fatty acids have chemical structures and fuel properties similar to those of hydrocarbons. Fatty acids with medium chain length and low unsaturation can be converted to biofuel. Third-generation biofuels are those obtained from algae, thus preserving agricultural land for human consumption.

To meet the strong societal demand for these various products, lipid production in plants and algae must be "optimized." The most sought-after phenotypic traits include oleaginous cells and tissues enriched in triacylglycerol (oil) or cells accumulating pigments (carotenoids). It is clear that the "optimization" of these systems requires a complete understanding of (i) their endogenous regulation and (ii) their integration within the metabolism as a whole. This will then allow the rational development of the most appropriate engineering strategies to modify and adapt these organisms for commercial and societal purposes.

This book includes comprehensive and authoritative reviews by leading experts on lipid metabolism in plants and algae. We thank them for their time and effort. The objective of the book is to bring together the recent and rapidly growing experimental information on this challenging and current topic. Some very recent fields in algal and plant lipid science are addressed here, providing key reviews that are missing in the current literature. In the book, the reader will find chapters on the physical structure of membranes, their composition, formation, and renewal. The book also includes chapters that deal with storage lipids, lipid droplets, and plastoglobules; their formations; and their roles in cell physiology and lipid metabolism. The dynamic mechanisms of acyl flux, fatty acid desaturation, and sterol biogenesis are also reviewed in detail. Finally, new information is presented on lipids as signaling molecules in plants and microalgae.

We believe that the book will be useful as a starting point for undergraduate and graduate students, as well as for researchers who wish to pursue specialized studies in this area.

Fabrice Rébeillé
Laboratoire de Physiologie Cellulaire et Végétale,
Université Grenoble Alpes, CEA, CNRS,
INRAE, Grenoble, France

Eric Maréchal
Laboratoire de Physiologie Cellulaire et Végétale,
Université Grenoble Alpes, CEA, CNRS,
INRAE, Grenoble, France

CHAPTER ONE

Biophysical properties of glycerolipids and their impact on membrane architecture and biology

Bolik Stéphanie[a,b], **Demé Bruno**[b], **and Jouhet Juliette**[a,*]
[a]Univ. Grenoble Alpes, CNRS, CEA, INRAE, IRIG, LPCV, 38000, Grenoble, France
[b]Large Scale Structures group, Institut Laue Langevin, Grenoble, France
*Corresponding author: e-mail address: juliette.jouhet@cea.fr

Contents

Abstract

Cellular membranes have a remarkable variety of lipids, and different organelles have different lipid compositions. Changes in lipid composition can alter the surface charge, thickness, and fluidity of a membrane—characteristics that affect, for example, photosynthesis efficiency or vesicular trafficking. Therefore, for proper organelle function, these parameters must be kept within an appropriate range and must be regulated. Due to their low mobility, plant and algae are dependent on their environment and face sudden changes such as light, temperature or osmotic variations, that will affect

Advances in Botanical Research, Volume 101
ISSN 0065-2296
https://doi.org/10.1016/bs.abr.2021.09.001

membrane features. This review focus on the physical and structural properties of glycerolipids and their impact on membrane specificities in response to environmental cues. Here, we present an overview of the methods that are currently used to establish biophysical membrane properties. We then describe the common glycerolipids present in plants and algae, with their characteristic and their distribution within cell membranes. In the light of the organelle lipid composition, we illustrate how membranes are able to sense and adapt their architecture to maintain their homeostasis and their properties in response to environmental stresses. Thanks to the improved techniques recently available to study membranes in their native context, we are now discovering that the regulation of membrane properties by lipids is far more complex and entangled than it was originally thought.

1. Introduction

The capacity of glycerolipids to self-organize in water is the physical basis for spontaneous membrane formation, and renders membranes virtually impermeable to polar solutes. This fundamental principle enabled the first cells to segregate their internal constituents from the external environment and also occurs within eukaryotic cells to form organelles. Whereas in principle a single lipid species is sufficient to generate a barrier to solute transfer, membrane lipids in eukaryotic cells are characterized by a remarkable structural diversity. In addition, the lipid composition and transbilayer arrangement among organelles show striking variations, and there is compelling evidence that the collective properties of the bulk lipids play a profound part in defining organelle identity and function (Holthuis & Menon, 2014).

Of particular interest are the lipid-induced changes in the membrane physical properties such as bilayer thickness, lipid packing density and surface charge that will shape organelle architecture and sort membrane protein (Bigay & Antonny, 2012; Singh & Mittal, 2016). In this review, we focus on glycerolipids that constitute the bulk of cell membrane. We present the biophysical approaches that can help us to determine the physicochemical properties of glycerolipids, the glycerolipid composition of membranes in plant and alga cells and the influence of environmental stresses on physical properties of membranes in vivo with three examples: thermic stress, osmotic stress and the evolution of plastid architecture.

2. Physicochemical properties and analysis methods

Lipid membrane domains are characterized by a distinctive protein and/or lipid composition. These domains confer specific properties to the membrane leading to original structure and function. Membranes are qualified as a mosaic fluid referring to the state of the lipidic phase. Lipid membranes are able to adjust to environmental parameters, thus maintaining the homeostasis of the cell (Martonosi, Kracke, Taylor, Dux, & Peracchia, 1985; Shinitzky, 1984). Moreover, lipids are able to form a bilayer structure, but also non-bilayer structures. These lipid conformations lead to specific physicochemical properties, essential for protein functions. This section describes some methods to explore lipid and fatty acid structures.

2.1 Membrane visualization techniques

The first microscopic observations described the plasma membrane only as a physical barrier between the cytoplasm and the outside of the cell. Scientists improved their knowledge about membranes and proposed some hypothesis on their structure and the presence of proteins around. The history of membrane discoveries is well described in the Lombard review (Lombard, 2014). A first complete model of bilayer was described in 1957 by Fernandez-Moran and Finean after their study of myelin sheath using electronic microscopy and X-rays diffraction (Fernandez-Moran & Finean, 1957). It was finally in 1972 that Singer and Nicolson described the fluid mosaic model as a lipid bilayer containing proteins more or less embedded in the membrane (Singer & Nicolson, 1972). This membrane model becomes a reference model for the next studies. Compared to the past, the resolution of membrane visualization techniques has improved significantly. In this section, we present a non-exhaustive list of biophysical techniques that are widely used to establish biophysical membrane properties.

2.1.1 Classical microscopy technique

As mentioned above, the first discoveries and studies about membranes have been made by electron microscopy. The resolution of microscopes varies a lot, depending on the technique used. For example, the resolution of the optical microscopy is 0.2 μm, while that for transmission electron microscopy is around 0.04 nm. In optical microscopy, the light is directed to a flatten object with a resolution that could not be achieved by eyes. By addition of a laser that emitted a specific wavelength, fluorescence microscopy permits to visualize a fluorescent dye in the cell. This technique can resolve the location of proteins or lipids inside a cell (Calvez, Jouhet, Vie, Durmort, & Zapun, 2019). Other techniques commonly used in biology are the confocal microscopy and the transmission electron microscopy (TEM). The first one can analyze cross sections without any impact of the light emitted outside the focal plan. The object appears in three dimensions. Confocal microscopy is usually coupled with fluorescence microscopy. The TEM technique consists of a very thin object placed under an electron beam. Then, the electronic picture is converted into an image. A 3D reconstruction of an object is possible after analysis by a focused ion beam-scanning electron microscopy (FIB-SEM) where the ionic beam cuts the object and the electron beam visualizes the cross section. Thanks to all these microscopy techniques, structures, contact sites of organelles, location of proteins can be visualized.

An example of application using microscopy techniques are the studies on chloroplast. The structure of the chloroplast was first discovered by Hugo von Mohl with optical microscopy (Staehelin, 2003). The electron microscopy techniques have been the principal tool for understanding thylakoid architecture and its functional organization. It is clearly possible to measure the size of thylakoids, count the number of grana, and measure the thickness of stroma thylakoids. Also, the two membranes of the chloroplast can be distinguished. Using 3D reconstructions, the structures inside the organelle and contacts between organelles can be understood. More recently, thanks to the laser confocal microscopy, Johnson and his collaborators have investigated the changes in the structural reorganization of photosynthetic membranes during the setup of photoprotective functions (Johnson, Brain, & Ruban, 2011).

2.1.2 Cryo-electron microscopy technique

The development of new sample preparation techniques such as cryo-electron microscopy (Cryo-EM) allows the possibility to establish structures at similar resolution than X-ray crystallography. Unlike preparation method for electron microscopy analysis requiring sample fixation in resin bed, samples analyzed by cryo-EM need to be cooled down into vitreous state before visualization (Lepault, 1985). The resolution of cryo-EM is better than 4 Å, higher than other electron microscopy techniques. Cryo-EM is widely used to study protein structure at atomic level into a bilayer membrane. It is possible to determine protein structure, liposome lamellarity, size, shape and ultrastructure (Mio & Sato, 2018). Demurtas and his collaborators used cryo-EM to visualize and understand the structure of the cubic lipid phase in the context of drug delivery system applications (Demurtas et al., 2015).

2.1.3 Freeze-fracture electron microscopy

Freeze-fracture electron microscopy is a method for studying the structure of biological samples at low and medium resolution (20 Å) (Gulik-Krzywicki, 1997). This technique is useful to investigate lipid structures, or protein insertion in the membrane. For example, it is possible to distinguish lamellar phase from the ripple phase, but only with ordered fatty acids. Indeed, with disordered acyl chains, the lipid structures are not well preserved after freezing. The Meyer and Richter's review describes some structures with or without proteins (Meyer & Richter, 2001). Cubic membrane architecture was also studied using this method (Delacroix, Gulik-Krzywicki, Mariani, & Luzzati, 1993).

Briefly, the preparation consists in freezing quickly the sample at very low temperature with liquid nitrogen (−170 °C) or liquid propane (−200 °C). Then, the ice block is fractured by a shock, always at low temperature and under pressure. The fracture happens along the line with the minimum resistance. For example, for a cell or a vesicle, two types of fractures exist: (1) the fracture can happen in the middle of the cell or vesicle, showing the inside of the sample, or (2) the fracture separates the bilayer leaflets in two monolayers, showing the center of the membrane. Proteins cannot be cut and are always intact (Shechter, 1984).

2.1.4 Atomic force microscopy (AFM)

The atomic force microscopy technique (AFM) is a very-high-resolution type of scanning probe microscopy with a resolution of a nanometer. An AFM generates images by scanning a small cantilever over the surface of a sample. The sharp tip on the end of the cantilever contacts the surface, bends the cantilever and transfers the signal by changing the amount of a laser light reflected by the cantilever into a photodiode. The AFM has three major abilities: force measurement, topographic imaging, and manipulation. It is capable of displaying the surface topology of protein complexes of hydrated membranes (Johnson et al., 2011; Kirchhoff, Lenhert, Buchel, Chi, & Nield, 2008; Sarkis et al., 2014; Wood et al., 2018). The organization of photosystem II (PSII) and cytochrome *b6f* complexes were also done by AFM (Johnson, Vasilev, Olsen, & Hunter, 2014). Nowadays, AFM is mainly used for protein structure into a membrane, but in the future this technique could progress to study the lipid organization.

2.2 Fatty acid phase transition measurement

The fluidity of the lipid bilayer influences many basic membrane functions. The lipid phase transition can affect the protein activity. Indeed, Ray et al. show that the activity of the phospholipase A_2 is mainly impacted by the lipid phase and less by the depth of insertion in the membrane (Ray, Scott, & Tatulian, 2007).

Membrane fluidity strongly depends on the lipid composition of the bilayer, on the length and the degree of unsaturation of fatty acids, on the temperature and on the state of the lipidic phase. Indeed, inside a membrane bilayer, hydrocarbon chains can adopt two main conformations: (1) an ordered conformation where fatty acids are rigid and stretched, it is called the gel phase ($L_{\beta'}$), or (2) a disordered conformation where fatty acids are liquid-crystalline, less stretched, the fluid phase (L_α). The rippled gel phase, between gel and fluid phases, is sometimes present and named $P_{\beta'}$. The phase structures are well described in the following review (Koynova & Tenchov, 2013). The lipid phase transition between gel and fluid phases is characterized by a specific temperature T_m, depending on the length and unsaturations of the carbon chains. This phase transition temperature can be measured using various physical techniques. On this part, a representative list of techniques is presented.

2.2.1 Differential scanning calorimetry (DSC)

Differential scanning calorimetry (DSC) is a relatively rapid, straightforward, and non-perturbing technique for studying the thermotropic phase behavior of hydrated lipid dispersions, or reconstituted lipid model of biological membranes (Demetzos, 2008; Lewis, Mannock, & McElhaney, 2007). It is a thermoanalytical technique that measures and compares the differences of energy (heat) between the sample and known reference. In other terms, the transition between gel to fluid conformation corresponds to a fusion of fatty acid chains, leading to the absorption of heat. Conversely, the transition between fluid to gel is characterized by a crystallization of fatty acid chains and heat release. The DSC measures the heat absorbed or released.

Sample and reference are sealed in an aluminum pan, and the temperature increases or decreases at a constant rate. In a typical thermogram, two main peaks are visible: the exothermic peak (fusion, heat released) and the endothermic peak (crystallization, heat absorbed). This result means that before the melting temperature, lipids are in gel phase, and beyond the crystallization peak, lipids are in fluid phase. In a lipid-protein system, the modification of the DSC peak area indicates that protein conformational

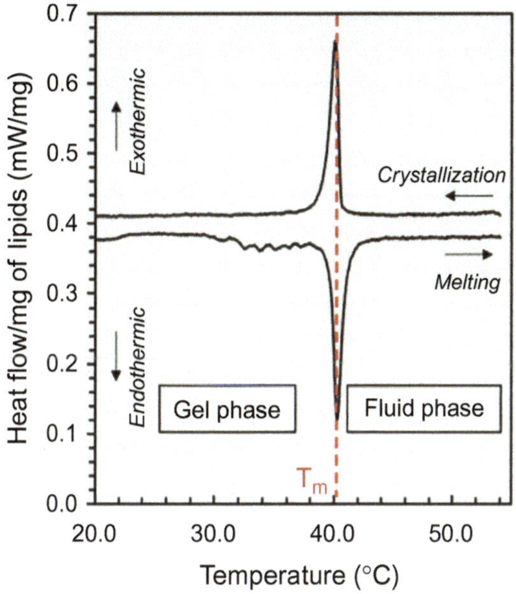

Fig. 1 DSC thermogram of DPPC in water. The thermogram shows the crystallization and melting peaks at the phase transition temperature. The phase transition between gel and fluid phase of DPPC is at 40 °C.

changes alter the nature of lipid–protein interactions (McElhaney, 1986). The following figure shows the DSC thermogram of dipalmitoylphosphatidylcholine (DPPC, PC 16:0/16:0) (Fig. 1).

2.2.2 Solid-state nuclear magnetic resonance (SS-NMR)

SS-NMR spectroscopy is used to study lipid dynamics in membrane models. Two types of NMR are used for lipid analysis: (1) ^{31}P NMR gives information on the polar head of phospholipid (Dufourc, Mayer, Stohrer, Althoff, & Kothe, 1992) and (2) ^{2}H NMR on deuterated lipids gives dynamics of the acyl chains (Davis, 1983). The shape of the ^{2}H spectrum is different between fatty acids in gel or fluid phase. Indeed, in gel phase, fatty acids are fully elongated and highly ordered leading to a wide spectrum, while disordered fluid fatty acids lead to a narrower spectrum (Furlan et al., 2020). The difference of value of the first moment M1 between the two phases determinates the phase transition temperature (Fig. 2). As an example, the influence of polyunsaturated fatty acids on membrane organization was widely studied with this technique (Shaikh, Kinnun, Leng, Williams, & Wassall, 2015).

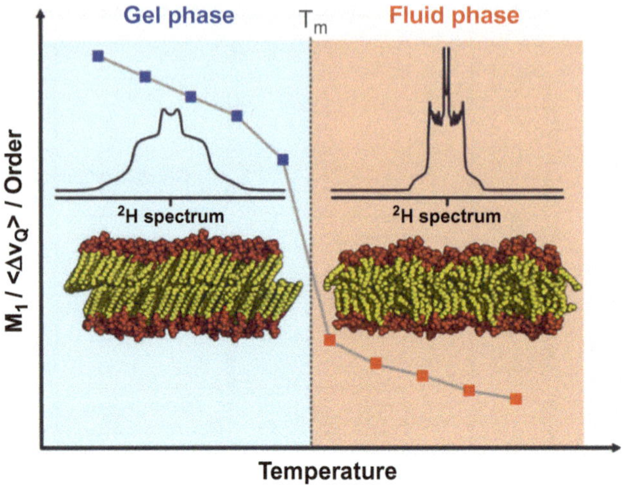

Fig. 2 The shape of the ^{2}H NMR spectrum of lipid membrane depending on the fatty acid crystalline state. The shape of the spectrum and the first moment M1 value determinate the lipid chain phase transition temperature T_m between gel and fluid. In gel phase, fatty acids are fully elongated and highly ordered leading to a wide spectrum, while disordered fluid fatty acid chains lead to a narrower spectrum (Furlan et al., 2020).

2.2.3 Electron spin resonance (ESR) spectroscopy

ESR spectroscopy is a non-destructive tool for studying membrane dynamics. It can measure both rotational and translational diffusion of membrane proteins and lipids. Briefly, ESR measures the transition between the electron spin energy levels. The sample is placed in the microwave resonator located between two magnet poles. The magnetic field of the studied molecule is scanned and the first derivative of the microwave radiation absorption is recorded. In order to detect phase transitions, the order parameter, $2A_{per}$, is plotted as a function of temperature. For example, Fig. 3 shows the pretransition between the gel and rippled phases at 28 °C, and the main phase transition at 40 °C for DPPC lipid (Pali & Pesti, 1996).

2.2.4 Fluorescence anisotropy

The measurement of fluorescence anisotropy is commonly used for the description of phase transitions (Kawski, 1993; van der Meer, van Hoeven, & van Blitterswijk, 1986; van Hoek, Vos, & Visser, 1987). As an example, by using this technique, Smutzer and Yeagle measured the phase transition temperature of a 1,2-dimyristoyl-sn-glycero-3-phosphocholine (DMPC) membrane with addition of various amounts of dehydroergosterol, a

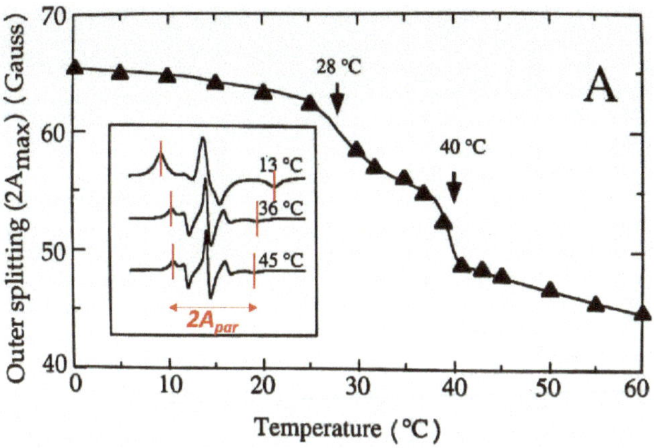

Fig. 3 Lipid phase transition using ESR. The plot of the spectral parameter $2A_{par}$ as a function of the temperature leads to the phase transition temperature of DPPC. Arrows indicate the pre-transition temperature at 28 °C, and the main transition at 40 °C. *Adapted from Pali T, Pesti M (1996) Phase transition of membrane lipids. In Manual on membrane lipids, R. Prasad (ed) pp. 80–93. Berlin: Springer.*

fluorescent molecule similar to cholesterol. They describe the sterol-rich domains thanks to this molecule (Smutzer & Yeagle, 1985). Biological membranes contain autofluorescent molecules, such as quinones, benzoic acid derivatives, alkaloids, etc. that can be used for biochemical, physiological or imaging studies (Donaldson, 2020). Different techniques exist, such as steady-state and time-resolved fluorescence anisotropy (SSFA and TRFA, respectively), fluorescence quenching (FQ) and inframolecular fluorescence energy transfer (IFET) (Esquembre, Ferrer, Gutierrez, Mallavia, & Mateo, 2007).

All lipids have their own phase diagram showing phase transition temperatures between each phase. This phase transition temperature is measured by the techniques described above. As described by Koynova and Tenchov, it is important to notice that the fatty acid chain length increases the fusion temperature, while the presence of unsaturation decreases the phase transition temperature. The polar head of lipids also has an impact on the phase transition of the lipid. In addition, the fusion temperature of fatty acid chains is dependent on the number of methyl groups, the position of the unsaturation on the carbon chain, and the environment such as pH (Koynova, Wang, & Macdonald, 2008). The following figure shows some examples of lipid phase transition temperatures depending on their fatty acid and polar head composition (Fig. 4).

A

Lipids	Phase transition temperature (gel to fluid) (°C)
PC 16:0/16:0	41.7
PC 18:0/18:0	54.5
PC 22:0/22:0	73.6
PC 16:0/16:1	30.0
PC 18:0/18:1	6.9
PC 18:0/18:3	-12.3
PE 16:0/16:0	64.4
DGDG 16:0/16:0	61.6

B

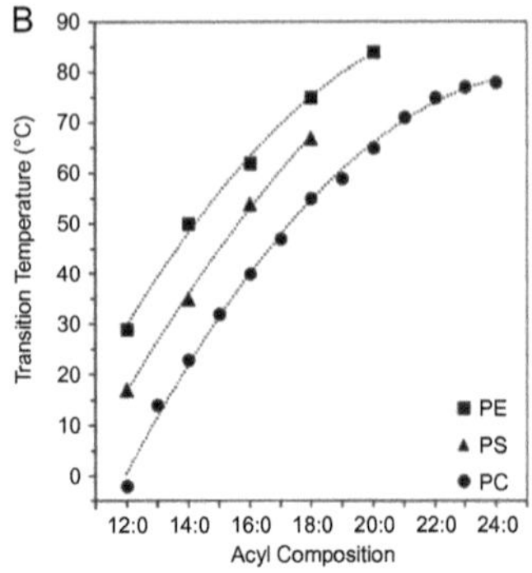

Fig. 4 Phase transition temperature depending on the fatty acid chain and lipid polar head. The length of fatty acid chains increases the fusion temperature while the unsaturation decreases it. The polar head also affects the temperature. (A) Phase transition temperature of lipids with different fatty acid chain lengths (Tenchov & Koynova, 2017). (B) Influence of fatty acid chains and polar heads on the phase transition temperature. *Panel (B) adapted from Avanti Polar Lipids data.*

2.3 Organized structure of membranes

As explained above, fatty acid chains can be frozen (gel phase) or liquid (fluid phase) inside membranes, leading to specific biophysical properties. Moreover, because lipids self-assemble to limit contacts between hydrophobic parts and water molecules, organized structures emerge (Fig. 5). First, lipids can adopt a structure composed of stacked bilayers (L_α) separated by water layers, whose thickness depends on the water content in the system. Second, the hexagonal conformation depends on the relative size of the polar heads compared to that of the fatty acid chains. If the volume occupied by the polar head is larger than the volume occupied by the fatty acid chains, which is mainly the case for lysolipids, a tubular hexagonal I (HI) conformation is formed with polar heads outside the tube and fatty acids inside. Conversely, if the volume occupied by the polar head is smaller than the volume occupied by fatty acid chains, an inverted hexagonal phase (HII) is formed, with the polar heads inside the tube and the fatty acids chains pointing outside. HII forming lipids are sometimes able to switch from a HII phase to a lamellar phase through an intermediate cubic phase (Fig. 5) by lowering the temperature (Tenchov & Koynova, 2012). All these phase transitions are spontaneous and reversible (Siegel & Tenchov, 2008). Most biological membranes are close to phase boundaries or in phase equilibrium between HII and L_α, allowing the formation of HII phase

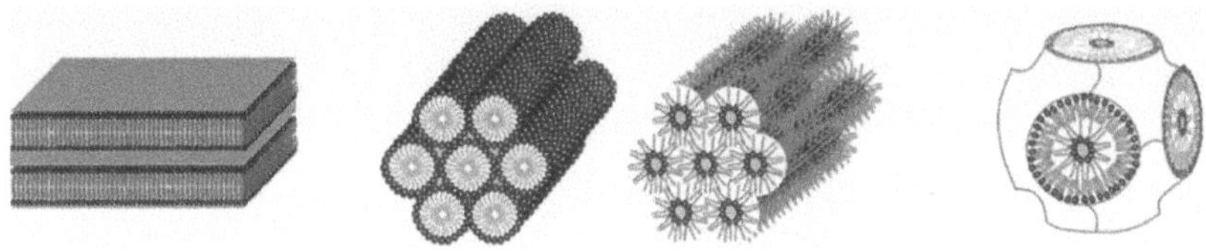

Fig. 5 Organized structures of lipids. Lipids self-organized in different structures, depending on the lipid composition and the water content in the system. Lamellar phase (L_α) is composed by stacked bilayers separated by a water layer. The hexagonal (HI) conformation is formed by lipids whose polar heads occupy more space than the fatty acid tail, i.e., lysolipids. This lipid can also form micelles. The inverted hexagonal (HII) phases are adopted by conic lipids whose polar heads occupy less space than the fatty acid tails, i.e., PE, PA, MGDG. In this structure, the polar heads are inside the tube containing water and fatty acid chains outside. The cubic structure is derived from HII forming lipids depending on the temperature and water amount. *Adapted from Jouhet J (2013) Importance of the hexagonal lipid phase in biological membrane organization. Frontiers in Plant Science 4: 494.*

nanodomains in the membrane (Jouhet, 2013). Biophysical techniques allow the study of these lipid structures. In this section, only a few of these techniques are presented.

2.3.1 X-ray and neutron diffraction analysis

X-ray diffraction and neutron diffraction are similar techniques, one using X-rays beams whereas the other is using neutron beams. They both provide information on the membrane organization, the bilayer and water layer thicknesses, and the bending rigidity. Samples are placed in a beam of thermal or cold neutrons to obtain a diffraction pattern that provides information on the structure of the material. Due to their different scattering properties, neutrons and X-rays provide complementary information: X-rays are suited for superficial analysis, strong X-rays from synchrotron radiation are suited for shallow depths or thin specimens, while neutrons having a higher penetration depth are suited for bulk samples.

As DSC technique, X ray and neutron diffractions can also determine the phase transition of lipids. A large diffraction peak around 0.46 nm characterizes disordered fatty acids, while ordered fatty acids are characterized by a smaller and thin diffraction peak, at 0.42 nm (Finean & Hutchinson, 1988; Lis & Quinn, 1991). A plot of the lipid peak position against the temperature shows the phase transition temperature. Many studies report lecithin bilayer characteristics, such as the conformation of the fatty acid chains and the phase transition temperature, using X-ray diffraction (Janiak, Small, & Shipley, 1979; Tardieu, Luzzati, & Reman, 1973). These bilayer parameters for synthetic lecithin permit to anticipate the structural behavior of natural lipids and other lipid classes.

Structure of membranes can be solved, and hypothesis on the role of lipids in the membrane can be formulated. Indeed, samples are single component or mixes of lipids, and so, depending on the concentration of each lipid class, membranes are not similarly structured. For example, the thylakoid-like lipid mixture, composed with MGDG/DGDG/PG/SQDG (respectively monogalactosyldiacylglycerol, digalactosyldiacylglycerol, phosphatidylglycerol, sulfoquinovosyldiacylglycerol), self-organizes in inverted hexagonal phase at low humidity and in lamellar phase at high humidity, as shown in the following figure (Fig. 6) (Demé, Cataye, Block, Marechal, & Jouhet, 2014).

One major advantage of neutron diffraction over X-ray diffraction is that the latter is rather insensitive to the presence of hydrogen (H) in a structure, whereas the nuclei ^1H and ^2H (i.e., Deuterium, D) are strong scatterers for neutrons. The greater scattering power of protons and deuterons means that the position of hydrogen in a crystal and its thermal motions can be

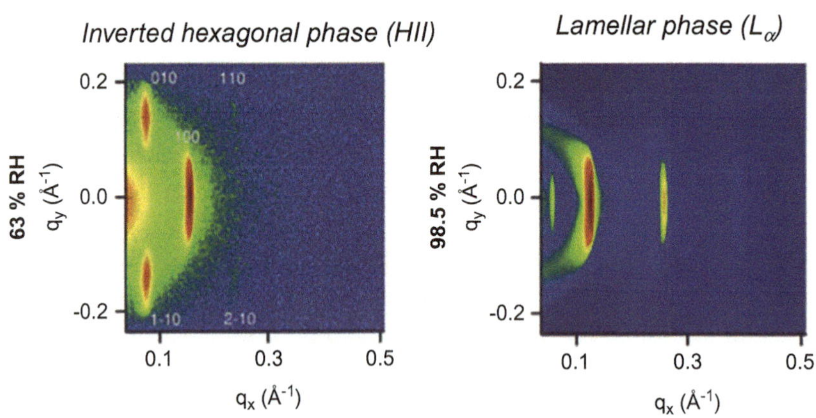

Fig. 6 Diffraction patterns of a thylakoid-like lipid mixture, obtained at the D16 diffractometer (ILL, Grenoble, France). The reconstituted membranes are composed of MGDG/DGDG/PG/SQDG (100/48/17/17 mol%) and measured in a dedicated chamber under controlled humidity. At low hydration (63% humidity, left) lipids self-organize in HII phase, while at high hydration (98.5% humidity right) lipids are in lamellar phase. *Adapted from Demé B, Cataye C, Block MA, Marechal E, Jouhet J (2014) Contribution of galactoglycerolipids to the 3-dimensional architecture of thylakoids. The FASEB Journal 28: 3373–83.*

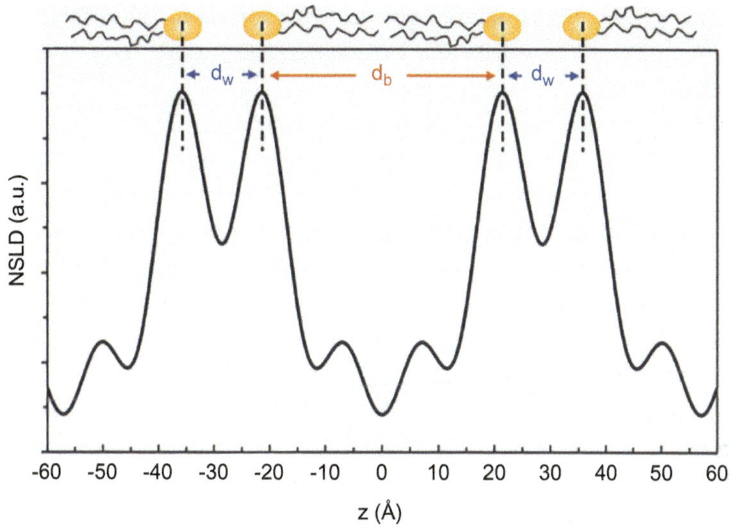

Fig. 7 Neutron scattering length density (NSLD) profile. Using neutron diffraction, the NLSD profile can be established by discrete Fourier analysis. The NSLD profile gives access to the bilayer thickness (d_b) and to the water layer thickness (d_w). At 8% D_2O solvent contrast, due to the high SLD of lipid headgroups, the maxima in the profile correspond to the polar head layer while the H-rich fatty acid chain ends correspond to the minima.

determined with greater precision. Using a 8% D_2O solvent contrast, the bilayer and water layer thickness can be calculated from the neutron scattering length density (NSLD) profile (Kucerka, Nieh, Pencer, Sachs, & Katsaras, 2009). As shown in the Fig. 7, at this specific solvent contrast and due to the H content of lipid molecules, the maxima correspond to the lipid polar heads, while the minima correspond to the center of the bilayer. Because the hydration can be increased in the humidity chamber, its impact on the membrane organization can be followed by calculating the NSLD profile for each humidity.

Other methods providing information on the lipid phase structure are X-ray diffraction (Luzzati & Husson, 1962) or ^{31}P NMR (Epand, D'Souza, Berno, & Schlame, 2015). They enable bilayer thickness calculation using similar method as neutron diffraction.

2.3.2 Membrane compressibility

The membrane compressibility can be estimated by looking at the Bragg peak profiles (or Bragg sheet) in neutron diffraction patterns, from which the bending rigidity and the compression modulus of the membrane are extracted. These two parameters give information on the physical properties of the membrane, especially the capacity of the bilayer stack to be compressed and bent.

Studying monolayer of lipids in Langmuir trough also gives information on the compressibility of the membrane. Many studies use this technique to see the impact of an addition of a protein or another lipid inside the monolayer. For example, Phan and Shin studied the effect of cardiolipin on membrane morphology. They conclude that the addition of cardiolipin increases membrane elasticity (Phan & Shin, 2015). Nintenberg et al. give insight into the reaction mechanism and binding properties of MGD1, the enzyme responsible for the bulk synthesis of monogalactosyldiacylglycerol (MGDG), the major lipid of photosynthetic membranes. In this paper, the authors aim to understand the role of phosphatidylglycerol (PG) in MGD1 binding. Indeed, a membrane of PG has a high affinity for MGD1 like MGDG membrane. Conversely, digalactosyldiacylglycerol (DGDG) membrane has a negative effect and tends to exclude the protein (Nitenberg et al., 2020). The membrane composition is crucial for the protein binding and activity.

2.3.3 Molecular dynamics simulations

Membranes can be modeled and their organization simulated using molecular dynamics simulation methods. Simulation of the repulsive and attractive charges, physical movements of atoms and molecules are generated in silico. A snapshot at the equilibrium of the system gives information on the lipid polar head and the water molecules orientation between two bilayers. This method is often used in parallel with other physical techniques. For example, molecular dynamics simulations on phosphatidylcholine (PC) and DGDG membranes confirmed the lamellar organization of these two lipids and allowed to calculate the attractive forces between two adjacent bilayers. It showed that DGDG bilayers are six times more attractive than PC bilayers (Kanduč et al., 2017).

Other physics techniques are available to study lipid membranes, such as small angle neutron and X-ray scattering (SANS and SAXS), reflectometry or circular dichroism. All these techniques can also investigate the protein mobility and insertion in the membrane.

By using all these techniques, biophysical properties of lipid membranes can be established. However, lipids are often classified depending on their chemical structures. In higher plants and algae, three main lipid classes coexist: glycerolipids, sphingolipids and isoprenoid lipids such as sterols and carotenoids. This chapter focuses on glycerolipids, the main lipid component in membranes, and their influence on membrane architecture. Because organelle membranes have a different lipid composition, they will display different biophysical properties and therefore different architectures.

3. Structure and distribution of glycerolipids

In plants and algae, glycerolipids consist of a glycerol backbone originating from glycerol-3-phosphate to which one to three fatty acids are linked by an ester bond. Glycerolipids containing two fatty acids may have a polar head, and they form the matrix of biological membranes. Glycerolipids containing three fatty acids are called triacylglycerols, or oils, and accumulate in the form of droplets within the cells (Fig. 8).

In plants and algae, less than a dozen classes of glycerolipids are sufficient to compose most biological membranes. Among them, phosphatidylcholine (PC) and phosphatidylethanolamine (PE), the major constituents of extra-plastidial membranes, are in the phospholipids class. The bulk of photosynthetic membranes is mainly composed by non-phosphorus galactolipids, monogalactosyldiacylglycerol (MGDG) and digalactosyldiacylglycerol (DGDG). In algae, there is a third type of membrane glycerolipids, called betaine lipids, which are non-phosphorous and localized in extra-plastidial membranes (Li-Beisson, Thelen, Fedosejevs, & Harwood, 2019). This section describes the structure and the cellular localisation of these glycerolipids.

The localization of each lipids is closely related to their synthesis place, such as the chloroplast, the endomembrane system and the mitochondria. In the environment, organisms are exposed to nutrients starvations, biotic and abiotic stresses, which often require modifications of their membrane

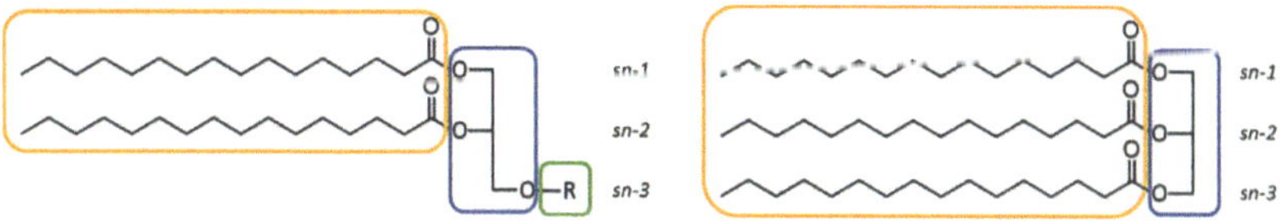

Fig. 8 Glycerolipids structures. Membrane glycerolipids (on the left) consist of a glycerol backbone (in blue) with two fatty acids (in yellow) esterified on the *sn-1* and *sn-2* positions and a variable polar head (in green) on the *sn-3* position. The triacylglycerol, present in lipid droplet or oil (on the right), contains three fatty acids esterified on the glycerol backbone.

lipid composition for surviving. Lipid trafficking exist between the three compartments thanks to vesicles or contact sites, but they are still not well understood.

In this section, the structure and localization of glycerolipids are mainly based on three biological models: *Arabidopsis thaliana* (Angiosperm), *Chlamydomonas reinhardtii* (Chlorophyta) and *Phaeodactylum tricornutum* (Heterokonta), respectively, the higher plant model, the green microalgae model and the diatom (brown microalgae) model. Indeed, these three models gather most of the literature on plant and alga glycerolipid knowledge.

3.1 Structure and biosynthesis of fatty acids

In plants and algae, fatty acids are carboxylic acids with even hydrocarbon chains of 14 up to 24 carbons and a terminal carboxyl group. In photosynthetic organisms, the synthesis of fatty acids takes place inside the stroma of chloroplasts. The characterization of the fatty acid synthesis pathways is well established in *Arabidopsis thaliana*. The extrapolation in algae, derived from primary (such as *Chlamydomonas reinhardtii*) or secondary endosymbiosis (such as *Phaeodactylum tricornutum*), helps to understand the fatty acid synthesis pathways in these organisms. However, many steps concerning the trafficking of fatty acids are still unknown. The fatty acid synthesis pathways and enzymes involved are not detailed in this chapter, but are well detailed in various reviews (Boudiere et al., 2014; Li-Beisson et al., 2019).

Fatty acids are diversified by their chain length, but also by the number of double bond in the carbon chain, also called unsaturation. In plants, membrane lipids contain mainly fatty acids with 16 or 18 carbons, while in microalgae, fatty acids up to 24 carbons are present. Table 1 lists some of the commonly occurring fatty acids in *Arabidopsis thaliana* and different microalgae.

On saturated fatty acids, desaturases can add unsaturations on specific positions in the carbon chain. Fatty acids can have one unsaturation (monounsaturated fatty acids), and up to six (polyunsaturated fatty acids). Two types of unsaturations exist, modifying the shape of the carbon chain and membrane fluidity (Fig. 9). As for saturated fatty acids, *trans*-unsaturation leads to a straight carbon chain. In vivo, the *trans*-configuration is present in one fatty acid, the 16:1Δ3, only in *sn-2* position of phosphatidylglycerol (PG) of thylakoid membrane (Dubacq & Tremolieres, 1983). All other double bonds in fatty acids are in *cis*-conformation, providing a rigid

Table 1 List of the most common fatty acids in higher plants and different microalgae.

		Angiosperm (plant)	Chlorophyta (green microalgae)		Heterokonta (brown microalgae)		Cyanobacteria
Fatty acids (mol%)	**Name**	***A. thaliana*[a]**	***C. reinhardtii*[b]**	***C. vulgaris*[c]**	***P. tricornutum*[a]**	***M. gaditana*[a]**	***S. maxima*[c]**
14:0	Myristic	0.07	–	0.7	4.3	2.5	–
16:0	Palmitic	17.8	22	14.4	15.2	34.0	35.8
16:1	Palmitoleic	3.7	7	4.0	22.6	32.0	0.9
16:2	Hexadecadienoic	–	2	5.3	3.5	0.7	4.6
16:3	Hexadecatrienoic	1.5	3	4.9	8.6	0.2	–
18:0	Stearic	0.5	2	1.6	2.8	1.8	1.5
18:1	Oleic *(cis)* Elaidic *(trans)*	26.2	16	17.6	0.9	3.2	5.0
18:2	Linoleic	12.2	8	12.0	2.7	1.7	16.3
18:3	Linolenic	38.2	25	15.8	1.4	0.4	18.2
18:4	Stearidonic	–	2	–	–	0.1	–
20:1	Eicosenoic	–	–	–	0.3	0.1	–
20:2	Eicosadienoic	–	–	–	0.4	1.1	0.6
20:4	Arachidonic	–	–	–	1.8	1.5	–
20:5	EPA	–	–	–	30.3	20.1	–
22:1	Erucic	–	–	–	–	0.4	–
22:5	Eicosapentaenoic	–	–	–	0.4	–	–
22:6	DHA	–	–	0.3	4.2	–	–
24:0	Lignoceric	–	–	0.2	0.7	–	0.6

[a](Jouhet et al., 2017).
[b](Giroud, Gerber, & Eichenberger, 1988).
[c](Otles & Pire, 2001).

Fatty acid composition in one plant, *Arabidopsis thaliana*, two green microalgae, *Chlamydomonas reinhardtii* and *Chlorella vulgaris*, two brown microalgae *Phaeodactylum tricornutum* and *Microchloropsis gaditana* and one cyanobacteria *Spirulina maxima*. The values are in mol%. DHA: decosahexaenoic acid, EPA: eicosapentaenoic acid.

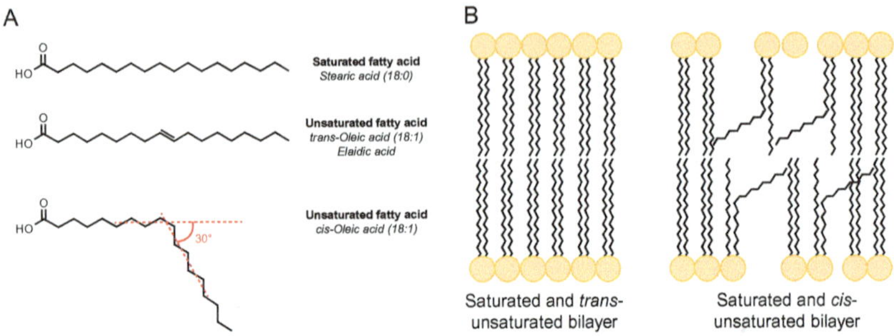

Fig. 9 Fatty acid and membrane structures. Fatty acids can be saturated or unsaturated (A). Two types of unsaturation exist: the *trans*-unsaturation extends the chain structure and the *cis*-unsaturation provides a rigid kink of 30° in the carbon chain. (B) Membranes containing lipids with saturated and *trans*-unsaturated fatty acids are more rigid than a membrane with *cis*-unsaturated fatty acids. The kink in the fatty acid chains maintains space between adjacent lipid molecules, influencing the membrane permeability.

kink of 30° in the hydrocarbon chain, solved by molecular dynamics on 1-palmitoyl-2-linoleoyl-sn-glycero-3-phosphocholine lipid (PLPC 16:0–18:2) (Hyvonen, Rantala, & Ala-Korpela, 1997). The length and degree of unsaturation of fatty acids determines their impact on the membrane biophysical properties: saturated fatty acids with short chains form viscous membranes, while unsaturated fatty acids form membranes that are more fluid (Ibarguren, Lopez, & Escriba, 2014; Maulucci et al., 2016).

Inoue and collaborators studied the impact of adding saturated or unsaturated fatty acids in a saturated bilayer of dipalmitoylphosphatidylcholine (DPPC). They concluded that the long chains of saturated and *trans*-monounsaturated fatty acids raise the main phase transition temperature of DPPC bilayer and rigidify the lipid acyl chain in gel phase, while *cis*-monounsaturated fatty acid exhibits a minimal perturbation on the physical properties of the bilayer (Inoue, Yanagihara, Misono, & Suzuki, 2001). Mouritsen writes that the enzymatic activity of desaturases and elongases can directly affect membrane fluidity by increasing the abundance of mono- and polyunsaturated fatty acids in membrane phospholipids (Mouritsen, 2005). In agreement, Maulucci et al. show that if the polyunsaturated fatty acid content increases, the membrane is more fluid, and vice versa (Maulucci et al., 2016). Moreover, if saturated fatty acids, with their straight tails, are compressed, they press on each other, making a dense and fairly rigid membrane. However, if unsaturated fatty acids are compressed, the kinks in their tails maintain some space between adjacent lipid molecules, thus influencing membrane permeability (de Mendoza & Pilon, 2019). Molecular dynamics simulations and biochemical measurements indicate that polyunsaturated fatty acids induce membrane curvature (Pinot et al., 2014; Risselada & Marrink, 2009). They also influence the membrane thickness because 22:6 chains have an average length of 8.2Å at 41 °C compared to 14.2Å for 18:1 chains (Fernandes, Castanho, & Garcia de la Torre, 2002).

Therefore, membranes adapt their fatty acid composition depending on their function and architecture. For example, organelles and vesicles, that generally display highly curved membranes, have an enrichment of polyunsaturated fatty acids that decreases the membrane rigidity (Takamori et al., 2006; Yang, Sugiura, Ikegami, Konishi, & Setou, 2012). Indeed, the lack of polyunsaturated fatty acids compromises vesicle formation (Tixier-Vidal, Picart, Loudes, & Bauman, 1986). Organisms are also capable to change the proportion of unsaturated fatty acids in their membranes in response to environmental stress, such as cold temperature (see next section).

Plants display fatty acids with two major chain lengths, C16 and C18, synthetized within the chloroplast. The desaturation of fatty acids on the glycerolipids lead to 16:3 from 16:0, and 18:3 from 18:1. Outside the chloroplast, only desaturation of fatty acids containing 18 carbons or more occurs (Heemskerk, Schmidt, Hammer, & Heinz, 1991). Glycerolipids esterified by two fatty acids on the position *sn-1* and *sn-2* of the glycerol-3-phosphate backbone form the phosphatidic acid (PA), first precursor of lipid biosynthesis. In plants, fatty acid chain length at the *sn-2* position indicates the place of synthesis for the diacylglycerol backbone. The starting point of the plastidial or "prokaryotic" pathway is the synthesis of PA in the chloroplast envelope by the sequential action of acyl-ACP/glycerol-3-phosphate acyltransferase and acyl-ACP/lysophosphatidic acid acyltransferase. These enzymes produce almost exclusively 18:1/16:0 PA (Frentzen, Heinz, McKeon, & Stumpf, 1983). This is in agreement with the finding that isolated chloroplasts (in which only the prokaryotic pathway can operate) synthesize PA, diacylglycerol, glycolipid and PG, which contain only C16 fatty acids at the *sn-2* position (Heinz & Roughan, 1983; Roughan, Holland, & Slack, 1980). However, the microsomal acyltransferase isoenzymes, which are responsible for PA synthesis in the endoplasmic reticulum, or "eukaryotic" pathway, give rise to lipids that contain a C18 fatty acid at the *sn-2* position and either C16 or C18 fatty acids at the *sn-1* position (Frentzen et al., 1983). Because both signatures are found in Arabidopsis galactolipids (C16 and C18 at the *sn-2* position), a traffic of diacylglycerol backbone occurs from the endoplasmic reticulum to the chloroplast. The trigalactosyldiacylglycerol transporter (TGD) is essential for this transport (Xu, Moellering, Muthan, Fan, & Benning, 2010). However, the nature of the transferred molecule, PA, DAG or PC, is still under debate (Block & Jouhet, 2015).

In microalgae, the fatty acid synthesis also takes place in the chloroplast but afterwards differences with land plants occur. While the fatty acid chain length in *Chlamydomonas reinhardtii* is similar to *A. thaliana*, plastidial lipids have a C16 fatty acid at the *sn-2* position and lack C18 fatty acid at the *sn-2* position, suggesting at first that there is no traffic from ER to chloroplast. However, Kim et al. identified a lysophosphatidic acid acyl transferase in the ER able to put a C16 fatty acid in *sn-2* position (Kim, Terng, Riekhof, Cahoon, & Cerutti, 2018). Moreover, TGD machinery is present in green algae (Warakanont et al., 2015), confirming the existence of a diacylglycerol backbone traffic from the ER to the plastid in these organisms. In addition, a plastidial Δ4 desaturase is present producing high amount of glycolipid 16:4 (Giroud et al., 1988; Zauner, Jochum, Bigorowski, & Benning, 2012), a typical feature from green algae.

Marine algae are rich in very-long chain polyunsaturated fatty acid (Li-Beisson et al., 2019). For example, 20:5 is found at the *sn-1* position in glycolipids in the diatom *Phaeodactylum tricornutum*, whereas the *sn-2* position hosts a C16 fatty acid, suggesting also a plastidial origin for the diacylglycerol backbone. However, the fatty acid 20:5 is not synthesized in the chloroplast but is obtained by the elongation and desaturation in the ER, probably from the 16:0 by the ω-pathway before its return to the chloroplast (Petroutsos et al., 2014; Smith et al., 2021). Nowadays, this traffic is still not known. At the opposite of *A. thaliana*, there is no evidence of 18:0 fatty acid synthesis in the plastid (Zulu, Zienkiewicz, Vollheyde, & Feussner, 2018) and 16:1 is produced by the desaturation of 16:0 inside the plastid that will be exported outside the chloroplast in other membrane lipids (Smith et al., 2021).

Glycerolipids are classified from the polar heads present at the *sn-3* position of the diacylglycerol backbone. The biosynthesis of glycerolipids occurs in two main organelles: the endoplasmic reticulum for phospholipid and betaine lipid biosynthesis, and the chloroplast for glycoglycerolipid and phosphatidylglycerol (PG) biosynthesis. Phosphatidylglycerol can also be synthesized in mitochondria, as well as diphosphatidylglycerol (DPG, also called cardiolipin). The biosynthesis of lipids is not described in details in this chapter, only the main steps are presented here and summarized in Fig. 10. More details about the enzymes involved in each step and their regulation are well described in the Li-Beisson's review (Li-Beisson et al., 2019).

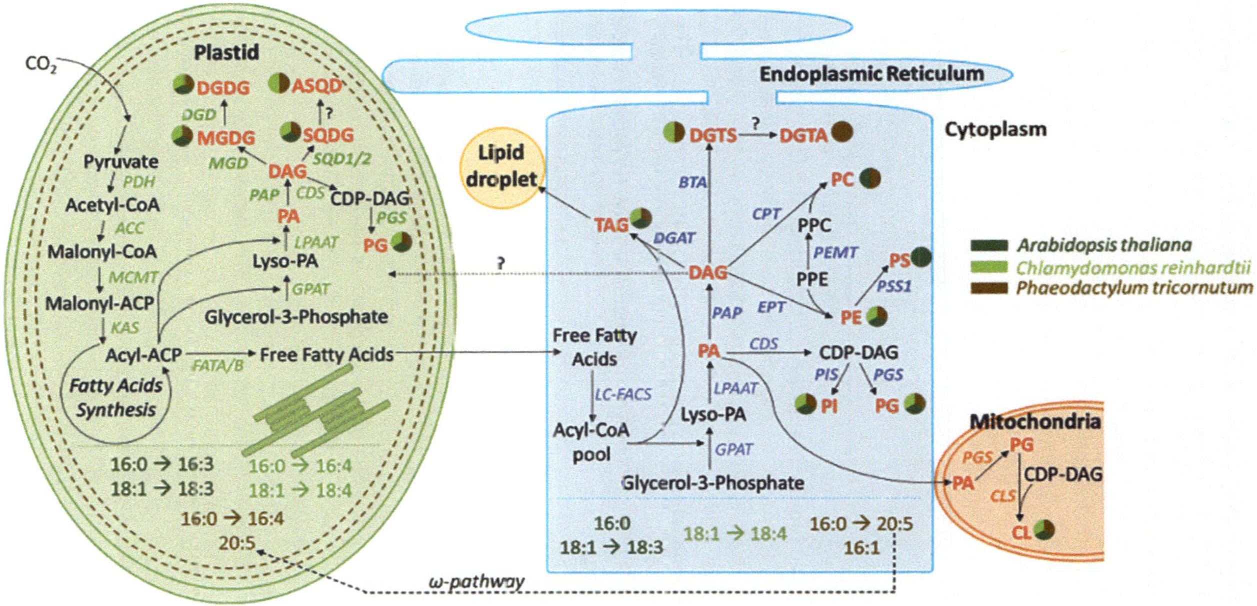

Fig. 10 See figure legend on next page.

3.2 Phospholipids

Phospholipids are the most abundant class of glycerolipids in non-plastidial membranes in plants. The polar head of phospholipids contains a phosphate group, and may have in addition, various alcohols linked to the phosphorous group leading to different phospholipids, as shown in Fig. 11.

In plants, the phospholipid synthesis takes place in the endoplasmic reticulum. As previously described, phosphatidic acid (PA) is the precursor of phospholipid synthesis. In plants, PA serves as substrate to synthesize CDP-diacylglycerol (CDG-DAG) by CDP-diacylglycerol synthase (CDS) or DAG by phosphatidic acid phosphatase (PAP) or phosphatidic hydrolase (PAH). CDP-DAG is used by phosphatidylinositol synthase (PIS) and

Fig. 10 Glycerolipids biosynthesis in higher plant *A. thaliana*, green microalgae *C. reinhardtii*, and brown microalgae *P. tricornutum*. The fatty acid synthesis occurs in chloroplast and fatty acid are exported in the endoplasmic reticulum (ER) for further desaturation and elongation. The glycolipid (galactolipids MGDG and DGDG, sulfolipid SQDG) synthesis takes place in the chloroplast. Acyl-SQDG (ASQD) synthesis is unknown, but probably also occurs in the chloroplast compartment. The phospholipid and betaine lipid synthesis occurs in the ER. Only the phosphatidylglycerol (PG) can be synthesized in the ER as well as in the chloroplast. The polyunsaturated fatty acids can be transported from the ER to the chloroplast through the DAG lipid pool using the TGD transporter. In *A. thaliana*, lipids containing a C16 signature at the *sn-2* position of the glycerol backbone are synthesized in the chloroplast and followed the "prokaryotic" pathway, while lipids containing a C18 signature at the *sn-2* position are synthesized in the ER and followed the "eukaryotic" pathway. In *P. tricornutum*, the return of elongated fatty acids from the ER to the chloroplast is called the ω-pathway. Lipids present in *A. thaliana* are in dark green, *C. reinhardtii* light green and *P. tricornutum* brown. The chloroplast of *A. thaliana* and *C. reinhardtii* is surrounded by two membranes (green ones), while the chloroplast of *P. tricornutum* is surrounded by four membranes (two green and two brown). The lipid structure and the fatty acid composition of each compartments are written on the bottom of each compartment representation. For an aesthetic reason, all enzymes involved in lipids synthesis are not placed inside the membranes. ACC: acetyl-CoA carboxylase, BTA: betaine lipid synthase, CPT: CDP-choline:DAG choline phosphotransferase, CL: cardiolipin, CLS: cardiolipin synthase, DGAT: acyl-CoA:DAG acyltransferase, DGD: DGDG synthase, EPT: CDP-ethanolamine: DAG ethanolamine phosphotransferase, FATA: fatty acid thioesterase A/B, GPAT: glycerol-3-phosphate acyl transferase, KAS: ketoacyl-ACP synthase, LC-FACS: long chain acyl-CoA synthetase, LPAAT: 1-acyl-glycerol-3-phosphoacyltransferase, MCMT: malonyl-CoA:ACP acyltransferase, MGD: MGDG synthase, PAP: phosphatidic acid phosphatase, PCT: phosphatidic acid cytidyl transferase, PDH: pyruvate dehydrogenase, PEMT: phosphatidylethanolamine N-transferase, PGS: phosphatidylglycerol synthase, PIS: phosphatidylinositol synthase, PPC: phosphocholine, PPE: phosphoethanolamine, PS: phosphatidylserine, PSS1: PS synthase SQD1/2: SQDG synthases.

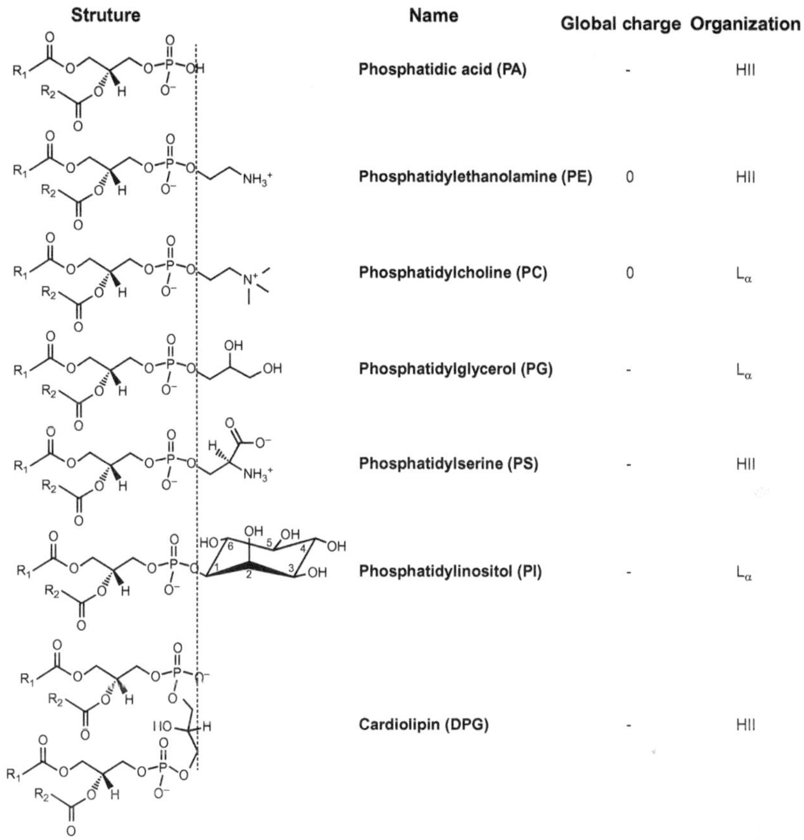

Fig. 11 Structure of the phosphoglycerolipids class. R_1, R_2 are the two fatty acids esterified on the *sn-1* and *sn-2* position of the glycerol backbone. The global charge of the lipid is shown: negative charge (−) or neutral charge (0). The organization into the membrane is also mentioned: inverted hexagonal (HII) or lamellar (L_α).

phosphatidylglycerol synthase (PGS) to form respectively phosphatidylinositol (PI) and phosphatidylglycerol (PG). DAG is the precursor of phosphatidylethanolamine (PE) and phosphatidylcholine (PC) through the Kennedy pathway (Dubots et al., 2012) with the respective enzyme activities CDP-ethanolamine:DAG ethanolamine phosphotransferase (EPT) and CDP-choline:DAG choline phosphotransferase (CPT). Because plants cannot find choline in their environment, the major pool of choline is synthesized by the trimethylation of phosphoethanolamine (PPE) by the phosphoethanolamine-*N*-methyltransferases (PEMTs) to form phosphocholine (PPC). Then, phosphocholine is esterified in *sn-3* position of the DAG backbone to form phosphatidylcholine (Nakamura, 2021). PG is the only phospholipid synthesized both in ER and plastid membranes. Quantitatively, PC is the most abundant phospholipid in photosynthetic organisms (Table 2). So far, in *A. thaliana*, almost all the enzymes involved in phospholipid synthesis have been identified and localized (Nakamura, 2021).

In plants, there is also a small amount of phosphatidylserine (PS). This lipid is synthesized from PE by the PS synthase (PSS) in the ER compartment. PS is mentioned as a signal lipid because it can interact with proteins present at low level, such as in the auxin signaling pathway (Colin & Jaillais, 2020). Finally, cardiolipin is a diphosphatidylglycerol lipid (DPG) containing four fatty acids bound to two glycerol moieties, which are linked by two phosphate groups to a third glycerol. In eukaryotes, cardiolipin is exclusively located in the inner mitochondrial membrane. It is synthesized by condensation of PG on CDP-DAG molecule, a reaction catalyzed by the cardiolipin synthase (CLS) (Katayama, Sakurai, & Wada, 2004; Zhou, Peisker, & Dormann, 2016).

In algae, genes responsible for phospholipid synthesis could be identified by sequence homology. Except for PG, that has also probably a dual location in ER and plastids, the phospholipid synthesis is predicted to be located in the ER. Therefore, by analogy, phospholipid synthesis pathways are supposed to be similar in plants and algae. However, *C. reinhardtii* lacks PC in its lipids composition, due to the absence of PEMT and PEAMT genes (Hirashima, Toyoshima, Moriyama, & Sato, 2018).

The physicochemical properties of phospholipids are described in Fig. 11. While phospholipids are mainly anionic due to the negative charge of the phosphate group, only PE and PC are zwitterionic because of the ammonium positive charge that compensates the phosphate negative charge. Due to differential ion concentrations between the inside and the outside of the cell compartments, membranes are polarized. In plants, there is an "electrostatic membrane territory" due to an electrostatic gradient from the plasma membrane to the tonoplast. For example, the plasma membrane is the most electronegative while the ER-derived compartments are neutral (Platre et al., 2018; Simon et al., 2016). The surface charges of membrane compartments are a key determinant to control the location of many proteins. Anionic lipids are major regulators of fundamental cellular processes, such as cell signaling, cell division, membrane trafficking, gene expression as described in the review by Noack and Jaillais (2020).

Thanks to neutron diffraction studies, the membrane organization of phospholipids were determined. The polar heads of PA and PE are small compared to their fatty acid volumes, leading to inverted hexagonal (HII)

Table 2 Lipid composition in higher plants and microalgae.

	Angiosperm (plant)	Chlorophyta (green microalgae)	Heterokonta (brown microalgae)	Ochrophyta (brown microalgae)	Cyanobacteria
Lipids (%)	***A. thaliana***[a]	***C. reinhardtii***[b]	***P. tricornutum***[c]	***M. gaditana***[c]	***S. platensis***[d]
MGDG	48.9	38.2	42.3	13.2	10.3
DGDG	14.5	14.8	8.5	5.8	6.44
SQDG	1.6	5	19.6	11.1	11.4
ASQD	–	–[e]	tr.	–	–
PG	8.6	7.7	6.8	5.3	25.9
PC	19.7	–	12.2	7.5	–
PE	6.7	6.1	1.1	2.5	–
PI	–	2.3	2.2	2.9	–
PS	–	–	–	–	–
DGTS	–	25.8	–	14.0	–
DGTA	–	–	4.5	–	–

[a](Kelly, Froehlich, & Dormann, 2003).
[b](Riekhof, Ruckle, Lydic, Sears, & Benning, 2003).
[c](Jouhet et al., 2017).
[d](Xue et al., 2002).
[e]ASQD was not analyzed in Xue et al. (2002) work but it is present in *C. Reinhardtii* (Riekhof et al., 2003).
Arabidopsis thaliana has six major lipids, three glycerolipids and three phospholipids. The composition in the green microalgae *Chlamydomonas reinhardtii* is similar to plant but DGTS replaces PC. The brown microalgae *Phaeodactylum tricornutum* and *Microchloropsis gaditana* have the same composition, with DGTA or DGTS, respectively. *Spirulina platensis* a cyanobacteria is only composed by four lipids: MGDG, DGDG, SQDG and PG. ASQD: acyl-sulfoquinovosyldiacylglycerol, DGDG: digalactosyldiacylglycerol, DGTA: 1,2-diaclyglyceryl-3-O-2′-(hydroxymethyl)-(*N*,*N*,*N*-trimethyl)-β-alanine, DGTS: 1,2-diacylglyceryl-3-O-4′-(*N*,*N*,*N*-trimethyl)-homoserine, MGDG: monogalactosyldiacylglycerol, PC: phosphatidylcholine, PE: phosphatidylethanolamine, PG: phosphatidylglycerol, PI: phosphatidylinositol, PS: phosphatidylserine, SQDG: sulfoquinovosyldiacylglycerol.

organization. DPG and PS organization is also hexagonal. PC, PG and PI are bilayer forming lipids. Bradshaw looked at the headgroup orientation of PI lipid in the membrane. Indeed, because PI-specific enzymes recognize this lipid, they investigated the conformation of PI headgroup. They concluded that the inositol ring in the headgroup extends perpendicular to the membrane surface (Bradshaw, Bushby, Giles, Saunders, & Reid, 1996).

3.3 Glycoglycerolipids

In the plant kingdom, glycoglycerolipids, including galactolipids, monogalactosyldiacylglycerol (MGDG), digalactosyldiacylglycerol (DGDG) and the sulfolipid sulfoquinovosyldiacylglycerol (SQDG), are prevalent components of photosynthetic membranes, representing around 80% of membrane lipids in chloroplasts (Block, Dorne, Joyard, & Douce, 1983). Their syntheses occur in the plastid envelope. DAG is the precursor of MGDG and SQDG. They are synthesized by MGDG synthases and SQDG synthases, respectively. The DAG precursor is synthesized within the chloroplast from PA by a PAP, and displays the plastid signature with a C16 fatty acid on the *sn-2* position. Galactolipids synthesized from this DAG share this signature. However, as indicated previously, a DAG backbone trafficking exists between the ER and the chloroplast, leading to a pool of MGDG and SQDG with C18 fatty acid on position *sn-2*, characteristic of the non-plastid pathway. MGDG synthases add a galactose from UDP-galactose on the *sn-3* position of the glycerol backbone in β-anomeric position (Kelly et al., 2003), while SQDG synthase adds a sulfoquinovose group on the *sn-3* position of DAG (Yu, Xu, & Benning, 2002). By glycosylation of MGDG by DGDG synthases, a second galactose is added in α-anomeric position. The position of DGDG polar head was also solved by neutron diffraction, it is oriented parallel to the plane of the bilayer such that the galactose moieties are tightly packed at the bilayer surface into a 0.8 nm thick polar layer (McDaniel, 1988).

Acyl-SQDG (ASQD) has been identified in some algae, such as *Chlorella vulgaris* (Morimoto, Murakami, Nagatsu, & Sakakibara, 1993), *C. reinhardtii* (Riekhof et al., 2003), *P. tricornutum* (Naumann et al., 2011) and dinoflagellate (personal observation). The ASQD structure is a SQDG backbone with an addition of a third fatty acid on the carbon 2 of the sulfoquinovose. ASQD location in the cell is not known but is probably chloroplastic because it derives from SQDG localized in this compartment. The third fatty acid on the sugar is usually the longest polyunsaturated fatty acid within the cell: 18:3 for *C. reinhardtii*, 20:5 for *P. tricornutum* and 22:6 for dinoflagellate (Riekhof et al., 2003). The location and the insertion of the third fatty acid on the membrane is not defined. We propose two hypotheses: (1) the third fatty acid is placed in the same membrane as the other two, resulting in a flattening of the polar head on the membrane, or (2) the third fatty acid binds an adjacent membrane, maintaining the two bilayers in contact. Preliminary neutron diffraction data are in favor of the first hypothesis (personal communication).

While galactolipids are uncharged, sulfolipids are negatively charged due to the sulfate group (Fig. 12). Because the polar head of MGDG is small, this lipid self-organizes in HII phase. DGDG and SQDG, with a larger polar head, are in lamellar organization. Many neutron diffraction studies have been done to understand galactolipid membrane organization. For example, it was shown that the architecture of thylakoid membrane depends on the MGDG/DGDG ratio. DGDG is important to keep the lamellar organization (Demé et al., 2014). Another study shows the influence of DGDG on membrane stacking in comparison with PC (Kanduč et al., 2017). However, the properties of SQDG and ASQD are not yet well understood and these lipids need further studies.

In plants, the envelope of the chloroplast is delimited by two membranes, the outer and the inner membranes, whose lipid composition has been

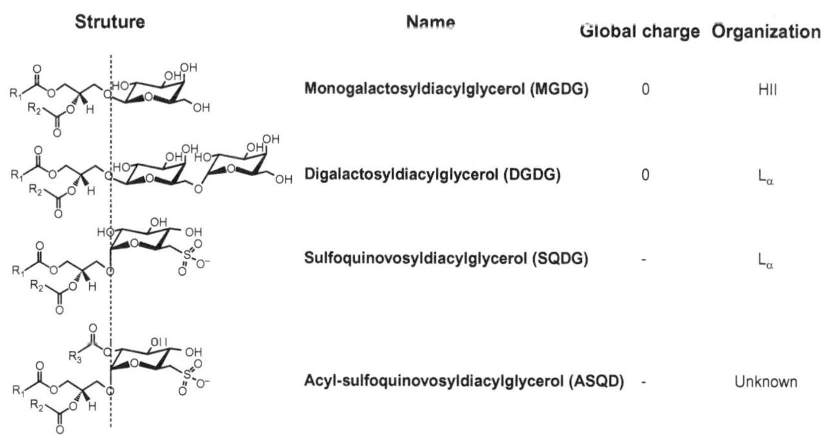

Fig. 12 Structure of the glycoglycerolipids class. R_1, R_2 are the two fatty acids esterified on the *sn-1* and *sn-2* position of the glycerol backbone. The global charge of the lipid is shown: negative charge (−) or neutral charge (0). The organization into the membrane is also mentioned: inverted hexagonal (HII) or lamellar (L_α). The organization of ASQD into the membrane is still unknown.

characterized. The inner envelope of chloroplasts and thylakoids contains MGDG, DGDG, SQDG and PG, while the outer envelope contains PC and PI in addition to MDGD, DGDG and PG (Block, Dorne, Joyard, & Douce, 1984; Douce, Holtz, & Benson, 1973; Douce & Joyard, 1990). In *C. reinhardtii*, the lipid composition of the chloroplast envelope and thylakoid membranes are similar to the plant chloroplast composition, but without PC and with a small amount of PE in the envelope. Betaine lipids (see below) are also present in the chloroplast envelope, but probably not in thylakoid membranes (Mendiola-Morgenthaler, Eichenberger, & Boschetti, 1985). In algae deriving from the secondary endosymbiosis event, such as *Phaeodactylum tricornutum*, the chloroplast is surrounded by four membranes whose lipid composition is unknown.

These two classes of lipids, phospholipids and glycolipids, are found in plant and alga membranes. In lower plants and algae, a third class of membrane glycerolipids is present, the betaine lipids that are absent in seed plants (angiosperms and gymnosperms) (Sato, 1992).

3.4 Betaine lipids

Betaine lipids are phosphate-free glycerolipids structurally related to phosphatidylcholine (PC) (Fig. 13). The presence of betaine lipids was first noted in a chrysophyte *Ochromonas danica* by Nichols and Appleby (1969).

Regarding evolution, betaine lipids are present exclusively in non-flowering plants, as well as lichen and fungi (Kunzler & Eichenberger, 1997). 1,2-diacylglyceryl-3-O-4′-(*N*,*N*,*N*-trimethyl)-homoserine (DGTS)

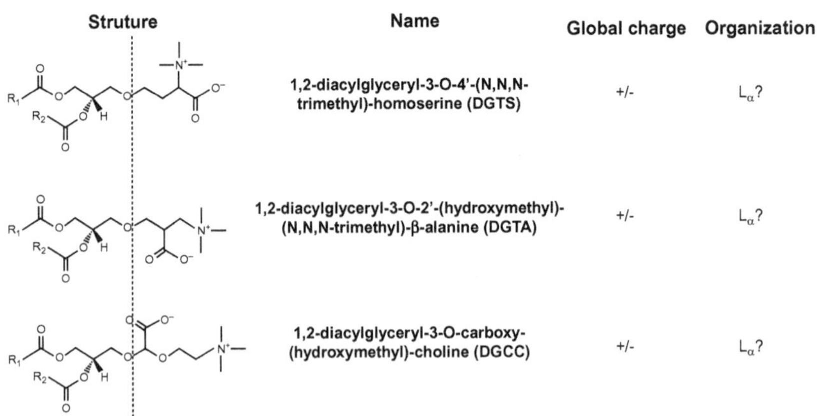

Fig. 13 Betaine lipids structures. R_1, R_2 are the two fatty acids esterified on the *sn-1* and *sn-2* position of the glycerol backbone. The three lipids are neutrally charged (+/−). The organization of the betaine lipids are still unknown but probably lamellar (L_α).

is found in many green algae, ferns, mosses, and in some algal species derived from secondary endosymbiosis. 1,2-diaclyglyceryl-3-O-2′-(hydroxymethyl)-(*N,N,N*-trimethyl)-β-alanine (DGTA) and 1,2-diacylglyceryl-3-O-carboxy-(hydroxymethyl)-choline (DGCC) are mostly found in brown algae such as *Phaeodactylum tricornutum* and *Thalassiosira pseudonana*, respectively (Canavate, Armada, Rios, & Hachero-Cruzado, 2016; Guschina & Harwood, 2006; Sato, 1992). DGCC is a common constituent of Haptophyceae (Kato, Sakai, Adachi, Ikemoto, & Sano, 1996). In green algae, a high amount of betaine lipids is often linked to a low amount or lack of PC (Kalisch, Dormann, & Holzl, 2016; Sato, 1992). For example, the green algae *Chlamydomonas reinhardtii* does not have PC but a high level of DGTS (Sakurai, Mori, & Sato, 2014).

The betaine lipids location is still elusive. However, Eichenberger studied the lipid composition of chloroplasts in many groups such as Chlorophyta, Bryophyta and Pteridophyta, and found that the amount of DGTS is insignificant, suggesting DGTS is an extraplastidial constituent (Eichenberger, Araki, & Müller, 1993). Also, the plasma membrane of *Dunaliella salina* contains a high concentration of DGTS (Sheffer, Fried, Gottlieb, Tietz, & Avron, 1986). The location of DGTS and DGTA by immunoelectron microscopy showed that these two betaine lipids are found in non-plastidial membranes (Kunzler, Eichenberger, & Radunz, 1997). First experiments made on *Pavlova lutheri* showed that DGCC is also found in non-plastid cell membranes, but further investigations are needed (Eichenberger & Gribi, 1997). All these works suggest that betaine lipids are located in non-plastidial membranes.

The polar parts of DGTS and DGTA are structural isomers (Fig. 13). However, the betaine lipid biosynthesis pathway is elucidated only for DGTS that occurs in the ER with the BTA1 enzyme (Riekhof, Sears, & Benning, 2005). Some researches were conducted to elucidate the biosynthesis pathway of DGTA that derives from DGTS (Vogel & Eichenberger, 1992) and DGCC (Kato et al., 1996), but the enzymes responsible for its synthesis are not yet known. The organization of betaine lipids in the membrane is also not known, but we can hypothesize that betaine lipids are bilayer forming lipids, alike PC. Indeed, PC and betaine lipids share similar chemical features: they are zwitterionic with a positive charge bared by a trimethylamine and a negative charge by a phosphate group (PC) or a carboxylic group (betaine lipids).

To summarize, by analogy with plants and by predicting the location of lipid synthesizing enzymes, some rules can be proposed: plastid membranes are mainly composed of glycoglycerolipids while phosphoglycerolipids and betaine lipids constitute non-plastidial membranes. The lipid composition of some organisms is shown in Table 2. The biosynthesis pathways of lipids in *A. thaliana, C. reinhardtii* and *P. tricornutum* are summarized in Fig. 10. These particular lipid compositions confer original properties to the compartment to which they belong.

4. Adaptation to a changing environment by rearranging membrane properties

4.1 Influence of fatty acid tail in membrane fluidity: Adaptation to thermal stress

As seen above, changing the fatty acid chain length and unsaturation has a huge impact on membrane properties. The acyl tails of plant and alga membrane lipids are enriched in long and unsaturated fatty acid chains and their saturation state heavily influences the viscosity of the hydrophobic membrane core (Harwood & Jones, 1989; Sarcina, Murata, Tobin, & Mullineaux, 2003). Temperature changes the physical properties of all molecules in the cell, thereby altering enzyme kinetics, protein binding, membrane fluidity, and protein folding. High temperatures lead to increased fluidity of the membrane. During chilling temperatures the opposite occurs, the rigidity of the membrane increases, leading to the loss of the bilayer structure and membrane leakiness during freezing (Thomashow, 1999). Regulation of the fluidity gets through the modulation of fatty acid desaturases (FAD) that are unstable at elevated temperature and overexpressed during cold stress (Niu & Xiang, 2018).

The study of FAD regulation from different organisms have revealed a conceptual convergence: (i) the physical properties of the membrane can activate a signal transduction pathway that controls the expression of FAD genes and (ii) unsaturated fatty acids, the product of desaturases, act as negative signaling molecules that turn the pathway off (Aguilar & de Mendoza, 2006). However, the perception of temperature stress is different depending on the temperature: heat perception is mediated by denaturated protein with Heat Shock Protein and light receptors such as phytochrom B (Lamers, van der Meer, & Testerink, 2020). Although studies have indicated that there are several ion channels that function as heat sensors in animal cells (Ramot, MacInnis, & Goodman, 2008; Yao, Liu, & Qin, 2011), the plasma membrane thermosensor in plants long remained unidentified. The existence of primary sensors for heat stress was revealed in the plasma membrane of *Physcomitrium patens,* which study supports the conclusion that heat-mediated calcium channels are probably regulated by membrane fluidity (Saidi, Finka, & Goloubinoff, 2011; Saidi et al., 2009). Further evidence of the role of these calcium channels as "thermosensors" in plants is needed.

Perception of cold stress is better understood. In cyanobacteria, the cold sensor Hik33 is involved in the activation of FADs. The structural characteristics of Hik33 suggest that cold stress might promote a conformational change of one of its transmembrane domain, which would trigger the dimerisation of Hik33 and its activation. However, the way in which Hik33 perceives cold-induced changes in membrane fluidity at the sub-molecular level remains to be clarified. Hik33 can be considered as a "multistress sensor" in Synechocystis because it can perceive also light, nutrient and hyperosmotic stresses (Mikami & Murata, 2003). Another histidine kinase, DesK, was identified as a cold sensor that regulates the cold inducible expression of the Δ5 desaturase in *Bacillus subtilis* (Aguilar, Hernandez-Arriaga, Cybulski, Erazo, & de Mendoza, 2001). Temperature sensing of DesK involves a built-in instability caused by a group of hydrophilic residues located near the N-terminus of the first transmembrane segment (Fig. 14). At lower temperature, the membrane is thicker due to a more ordered packing of lipids, and these residues are buried in the lipid phase, triggering the kinase conformation of DesK. At higher temperatures, the membrane becomes thinner with more disordered lipids, and hydrophilic residues are pushed toward the cytoplasm, promoting the required conformational changes to switch to the phosphatase conformation (Inda et al., 2014).

In plants, there is no evidence of membrane fluidity change triggering cold signaling. Two sensors were identified, COLD1 in *Oriza sativa* and the protein kinase OST1 in *Arabidopsis thaliana,* but the perception mechanisms are not known. COLD1 interacts with the protein RGA1 during cold temperatures, resulting in increased GTPase activity, which activates, in turn, an unknown calcium influx channel. OST1 is activated upon myristoylation of its interactor EGR2 in response to cold temperatures (Lamers et al., 2020).

Therefore, plants and algae adapt their membrane composition to face changing temperatures. In plants, during both heat and cold stress, TAG content is increased serving as a storage of polyunsaturated lipids. Indeed, heat stress reduces the quantities of chloroplast glycerolipids containing 18:3 acyl chains but increased 18:3-containing PC, DAG, and TAG. TAG is proposed to serve as acyl carrier for lipid turnover and β-oxidation

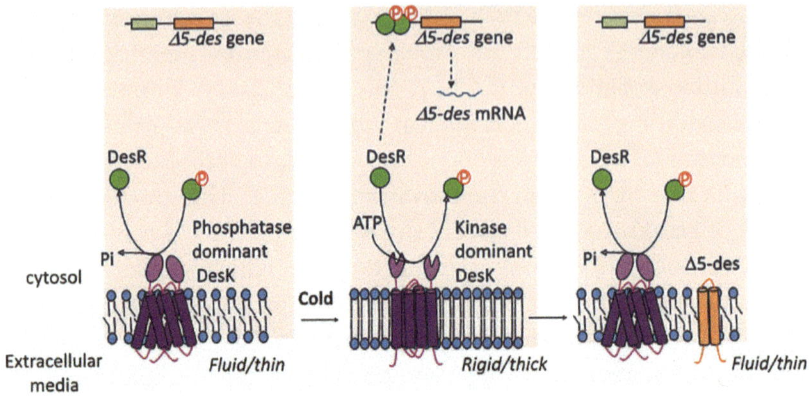

Fig. 14 Influence of membrane thickness and rigidity on the thermosensor Desk. The thermosensor DesK is composed of transmembrane domains and a cytosolic histidine kinase/phosphatase domain acting at the top of a regulatory cascade controlling the synthesis of unsaturated fatty acids in *Bacillus subtilis*. Following a drop in temperature, the membrane spans sense an increase in bilayer thickness due to an increased ordering of the lipid acyl chains by the burial of hydrophilic residues. This promotes a kinase dominant state of DesK, which autophosphorylates and then transfers the phosphate group to DesR. P-DesR activates transcription of the *des* gene coding for a Δ5-desaturase. Δ5-Des desaturates the acyl chains of membrane phospholipids, increasing membrane fluidity. This change promotes the phosphatase-dominant state of DesK, leading to DesR dephosphorylation and turning off transcription of the *des* gene. *Adapted from Holthuis JC, Menon AK (2014) Lipid landscapes and pipelines in membrane homeostasis. Nature 510: 48–57.*

of chloroplast-derived fatty acids (Higashi, Okazaki, Myouga, Shinozaki, & Saito, 2015). HIL1 (Heat Induce Lipase 1) encodes a chloroplastic lipase that releases 18:3 in the *sn-1* position from MGDG. HIL1 is activated by heat stress, suggesting it has an important role in the lipid remodeling process induced by heat stress in plants (Higashi et al., 2018). On the opposite, desaturase activity is stimulated during cold conditions to increase desaturation of lipids in the lipid bilayer of chloroplast and other membranes to maintain membrane fluidity (Lamers et al., 2020). The hydrophobic fatty acyl chains might also be shortened to maintain the membrane fluidity (Barrero-Sicilia, Silvestre, Haslam, & Michaelson, 2017; Pittera et al., 2018).

In eukaryotes, lipid species that increase membrane viscosity—sterols and sphingolipids—are depleted from energy-transducing membranes, i.e., chloroplast and mitochondria (van Meer, Voelker, & Feigenson, 2008). Therefore, the level of fatty acid unsaturation drives most of the membrane fluidity. Cold and heat stress modulating fatty acid unsaturation have therefore an influence on the activity of these organelles. For instance, an increase in thylakoid fatty acid saturation inhibits the rate of linear photosynthetic electron transport (Oquist, 1982; Vigh, Horvath, van Hasselt, & Kuiper, 1985), whereas a decrease in fatty acid saturation in mitochondria membranes is connected to an increase of quinone diffusion during respiration (Budin et al., 2018).

To summarize, membrane fluidity is affected by temperature. A homeostasis of the membrane viscosity and fluidity is necessary for the survival of plants and algae and to preserve metabolic reactions occurring in the membrane, such as photosynthesis or respiration. Fatty acid desaturases are therefore regulated to preserve these properties. However, as indicated earlier, membranes contain an equilibrium of HII and bilayer forming lipids and this equilibrium must be preserved for membrane integrity.

4.2 Homeostasis of the membrane tension and curvature: Effect of mechanical stress

The various organelles (ER, Golgi, endosome, plasma membrane) associated with the secretory pathway have differences in lipid composition that help shaping their specialized tasks. For example, they do not have the same membrane thickness throughout the secretory pathway, ER membrane being thinner than plasma membrane. This implies that transmembrane domains of ER proteins contain two amino acids less than transmembrane domains of plasma membrane proteins (Singh & Mittal, 2016). Therefore, the collective action of bulk lipids determines organelle function by influencing generic physical membrane parameters such as fluidity, thickness, lipid packing density and surface charges, despite the fact that specific lipids are also able to modulate locally the physical properties of membranes (Noack & Jaillais, 2020). For proper organelle function, these physical parameters must be kept within an appropriate range (Holthuis & Menon, 2014).

We already saw that freezing causes the formation of non-bilayer structures, such as the HII phase in the plasma and chloroplast membranes, inducing in turn membrane lesions and damaging the cell membrane integrity (Barrero-Sicilia et al., 2017). In *Arabidopsis thaliana*, during freezing, the chloroplast enzyme SFR2 is activated in addition to the diacylglycerol acyltransferase DGAT1 to remove MGDG and to produce TAG, thus preventing membrane disruption by removing HII forming lipids (Arisz et al., 2018; Moellering, Muthan, & Benning, 2010). SFR2 is activated by acidification of the cytosol, a process that occurs during freezing

(Barnes, Benning, & Roston, 2016; Barnes, Elowsky, & Roston, 2019), but the mechanism of perception of this acidification is not known.

Perception mechanism of hypo-osmotic or hyper-osmotic stresses is starting to emerge. These stresses increase membrane surface tension produced either by osmotic swelling of the cell or by shrinking of the plasma membrane prevented by the attachment of the membrane to the cell wall. How plant and alga cells are able to distinguish between these two stresses is not fully understood but they perceive membrane property changes and trigger cell signals to recover their initial properties. The primary sensors that mediate such rapid responses to mechanical changes are ion channels and they are called mechanosensors (Kefauver, Ward, & Patapoutian, 2020).

In plants, four kinds of mechanosensors have been identified so far: MscS, Piezo, OSCA and MCA. No study was done on algae so far. MscS, for mechanosensitive channel small conductance, is a family widely distributed from bacteria to animals, yeast and plants. Land plants encode 10 MscS-like (MSL) genes that are grouped into three categories on the basis of their subcellular location (Hamilton, Schlegel, & Haswell, 2015). Group I and group II are expressed in mitochondria and plastids, where they have an osmoregulatory role. *For example,* MSL2 and MSL3 control plastid size, shape, and perhaps division during normal plant development by altering ion flux in response to changes in membrane tension (Haswell & Meyerowitz, 2006). During hypo-osmotic stress, such as flooding or rehydration, they prevent epidermis plastid lysis allowing the efflux of osmolytes from the plastid into the cytosol (Veley, Marshburn, Clure, & Haswell, 2012). Group III reside at the plasma membrane, where their roles remain an active area of research. In *Arabidopsis thaliana*, MSL8 and MSL10 have been shown to be mechanosensitive ion channels, with roles in pollen survival, stress-induced cell death and cell swelling in seeds (Basu & Haswell, 2020; Basu, Shoots, & Haswell, 2020; Hamilton & Haswell, 2017). Recent structures of MSL1 reveal a multimeric architecture shared with bacterial MscS. An amphipathic helix on the cytosol leaflet is bound to several lipids that may be important for channel opening in response to mechanical stimuli (Rasmussen, Flegler, Rasmussen, & Bottcher, 2019; Reddy, Bavi, Lu, Park, & Perozo, 2019) (Fig. 15A). It is proposed that when cells are exposed to hypo-osmotic shock, the resultant cell swelling induces an increase in membrane tension that will move the amphipathic helixes and increase the probability of opening the channel. When open, these channels allow ions and other small osmolytes to flow out of the cell, reducing internal osmotic pressure and protecting the membrane from lysis (Hamilton & Haswell, 2017).

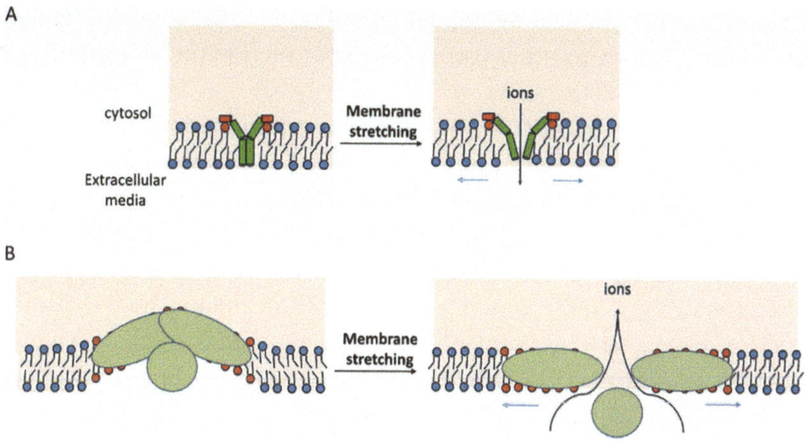

Fig. 15 "Force from lipids" models of mechanosensing (A) The dragging model. Lipids interact with an amphipathic helix and drag it outwards upon membrane expansion. MscS, for example, has an amphipathic helix on the internal leaflet (red) that drives a tilt to the pore-lining helix (green) as it is "dragged" outward under tension opening the channel to allow the efflux of ion. (B) Membrane dome model. Channel curvature within the membrane induced by the mechanosensor stores energy. During mechanic stress inducing stretching of the membrane, PIEZOs, for example, expand and flatten, gating the pore. *Adapted from Kefauver JM, Ward AB, Patapoutian A (2020) Discoveries in structure and physiology of mechanically activated ion channels. Nature 587: 567–576.*

PIEZOs are large trimeric proteins with a triskelion or three-blade propeller architecture. The three blade domains extend outward within the lipid bilayer and an extracellular cap domain resides below the central pore (Fig. 15B). PIEZOs have an unusually large number of transmembrane domain per protomer; 38 transmembrane helices per subunit are resolved in the mammalian PIEZO2 structure (Wang et al., 2019), matching previous membrane topology predictions for PIEZO1 (Coste et al., 2015). Amphipathic helices are known to sense and/or induce membrane curvature (Saotome et al., 2018). The curved form of PIEZOs could cause a local distortion to the native cell membrane or result in its preferred location to a membrane domain of similar curvature. Membrane tension would flatten the piezo trimer and open the cation channel. Most plant species harbor a single Piezo protein but it is absent in Stramenopiles (brown algae) (Cox, Bavi, & Martinac, 2017). It is involved in limiting the systemic movement of plant viruses (Zhang et al., 2019) and is required for proper root penetration in compacted environments creating mechanical stresses (Mousavi et al., 2021). AtPIEZO protein might directly alleviate mechanical pressure in columella cells by protecting cell wall integrity and/or by transducing Ca^{2+} signals to other parts of the root such as the elongation zone (Mousavi et al., 2021).

Genes coding for *MCA* (*mid1*-complementing activity) orthologs are found exclusively in the plant kingdom. *Arabidopsis thaliana* contain two genes that have redundant functions. MCA mediates Ca^{2+} influx upon mechanical stimulation, such as hypo-osmotic shock and membrane stretch, and facilitates, alike PIEZO, penetration of the root in a hard medium (Nakagawa et al., 2007). In contrast to PIEZO proteins, the polypeptides of MCA have a single transmembrane segment and assemble as a homotetramer to form a Ca^{2+}-permeable channel (Kamano et al., 2015; Shigematsu et al., 2014). The mechanism of mechanoperception is not known.

The last known family, OSCA (osmosensing calcium antiporter)/TMEM63 (transmembrane protein 63) proteins, constitutes the largest family of mechanosensitive channels. Initially described as hyperosmolarity sensors in *A. thaliana*, OSCA proteins have recently been shown to be pore-forming, inherently mechanosensitive channels that are conserved across plants and animals (Kefauver et al., 2020). OSCAs are stretch-activated at a high threshold compared to PIEZO channels (Zhang et al., 2018). In plants, mutations in OSCA1.1 reduce the influx of cytosolic Ca^{2+} in response to hyperosmolarity stress. Mutations prevent Ca^{2+} signal to be generated and funneled to downstream events, such as the activation of ASK1 protein kinase, ABA accumulation and stomatal closure, leading to reduction of water loss (Yuan et al., 2014). OSCA may act as a stretch sensor that coordinates the movements of the transmembrane helices, similar to the model for PIEZO mechanosensing (Liu, Wang, & Sun, 2018; Maity et al., 2019).

At present, two basic models are used to describe the gating of mechanosensitive channels by mechanical force. The "force-from-lipids" model favors the view that force transmitted through the lipid bilayer directly gates the channels, whereas the "force-from-filament" model favors the notion of channel gating via cytoskeleton or cell wall tethers that are directly connected to the mechanosensitive channels. The force-from filament model is not developed here because it does not involve lipids. Force-from-lipids was validated for MscS, Piezo and OSCA from liposome or artificial bilayer experiments where no cytoskeleton or cell wall is present (Kefauver et al., 2020). Force-from-lipids manifests itself in many ways that are not mutually exclusive. Two main examples are hydrophobic mismatch and bilayer curvature. Given that the lipid bilayer is practically incompressible, membrane stretching thins the lipid bilayer, which results in hydrophobic mismatch between the hydrophobic length of the transmembrane domain of a channel and the bilayer. To minimize the mechanical strain on the bilayer, two hypothesizes are proposed: (1) lipids will move laterally in the bilayer to surround the channel with lipids of matching size and shape, which may influence helix tilt, and thus mechanosensitive channel gating, by shifting the equilibrium between open and closed conformations; (2) lipids bound to an amphipathic helix connected to the pore will move in the membrane away from the pore, dragging the amphipathic helix and straightening the channel as the membrane equilibrates under tension. This will favor the opening of the channel (Fig. 15A) (Hamill & Martinac, 2001). A different type of membrane distortion is a change in curvature (Fig. 15B). Channel activity can be modulated by the addition of conical lipids or amphipathic molecules that tend to bend membranes (Martinac et al., 2018). For example, the addition of lysoPC that will tend to flatten the curved membrane is able to activate OSCA (Zhang et al., 2018). Furthermore, the intrinsic curvature of PIEZOs will favor membrane bending and when tension is applied to the membrane by an atomic force microscopy cantilever, PIEZO1 protein reconstituted in planar membranes undergoes substantial area expansion implying the opening of the channel (Lin et al., 2019). Of course, these structural and mechanistic similarities exist in other channels, where instead of force-from-lipids, the gating stimuli can be membrane potential, pH, or ligand binding. Therefore, not only mechanosensors are sensitive to force-from-lipids. For example, water permeability of individual aquaporin channels depends on composition, thickness, and elasticity of the lipid bilayer (Tong, Briggs, & McIntosh, 2012).

Hydrophobic mismatch can also be used to regulate lipid synthesis and to maintain membrane homeostasis. The most studied regulation by hydrophobic mismatch is the balance between PE and PC that is controlled by the cytidine triphosphate (CTP): phosphocholine cytidylyltransferase (CCT). This enzyme is the limiting step for PC synthesis; it transfers CTP to phosphocholine to form CDP-choline which is thereafter transferred to a DAG to form PC. CCT contains an amphipathic helix—called domain M—that silences the activity of the enzyme in its soluble form (Lee, Taneva, Holland, Tieleman, & Cornell, 2014). In the presence of PC

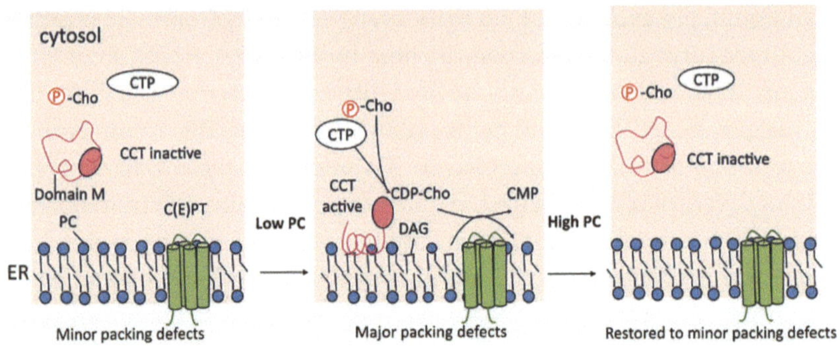

Fig. 16 Control of the CTP:phosphocholine cytidylyltransferase (CCT) activity by membrane defect. CCT, the rate-limiting enzyme in PC biosynthesis, interconverts between an inactive soluble and active membrane bound form in response to changes in membrane lipid packing density. An amphipathic helix, called domain M, silences the activity of the enzyme in its soluble form. Lipid packing defects arising from increased levels of conical lipids (for example, DAG or PE) are sensed by domain M, creating a hydrophobic face for membrane binding. Association of domain M with membranes relieves self-inhibition, increasing the affinity of CCT for its substrate CTP. This then enhances biosynthesis of PtdCho to prevent transition of the membrane into a porous state. C(*E*)PT, choline/ethanolaminephosphotransferase. *Adapted from Holthuis JC, Menon AK (2014) Lipid landscapes and pipelines in membrane homeostasis. Nature 510: 48–57.*

bilayers with a high content of HII forming lipids such as DAG or PE, domain M undergoes a conformational switch that creates a hydrophobic surface for membrane binding (Attard, Templer, Smith, Hunt, & Jackowski, 2000) (Fig. 16). Association with membranes relieves self-inhibition and drastically increases the affinity of CCT for its substrate CTP. By contrast, membrane association and activity of CCT is reduced when DAG/PC or PE/PC ratios are low. Thus, CCT seems to act as a general sensor of lipid packing defects that signals a demand for PC biosynthesis to allow membrane expansion or to prevent its transition into a porous state (Holthuis & Menon, 2014). Consistent with this model, in mammals, CCT is recruited to lipid droplets in the growing phase to allow production of more PC molecules to surround the expanding oily core. It will be interesting to investigate the BTA1 enzyme involved in betaine lipid synthesis to see if they share the same kind of regulation.

Finally, to illustrate the influence of glycerolipid properties on membrane architecture, we describe in the next paragraph the changes occurring in one organelle, the chloroplast, through the life cycle of the land plant, and architectural differences existing between different photosynthetic organisms.

4.3 Plastid lipid composition has an influence on plastid architecture and therefore on plastid compartmentalization and function

Plastids are typical organelles present in photosynthetic eukaryotes. They derived from the endosymbiosis of a cyanobacteria and are conserved through the evolution up to higher plants and through the alga kingdoms. They have a unique lipid composition that is conserved throughout all types of plastids whereas they can adopt different membrane architecture (Fig. 17). One hypothesis for the conservation of the plastid lipid composition is their specific biophysical properties.

As indicated earlier, plastid membranes are mainly composed of glycolipids MGDG, DGDG, SQDG and one phospholipid PG with a particular fatty acid in the *sn-2* position, a 16:1Δ3 in *trans*-conformation. The fact that 16:1 t-PG is only associated with the photosynthetic machinery and conserved throughout the evolution (Nichols, Harris, & James, 1965; Roughan & Boardman, 1972), suggests a crucial structural or functional role in eukaryotic photosynthesis (Boudiere et al., 2014; Wada & Murata, 2007). However, plants devoid of 16:1 t-PG are growing normally with minor defects in photosynthesis (Browse, McCourt, & Somerville, 1985; Horn, Smith, Clark, Froehlich, & Benning, 2020; McCourt, Browse, Watson, Arntzen, & Somerville, 1985), leaving the conservation of 16:1 t-PG as an unsolved mystery. As described above, MGDG and DGDG are neutral lipids, the first one forming HII phases whereas the second one bilayer structures. As a functional consequence, DGDG forms and stabilizes membrane bilayers, whereas MGDG is likely to favor curvature of thylakoids (Gounaris, Sen, Brain, Quinn, & Williams, 1983). On the opposite, SQDG and PG are anionic lipid and are both bilayer forming lipid. SQDG is able to replace PG during phosphate starvation indicating that both lipids share common properties. All these lipids associate specifically with protein complexes including photosystems, explaining why they might be conserved through the evolution (Boudiere et al., 2014).

In angiosperms, plastids can adopt different architectures (Fig. 17A). They start in meristem tissues as proplastids. Proplastids are small plastids, about 1 μm in diameter, and less complex at the structural level than other plant plastids. They are colorless, can change their morphology and may content a variable amount of tubular-like internal membranous compartments, as well as starch depots. By division and differentiation, they give rise to the rest of the plant plastids. In hypocotyls and roots of growing seedlings, proplastids usually develop into a number of colorless plastids commonly

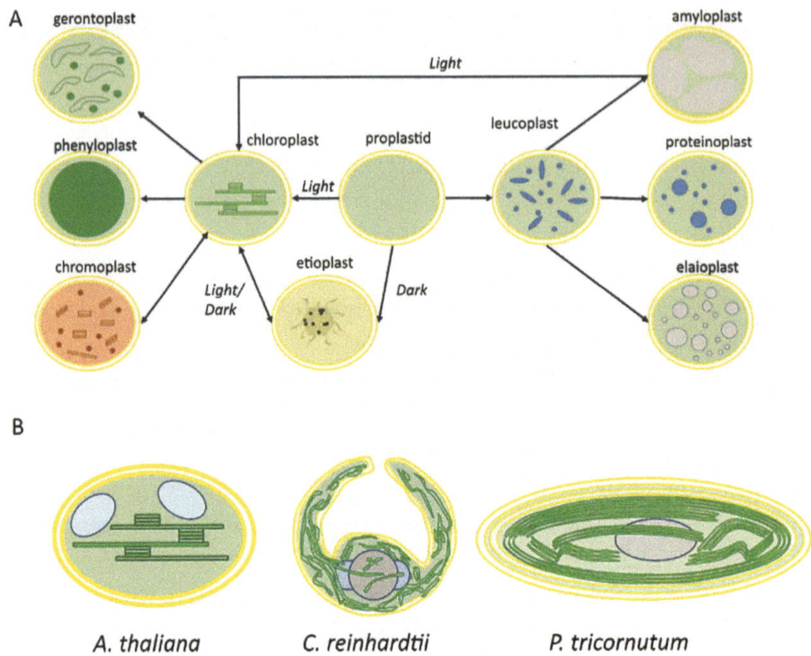

Fig. 17 Diversity and differentiation of plastids depending on tissue type, developmental stages and organisms. (A) Diversity of plastids in land plants. Arrows indicate transitions between different plastid types. The proplastid is the progenitor plastid, point of differentiation, developing into either etioplast or chloroplast depending on exposure to darkness or light, respectively. Proplastids are also able to develop into the intermediate plastid, the leucoplast, that may further differentiate into either amylo-, proteino- or elaioplasts, accumulating starch, proteins or oils, respectively. Chloroplasts can also further differentiate depending on environmental stimuli and plant cell types into chromoplasts, phenyloplasts and the dying gerontoplast, in which resources are recycled and redistributed. Chromoplast and phenyloplast accumulate phytochemicals such as carotenoids and phenyl-propanoids, respectively. (B) Architectures of thylakoid membranes. Models of the diverse structures of thylakoid membranes (green) in *Arabidopsis thaliana*, *Chlamydomonas reinhardtii* and *Phaeodactylum tricornutum*. Pyrenoids are shown in gray, starch in blue, envelope membranes are indicated in yellow, thylakoid in green. *Panel (A) adapted from Knudsen C, Gallage NJ, Hansen CC, Moller BL, Laursen T (2018) Dynamic metabolic solutions to the sessile life style of plants. Natural Product Reports 35: 1140–1155. Panel (B) adapted from Rast A, Heinz S, Nickelsen J (2015) Biogenesis of thylakoid membranes. Biochimica et Biophysica Acta 1847: 821–30.*

summarized under the term leucoplasts. This group does not develop a network of internal membranes and therefore are delimitated only by the plastid envelope. It comprises amyloplasts, proteinoplast, and elaioplasts. In later stages, they may also develop in other parts of the plant. These colorless plastids develop even if tissues are exposed to light (Liebers,

Chevalier, Blanvillain, & Pfannschmidt, 2018). Amyloplasts can reversibly differentiate into photosynthetic chloroplasts when exposed to light, as exampled by the greening of potato tubers. In contrast, differentiation of proplastids into elaioplasts and proteinoplasts is non-reversible (Knudsen, Gallage, Hansen, Moller, & Laursen, 2018). Elaioplasts play a key role in storage and biosynthesis of lipids and oils, including terpenoids, which in citrus fruits are then exported into secretory pockets. Proteinoplasts (or proteoplasts) are defined as the site of protein and lipoprotein storage in the plant cell, and these plastids do not apparently encompass biosynthesis or storage roles for phytochemicals (Knudsen et al., 2018).

In the dark, seedlings follow a developmental program called skotomorphogenesis (Solymosi & Schoefs, 2010). In cotyledons of such seedlings, proplastids develop into etioplasts. Etioplasts are characteristic for this developmental program and represent an intermediate stand-by state of chloroplast formation. They do not develop a thylakoid membrane system, but paracrystalline structures, i.e., three dimensional lattice membrane structures called prolamellar bodies (PLBs) from which flattened lamellar prothylakoids (PTs) are radiated. PLBs are composed of regular arrangements of NADPH, the enzyme protochlorophyllide-oxido-reductase (POR), the chlorophyll precursor protochlorophyllide (Pchlide) and galactolipids (Bastien et al., 2016).

The paracrystalline PLB structure is identified by its specific composition, in which polar lipids supposedly play a part in the formation and maintenance of such unique cubic arrangements. The lipid composition of PLBs is similar to that of the thylakoid membrane: in PLBs of wheat etioplasts, MGDG and DGDG account for approximately 50% and 30%, respectively, of total glycerolipids but with a slight increase of the MGDG/DGDG ratio compared with thylakoids (Selstam & Sandelius, 1984). This increase of the ratio MGDG/DGDG in PLB was supposed to be the main driver in favor of cubic arrangement of the membrane instead of bilayer organization. Such hypothesis was mainly based on the known roles of the conically shaped MGDG and the bilayer-forming DGDG in the bending of etioplast membranes (Demé et al., 2014). Expression of a dexamethasone inducible artificial microRNA targeting MGD1 in dark-grown Arabidopsis seedlings reduced MGD1 mRNA level by 70% as compared with the control, which resulted in 36% loss of MGDG without affecting DGDG content (Fujii, Kobayashi, Nagata, Masuda, & Wada, 2017). The decrease of MGDG in etioplasts only slightly changed the lattice structure of PLBs, whereas it strongly impaired the elongation of

PTs and the formation and oligomerisation of Pchlide–POR complexes in membranes. On the opposite, in the dark-grown dgd1 mutant, DGDG content was reduced by 80% as compared with the wild type and the lattice structure of PLBs was strongly disordered while PTs were severely underdeveloped (Fujii, Kobayashi, Nagata, Masuda, & Wada, 2018). These data demonstrate the essential roles of galactolipids in the formation of internal membranes and membrane-associated processes in etioplasts but the ratio MGDG/DGDG is not the only actor of the cubic architecture of PLB.

More generally, PLB, similar to other cubic membranes in nature such as alveolar surface of mammalian lungs, retina cells, cuticle of butterfly, etc., separates aqueous phases into a two-channel system corresponding, respectively, to the future stroma and lumen compartment. This enables different molecular composition and, therefore, function (Mezzenga et al., 2019). The size of the water channels, controlled by the lipid and pigment composition that will drive the scale of the structure, can exclude or authorize the location of certain molecules on one particular side of the membrane (Bykowski et al., 2020). In such way, we can directly link the ultrastructural features of PLBs and PTs, length-scale in particular, with new possible biological functions, such as the segregation of proteins in different membrane domains. Tomographic reconstructions show that ATP synthase monomers are enriched in the PTs, whereas plastid ribosomes and POR are in the tubular lattice (Floris & Kuhlbrandt, 2021).

With light exposure, etioplasts rapidly differentiate into chloroplasts to allow photo-autotrophic growth. The differentiation from etioplasts to chloroplasts after light exposure involves the dynamic transformation of PLBs to thylakoids, which is accompanied by changes in pigment compositions from Pchlide to chlorophyll and accumulation of other photosynthetic components including photosystem (PS) I, PSII, and the light-harvesting complex (Fujii, Wada, & Kobayashi, 2019). The dynamic conversion of etioplasts into chloroplasts could operate, at least partly, via a cubic-to-lamellar lipid-phase transition. This phase transition is a spontaneous process and the released energy could be used as a driving force to change the plastid inner architecture (Demé et al., 2014; Pipitone et al., 2021).

Chloroplasts are the most studied plastid because they contain the photosynthetic membranes, the thylakoids. However, thylakoid architecture is highly different depending on light intensity and quality, and also on the organism it belongs to (Fig. 17B). In angiosperms, thylakoids comprise a sum of stroma lamellae, which turns around the grana stacks as a right-handed helix, connecting individual grana disks by narrow membrane protrusions (Daum et al., 2014). According to the helical model, the granum has a roughly cylindrical shape comprising a variable number of stacked granal thylakoids with a flat discoid shape and typical diameters in the range of 400–600 nm (Lambrev & Akhtar, 2019). In *Chlamydomonas reinhardtii*, thylakoids were most commonly found in stacks of variable length but uniform spacing, with a number of thylakoids per stack ranging from 2 to 15, with a median of 3 thylakoids (Engel et al., 2015). In the stroma, in contrast to the disordered arrangement of Rubisco in higher plants, this enzyme is concentrated in membrane-less organelles called pyrenoids, a highly dynamic liquid-like body packed with Rubisco and traversed by thylakoid membranes (Freeman Rosenzweig et al., 2017). In *Phaeodactylum tricornutum*, the 3D structure of the photosynthetic membranes show the presence of three parallel layers of stacked thylakoids, connected between them, as well as the presence of a pyrenoid (Flori et al., 2017; Pyszniak & Gibbs, 1992). Why these organisms display different membrane organizations whereas their constituting lipids and proteins are conserved? One explanation could be the different proportion of lipid classes, or differences in the fatty acid composition or pigment content such as chlorophyll or carotenoid. For example, diatoms are richer in SQDG than green algae and land plants, contain very long polyunsaturated fatty acid chains and their photosystems comprise chlorophyll *c* and fucoxanthin instead of chlorophyll *a* and xanthophyll (Tables 1 and 2).

As a common feature, in most photosynthetic eukaryotes, the thylakoid membranes are subdivided into appressed regions that face adjacent membranes within thylakoid stacks and non-appressed regions that freely face the stroma. In both plants and algae, as in etioplast, membrane architecture might be used to isolate proteins. Indeed, PSII and PSI appear to be segregated to the appressed and non-appressed membranes, respectively (Flori et al., 2017; Goodenough & Staehelin, 1971; Pribil, Labs, & Leister, 2014; Wietrzynski & Engel, 2021; Wollman, Olive, Bennoun, & Recouvreur, 1980). This lateral heterogeneity is believed to coordinate photosynthesis by concentrating different reactions within specialized membrane domains, while the redistribution of light-harvesting complex II (LHCII) antennas between these domains may enable adaptation to changing environmental conditions (Minagawa & Tokutsu, 2015; Nawrocki, Santabarbara, Mosebach, Wollman, & Rappaport, 2016).

The stacking of thylakoids to form grana or appressed membranes depends on several parameters that are still not completely understood.

Forces that sustain thylakoid stacking were listed by (Chow, Kim, Horton, & Anderson, 2005). The LHCII of PSII is supposed to drive the structural dynamics of grana stacking in higher plants; phosphorylations of LHCII and PSII are involved in state transition implying stacking and destacking of thylakoids (Tikkanen & Aro, 2012). However, LHCII is apparently not the only one factor driving thylakoid stacking. Galactolipids, mainly DGDG, favor also membrane stacking by forming hydrogen bonds between the galactose polar heads screening repulsive electrostatic contribution of the charged lipids PG and SQDG (Demé et al., 2014; Kanduč et al., 2017).

Formation of thylakoids needs an intense traffic of material from the inner membrane to the thylakoid network. Vesicles accumulating in the stroma near the chloroplast envelope could organize such a traffic. Indeed, a study uncovered the presence of such vesicles in mature chloroplasts, proplastids, etioplasts and under a variety of conditions, supporting the hypothesis of vesicle-mediated transport (Lindquist, Solymosi, & Aronsson, 2016). However, other mechanisms using the non-bilayer properties of MGDG could also participate, such as flattened invaginations from the inner membrane and HII intermediate structures (Bastien et al., 2016; Garab, Ughy, & Goss, 2016).

Maintaining a constant MGDG/DGDG ratio in thylakoid membranes (at least under standard growth conditions) seems crucial for the stability and functional integrity of these membranes (Dormann & Benning, 2002). Signatures of non-bilayer phases in thylakoid were found by ^{31}P NMR and by the fluorescence of the lipid-phase-sensitive dye merocyanine-540 (Krumova, Dijkema, et al., 2008; Krumova, Koehorst, et al., 2008). More recently, higher-resolution NMR data have identified on the stromal and lumenal side of thylakoid membranes the coexistence of lamellar and non-lamellar phases, tentatively assigned to inverted hexagonal HII and isotropic phases, respectively (Garab et al., 2017). This HII/Lα coexistence would for instance explain the influence of MGDG on non-photochemical quenching (NPQ) via the violaxanthin de-epoxidase activity (Schaller et al., 2010), because this enzyme requires inverted hexagonal lipid structures for its activity (Jahns, Latowski, & Strzalka, 2009).

In angiosperms, during the course of plant development, chloroplasts are matured in gerontoplast, chromoplast or phenyloplast (Fig. 17A). In senescing leaves, gerontoplasts are generated as the last ontogeny stage of chloroplasts. These plastids no longer harbor functional DNA. How exactly the degradation of the thylakoid membrane systems occurs is not fully understood, but during degradation more and bigger lipid-rich micro-compartment, called plastoglobuli, are formed. In fleshy fruits such as tomatoes, ripening is associated with differentiation of green fruit chloroplasts into ripe fruit chromoplasts. Chromoplasts are recognized by massive accumulation of carotenoid pigments that give the red, orange and yellow colors to the plant structures. Upon maturity, the concentration of carotenoids, stored in the plastoglobuli, increases and may result in the formation of carotenoid crystals. Recent studies on re-differentiated chloroplasts demonstrated that plastoglobuli may serve as biosynthetic units as well as storage sites of phenylpropanoids. Based on studies in the vanilla pod that focused on their ability to accumulate vanillin glucoside, a phenylpropanoid derived defense compound, these plastids were named phenyloplasts. The transition from chloroplasts to phenyloplasts proceeds by a decline of photosynthetic capacity, thylakoid lamellae and chlorophyll content, associated with accumulation of high concentrations of phenylpropanoid-derived glucosides (Knudsen et al., 2018). Plastoglobuli are also present in chloroplasts but they are much smaller than in phenyloplasts or chromoplasts. In contrast to the thylakoid membrane, plastoglobuli are composed of neutral lipids (triacylglycerol, phytol esters, and free fatty acids), prenylquinones (α-tocopherol, plastoquinone, plastochromanol, and phylloquinone), as well as carotenoids—the last category being especially prevalent in the plastoglobuli of chromoplasts (van Wijk & Kessler, 2017).

A possible function of plastoglobuli that has not been considered so far is that they serve as dynamic lipid reservoirs for thylakoid membranes. Indeed, plastoglobuli sizes vary and they contain thylakoid lipids at their surface (van Wijk & Kessler, 2017). A challenge on this hypothesis relies on the observation that plastoglobuli monolayers are connected only with the outer thylakoid membrane leaflet (Kirchhoff, 2019). Because thylakoid membranes most likely do not contain flippases (Jouhet, Marechal, & Block, 2007), the question arises of how the inner leaflet can profit from a dynamic lipid exchange mechanism between plastoglobuli and the outer leaflet. A possibility is the formation of non-bilayer hexagonal II phases that can act as lipid carousels between two lipid monolayers in thylakoid membranes (Garab et al., 2016). In addition to the "lipid-exchange" mechanism to maintain the thylakoid lipid/protein ratio, plastoglobuli also might adjust the thylakoid lipid content by remobilization of lipids (van Wijk & Kessler, 2017).

To summarize, by adapting lipid class distribution, plastids are able to modulate their membrane architecture, allowing the formation of membrane domains and favoring protein or pigment segregation. All these modulation needs to be highly dynamic to face sudden change in the environment such as light intensity, temperature variation or osmotic stress. How plants and algae transduce the perception of these changes to modify their lipid metabolism and membrane composition remains a huge and challenging question.

Acknowledgment

This work was supported by ANR-DFG ANR-18-CE92-0015, by ILL PhD funding SPEC-2-2018, by the French National Research Agency in the framework of the "Investissements d'avenir" program Glyco@Alps (ANR-15-IDEX-02) and the Labex GRAL, financed within the University Grenoble Alpes graduate school (Ecoles Universitaires de Recherche) CBH-EUR-GS (ANR-17-EURE-0003).

References

Aguilar, P. S., & de Mendoza, D. (2006). Control of fatty acid desaturation: A mechanism conserved from bacteria to humans. *Molecular Microbiology*, *62*, 1507–1514.

Aguilar, P. S., Hernandez-Arriaga, A. M., Cybulski, L. E., Erazo, A. C., & de Mendoza, D. (2001). Molecular basis of thermosensing: A two-component signal transduction thermometer in Bacillus subtilis. *The EMBO Journal*, *20*, 1681–1691.

Arisz, S. A., Heo, J. Y., Koevoets, I. T., Zhao, T., van Egmond, P., Meyer, A. J., et al. (2018). Diacylglycerol acyltransferase1 contributes to freezing tolerance. *Plant Physiology*, *177*, 1410–1424.

Attard, G. S., Templer, R. H., Smith, W. S., Hunt, A. N., & Jackowski, S. (2000). Modulation of CTP:Phosphocholine cytidylyltransferase by membrane curvature elastic stress. *Proceedings of the National Academy of Sciences of the United States of America*, *97*, 9032–9036.

Barnes, A. C., Benning, C., & Roston, R. L. (2016). Chloroplast membrane remodeling during freezing stress is accompanied by cytoplasmic acidification activating SENSITIVE TO FREEZING2. *Plant Physiology*, *171*, 2140–2149.

Barnes, A. C., Elowsky, C. G., & Roston, R. L. (2019). An Arabidopsis protoplast isolation method reduces cytosolic acidification and activation of the chloroplast stress sensor SENSITIVE TO FREEZING 2. *Plant Signaling & Behavior*, *14*, 1629270.

Barrero-Sicilia, C., Silvestre, S., Haslam, R. P., & Michaelson, L. V. (2017). Lipid remodelling: Unravelling the response to cold stress in Arabidopsis and its extremophile relative Eutrema salsugineum. *Plant Science*, *263*, 194–200.

Bastien, O., Botella, C., Chevalier, F., Block, M. A., Jouhet, J., Breton, C., et al. (2016). New insights on thylakoid biogenesis in plant cells. *International Review of Cell and Molecular Biology*, *323*, 1–30.

Basu, D., & Haswell, E. S. (2020). The mechanosensitive Ion Channel MSL10 potentiates responses to cell swelling in Arabidopsis seedlings. *Current Biology*, *30*, 2716–2728 (e6).

Basu, D., Shoots, J. M., & Haswell, E. S. (2020). Interactions between the N- and C-termini of the mechanosensitive ion channel AtMSL10 are consistent with a three-step mechanism for activation. *Journal of Experimental Botany*, *71*, 4020–4032.

Bigay, J., & Antonny, B. (2012). Curvature, lipid packing, and electrostatics of membrane organelles: Defining cellular territories in determining specificity. *Developmental Cell*, *23*, 886–895.

Block, M. A., Dorne, A. J., Joyard, J., & Douce, R. (1983). Preparation and characterization of membrane fractions enriched in outer and inner envelope membranes from spinach chloroplasts. I. Electrophoretic and immunochemical analyses. *The Journal of Biological Chemistry*, *258*, 13273–13280.

Block, M. A., Dorne, A., Joyard, J., & Douce, R. (1984). Preparation and characterization of membrane fractions enriched in outer and inner envelope membranes from spinach chloroplasts. In C. Sybesma (Ed.), *Advances in photosynthesis research* (pp. 27–30). Dordrecht: Springer.

Block, M. A., & Jouhet, J. (2015). Lipid trafficking at endoplasmic reticulum-chloroplast membrane contact sites. *Current Opinion in Cell Biology, 35*, 21–29.

Boudiere, L., Michaud, M., Petroutsos, D., Rebeille, F., Falconet, D., Bastien, O., et al. (2014). Glycerolipids in photosynthesis: Composition, synthesis and trafficking. *Biochimica et Biophysica Acta, 1837*, 470–480.

Bradshaw, J. P., Bushby, R. J., Giles, C. C., Saunders, M. R., & Reid, D. G. (1996). Neutron diffraction reveals the orientation of the headgroup of inositol lipids in model membranes. *Nature Structural Biology, 3*, 125–127.

Browse, J., McCourt, P., & Somerville, C. R. (1985). A mutant of Arabidopsis lacking a chloroplast-specific lipid. *Science, 227*, 763–765.

Budin, I., de Rond, T., Chen, Y., Chan, L. J. G., Petzold, C. J., & Keasling, J. D. (2018). Viscous control of cellular respiration by membrane lipid composition. *Science, 362*, 1186–1189.

Bykowski, M., Mazur, R., Buszewicz, D., Szach, J., Mostowska, A., & Kowalewska, L. (2020). Spatial Nano-morphology of the Prolamellar body in etiolated Arabidopsis thaliana plants with disturbed pigment and Polyprenol composition. *Frontiers in Cell and Development Biology, 8*, 586628.

Calvez, P., Jouhet, J., Vie, V., Durmort, C., & Zapun, A. (2019). Lipid phases and cell geometry during the cell cycle of Streptococcus pneumoniae. *Frontiers in Microbiology, 10*, 351.

Canavate, J. P., Armada, I., Rios, J. L., & Hachero-Cruzado, I. (2016). Exploring occurrence and molecular diversity of betaine lipids across taxonomy of marine microalgae. *Phytochemistry, 124*, 68–78.

Chow, W. S., Kim, E. H., Horton, P., & Anderson, J. M. (2005). Granal stacking of thylakoid membranes in higher plant chloroplasts: The physicochemical forces at work and the functional consequences that ensue. *Photochemical & Photobiological Sciences, 4*, 1081–1090.

Colin, L. A., & Jaillais, Y. (2020). Phospholipids across scales: Lipid patterns and plant development. *Current Opinion in Plant Biology, 53*, 1–9.

Coste, B., Murthy, S. E., Mathur, J., Schmidt, M., Mechioukhi, Y., Delmas, P., et al. (2015). Piezo1 ion channel pore properties are dictated by C-terminal region. *Nature Communications, 6*, 7223.

Cox, C. D., Bavi, N., & Martinac, B. (2017). Origin of the force: The force-from-lipids principle applied to piezo channels. *Current Topics in Membranes, 79*, 59–96.

Daum, B., Quax, T. E., Sachse, M., Mills, D. J., Reimann, J., Yildiz, O., et al. (2014). Self-assembly of the general membrane-remodeling protein PVAP into sevenfold virus-associated pyramids. *Proceedings of the National Academy of Sciences of the United States of America, 111*, 3829–3834.

Davis, J. H. (1983). The description of membrane lipid conformation, order and dynamics by 2H-NMR. *Biochimica et Biophysica Acta, 737*, 117–171.

de Mendoza, D., & Pilon, M. (2019). Control of membrane lipid homeostasis by lipid-bilayer associated sensors: A mechanism conserved from bacteria to humans. *Progress in Lipid Research, 76*, 100996. https://doi.org/10.1016/j.plipres.2019.100996.

Delacroix, H., Gulik-Krzywicki, T., Mariani, P., & Luzzati, V. (1993). Freeze-fracture electron microscope study of lipid systems. The cubic phase of space group Pm3n. *Journal of Molecular Biology, 229*, 526–539.

Demé, B., Cataye, C., Block, M. A., Marechal, E., & Jouhet, J. (2014). Contribution of galactoglycerolipids to the 3-dimensional architecture of thylakoids. *The FASEB Journal, 28*, 3373–3383.

Demetzos, C. (2008). Differential scanning calorimetry (DSC): A tool to study the thermal behavior of lipid bilayers and liposomal stability. *Journal of Liposome Research, 18*, 159–173.

Demurtas, D., Guichard, P., Martiel, I., Mezzenga, R., Hebert, C., & Sagalowicz, L. (2015). Direct visualization of dispersed lipid bicontinuous cubic phases by cryo-electron tomography. *Nature Communications, 6*, 8915.

Donaldson, L. (2020). Autofluorescence in plants. *Molecules, 25*, 2393.

Dormann, P., & Benning, C. (2002). Galactolipids rule in seed plants. *Trends in Plant Science, 7*, 112–118.

Douce, R., Holtz, R. B., & Benson, A. A. (1973). Isolation and properties of the envelope of spinach chloroplasts. *The Journal of Biological Chemistry, 248*, 7215–7222.

Douce, R., & Joyard, J. (1990). Biochemistry and function of the plastid envelope. *Annual Review of Cell Biology, 6*, 173–216.

Dubacq, J. P., & Tremolieres, A. (1983). Occurrence and function of Phosphatidylglycerol Containing Delta-3-trans-Hexadecenoic acid in photosynthetic lamellae. *Physiol Veg, 21*, 293–312.

Dubots, E., Botte, C., Boudiere, L., Yamaryo-Botte, Y., Jouhet, J., Marechal, E., et al. (2012). Role of phosphatidic acid in plant galactolipid synthesis. *Biochimie, 94*, 86–93.

Dufourc, E. J., Mayer, C., Stohrer, J., Althoff, G., & Kothe, G. (1992). Dynamics of phosphate head groups in biomembranes. Comprehensive analysis using phosphorus-31 nuclear magnetic resonance lineshape and relaxation time measurements. *Biophysical Journal, 61*, 42–57.

Eichenberger, W., Araki, S., & Müller, D. G. (1993). Betaine lipids and phospholipids in brown algae. *Phytochemistry, 34*, 1323–1333.

Eichenberger, W., & Gribi, C. (1997). Lipids of Pavlova lutheri: Cellular site and metabolic role of DGCC. *Phytochemistry, 45*, 1561–1567.

Engel, B. D., Schaffer, M., Kuhn Cuellar, L., Villa, E., Plitzko, J. M., & Baumeister, W. (2015). Native architecture of the Chlamydomonas chloroplast revealed by in situ cryo-electron tomography. *eLife, 4*, e04889.

Epand, R. M., D'Souza, K., Berno, B., & Schlame, M. (2015). Membrane curvature modulation of protein activity determined by NMR. *Biochimica et Biophysica Acta, 1848*, 220–228.

Esquembre, R., Ferrer, M. L., Gutierrez, M. C., Mallavia, R., & Mateo, C. R. (2007). Fluorescence study of the fluidity and cooperativity of the phase transitions of zwitterionic and anionic liposomes confined in sol-gel glasses. *The Journal of Physical Chemistry. B, 111*, 3665–3673.

Fernandes, M. X., Castanho, M. A., & Garcia de la Torre, J. (2002). Brownian dynamics simulation of the unsaturated lipidic molecules oleic and docosahexaenoic acid confined in a cellular membrane. *Biochimica et Biophysica Acta, 1565*, 29–35.

Fernandez-Moran, H., & Finean, J. B. (1957). Electron microscope and low-angle x-ray diffraction studies of the nerve myelin sheath. *The Journal of Biophysical and Biochemical Cytology, 3*, 725–748.

Finean, J. B., & Hutchinson, A. L. (1988). X-ray diffraction studies of lipid phase transitions in cholesterol-rich membranes at sub-zero temperatures. *Chemistry and Physics of Lipids, 46*, 63–71.

Flori, S., Jouneau, P. H., Bailleul, B., Gallet, B., Estrozi, L. F., Moriscot, C., et al. (2017). Plastid thylakoid architecture optimizes photosynthesis in diatoms. *Nature Communications, 8*, 15885.

Floris, D., & Kuhlbrandt, W. (2021). Molecular landscape of etioplast inner membranes in higher plants. *Nature Plants, 7*, 514–523.

Freeman Rosenzweig, E. S., Xu, B., Kuhn Cuellar, L., Martinez-Sanchez, A., Schaffer, M., Strauss, M., et al. (2017). The eukaryotic CO2-concentrating organelle is liquid-like and exhibits dynamic reorganization. *Cell, 171*, 148–162 (e19).

Frentzen, M., Heinz, E., McKeon, T. A., & Stumpf, P. K. (1983). Specificities and selectivities of glycerol-3-phosphate acyltransferase and monoacylglycerol-3-phosphate acyltransferase from pea and spinach chloroplasts. *European Journal of Biochemistry, 129*, 629–636.

Fujii, S., Kobayashi, K., Nagata, N., Masuda, T., & Wada, H. (2017). Monogalactosyldiacylglycerol facilitates synthesis of photoactive Protochlorophyllide in Etioplasts. *Plant Physiology, 174*, 2183–2198.

Fujii, S., Kobayashi, K., Nagata, N., Masuda, T., & Wada, H. (2018). Digalactosyldiacylglycerol is essential for Organization of the Membrane Structure in Etioplasts. *Plant Physiology, 177*, 1487–1497.

Fujii, S., Wada, H., & Kobayashi, K. (2019). Role of Galactolipids in plastid differentiation before and after light exposure. *Plants (Basel), 8*, 357.

Furlan, A. L., Laurin, Y., Botcazon, C., Rodriguez-Moraga, N., Rippa, S., Deleu, M., et al. (2020). Contributions and limitations of biophysical approaches to study of the interactions between amphiphilic molecules and the plant plasma membrane. *Plants (Basel), 9*, 648.

Garab, G., Ughy, B., & Goss, R. (2016). Role of MGDG and non-bilayer lipid phases in the structure and dynamics of chloroplast thylakoid membranes. *Sub-Cellular Biochemistry, 86*, 127–157.

Garab, G., Ughy, B., Waard, P., Akhtar, P., Javornik, U., Kotakis, C., et al. (2017). Lipid polymorphism in chloroplast thylakoid membranes - as revealed by (31) P-NMR and time-resolved merocyanine fluorescence spectroscopy. *Scientific Reports, 7*, 13343.

Giroud, C., Gerber, A., & Eichenberger, W. (1988). Lipids of Chlamydomonas-Reinhardtii analysis of molecular species and intracellular site(S) of biosynthesis. *Plant and Cell Physiology, 29*, 587–595.

Goodenough, U. W., & Staehelin, L. A. (1971). Structural differentiation of stacked and unstacked chloroplast membranes. Freeze-etch electron microscopy of wild-type and mutant strains of Chlamydomonas. *The Journal of Cell Biology, 48*, 594–619.

Gounaris, K., Sen, A., Brain, A. P. R., Quinn, P. J., & Williams, W. P. (1983). Phase-separation of non-bilayer lipids in thylakoid membranes of chloroplasts. *Biophysical Journal, 41*, A315.

Gulik-Krzywicki, T. (1997). Freeze-fracture transmission electron microscopy. *Current Opinion in Colloid & Interface Science, 2*, 137–144.

Guschina, I. A., & Harwood, J. L. (2006). Lipids and lipid metabolism in eukaryotic algae. *Progress in Lipid Research, 45*, 160–186.

Hamill, O. P., & Martinac, B. (2001). Molecular basis of mechanotransduction in living cells. *Physiological Reviews, 81*, 685–740.

Hamilton, E. S., & Haswell, E. S. (2017). The tension-sensitive ion transport activity of MSL8 is critical for its function in pollen hydration and germination. *Plant & Cell Physiology, 58*, 1222–1237.

Hamilton, E. S., Schlegel, A. M., & Haswell, E. S. (2015). United in diversity: Mechanosensitive ion channels in plants. *Annual Review of Plant Biology, 66*, 113–137.

Harwood, J. L., & Jones, A. L. (1989). Lipid Metabolism in Algae. *Advances in Botanical Research, 16*, 1–53.

Haswell, E. S., & Meyerowitz, E. M. (2006). MscS-like proteins control plastid size and shape in Arabidopsis thaliana. *Current Biology, 16*, 1–11.

Heemskerk, J. W., Schmidt, H., Hammer, U., & Heinz, E. (1991). Biosynthesis and desaturation of prokaryotic galactolipids in leaves and isolated chloroplasts from spinach. *Plant Physiology, 96*, 144–152.

Heinz, E., & Roughan, P. G. (1983). Similarities and differences in lipid metabolism of chloroplasts isolated from 18:3 and 16:3 plants. *Plant Physiology, 72*, 273–279.

Higashi, Y., Okazaki, Y., Myouga, F., Shinozaki, K., & Saito, K. (2015). Landscape of the lipidome and transcriptome under heat stress in Arabidopsis thaliana. *Scientific Reports*, *5*, 10533.

Higashi, Y., Okazaki, Y., Takano, K., Myouga, F., Shinozaki, K., Knoch, E., et al. (2018). HEAT INDUCIBLE LIPASE1 remodels Chloroplastic Monogalactosyldiacylglycerol by liberating alpha-linolenic acid in Arabidopsis leaves under heat stress. *Plant Cell*, *30*, 1887–1905.

Hirashima, T., Toyoshima, M., Moriyama, T., & Sato, N. (2018). Evolution of the phosphatidylcholine biosynthesis pathways in green algae: Combinatorial diversity of methyltransferases. *Journal of Molecular Evolution*, *86*, 68–76.

Holthuis, J. C., & Menon, A. K. (2014). Lipid landscapes and pipelines in membrane homeostasis. *Nature*, *510*, 48–57.

Horn, P. J., Smith, M. D., Clark, T. R., Froehlich, J. E., & Benning, C. (2020). PEROXIREDOXIN Q stimulates the activity of the chloroplast 16:1(Delta3trans) FATTY ACID DESATURASE4. *The Plant Journal*, *102*, 718–729.

Hyvonen, M. T., Rantala, T. T., & Ala-Korpela, M. (1997). Structure and dynamic properties of diunsaturated 1-palmitoyl-2-linoleoyl-sn-glycero-3-phosphatidylcholine lipid bilayer from molecular dynamics simulation. *Biophysical Journal*, *73*, 2907–2923.

Ibarguren, M., Lopez, D. J., & Escriba, P. V. (2014). The effect of natural and synthetic fatty acids on membrane structure, microdomain organization, cellular functions and human health. *Biochimica et Biophysica Acta*, *1838*, 1518–1528.

Inda, M. E., Vandenbranden, M., Fernandez, A., de Mendoza, D., Ruysschaert, J. M., & Cybulski, L. E. (2014). A lipid-mediated conformational switch modulates the thermosensing activity of DesK. *Proceedings of the National Academy of Sciences of the United States of America*, *111*, 3579–3584.

Inoue, T., Yanagihara, S., Misono, Y., & Suzuki, M. (2001). Effect of fatty acids on phase behavior of hydrated dipalmitoylphosphatidylcholine bilayer: Saturated versus unsaturated fatty acids. *Chemistry and Physics of Lipids*, *109*, 117–133.

Jahns, P., Latowski, D., & Strzalka, K. (2009). Mechanism and regulation of the violaxanthin cycle: The role of antenna proteins and membrane lipids. *Biochimica et Biophysica Acta*, *1787*, 3–14.

Janiak, M. J., Small, D. M., & Shipley, G. G. (1979). Temperature and compositional dependence of the structure of hydrated dimyristoyl lecithin. *The Journal of Biological Chemistry*, *254*, 6068–6078.

Johnson, M. P., Brain, A. P., & Ruban, A. V. (2011). Changes in thylakoid membrane thickness associated with the reorganization of photosystem II light harvesting complexes during photoprotective energy dissipation. *Plant Signaling & Behavior*, *6*, 1386–1390.

Johnson, M. P., Vasilev, C., Olsen, J. D., & Hunter, C. N. (2014). Nanodomains of cytochrome b6f and photosystem II complexes in spinach grana thylakoid membranes. *Plant Cell*, *26*, 3051–3061.

Jouhet, J. (2013). Importance of the hexagonal lipid phase in biological membrane organization. *Frontiers in Plant Science*, *4*, 494.

Jouhet, J., Lupette, J., Clerc, O., Magneschi, L., Bedhomme, M., Collin, S., et al. (2017). LC-MS/MS versus TLC plus GC methods: Consistency of glycerolipid and fatty acid profiles in microalgae and higher plant cells and effect of a nitrogen starvation. *PLoS One*, *12*, e0182423.

Jouhet, J., Marechal, E., & Block, M. A. (2007). Glycerolipid transfer for the building of membranes in plant cells. *Progress in Lipid Research*, *46*, 37–55.

Kalisch, B., Dormann, P., & Holzl, G. (2016). DGDG and glycolipids in plants and algae. *Sub-Cellular Biochemistry*, *86*, 51–83.

Kamano, S., Kume, S., Iida, K., Lei, K. J., Nakano, M., Nakayama, Y., et al. (2015). Transmembrane topologies of Ca2+–permeable mechanosensitive channels MCA1 and MCA2 in Arabidopsis thaliana. *The Journal of Biological Chemistry*, *290*, 30901–30909.

Kanduč, M., Schlaich, A., de Vries, A. H., Jouhet, J., Marechal, E., Deme, B., et al. (2017). Tight cohesion between glycolipid membranes results from balanced water-headgroup interactions. *Nature Communications, 8*, 14899.

Katayama, K., Sakurai, I., & Wada, H. (2004). Identification of an Arabidopsis thaliana gene for cardiolipin synthase located in mitochondria. *FEBS Letters, 577*, 193–198.

Kato, M., Sakai, M., Adachi, K., Ikemoto, H., & Sano, H. (1996). Distribution of betaine lipids in marine algae. *Phytochemistry, 42*, 1341–1345.

Kawski, A. (1993). Fluorescence anisotropy - theory and applications of rotational depolarization. *Critical Reviews in Analytical Chemistry, 23*, 459–529.

Kefauver, J. M., Ward, A. B., & Patapoutian, A. (2020). Discoveries in structure and physiology of mechanically activated ion channels. *Nature, 587*, 567–576.

Kelly, A. A., Froehlich, J. E., & Dormann, P. (2003). Disruption of the two digalactosyldiacylglycerol synthase genes DGD1 and DGD2 in Arabidopsis reveals the existence of an additional enzyme of galactolipid synthesis. *Plant Cell, 15*, 2694–2706.

Kim, Y., Terng, E. L., Riekhof, W. R., Cahoon, E. B., & Cerutti, H. (2018). Endoplasmic reticulum acyltransferase with prokaryotic substrate preference contributes to triacylglycerol assembly in Chlamydomonas. *Proceedings of the National Academy of Sciences of the United States of America, 115*, 1652–1657.

Kirchhoff, H. (2019). Chloroplast ultrastructure in plants. *The New Phytologist, 223*, 565–574.

Kirchhoff, H., Lenhert, S., Buchel, C., Chi, L., & Nield, J. (2008). Probing the organization of photosystem II in photosynthetic membranes by atomic force microscopy. *Biochemistry, 47*, 431–440.

Knudsen, C., Gallage, N. J., Hansen, C. C., Moller, B. L., & Laursen, T. (2018). Dynamic metabolic solutions to the sessile life style of plants. *Natural Product Reports, 35*, 1140–1155.

Koynova, R., & Tenchov, B. (2013). Recent patents on nonlamellar liquid crystalline lipid phases in drug delivery. *Recent Patents on Drug Delivery & Formulation, 7*, 165–173.

Koynova, R., Wang, L., & Macdonald, R. C. (2008). Cationic phospholipids forming cubic phases: Lipoplex structure and transfection efficiency. *Molecular Pharmaceutics, 5*, 739–744.

Krumova, S. B., Dijkema, C., de Waard, P., Van As, H., Garab, G., & van Amerongen, H. (2008). Phase behavior of phosphatidylglycerol in spinach thylakoid membranes as revealed by 31P-NMR. *Biochimica et Biophysica Acta, 1778*, 997–1003.

Krumova, S. B., Koehorst, R. B., Bota, A., Pali, T., van Hoek, A., Garab, G., et al. (2008). Temperature dependence of the lipid packing in thylakoid membranes studied by time- and spectrally resolved fluorescence of Merocyanine 540. *Biochimica et Biophysica Acta, 1778*, 2823–2833.

Kucerka, N., Nieh, M. P., Pencer, J., Sachs, J. N., & Katsaras, J. (2009). What determines the thickness of a biological membrane. *General Physiology and Biophysics, 28*, 117–125.

Kunzler, K., & Eichenberger, W. (1997). Betaine lipids and zwitterionic phospholipids in plants and fungi. *Phytochemistry, 46*, 883–892.

Kunzler, K., Eichenberger, W., & Radunz, A. (1997). Intracellular localization of two betaine lipids by cell fractionation and immunomicroscopy. *Zeitschrift fuer Naturforschung, C: Journal of Biosciences, 52*, 487–495.

Lambrev, P. H., & Akhtar, P. (2019). Macroorganisation and flexibility of thylakoid membranes. *The Biochemical Journal, 476*, 2981–3018.

Lamers, J., van der Meer, T., & Testerink, C. (2020). How plants sense and respond to stressful environments. *Plant Physiology, 182*, 1624–1635.

Lee, J., Taneva, S. G., Holland, B. W., Tieleman, D. P., & Cornell, R. B. (2014). Structural basis for autoinhibition of CTP:Phosphocholine cytidylyltransferase (CCT), the regulatory enzyme in phosphatidylcholine synthesis, by its membrane-binding amphipathic helix. *The Journal of Biological Chemistry, 289*, 1742–1755.

Lepault, J. (1985). Cryo-electron microscopy of helical particles TMV and T4 polyheads. *Journal of Microscopy, 140*, 73–80.

Lewis, R. N., Mannock, D. A., & McElhaney, R. N. (2007). Differential scanning calorimetry in the study of lipid phase transitions in model and biological membranes: Practical considerations. *Methods in Molecular Biology, 400*, 171–195.

Li-Beisson, Y., Thelen, J. J., Fedosejevs, E., & Harwood, J. L. (2019). The lipid biochemistry of eukaryotic algae. *Progress in Lipid Research, 74*, 31–68.

Liebers, M., Chevalier, F., Blanvillain, R., & Pfannschmidt, T. (2018). PAP genes are tissue- and cell-specific markers of chloroplast development. *Planta, 248*, 629–646.

Lin, Y. C., Guo, Y. R., Miyagi, A., Levring, J., MacKinnon, R., & Scheuring, S. (2019). Force-induced conformational changes in PIEZO1. *Nature, 573*, 230–234.

Lindquist, E., Solymosi, K., & Aronsson, H. (2016). Vesicles are persistent features of different plastids. *Traffic, 17*, 1125–1138.

Lis, L. J., & Quinn, P. J. (1991). The application of synchrotron X-radiation for the study of phase transitions in lipid model membrane systems. *Journal of Applied Crystallography, 24*, 48–60.

Liu, X., Wang, J., & Sun, L. (2018). Structure of the hyperosmolality-gated calcium-permeable channel OSCA1.2. *Nature Communications, 9*, 5060.

Lombard, J. (2014). Once upon a time the cell membranes: 175 years of cell boundary research. *Biology Direct, 9*, 32.

Luzzati, V., & Husson, F. (1962). The structure of the liquid-crystalline phasis of lipid-water systems. *The Journal of Cell Biology, 12*, 207–219.

Maity, K., Heumann, J. M., McGrath, A. P., Kopcho, N. J., Hsu, P. K., Lee, C. W., et al. (2019). Cryo-EM structure of OSCA1.2 from Oryza sativa elucidates the mechanical basis of potential membrane hyperosmolality gating. *Proceedings of the National Academy of Sciences of the United States of America, 116*, 14309–14318.

Martinac, B., Bavi, N., Ridone, P., Nikolaev, Y. A., Martinac, A. D., Nakayama, Y., et al. (2018). Tuning ion channel mechanosensitivity by asymmetry of the transbilayer pressure profile. *Biophysical Reviews, 10*, 1377–1384.

Martonosi, A., Kracke, G., Taylor, K. A., Dux, L., & Peracchia, C. (1985). The regulation of the Ca2+ transport activity of sarcoplasmic reticulum. *Society of General Physiologists Series, 39*, 57–85.

Maulucci, G., Cohen, O., Daniel, B., Sansone, A., Petropoulou, P. I., Filou, S., et al. (2016). Fatty acid-related modulations of membrane fluidity in cells: Detection and implications. *Free Radical Research, 50*, S40–S50.

McCourt, P., Browse, J., Watson, J., Arntzen, C. J., & Somerville, C. R. (1985). Analysis of photosynthetic antenna function in a mutant of Arabidopsis thaliana (L.) lacking trans-Hexadecenoic acid. *Plant Physiology, 78*, 853–858.

McDaniel, R. V. (1988). Neutron diffraction studies of digalactosyldiacylglycerol. *Biochimica et Biophysica Acta, 940*, 158–164.

McElhaney, R. N. (1986). Differential scanning calorimetric studies of lipid-protein interactions in model membrane systems. *Biochimica et Biophysica Acta, 864*, 361–421.

Mendiola-Morgenthaler, L., Eichenberger, W., & Boschetti, A. (1985). Isolation of chloroplast envelopes from Chlamydomonas. Lipid and polypeptide composition. *Plant Science, 41*, 97–104.

Meyer, H. W., & Richter, W. (2001). Freeze-fracture studies on lipids and membranes. *Micron, 32*, 615–644.

Mezzenga, R., Seddon, J. M., Drummond, C. J., Boyd, B. J., Schroder-Turk, G. E., & Sagalowicz, L. (2019). Nature-inspired design and application of Lipidic Lyotropic liquid crystals. *Advanced Materials, 31*, e1900818.

Mikami, K., & Murata, N. (2003). Membrane fluidity and the perception of environmental signals in cyanobacteria and plants. *Progress in Lipid Research, 42*, 527–543.

Minagawa, J., & Tokutsu, R. (2015). Dynamic regulation of photosynthesis in Chlamydomonas reinhardtii. *The Plant Journal, 82*, 413–428.

Mio, K., & Sato, C. (2018). Lipid environment of membrane proteins in cryo-EM based structural analysis. *Biophysical Reviews, 10*, 307–316.

Moellering, E. R., Muthan, B., & Benning, C. (2010). Freezing tolerance in plants requires lipid remodeling at the outer chloroplast membrane. *Science, 330*, 226–228.

Morimoto, T., Murakami, N., Nagatsu, A., & Sakakibara, J. (1993). Studies on Glycolipids .7. Isolation of 2 New Sulfoquinovosyl Diacylglycerols from the Green-Alga Chlorella-Vulgaris. *Chemical & Pharmaceutical Bulletin, 41*, 1545–1548.

Mouritsen, O. G. (2005). Lively lipids provide for function. In T. F. Collection (Ed.), *Life— As a matter of fat: The emerging science of Lipidomics* (pp. 129–136). Berlin, Heidelberg: Springer Berlin Heidelberg.

Mousavi, S. A. R., Dubin, A. E., Zeng, W. Z., Coombs, A. M., Do, K., Ghadiri, D. A., et al. (2021). PIEZO ion channel is required for root mechanotransduction in Arabidopsis thaliana. *Proceedings of the National Academy of Sciences of the United States of America, 118*, e2102188118.

Nakagawa, Y., Katagiri, T., Shinozaki, K., Qi, Z., Tatsumi, H., Furuichi, T., et al. (2007). Arabidopsis plasma membrane protein crucial for Ca2+ influx and touch sensing in roots. *Proceedings of the National Academy of Sciences of the United States of America, 104*, 3639–3644.

Nakamura, Y. (2021). Headgroup biosynthesis of phosphatidylcholine and phosphatidylethanolamine in seed plants. *Progress in Lipid Research, 82*, 101091.

Naumann, I., Klein, B. C., Bartel, S. J., Darsow, K. H., Buchholz, R., & Lange, H. A. (2011). Identification of sulfoquinovosyldiacyglycerides from Phaeodactylum tricornutum by matrix-assisted laser desorption/ionization QTrap time-of-flight hybrid mass spectrometry. *Rapid Communications in Mass Spectrometry, 25*, 2517–2523.

Nawrocki, W. J., Santabarbara, S., Mosebach, L., Wollman, F. A., & Rappaport, F. (2016). State transitions redistribute rather than dissipate energy between the two photosystems in Chlamydomonas. *Nature Plants, 2*, 16031.

Nichols, B. W., & Appleby, R. S. (1969). The distribution and biosynthesis of arachidonic acid in algae. *Phytochemistry, 8*, 1907–1915.

Nichols, B. W., Harris, P., & James, A. T. (1965). The biosynthesis of trans-delta-3-hexadecenoic acid by chlorella vulgaris. *Biochemical and Biophysical Research Communications, 21*, 473–479.

Nitenberg, M., Makshakova, O., Rocha, J., Perez, S., Marechal, E., Block, M. A., et al. (2020). Mechanism of activation of plant monogalactosyldiacylglycerol synthase 1 (MGD1) by phosphatidylglycerol. *Glycobiology, 30*, 396–406.

Niu, Y., & Xiang, Y. (2018). An overview of biomembrane functions in plant responses to high-temperature stress. *Frontiers in Plant Science, 9*, 915.

Noack, L. C., & Jaillais, Y. (2020). Functions of anionic lipids in plants. *Annual Review of Plant Biology, 71*, 71–102.

Oquist, G. (1982). Seasonally induced changes in acyl lipids and fatty acids of chloroplast thylakoids of Pinus silvestris: A correlation between the level of unsaturation of monogalactosyldiglyceride and the rate of electron transport. *Plant Physiology, 69*, 869–875.

Otles, S., & Pire, R. (2001). Fatty acid composition of chlorella and spirulina microalgae species. *Journal of AOAC International, 84*, 1708–1714.

Pali, T., & Pesti, M. (1996). Phase transition of membrane lipids. In R. Prasad (Ed.), *Manual on membrane lipids* (pp. 80–93). Berlin: Springer.

Petroutsos, D., Amiar, S., Abida, H., Dolch, L. J., Bastien, O., Rebeille, F., et al. (2014). Evolution of galactoglycerolipid biosynthetic pathways- -from cyanobacteria to primary plastids and from primary to secondary plastids. *Progress in Lipid Research, 54*, 68–85.

Phan, M. D., & Shin, K. (2015). Effects of cardiolipin on membrane morphology: A Langmuir monolayer study. *Biophysical Journal, 108*, 1977–1986.

Pinot, M., Vanni, S., Pagnotta, S., Lacas-Gervais, S., Payet, L. A., Ferreira, T., et al. (2014). Lipid cell biology. Polyunsaturated phospholipids facilitate membrane deformation and fission by endocytic proteins. *Science, 345*, 693–697.

Pipitone, R., Eicke, S., Pfister, B., Glauser, G., Falconet, D., Uwizeye, C., et al. (2021). A multifaceted analysis reveals two distinct phases of chloroplast biogenesis during de-etiolation in Arabidopsis. *eLife, 10*, e62709.

Pittera, J., Jouhet, J., Breton, S., Garczarek, L., Partensky, F., Marechal, E., et al. (2018). Thermoacclimation and genome adaptation of the membrane lipidome in marine Synechococcus. *Environmental Microbiology, 20*, 612–631.

Platre, M. P., Noack, L. C., Doumane, M., Bayle, V., Simon, M. L. A., Maneta-Peyret, L., et al. (2018). A combinatorial lipid code shapes the electrostatic landscape of plant Endomembranes. *Developmental Cell, 45*, 465–480 (e11).

Pribil, M., Labs, M., & Leister, D. (2014). Structure and dynamics of thylakoids in land plants. *Journal of Experimental Botany, 65*, 1955–1972.

Pyszniak, A. M., & Gibbs, S. P. (1992). Immunocytochemical localization of photosystem I and the fucoxanthin-chlorophylla/c light-harvesting complex in the diatomPhaeodactylum tricornutum. *Protoplasma*, 208–217.

Ramot, D., MacInnis, B. L., & Goodman, M. B. (2008). Bidirectional temperature-sensing by a single thermosensory neuron in C. elegans. *Nature Neuroscience, 11*, 908–915.

Rasmussen, T., Flegler, V. J., Rasmussen, A., & Bottcher, B. (2019). Structure of the Mechanosensitive Channel MscS embedded in the membrane bilayer. *Journal of Molecular Biology, 431*, 3081–3090.

Ray, S., Scott, J. L., & Tatulian, S. A. (2007). Effects of lipid phase transition and membrane surface charge on the interfacial activation of phospholipase A2. *Biochemistry, 46*, 13089–13100.

Reddy, B., Bavi, N., Lu, A., Park, Y., & Perozo, E. (2019). Molecular basis of force-from-lipids gating in the mechanosensitive channel MscS. *eLife, 8*, e50486.

Riekhof, W. R., Ruckle, M. E., Lydic, T. A., Sears, B. B., & Benning, C. (2003). The sulfolipids 2'-O-acyl-sulfoquinovosyldiacylglycerol and sulfoquinovosyldiacylglycerol are absent from a Chlamydomonas reinhardtii mutant deleted in SQD1. *Plant Physiology, 133*, 864–874.

Riekhof, W. R., Sears, B. B., & Benning, C. (2005). Annotation of genes involved in glycerolipid biosynthesis in Chlamydomonas reinhardtii: Discovery of the betaine lipid synthase BTA1Cr. *Eukaryotic Cell, 4*, 242–252.

Risselada, H. J., & Marrink, S. J. (2009). Curvature effects on lipid packing and dynamics in liposomes revealed by coarse grained molecular dynamics simulations. *Physical Chemistry Chemical Physics, 11*, 2056–2067.

Roughan, P. G., & Boardman, N. K. (1972). Lipid composition of pea and bean leaves during chloroplast development. *Plant Physiology, 50*, 31–34.

Roughan, P. G., Holland, R., & Slack, C. R. (1980). The role of chloroplasts and microsomal fractions in polar-lipid synthesis from [1-14C]acetate by cell-free preparations from spinach (Spinacia oleracea) leaves. *The Biochemical Journal, 188*, 17–24.

Saidi, Y., Finka, A., & Goloubinoff, P. (2011). Heat perception and signalling in plants: A tortuous path to thermotolerance. *The New Phytologist, 190*, 556–565.

Saidi, Y., Finka, A., Muriset, M., Bromberg, Z., Weiss, Y. G., Maathuis, F. J., et al. (2009). The heat shock response in moss plants is regulated by specific calcium-permeable channels in the plasma membrane. *Plant Cell, 21*, 2829–2843.

Sakurai, K., Mori, N., & Sato, N. (2014). Detection and characterization of phosphatidylcholine in various strains of the genus Chlamydomonas (Volvocales, Chlorophyceae). *Journal of Plant Research, 127*, 641–650.

Saotome, K., Murthy, S. E., Kefauver, J. M., Whitwam, T., Patapoutian, A., & Ward, A. B. (2018). Structure of the mechanically activated ion channel Piezo1. *Nature, 554*, 481–486.

Sarcina, M., Murata, N., Tobin, M. J., & Mullineaux, C. W. (2003). Lipid diffusion in the thylakoid membranes of the cyanobacterium Synechococcus sp.: Effect of fatty acid desaturation. *FEBS Letters, 553*, 295–298.

Sarkis, J., Rocha, J., Maniti, O., Jouhet, J., Vie, V., Block, M. A., et al. (2014). The influence of lipids on MGD1 membrane binding highlights novel mechanisms for galactolipid biosynthesis regulation in chloroplasts. *The FASEB Journal, 28*, 3114–3123.

Sato, N. (1992). Betaine lipids. *Botanical Magazine Tokyo, 105*, 185–197.

Schaller, S., Latowski, D., Jemiola-Rzeminska, M., Wilhelm, C., Strzalka, K., & Goss, R. (2010). The main thylakoid membrane lipid monogalactosyldiacylglycerol (MGDG) promotes the de-epoxidation of violaxanthin associated with the light-harvesting complex of photosystem II (LHCII). *Biochimica et Biophysica Acta, 1797*, 414–424.

Selstam, E., & Sandelius, A. S. (1984). A comparison between Prolamellar bodies and Prothylakoid membranes of Etioplasts of dark-grown wheat concerning lipid and polypeptide composition. *Plant Physiology, 76*, 1036–1040.

Shaikh, S. R., Kinnun, J. J., Leng, X., Williams, J. A., & Wassall, S. R. (2015). How polyunsaturated fatty acids modify molecular organization in membranes: Insight from NMR studies of model systems. *Biochimica et Biophysica Acta, 1848*, 211–219.

Shechter, E. (1984). *Membranes biologiques: structure, transports, bioénergétique*. (Masson (ed) Paris).

Sheffer, M., Fried, A., Gottlieb, H. E., Tietz, A., & Avron, M. (1986). Lipid composition of the plasma-membrane of the halotolerant alga, Dunaliella salina. *Biochimica et Biophysica Acta - Biomembranes, 857*, 165–172.

Shigematsu, H., Iida, K., Nakano, M., Chaudhuri, P., Iida, H., & Nagayama, K. (2014). Structural characterization of the mechanosensitive channel candidate MCA2 from Arabidopsis thaliana. *PLoS One, 9*, e87724.

Shinitzky, M. (1984). Membrane fluidity in malignancy. Adversative and recuperative. *Biochimica et Biophysica Acta, 738*, 251–261.

Siegel, D. P., & Tenchov, B. G. (2008). Influence of the lamellar phase unbinding energy on the relative stability of lamellar and inverted cubic phases. *Biophysical Journal, 94*, 3987–3995.

Simon, M. L., Platre, M. P., Marques-Bueno, M. M., Armengot, L., Stanislas, T., Bayle, V., et al. (2016). A PtdIns(4)P-driven electrostatic field controls cell membrane identity and signalling in plants. *Nature Plants, 2*, 16089.

Singer, S. J., & Nicolson, G. L. (1972). The fluid mosaic model of the structure of cell membranes. *Science, 175*, 720–731.

Singh, S., & Mittal, A. (2016). Transmembrane domain lengths serve as signatures of organismal complexity and viral transport mechanisms. *Scientific Reports, 6*, 22352.

Smith, R., Jouhet, J., Gandini, C., Nekrasov, V., Marechal, E., Napier, J. A., et al. (2021). Plastidial acyl carrier protein Delta9 desaturase modulates eicosapentaenoic acid biosynthesis and triacylglycerol accumulation in Phaeodactylum tricornutum. *The Plant Journal*.

Smutzer, G., & Yeagle, P. L. (1985). A fluorescence anisotropy study on the phase behavior of dimyristoylphosphatidylcholine/cholesterol mixtures. *Biochimica et Biophysica Acta, 814*, 274–280.

Solymosi, K., & Schoefs, B. (2010). Etioplast and etio-chloroplast formation under natural conditions: The dark side of chlorophyll biosynthesis in angiosperms. *Photosynthesis Research, 105*, 143–166.

Staehelin, L. A. (2003). Chloroplast structure: From chlorophyll granules to supra-molecular architecture of thylakoid membranes. *Photosynthesis Research, 76*, 185–196.

Takamori, S., Holt, M., Stenius, K., Lemke, E. A., Gronborg, M., Riedel, D., et al. (2006). Molecular anatomy of a trafficking organelle. *Cell, 127*, 831–846.

Tardieu, A., Luzzati, V., & Reman, F. C. (1973). Structure and polymorphism of the hydrocarbon chains of lipids: A study of lecithin-water phases. *Journal of Molecular Biology, 75*, 711–733.

Tenchov, B., & Koynova, R. (2012). Cubic phases in membrane lipids. *European Biophysics Journal, 41*, 841–850.

Tenchov, B., & Koynova, R. (2017). Cubic phases in phosphatidylethanolamine dispersions: Formation, stability and phase transitions. *Chemistry and Physics of Lipids, 208*, 65–74.

Thomashow, M. F. (1999). PLANT COLD ACCLIMATION: Freezing tolerance genes and regulatory mechanisms. *Annual Review of Plant Physiology and Plant Molecular Biology, 50*, 571–599.

Tikkanen, M., & Aro, E. M. (2012). Thylakoid protein phosphorylation in dynamic regulation of photosystem II in higher plants. *Biochimica et Biophysica Acta, 1817*, 232–238.

Tixier-Vidal, A., Picart, R., Loudes, C., & Bauman, A. F. (1986). Effects of polyunsaturated fatty acids and hormones on synaptogenesis in serum-free medium cultures of mouse fetal hypothalamic cells. *Neuroscience, 17*, 115–132.

Tong, J., Briggs, M. M., & McIntosh, T. J. (2012). Water permeability of aquaporin-4 channel depends on bilayer composition, thickness, and elasticity. *Biophysical Journal, 103*, 1899–1908.

van der Meer, B. W., van Hoeven, R. P., & van Blitterswijk, W. J. (1986). Steady-state fluorescence polarization data in membranes. Resolution into physical parameters by an extended Perrin equation for restricted rotation of fluorophores. *Biochimica et Biophysica Acta, 854*, 38–44.

van Hoek, A., Vos, K., & Visser, A. J. W. G. (1987). Ultrasensitive time-resolved polarized fluorescence spectroscopy as a tool in biology and medicine. *IEEE Journal of Quantum Electronics, 23*, 1812–1820.

van Meer, G., Voelker, D. R., & Feigenson, G. W. (2008). Membrane lipids: Where they are and how they behave. *Nature Reviews. Molecular Cell Biology, 9*, 112–124.

van Wijk, K. J., & Kessler, F. (2017). Plastoglobuli: Plastid microcompartments with integrated functions in metabolism, plastid developmental transitions, and environmental adaptation. *Annual Review of Plant Biology, 68*, 253–289.

Veley, K. M., Marshburn, S., Clure, C. E., & Haswell, E. S. (2012). Mechanosensitive channels protect plastids from hypoosmotic stress during normal plant growth. *Current Biology, 22*, 408–413.

Vigh, L., Horvath, I., van Hasselt, P. R., & Kuiper, P. J. (1985). Effect of frost hardening on lipid and fatty acid composition of chloroplast thylakoid membranes in two wheat varieties of contrasting hardiness. *Plant Physiology, 79*, 756–759.

Vogel, G., & Eichenberger, W. (1992). Betaine lipids in lower plants. Biosynthesis of DGTS and DGTA in Ochromonas danica (Chrysophyceae) and the possible role of DGTS in lipid metabolism. *Plant and Cell Physiology, 33*, 427–436.

Wada, H., & Murata, N. (2007). The essential role of phosphatidylglycerol in photosynthesis. *Photosynthesis Research, 92*, 205–215.

Wang, L., Zhou, H., Zhang, M., Liu, W., Deng, T., Zhao, Q., et al. (2019). Structure and mechanogating of the mammalian tactile channel PIEZO2. *Nature, 573*, 225–229.

Warakanont, J., Tsai, C. H., Michel, E. J., Murphy, G. R., 3rd, Hsueh, P. Y., Roston, R. L., et al. (2015). Chloroplast lipid transfer processes in Chlamydomonas reinhardtii involving a TRIGALACTOSYLDIACYLGLYCEROL 2 (TGD2) orthologue. *The Plant Journal, 84*, 1005–1020.

Wietrzynski, W., & Engel, B. D. (2021). Chlorophyll biogenesis sees the light. *Nature Plants, 7*, 380–381.

Wollman, F. A., Olive, J., Bennoun, P., & Recouvreur, M. (1980). Organization of the photosystem II centers and their associated antennae in the thylakoid membranes: A comparative ultrastructural, biochemical, and biophysical study of Chlamydomonas wild type and mutants lacking in photosystem II reaction centers. *The Journal of Cell Biology*, *87*, 728–735.

Wood, W. H. J., MacGregor-Chatwin, C., Barnett, S. F. H., Mayneord, G. E., Huang, X., Hobbs, J. K., et al. (2018). Dynamic thylakoid stacking regulates the balance between linear and cyclic photosynthetic electron transfer. *Nature Plants*, *4*, 116–127.

Xu, C., Moellering, E. R., Muthan, B., Fan, J., & Benning, C. (2010). Lipid transport mediated by Arabidopsis TGD proteins is unidirectional from the endoplasmic reticulum to the plastid. *Plant & Cell Physiology*, *51*, 1019–1028.

Xue, C. H., Hu, Y. Q., Saito, H., Zhang, Z. H., Li, Z. J., Cai, Y. P., et al. (2002). Molecular species composition of glycolipids from Sprirulina platensis. *Food Chemistry*, *77*, 9–13.

Yang, H. J., Sugiura, Y., Ikegami, K., Konishi, Y., & Setou, M. (2012). Axonal gradient of arachidonic acid-containing phosphatidylcholine and its dependence on actin dynamics. *The Journal of Biological Chemistry*, *287*, 5290–5300.

Yao, J., Liu, B., & Qin, F. (2011). Modular thermal sensors in temperature-gated transient receptor potential (TRP) channels. *Proceedings of the National Academy of Sciences of the United States of America*, *108*, 11109–11114.

Yu, B., Xu, C., & Benning, C. (2002). Arabidopsis disrupted in SQD2 encoding sulfolipid synthase is impaired in phosphate-limited growth. *Proceedings of the National Academy of Sciences of the United States of America*, *99*, 5732–5737.

Yuan, F., Yang, H., Xue, Y., Kong, D., Ye, R., Li, C., et al. (2014). OSCA1 mediates osmotic-stress-evoked Ca2+ increases vital for osmosensing in Arabidopsis. *Nature*, *514*, 367–371.

Zauner, S., Jochum, W., Bigorowski, T., & Benning, C. (2012). A cytochrome b5-containing plastid-located fatty acid desaturase from Chlamydomonas reinhardtii. *Eukaryotic Cell*, *11*, 856–863.

Zhang, Z., Tong, X., Liu, S. Y., Chai, L. X., Zhu, F. F., Zhang, X. P., et al. (2019). Genetic analysis of a piezo-like protein suppressing systemic movement of plant viruses in Arabidopsis thaliana. *Scientific Reports*, *9*, 3187.

Zhang, M., Wang, D., Kang, Y., Wu, J. X., Yao, F., Pan, C., et al. (2018). Structure of the mechanosensitive OSCA channels. *Nature Structural & Molecular Biology*, *25*, 850–858.

Zhou, Y., Peisker, H., & Dormann, P. (2016). Molecular species composition of plant cardiolipin determined by liquid chromatography mass spectrometry. *Journal of Lipid Research*, *57*, 1308–1321.

Zulu, N. N., Zienkiewicz, K., Vollheyde, K., & Feussner, I. (2018). Current trends to comprehend lipid metabolism in diatoms. *Progress in Lipid Research*, *70*, 1–16.

CHAPTER TWO

Plant and algal galactolipids: Their function, biosynthesis and evolution

Mie Shimojima* and Hiroyuki Ohta
School of Life Science and Technology, Tokyo Institute of Technology, Yokohama, Japan
*Corresponding author: e-mail address: shimojima.m.aa@m.titech.ac.jp

Contents

Abstract

Two galactolipids, namely monogalactosyldiacylglycerol (MGDG) and digalactosyldiacylglycerol (DGDG), are the most abundant lipids in plastid membranes of plant vegetative tissues. MGDG synthesis is carried out by type A and B MGDG synthases, which are widely conserved in seed plants. Type A MGDG synthase produces the bulk of MGDG for thylakoid membranes of chloroplasts in seed plants. Type B MGDG synthase primarily supplies MGDG as a substrate for DGDG synthesis —especially in response to abiotic stress. Under phosphate-starved conditions, extraplastidial membranes lack sufficient phospholipids. To compensate for this deficiency, the DGDG produced by the type B MGDG synthase pathway is transported out of chloroplasts and becomes incorporated in extraplastidial membranes. In addition to MGDG synthase and DGDG synthase, the third enzyme of galactolipid synthesis, namely galactolipid:galactolipid galactosyltransferase, is involved in freezing tolerance. The various galactolipid synthases seem to be regulated in response to diverse abiotic stresses, but many aspects remain unclear. Considering that galactolipid synthesis is conserved among land plants and algae, the findings obtained from algal galactolipid research will further

Advances in Botanical Research, Volume 101
ISSN 0065-2296
https://doi.org/10.1016/bs.abr.2021.09.007

our understanding of galactolipid function. Here, we review the history and progress of galactolipid research in seed plants as well as the few, but very recent, advances achieved with algae.

1. Introduction

The two galactolipids monogalactosyldiacylglycerol (MGDG) and digalactosyldiacylglycerol (DGDG) are the most abundant lipids in oxygen-evolving photosynthetic organisms such as cyanobacteria (Fig. 1) (Douce & Joyard, 1980; Wada & Murata, 1989) and in the thylakoid membranes of chloroplasts.

Indeed, these two galactolipids comprise ~50% of membrane glycerolipids of plant leaves and ~80% of that of chloroplasts (Block, Dorne, Joyard, & Douce, 1983). Notably, cyanobacteria and chloroplasts have similar membrane-glycerolipid compositions, perhaps because the endosymbiotic origin of chloroplasts is an ancient cyanobacteria (Reyes-Prieto, Weber, & Bhattacharya, 2007). However, the galactolipid biosynthesis pathways differ substantially between plants and cyanobacteria. In cyanobacteria, MGDG is produced from glucose transfer to diacylglycerol (DAG) followed by epimerization of the glucose to galactose (Awai et al., 2006; Awai, Ohta, & Sato, 2014; Feige, Heinz, Wrage, Cochems, & Ponzelar, 1980; Omata & Murata, 1986; Sato & Murata, 1982a, 1982b, 1982c). In contrast, plants produce MGDG via a one-step reaction in which

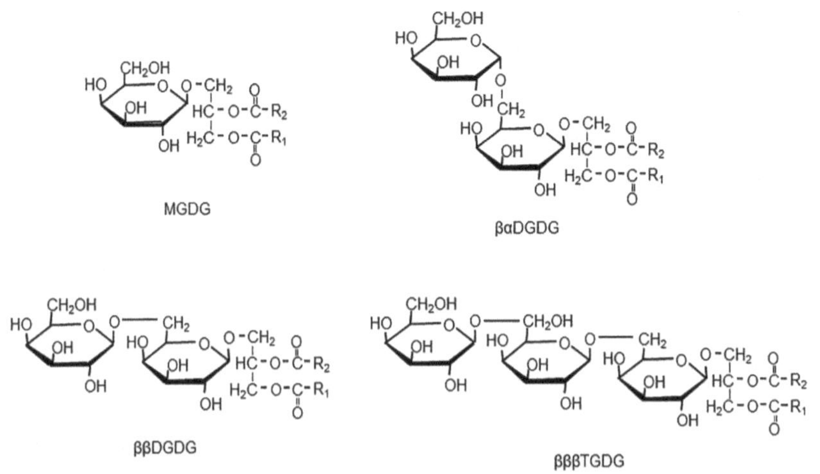

Fig. 1 Chemical structures of galactolipids in plants. MGDG, monogalactosyldiacylglycerol; DGDG, digalactosyldiacylglycerol; TGDG, trigalactosyldiacylglycerol; TeGDG, tetragalactosyldiacylglycerol.

the galactose moiety of UDP-galactose is transferred to DAG (Douce & Joyard, 1980). Because galactolipids are produced by a wide variety of photosynthetic organisms, it is not surprising that one of the principal roles for MGDG and DGDG is their involvement in photosynthesis as components of the photosynthetic apparatus. Indeed, MGDG is a component of both the photosystem I (PSI) (Jordan et al., 2001; Qin, Suga, Kuang, & Shen, 2015) and PSII complexes (Guskov et al., 2009; Loll, Kern, Saenger, Zouni, & Biesiadka, 2005; Sakurai et al., 2006; Umena, Kawakami, Shen, & Kamiya, 2011). MGDG is likely involved in the dimerization of the two PSII core complexes (Kansy, Wilhelm, & Goss, 2014) and energy transfer from the light-harvesting complex of PSII (LHCII) to PSII core complexes (Schaller et al., 2011, 2010; Zhou et al., 2009). DGDG exists in the core of LHCII (Liu et al., 2004) and PSII (Guskov et al., 2009; Loll et al., 2005; Sakurai et al., 2006; Umena et al., 2011) and is also involved in LHCII trimerization (Liu et al., 2004). Notably, MGDG alone cannot form a membrane-lipid bilayer owing to the unique shape and small size of the polar head. However, the proportion of MGDG can affect the propensity for photosynthetic membranes to form a bilayer, which can affect membrane fluidity (Demé, Cataye, Block, Maréchal, & Jouhet, 2014; Garab, 2016; Pribil, Labs, & Leister, 2014; Webb & Green, 1991). By contrast, DGDG promotes bilayer formation and thus is involved in the stacking of thylakoid membranes (Demé et al., 2014; Garab, 2016).

The functions, biosynthetic pathways (and their regulation), and evolution of galactolipids in plants have long been the subject of research. In addition to their roles in the photosynthetic apparatus, galactolipids have many other essential roles in plants. The initial breakthrough in our understanding of galactolipid functions was the identification of a gene involved in MGDG synthesis in cucumber (Shimojima et al., 1997), followed by the identification of two types of MGDG synthases in Arabidopsis (Awai et al., 2001). These findings allowed in vitro analyses of the corresponding enzyme-coding genes and genetic approaches using transgenic plants for understanding the mechanism of MGDG synthesis. Subsequent identification of two DGDG synthases (DGDs) in Arabidopsis and the processive galactosyltransferase, galactolipid:galactolipid galactosyltransferase (GGGT), completed the list of the major genes responsible for galactolipid synthesis (Dörmann, Balbo, & Benning, 1999; Dörmann, Hoffmann-Benning, Balbo, & Benning, 1995; Kelly & Dörmann, 2002; Kelly, Froehlich, & Dörmann, 2003; Moellering, Muthan, & Benning, 2010). Analyses of numerous Arabidopsis galactolipid synthase mutants revealed novel functions

Table 1 Distribution of galactolipid synthases (Hori et al., 2016)

	Lineage	MGDG synthase	DGDG synthase	Oligogalactolipid synthase
Cyanobacteria	Cyanobacteria	MgdA and MgdE	DgdA	NF
Algae	Glaucophytes[a]	MGD	DGD (LS), DgdA	NF
	Cyanidiophytina	MGD	DgdA	NF
	Rhodophytina	MGD	DGD	NF
	Stramenopiles	MGD	DGD	NF
	Green algae	MGD	DGD	NF
	Chlorarachniophytes[a]	MGD	DGD	NF
	Charophytes[a]	MGD	DGD	GGGT
Land plants	Moss, Fern	MGD	DGD	GGGT
	Seed plants	MGD (type A, type B)	DGD1, DGD2	GGGT (SFR2)

[a]Data from a single organism.

for galactolipids, and the regulatory mechanisms have since been unraveled. The findings obtained with Arabidopsis enabled galactolipid research to expand to other seed plants and algae (Table 1). These researches clarified that most galactolipid functions in plastids are essential and cannot be substituted by other glycerolipids. Moreover, the roles of galactolipids are not only limited to physiological events in plastids. In extraplastidial membranes, for example, phosphatidylcholine (PC) is replaced with DGDG under phosphate-starved growth conditions (Andersson, Larsson, Tjellström, Liljenberg, & Sandelius, 2005; Andersson, Stridh, Larsson, Liljenberg, & Sandelius, 2003; Benning & Ohta, 2005; Härtel & Benning, 2000; Härtel, Dormann, & Benning, 2000; Härtel et al., 1998; Jouhet et al., 2004; Jouhet, Maréchal, & Block, 2007; Klaus et al., 2002). Here, we discuss the function of galactolipids and the regulation of galactolipid biosynthesis in plants and algae. We refer readers to other published and highly regarded reviews that cover the details of galactolipid structure and function (Boudière et al., 2012; Dörmann & Benning, 2002; Fujii, Wada, & Kobayashi, 2019; Garab, Ughy, & Goss, 2016; Kobayashi, 2016;

Kobayashi et al., 2009; Kobayashi & Wada, 2016; Li & Yu, 2018; Lyu, Gao, & Guo, 2021; Nakamura, 2013; Petroutsos et al., 2014; Rocha et al., 2018; Shimojima & Ohta, 2011).

2. Synthesis and function of galactolipids in plants

2.1 MGDG in seed plants

The last step of MGDG synthesis is catalyzed by MGDG synthase, which transfers the galactose moiety from UDP-galactose to DAG to produce MGDG (Fig. 2) (Douce & Joyard, 1980).

Plant MGDG synthesis was first demonstrated in the plastid envelope membrane (Douce, 1974), and thus early biochemical studies mainly

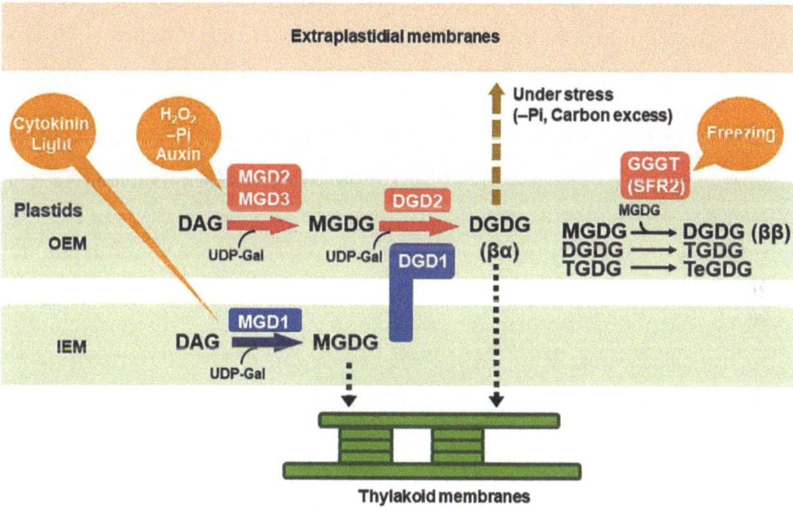

Fig. 2 Biosynthesis of galactolipids in Arabidopsis leaves. Extraplastidial membranes include plasma membranes, mitochondrial membranes and vacuolar membranes. The factors/stresses which upregulate each galactolipid synthase were indicated in the speech bubbles. Broken arrows represent the movement of the lipids. Black filled broken arrows represent enzymatic reactions activated under normal growth conditions and white/red arrows represent enzymatic reactions activated under phosphate (Pi) starvation or carbon-excess conditions. *Abbreviations*: MGD1, 2, 3, monogalactosyldiacylglycerol synthase; DGD1, 2, digalactosyldiacylglycerol synthase; GGGT, galactolipid: galactolipid galactosyltransferase; MGDG, monogalactosyldiacylglycerol; DGDG, digalactosyldiacylglycerol; TGDG, trigalactosyldiacylglycerol; TeGDG, tetragalactosyldiacylglycerol; DAG, diacylglycerol; OEM, outer envelope membranes; IEM, inner envelope membranes.

utilized chloroplast envelope membranes isolated from spinach (Heemskerk, Bögemann, Helsper, & Wintermans, 1988; Heemskerk, Jacobs, Scheijen, Helsper, & Wintermans, 1987; Maréchal, Block, Joyard, & Douce, 1994a, 1994b; Maréchal, Miège, Block, Douce, & Joyard, 1995). However, the exact site of biosynthesis in the envelope membrane remained controversial, i.e., it appeared to occur between the inner envelope membranes in spinach (*Spinacia oleracea*) but in the outer envelope membranes in pea (*Pisum sativum*) (Block et al., 1983; Cline & Keegstra, 1983). Biochemical approaches for identification of MGDG synthase were initially unsuccessful owing to its low abundance in envelope membranes and the instability of MGDG synthase activity, which became undetectable midway through the purification process (Maréchal, Block, Joyard, & Douce, 1991; Slabas, 1997; Teucher & Heinz, 1991). A breakthrough came when the microsomal fraction of cucumber cotyledons was used for MGDG synthase purification (Ohta, Shimojima, Arai, et al., 1995; Shimojima et al., 1997).

The identification of the first MGDG synthase from cucumber (*Cucumis sativus*) contributed greatly to the identification and functional characterization of other plant MGDG synthases (Miège et al., 1999). In *Arabidopsis thaliana*, three enzymes for MGDG synthase were then identified and designated as type A (AtMGD1) and type B (AtMGD2, AtMGD3) based on their similarity of the amino acid sequence and localization in the envelope membrane (Fig. 2) (Awai et al., 2001). MGD1 is localized in the inner envelope membrane, presumptively facing the intermembrane space of the envelope membrane both in *S. oleracea* and *A. thaliana* (Miège et al., 1999; Vojta, Soll, & Bölter, 2007; Xu, Fan, Froehlich, Awai, & Benning, 2005). However, Li et al. (2011) reported that a UDP-glucose epimerase, which produces UDP-galactose in the stroma of chloroplasts, is crucial for galactolipid synthesis in rice. On the other hand, Zhang et al. (2020) found that the rice mutant of cytosol-localized UDP-glucose epimerase 2 exhibits dwarf phenotype and pale leaves with a marked decrease in galactolipids and chlorophyll. Thus, the pathway by which UDP-galactose is provided for the MGDG biosynthesis remains uncertain. AtMGD1 is highly expressed in photosynthetic organs and mediates the bulk synthesis of MGDG in thylakoid membranes. In vitro analyses of recombinant AtMGDs clearly indicated that AtMGD1 has the highest activity and DAG affinity among the three MGDs of Arabidopsis (Awai et al., 2001). Analysis of the crystal structure of the catalytic domain of AtMGD1 revealed the presence of an unstructured sequence at the N terminus that may anchor AtMGD1 to the membrane and is essential for binding the substrate DAG (Rocha et al., 2013, 2016). As observed during partial purification of MGDG synthase from cucumber, AtMGD1 activity requires association with anionic lipids like phosphatidic acid (PA) and phosphatidylglycerol (PG), although PA and PG likely have distinct binding sites on AtMGD1 (Covés, Joyard, & Douce, 1988; Dubots et al., 2010; Ohta, Shimojima, Arai, et al., 1995; Rocha et al., 2016). Moreover, Sarkis et al. (2014) suggested that MGDG helps AtMGD1 bind to membranes, whereas DGDG releases AtMGD1 from the membranes. These findings indicate that the ratio of MGDG to DGDG in the membranes is important for proper MGD1 localization and activity.

The Arabidopsis type B MGDG synthases AtMGD2 and AtMGD3 are mainly expressed in non-photosynthetic tissues (Awai et al., 2001; Kobayashi, Awai, et al., 2009). AtMGD2 and AtMGD3 are localized in the outer side of the outer envelope membrane (Awai et al., 2001). Biochemical analyses using recombinant AtMGD1, AtMGD2 and AtMGD3 expressed in *Escherichia coli* revealed that all three enzymes utilize DAG and UDP-galactose as substrates, although their preference for various DAG species differs (Awai et al., 2001). Recombinant MGD1 (type A) can utilize both 18:2/18:2 DAG and 18:1/16:0 DAG equally (Awai et al., 2001), and this was also observed with spinach MGDG synthase (Maréchal et al., 1994a, 1994b; Miège et al., 1999). On the other hand, recombinant MGD2 and MGD3 (type B) prefer 18:2/18:2 DAG over 18:1/16:0 DAG (Awai et al., 2001), suggesting that the type B enzymes have substrate preference for DAGs supplied from the endoplasmic reticulum (ER) (Awai et al., 2001). Thus, the DAG substrate preferences of type A and B MGDG synthases may be similar among seed plants. However, seed plants are categorized as "16:3" or "18:3" plants based on differences between their biosynthetic pathways: 16:3 plants have both eukaryotic (ER) and prokaryotic (chloroplast) pathways, whereas 18:3 plants have only the eukaryotic (ER) pathway (Roughan & Slack, 1982). Although genes for both type A and B MGDG synthases are conserved in angiosperms (Hori et al., 2016; Kobayashi, Nakamura, & Ohta, 2009; Yuzawa et al., 2012), additional biochemical data for MGDG biosynthesis among 16:3 and 18:3 plants will be needed to support the functional generality of the enzymes in higher plants.

MGDG biosynthesis is regulated differently with respect to type A and type B MGDG synthases. Type A MGDG synthase predominates in photosynthetic organs, where it contributes to chloroplast development. Upon irradiation with visible light, the abundance of thylakoid membranes increases in plastids, with a concomitant increase in both MGDG synthase activity and MGDG and DGDG production/accumulation (Ohta, Shimojima, Ookata, et al., 1995). MGDG synthesis that is induced by illumination is mediated by type A MGDG synthase and is regulated by cytokinin signaling (Kobayashi et al., 2012; Kobayashi, Masuda, Takamiya, & Ohta, 2006; Yamaryo et al., 2003). Biochemical analyses using spinach revealed that the reducing agent dithiothreitol stabilizes type A MGDG synthase activity (Maréchal et al., 1991; Teucher & Heinz, 1991), implying that cysteine residues must remain in a reduced state (Maréchal et al., 1991; Teucher & Heinz, 1991). Later, use of recombinant type A MGDG synthase confirmed that the cysteine thiol groups are essential for type A MGDG synthase activity (Miège et al., 1999; Yamaryo, Motohashi, Takamiya, Hisabori, & Ohta, 2006). Thioredoxin is likely the reductant for the activation of type A MGDG synthase in chloroplasts, but it remains unclear which type of thioredoxin actually interacts with type A MGDG synthase because type A MGDG synthase localizes on the intermembrane surface of the envelope membrane. As noted above, phospholipids such as PA and PG enhance the activity of type A MGDG synthase (Covés et al., 1988; Dubots et al., 2010; Ohta, Shimojima, Arai, et al., 1995). Although mature chloroplasts are not observed, some non-photosynthetic tissues contain abundant galactolipids (Camara, Bardat, Dogbo, Brangeon, & Monéger, 1983; Kleinig & Liedvogel, 1978). In Arabidopsis, MGD2 is highly expressed in flowers and likewise MGD3 in roots as well as in young seedlings, although MGD3 expression decreases during seedling maturation (Awai et al., 2001). In analyses using a β-glucuronidase (GUS) reporter gene, both MGD2-GUS and MGD3-GUS were expressed in pollen tubes (Kobayashi, Awai, Takamiya, & Ohta, 2004). In *Petunia hybrida*, DGDG was found to be the most abundant galactolipid in flowers, and MGDG synthase activity increased during flower development (Murphy, 1982; Nakamura et al., 2003). In *Lilium longiflorum*, galactolipids accumulated during pollen tube growth (Nakamura, Kobayashi, & Ohta, 2009). All these results suggest that type B MGDG synthases play a role in flower development. However, Arabidopsis double-knockout mutants of MGD2 and MGD3 were found not to be defective for pollen tube growth (Kobayashi, Awai, et al., 2009). Thus, any lack of type B MGDG synthase might be compensated by type A MGDG synthase. The most well-analyzed function of type B MGDG synthases is MGDG synthesis induced by phosphate starvation. Phosphate is essential for plant growth, but it tends to run short in soil because inorganic phosphate forms insoluble salts that are unavailable for absorption by roots. Phosphate-deprived plants undergo membrane-lipid remodeling, in which phospholipids are replaced by glycolipids. During phosphate starvation, PC in the extraplastidial membranes is degraded and replaced by DGDG that has been exported from chloroplasts (Boudière et al., 2014; Jouhet et al., 2004; Kelly et al., 2003; Tjellström, Andersson, Larsson, & Sandelius, 2008). Phospholipid degradation supplies phosphate for certain other essential biological processes and thus enables plants to survive under this type of stress. The lack of type B MGDG synthases in Arabidopsis does not affect the growth, galactolipid content, or fatty acid composition of cellular membranes under normal growth conditions. However, the double-knockout mutant *mgd2mgd3* of Arabidopsis had decreased DGDG content along with growth inhibition under phosphate starvation (Kobayashi, Awai, et al., 2009), indicating that type B MGDG synthases synthesize MGDG for DGDG synthesis. The exchange between DGDG and PC is possible because both lipids are bilayer-forming lipids. Recent analyses revealed that type B MGDG synthases are also involved in tolerance to acid stress under phosphate starvation (Murakawa, Ohta, & Shimojima, 2019), indicating that type B MGDG synthases may have other important roles in plant growth. In addition, AtMGD3 enhances Arabidopsis plant growth in the presence of exogenously supplied sucrose by increasing the DGDG content in extraplastidial membranes, plausibly for supporting the expansion of cell membranes (Murakawa et al., 2014). It must also be noted that type B MGDG synthase can compensate for the function of type A MGDG synthase under phosphate starvation (Kobayashi et al., 2013). Given that type A MGDG synthase and type B MGDG synthase localize differently in the envelope membrane, have different substrate preference, and are regulated by different factors, it remains to be clarified how they can substitute for each other.

In Arabidopsis, the transcriptional regulation of galactolipid and chlorophyll biosynthesis is coordinated during chloroplast biogenesis (Kobayashi et al., 2014) along with the assembly of photosynthetic complexes and thylakoid membranes (Kobayashi et al., 2013). Inducible knockdown of MGD1 in Arabidopsis revealed that galactolipids are necessary for organelle differentiation in cotyledons (Fujii, Kobayashi, Nakamura, & Wada, 2014). Although MGDG is required for the formation and function of thylakoid membranes in chloroplasts, the roles of MGDG in prolamellar body formation and etioplast-to-chloroplast development remain unknown.

Recently, however, studies have shown that MGDG biosynthesis at an early germination stage is particularly important for protochlorophyllide accumulation. During the initial stage of etioplast development, MGDG may serve as a lipid matrix for protochlorophyllide biosynthesis and the formation of complexes comprising protochlorophyllide and light-dependent NADPH:protochlorophyllide oxidoreductase, by which chlorophyll is immediately formed after illumination. Furthermore, analyses of Arabidopsis MGD1 knockdown mutants produced by an artificial microRNA targeting type A MGDG synthase (amiR-MGD1) and DGDG-deficient mutants (*dgd1*) suggested that the MGDG-to-DGDG ratio may substantively affect the migration of prolamellar bodies to the thylakoid membrane during the etioplast-to-chloroplast transition (Fujii, Nagata, Masuda, Wada, & Kobayashi, 2019; Ohta, Shimojima, Ookata, et al., 1995). Details concerning the role of the galactolipids in this regard are summarized in recent reviews (Fujii, Wada, & Kobayashi, 2019; Kobayashi & Wada, 2016).

The expression of all types of MGDG synthase and the corresponding galactolipid levels change in response to various environmental stresses. In Arabidopsis, expression of *MGD2* and *MGD3* is induced by hydrogen peroxide that is largely produced via conversion of superoxide anion (O_2^-) by cellular superoxide dismutase (Xu, Moellering, Fan, & Benning, 2008). Expression of *Oryza sativa MGD*, which is another type of MGDG synthase, increases upon submergence, drought, or high salinity (Qi et al., 2004). Drought stress can affect the expression of *MGD1*, *MGD2*, *DGD1* and *DGD2* in cowpea leaves in a cultivar-dependent manner (Torres-Franklin, Gigon, de Melo, Zuily-Fodil, & Pham-Thi, 2007). Salinity can also decrease the abundance of chloroplast lipids and reduce galactolipid synthesis in cell suspensions of olive, oilseed rape, and *Cantharanthus roseus* (Elkahoui, Smaoui, Zarrouk, Ghrir, & Limam, 2004; Harwood, 1984; Najine, Marzouk, & Cherif, 1995; Zarrouk, Seqqat-Dakhma, & Cherif, 1995). At*MGD1* expression in shoots increases after wounding, and likewise in roots during drought or salt stress (Kobayashi, Awai, et al., 2009). Chloroplast-derived MGDG decreases slightly in Arabidopsis exposed to elevated levels of ozone (Afitlhile, Worthington, Banigan, Jirik, & Hildebrand, 2021). Thus, expression of *MGDs* may be affected by many kinds of environmental stresses, but the physiological relevance remains unclear.

To clarify the direct involvement of MGDG synthases in stress tolerance, several research groups analyzed MGD knockouts and MGD-overexpressing transgenic plants. Transgenic tobacco plants overexpressing

O. sativa MGD (OsMGD) showed enhanced tolerance to salinity (Wang et al., 2014). Overexpression of OsMGD also enhanced aluminum tolerance in tobacco (Zhang et al., 2016), and an Arabidopsis *mgd* mutant was hypersensitive to aluminum stress (Liu et al., 2020), clearly indicating the involvement of MGDs in aluminum stress. However, a change in the galactolipid profile and content can affect the abundance of other metabolites and macromolecules. Moreover, a change in galactolipid content can affect jasmonic acid production. For example, an increase in the MGDG: DGDG ratio induced overproduction of jasmonic acid and also altered chloroplast shape (Yu, Lin, & Li, 2020). Exposure to nitric oxide leads to a decrease in MGDG in plastids via posttranslational inhibition of MGDG synthase activity, followed by accumulation of triacylglycerol (TAG) (Dolch, Lupette, et al., 2017). Notably, the head-group of chloroplast membrane lipids can be acylated. Approximately 50 years ago, an MGDG was discovered that had a third acyl group esterified at the carbon of the galactose moiety (acMGDG) in spinach leaf homogenates (Myhre, 1968) and wheat flour (Heinz & Tulloch, 1969). Among more than 100 acylated chloroplast lipids identified to date, MGDGs containing esters of the oxylipin 12 oxo-phytodienoic acid or dinor-12-oxo-phytodienoic acid were identified and named arabidopside E, a species of the plant metabolite class called arabidopsides (Andersson et al., 2006; Buseman et al., 2006; Hisamatsu, Goto, Hasegawa, & Shigemori, 2003; Hisamatsu, Goto, Sekiguchi, Hasegawa, & Shigemori, 2005). Although arabidopsides were initially thought to exist only in Arabidopsis, later they were also found in certain other Brassicaceae species, lemon balm *Melissa officinalis*, Mexican Morning glory *Ipomoea tricolor*, and creeping thistle *Cirsium arvense* (Hartley, Eschen, Horwood, Gange, & Hill, 2015; Nilsson et al., 2015; Ohashi, Ito, Okada, & Sakagami, 2005; Zábranská, Vrkoslav, Sobotníková, & Cvačka, 2012). Although the functions of acMGDG remain unknown, MGDG accumulation might also affect acMGDG abundance. Advances made in this regard have been summarized by Song, Zoong Lwe, Wickramasinghe, and Welti (2021).

2.2 DGDG in seed plants

The first DGD gene was identified in Arabidopsis after discovery of the DGDG mutant, *dgd1* (Dörmann et al., 1999, 1995), in which the DGDG content was reduced by 90% compared with that of wild type, suggesting that DGD1 is the predominant enzyme for DGDG synthesis in Arabidopsis (Dörmann et al., 1995). The subsequent discovery of DGD2 accounted for the residual 10% of DGDG seen in *dgd1* plants (Kelly & Dörmann, 2002). *dgd1* mutant plants are stunted, owing mainly to the hyperaccumulation of jasmonic acids and their biosynthetic intermediates including arabidopsides. As described above, this may be mainly associated with changes in the MGDG:DGDG ratio in plastids. Although both DGD1 and DGD2 are associated with the outer envelope membrane of plastids, their protein structures differ markedly—especially at the N-terminus. The N-terminal sequence of DGD1 does not contribute to enzymatic activity but is required for localization to the outer envelope membrane (Froehlich, Benning, & Dörmann, 2001). Later work showed that this N-terminal sequence is important for galactolipid transfer between envelope membranes (Kalisch, Dörmann, & Hölzl, 2016; Kelly et al., 2016). The N-terminal sequence may also bind PA, after which it may promote membrane fusion (Kelly et al., 2016). Szpryngiel et al. (2011) found that DGD2 interacts with membranes via binding of its N-terminal region to negatively charged lipids. The C-terminus of DGD2 serves as a lipid-sensing switch that activates DGD2 activity (Szpryngiel & Mäler, 2016). *DGD2* expression is induced upon phosphate starvation (Kelly & Dörmann, 2002). Under normal growth conditions, wild-type and *dgd2* knockout mutant plants have similar growth, photosynthetic activity, and glycerolipid composition; however, under phosphate starvation, *dgd2* plants represses the DGDG accumulation that normally occurs under this stress (Kelly et al., 2003). Moreover, the fatty acid composition of DGDG indicates that MGD2 and MGD3 cooperate with DGD2 to increase the supply of DGDG under phosphate starvation (Benning & Ohta, 2005; Kobayashi, Awai, et al., 2009).

The nitrogen-fixing rhizobia are surrounded by a peribacteroid membrane in the nodules of legumes, which is critical for the exchange of ammonium and nutrients between the bacteria and cells of the host plant. DGDG also accumulates in the peribacteroid membrane of nitrogen-fixing nodules of soybean and lotus (Gaude et al., 2004). Accumulation of the nonphosphate-containing membrane-lipid DGDG in the peribacteroid membrane may be important for phosphate conservation for later use in nodule formation (Gaude et al., 2004). The preference for galactose over glucose in glycolipids that form thylakoid membranes in plant chloroplasts remains unknown. The bacterial glucosyltransferase from *Chloroflexus aurantiacus* was introduced into Arabidopsis *dgd1* and the double-mutant *dgd1dgd2* to produce a glucose-containing lipid (Hölzl et al., 2006). Subsequent analyses demonstrated that the substitution of the second galactose with glucose in DGDG did not substantively affect plant growth but decreased the capacity for chloroplast membrane assembly and proliferation, followed by the reduction of the maximal photosynthetic efficiency (Hölzl & Dörmann, 2007; Hölzl et al., 2009, 2006; Kalisch et al., 2016).

2.3 Oligogalactolipids in seed plants

The common feature of the MGDG synthases (MGD1, MGD2, MGD3) and DGDG synthases (DGD1, DGD2) is that UDP-galactose serves as the galactose donor for galactosylation. However, plants also have a processive galactosyltransferase (GGGT) that transfers galactose from one MGDG to another MGDG to produce DGDG. Each of trigalactosyldiacylglycerol (TGDG) and tetragalactosyldiacylglycerol (TeGDG) can be synthesized in the same manner (Fig. 2) (van Besouw & Wintermans, 1978). However, the enzyme corresponding to the reaction remained unknown for many years (Benning & Ohta, 2005). The existence of GGGT was confirmed by the analysis of *dgd1dgd2* (Kelly et al., 2003). The *dgd1dgd2* mutants retained slight DGDG synthase activity, and TGDG and tetragalactosyldiacylglycerol were observed in this mutant (Kelly et al., 2003). Later, Moellering et al. (2010) found that *SENSITIVE TO FREEZING 2* (*SFR2*) is the gene encoding GGGT in Arabidopsis. SFR2 is not crucial for growth under normal conditions, but the *sfr2* knockout mutant is hypersensitive to freezing stress (Fourrier et al., 2008; Moellering et al., 2010; Thorlby, Fourrier, & Warren, 2004). The molecular mechanism of freezing tolerance mediated by SFR2 has been suggested to be via lipid remodeling, i.e., by increasing the ratio of bilayer-forming lipids to nonbilayer-forming lipids to avoid destructive fusion of two membranes (Moellering & Benning, 2011). SFR2 is classified as a family I glycosyl hydrolase, but has no hydrolase activity. Roston, Wang, Kuhn, and Benning (2014) investigated the relationship between its structure and function using biochemical methods and identified three essential regions required for galactosyltransferase activity. GGGT is activated by divalent cations (e.g., Mg^{2+}, Ca^{2+} and Ba^{2+}) as well as free fatty acids added to the assay (Barnes, Benning, & Roston, 2016; Heemskerk, Bögemann, & Wintermans, 1983; Sakaki, Kondo, & Yamada, 1990a). Oligogalactolipid formation is also associated with several abiotic stresses including ozone fumigation, high salinity and desiccation (Barnes et al., 2016; Sakaki, Kondo, & Yamada, 1990b; Wang, Hersh, & Benning, 2016). Interestingly, membranes of the terrestrial alga *Klebsormidium nitens* contain oligogalactolipids, and the content is >3% in the membrane-lipid fraction, a proportion similar to that measured for the other major membrane lipids such as PC and PG (Hori et al., 2016). Indeed, the *K. nitens* genome encodes two homologues of Arabidopsis *SFR2* (Hori et al., 2014; Hori et al., 2016). It was recently postulated that the presence of oligogalactolipids in membranes was one of the key factors contributing to the adaptation of ancient aquatic plants to terrestrial environments (Gasulla, García-Plazaola, López-Pozo, & Fernández-Marín, 2019; Hori et al., 2016).

2.4 Galactolipids in algae

The biosynthesis of each of MGDG and DGDG has diverged during the evolution of land plants; for example, enzyme-coding genes of seed plants have undergone duplication and functional specialization (Hori et al., 2016). A specialized biosynthesis pathway comprising MGD2/3 and DGD2 rather than pathways for biosynthesis of the other major galactolipids for thylakoid assembly in chloroplasts may have evolved for adaptation to phosphate-deficient conditions (Yuzawa et al., 2012). This may have proved advantageous for adapting to severe nutrient deficiencies that sometimes occur in terrestrial environments (Hori et al., 2016). Apart from these functional divergences of galactolipid biosynthesis pathways during land colonization, several secondary endosymbionts possess different pathways for galactolipid biosynthesis. For example, the plastid genome of red algae acquired a DGDG biosynthesis gene (*DgdA*) from a primordial cyanobacterium. Indeed, DgdA in both cyanobacteria and red algae has two glycosyltransferase (GT) domains, i.e., GT1 and GT4. This structural feature differs from land-plant DGD1 and DGD2, each of which has only GT4 (Sato & Awai, 2016). Interestingly, certain red algae such as *Porphyridium purpureum* lack *DgdA* in the plastid genome but have *DGD1* in the nuclear genome (Bhattacharya et al., 2013; Tajima et al., 2014).

Galactolipid biosynthesis in algae is dependent on the glycerolipid biosynthesis pathway in the ER, a scenario that differs from that of land plants. In fact, various algal galactolipids contain very-long-chain polyunsaturated fatty acids that are elongated and desaturated in the ER. Typically, eicosapentaenoic acid (EPA) and docosahexaenoic acid (DHA) are the major fatty acids of galactolipids, and the pathway that provides these omega 3 polyunsaturated fatty acids is called the omega 3 pathway. The heterokont and secondary symbiont *Nannochloropsis* contains EPA particularly in galactolipids and diacylglyceryltrimethylhomoserine (DGTS)

(Iwai, Hori, Sasaki-Sekimoto, Shimojima, & Ohta, 2015; Murakami, Nobusawa, Hori, Shimojima, & Ohta, 2018; Simionato et al., 2013). When DGTS biosynthesis is knocked out, the proportion of EPA in galactolipids increases. If DGTS is overproduced, however, EPA abundance in galactolipids decreases, suggesting that galactolipids and DGTS are the two major pools of EPA. However, the mechanism by which ER-born EPA is transported to plastids is unknown. A hint is offered by the fact that knockout of a gene encoding elongase results in a specific reduction of the proportion of free EPA, thereby causing a significant reduction of MGDG in *Nannochloropsis* (Dolch, Rak, et al., 2017). This result also suggests that EPA is a major and important constituent of MGDG, and thus its reduction strongly affects MGDG level in *Nannochloropsis*. However, the role of EPA in MGDG remains to be clarified.

3. Evolution of galactolipid synthesis

3.1 Origin and divergence of MGDG synthases

An ancient cyanobacterium is thought to be the origin of plant chloroplasts. Indeed, the typical membrane lipid composition of cyanobacteria is similar to that of chloroplasts, composed mainly of MGDG, DGDG, sulfoquinovosyldiacylglycerol, and PG. However, their MGDG biosynthesis pathways are not similar. In plants, MGDG is synthesized by a one-step reaction, i.e., galactose transfer from UDP-galactose to DAG. In cyanobacteria, however, MGDG is synthesized by a two-step reaction, i.e., glucose transfer from UDP-glucose to DAG to produce monoglucosyldiacylglycerol with concomitant epimerization of the glucose (Sato & Murata, 1982b). Indeed, the homologous gene of plant MGDG synthase has not been found in cyanobacterial genomes. The subsequent identification of the cyanobacterial MGDG synthases monoglucosyldiacylglycerol synthase (MgdA) and glucolipid epimerase (MgdE) clearly demonstrated that the biosynthetic pathways for MGDG differ between cyanobacteria and plants (Awai et al., 2006, 2001; Sato & Murata, 1982b). MgdA is conserved in cyanobacteria, whereas MgdE is not found in certain cyanobacterial species. Given that MGDG is produced in all cyanobacteria, another epimerase might exist for the epimerization step. The two-step MGDG synthesis can be experimentally replaced with the aforementioned one-step MGDG synthesis in the cyanobacteria *Synechocystis* sp. PCC 6803 (Yuzawa et al., 2014). Both the growth and photosynthetic capacity of the transformants were comparable with those aspects of wild-type cells. Low temperature, however, retarded the growth of the transformants, indicating that monoglucosyldiacylglycerol might be important in adaptation to cold stress (Yuzawa et al., 2014). DGDG is dispensable in *Synechocystis* sp. PCC 6803 under optimal growth conditions but is required for maximal photosynthetic growth under phosphate starvation (Awai, Watanabe, Benning, & Nishida, 2007). A mutant of *Synechocystis* sp. PCC 6803 that lacks both MGDG and DGDG but accumulates monoglucosyldiacylglycerol retained the capacity to assemble thylakoid membranes and had maximal photosynthetic activity (Awai et al., 2014). This result clearly indicates that oxygenic photosynthesis does not necessarily require galactolipids. However, in another species of cyanobacteria, namely *Synechococcus elongatus* PCC 7942, DGDG is essential and cannot be compensated by other membrane lipids (Maida & Awai, 2016). Although the existence of DGDG itself is crucial for growth, the DGDG biosynthesis pathway can be engineered; a mutant of *S. elongatus* PCC 7942 was produced in which the cyanobacterial pathway for galactolipid synthesis was replaced with the plant-type pathway, and the mutant could grow photoautotrophically, suggesting that at least certain cyanobacterial species can be engineered to utilize alternative galactolipid synthesis pathways (Apdila, Inoue, Shimojima, & Awai, 2020). For more detailed discussion of the research and advances pertaining to galactolipid synthesis in cyanobacteria, we recommend several recent reviews (Apdila & Awai, 2018; Awai, 2016; Sato & Awai, 2016).

Because the pathways for galactolipid biosynthesis differ substantially between cyanobacteria and plants, the origin of plant MGDG synthase has been the subject of intensive research. Phylogenetic analyses using plant MGDG synthase and enzyme assays for the homologous proteins expressed in *E. coli* suggested that plant MGDG synthase was derived from the filamentous anoxygenic phototroph *Chloroflexi* through horizontal gene transfer, which occurred independently of the endosymbiosis of cyanobacteria (Yuzawa et al., 2012). Moreover, the importance of MGDG in anoxygenic photosynthetic bacteria has also been revealed, in that certain species of these bacteria contain MGDG in the chlorosome, which is an ellipsoid organelle with an interior of self-organized bacteriochlorophyll aggregates (Pedersen, Linnanto, Frigaard, Nielsen, & Miller, 2010). Thus, it has been suggested that MGDG also has an indispensable role in anoxygenic photosynthetic bacteria. The MGDG synthase gene was later identified in the green sulfur bacterium *Chlorobaculum tepidum*, and analysis of the knockout mutant revealed that MGDG is essential for chlorosome assembly (Masuda et al., 2011). All these findings suggest that MGDG is necessary for the viability of all photosynthetic organisms, although the biosynthetic machineries have been independently acquired and diverged since that acquisition.

Phylogenetic analyses have suggested that type B MGDG synthases were acquired 320 million years ago, immediately after the emergence of Spermatophyta (seed plants) (Yuzawa et al., 2012). Indeed, type B MGDG synthase has been documented only in seed plants and not in other land plants such as mosses and ferns (Hori et al., 2016). Given that nutrient shortages may have occurred frequently after plants transitioned to land, the evolution of type B MGDG synthases might have been crucial for their growth. Moreover, environmental stresses often occur simultaneously. In Arabidopsis, for example, lipid remodeling mediated by type B MGDG synthase in roots was shown to be important for survival during phosphate deficiency specifically under acidic conditions (Murakawa et al., 2019). Thus, upregulated expression of type B MGDG synthase often correlates with environmental changes, suggesting its involvement in tolerating multiple stressors and a rationale for why seed plants acquired type B MGDG synthase.

3.2 Divergence of DGDG synthases

As noted above, seed plants produce two DGDG synthases, namely DGD1 and DGD2. On the other hand, in cyanobacteria and Cyanidiophytina, the particular DGDG synthase DgdA evolved independently (Awai et al., 2007; Sakurai, Mizusawa, Wada, & Sato, 2007). The amino acid sequence of DgdA is similar to that of sqdX/SQD2 (sulfoquinovosyldiacylglycerol synthase). Indeed, phylogenetic analysis of DgdA and SqdX/SQD2 revealed that DgdA was derived from an sqdX-like gene before the emergence of cyanobacteria (Sato & Awai, 2016). One candidate DGD has been identified in each of Rhodophytina, Stramenopiles, green algae, *Bigelowiella natans*, and *K. nitens*, and one low-similarity candidate DGD was identified in *Cyanophora paradoxa* (Maida & Awai, 2016). Thus, DgdA and DGD may have initially coexisted, and DGD was later utilized for DGDG synthesis in the green-algae lineage (Hori et al., 2016).

DGD1 and DGD2 are ubiquitous among seed plants, suggesting that their distinct roles have been evolutionarily conserved. Phylogenetic analyses suggested that these two DGDs diverged before the emergence of Amborella (Hori et al., 2016), but biochemical and molecular biological analyses of MGDs and DGDs in gymnosperms and ferns are required to confirm this hypothesis.

3.3 Divergence of oligogalactolipid synthases

As noted above, TGDG accumulates in seed plants, but it also accumulates in Oscillatoria, Tolypothrix (cyanobacteria) (Zepke, Heinz, Radunz, Linscheid, & Pesch, 2004), and Chlorella (Chlorophyta) (Benson, Wiser, Ferrari, & Miller, 1958; Harwood, 1998), yet SFR2 homologues have been found only in land plants and charophyte algae (Hori et al., 2016). In *K. nitens*, TGDG accumulates even under normal growth conditions (Hori et al., 2016), whereas it is hardly observed in Arabidopsis under normal conditions (Moellering et al., 2010). The corresponding gene for TGDG synthase and the reason for TGDG accumulation in *K. nitens* remain unknown, but analysis of the biosynthesis pathway might help determine how ancient aquatic plants adapted to terrestrial environments. As described above, the evolution of oligogalactolipids may be one of the key factors underlying the colonization of land environments by plants (Gasulla et al., 2019; Hori et al., 2016).

4. Perspectives

Progress in galactolipid research in plants has been considerable since the first discovery of MGDG and DGDG. The functions, biosynthetic pathways, and evolution of galactolipids have long been analyzed by numerous number of research groups. There are still many issues to be addressed by extensive research, but one of the most intriguing research topics might be regulation of galactolipid catabolism and recycling in plants and algae.

Galactolipid-specific lipases have been reported in several plants and microalgae. In Chlamydomonas, for example, the galactolipase PGD1 is involved in TAG biosynthesis (Li et al., 2012). PGD1 specifically releases the *sn*-1-attached fatty acid from *de novo* synthesized MGDG. Upon this hydrolysis reaction, 18:1(9) is mainly released from the de novo MGDG and supplied for TAG synthesis, particularly during nitrate deprivation. In seed plants, PC is the main membrane phospholipid that undergoes changes in fatty-acid composition. The resulting diacylglyceryl backbones are further supplied to the ER pathway for galactolipid and TAG biosynthesis. This is known as PC editing (Bates et al., 2012). However, *Chlamydomonas reinhardtii* lacks PC, and thus the galactolipid MGDG may be the major source of fatty acids for TAG biosynthesis. Indeed, TAG levels are greatly reduced in the *pgd1* mutant of *C. reinhardtii*. PGD1-mediated hydrolysis of MGDG produces the lysogalactolipid monogalactosylmonoacylglycerol (MGMG) along with free fatty acids. Iwai et al. (2021) reported that the *C. reinhardtii* acyltransferase CrLAT1, which is a homologue of Arabidopsis lysophosphatidylcholine acyltransferase, mediates the recycling of MGMG to produce MGDG. MGMG level increases in CrLAT knockdown mutants and decreases in CrLAT overexpressors. Interestingly, *C. reinhardtii* growth was compromised in both knockdown and overexpressor lines, suggesting that an appropriate balance between MGDG and MGMG is required for maximal cell growth. This conversion of MGMG to MGDG is another pathway for galactolipid synthesis specifically found in *C. reinhardtii*. *CrGH1*, which encodes galactosyl hydrolase, was also found in *Chlamydomonas* by data mining of the previously reported oil body proteome (Gu et al., 2021). Although its in vivo substrate is unclear, the *CrGH1*-deficient mutant had increased galactolipids and decreased TAG under high-light conditions. This result suggests that this enzyme is involved in conversion of galactolipids to TAG under high-light conditions in *Chlamydomonas*.

It remains unknown whether a similar MGDG recycling system exists in land plants. The galactolipase HEAT INDUCIBLE LIPASE1 (HIL1) is involved in breakdown of MGDG, which also produces a single free fatty acid and MGMG (Higashi et al., 2018). HIL1 functions in membrane remodeling during heat acclimation, during which polyunsaturated fatty acid levels are reduced in chloroplasts. Upon heating to 38 °C, *sn*-1-linked 18:3 is preferentially liberated from MGDG by HIL1. This reaction decreases the total levels of highly polyunsaturated MGDG species and may help to decrease fluidity of the chloroplast membrane under heat conditions. HIL1 also produces MGMG along with a fatty acid at the *sn*-2 position, but the function and fate of this MGMG remains to be clarified. Because HIL1 localizes to the chloroplast stroma, another turnover system for lysogalactolipids that differs from that of PGD1 and LAT1 in *Chlamydomonas* may exist inside chloroplasts in seed plants. We hope that many questions regarding galactolipid catabolism and recycling will be answered in the future studies.

Acknowledgments

We apologize to researchers whose work we may have overlooked. Our recent work on plant and algal galactolipids was supported in part by the Core Research for Evolutional Science and Technology (CREST) program of Japan Science and Technology Agency (JST); by Grant-in-Aid for Scientific Research (A)18H03941, (B)15H04393, (C) 18K05371; by Grants-in-Aid for Scientific Research on Innovative Areas (17H06417) from the Ministry of Education, Culture, Sports, Science and Technology of Japan (MEXT) and the Japan Society for the Promotion of Science (JSPS).

References

Afitlhile, M., Worthington, R., Banigan, E., Jirik, J., & Hildebrand, D. (2021). Ozone-induced lipid changes in the wildtype and *toc*132*toc*120 heterozygote mutant of *Arabidopsis thaliana*. *Plant Physiology and Biochemistry, 158*, 13–20. https://doi.org/10.1016/j.plaphy.2020.11.050.

Andersson, M. X., Hamberg, M., Kourtchenko, O., Brunnström, A., McPhail, K. L., Gerwick, W. H., et al. (2006). Oxylipin profiling of the hypersensitive response in *Arabidopsis thaliana*. Formation of a novel oxo-phytodienoic acid-containing galactolipid, arabidopside E. *Journal of Biological Chemistry, 281*(42), 31528–31537. https://doi.org/10.1074/jbc.M604820200.

Andersson, M. X., Larsson, K. E., Tjellström, H., Liljenberg, C., & Sandelius, A. S. (2005). Phosphate-limited oat. The plasma membrane and the tonoplast as major targets for phospholipid-to-glycolipid replacement and stimulation of phospholipases in the plasma membrane. *Journal of Biological Chemistry, 280*(30), 27578–27586. https://doi.org/10.1074/jbc.M503273200.

Andersson, M. X., Stridh, M. H., Larsson, K. E., Liljenberg, C., & Sandelius, A. S. (2003). Phosphate-deficient oat replaces a major portion of the plasma membrane phospholipids with the galactolipid digalactosyldiacylglycerol. *FEBS Letters, 537*(1–3), 128–132. https://doi.org/10.1016/s0014-5793(03)00109-1.

Apdila, E. T., & Awai, K. (2018). Configuration of the sugar head of glycolipids in thylakoid membranes. *Genes & Genetic Systems, 92*(5), 235–242. https://doi.org/10.1266/ggs.17-00047.

Apdila, E. T., Inoue, S., Shimojima, M., & Awai, K. (2020). Complete replacement of the galactolipid biosynthesis pathway with a plant-type pathway in the cyanobacterium *Synechococcus elongatus* PCC 7942. *Plant and Cell Physiology, 61*(9), 1661–1668. https://doi.org/10.1093/pcp/pcaa090.

Awai, K. (2016). Thylakoid development and Galactolipid synthesis in cyanobacteria. *Sub-Cellular Biochemistry, 86*, 85–101. https://doi.org/10.1007/978-3-319-25979-6_4.

Awai, K., Kakimoto, T., Awai, C., Kaneko, T., Nakamura, Y., Takamiya, K., et al. (2006). Comparative genomic analysis revealed a gene for monoglucosyldiacylglycerol synthase, an enzyme for photosynthetic membrane lipid synthesis in cyanobacteria. *Plant Physiology, 141*(3), 1120–1127. https://doi.org/10.1104/pp.106.082859.

Awai, K., Maréchal, E., Block, M. A., Brun, D., Masuda, T., Shimada, H., et al. (2001). Two types of MGDG synthase genes, found widely in both 16:3 and 18:3 plants, differentially mediate galactolipid syntheses in photosynthetic and nonphotosynthetic tissues in *Arabidopsis thaliana*. *Proceedings of the National Academy of Sciences of the United States of America, 98*(19), 10960–10965. https://doi.org/10.1073/pnas.181331498.

Awai, K., Ohta, H., & Sato, N. (2014). Oxygenic photosynthesis without galactolipids. *Proceedings of the National Academy of Sciences of the United States of America, 111*(37), 13571–13575. https://doi.org/10.1073/pnas.1403708111.

Awai, K., Watanabe, H., Benning, C., & Nishida, I. (2007). Digalactosyldiacylglycerol is required for better photosynthetic growth of *Synechocystis* sp. PCC6803 under phosphate limitation. *Plant and Cell Physiology, 48*(11), 1517–1523. https://doi.org/10.1093/pcp/pcm134.

Barnes, A. C., Benning, C., & Roston, R. L. (2016). Chloroplast membrane remodeling during freezing stress is accompanied by cytoplasmic acidification activating SENSITIVE TO FREEZING2. *Plant Physiology, 171*(3), 2140–2149. https://doi.org/10.1104/pp.16.00286.

Bates, P. D., Fatihi, A., Snapp, A. R., Carlsson, A. S., Browse, J., & Lu, C. (2012). Acyl editing and headgroup exchange are the major mechanisms that direct polyunsaturated fatty acid flux into triacylglycerols. *Plant Physiology, 160*(3), 1530–1539. https://doi.org/10.1104/pp.112.204438.

Benning, C., & Ohta, H. (2005). Three enzyme systems for galactoglycerolipid biosynthesis are coordinately regulated in plants. *Journal of Biological Chemistry, 280*(4), 2397–2400. https://doi.org/10.1074/jbc.R400032200.

Benson, A. A., Wiser, R., Ferrari, R. A., & Miller, J. A. (1958). Photosynthesis of galactolipids. *Journal of the American Chemical Society, 80*(17), 4740. https://doi.org/10.1021/ja01550a079.

Bhattacharya, D., Price, D. C., Chan, C. X., Qiu, H., Rose, N., Ball, S., et al. (2013). Genome of the red alga *Porphyridium purpureum. Nature Communications, 4*, 1941. https://doi.org/10.1038/ncomms2931.

Block, M. A., Dorne, A. J., Joyard, J., & Douce, R. (1983). Preparation and characterization of membrane fractions enriched in outer and inner envelope membranes from spinach chloroplasts. II. Biochemical characterization. *Journal of Biological Chemistry, 258*(21), 13281–13286.

Boudière, L., Botté, C. Y., Saidani, N., Lajoie, M., Marion, J., Bréhélin, L., et al. (2012). Galvestine-1, a novel chemical probe for the study of the glycerolipid homeostasis system in plant cells. *Molecular BioSystems, 8*(8), 2023–2035. https://doi.org/10.1039/c2mb25067e.

Boudière, L., Michaud, M., Petroutsos, D., Rébeillé, F., Falconet, D., Bastien, O., et al. (2014). Glycerolipids in photosynthesis: Composition, synthesis and trafficking. *Biochimica et Biophysica Acta, 1837*(4), 470–480. https://doi.org/10.1016/j.bbabio.2013.09.007.

Buseman, C. M., Tamura, P., Sparks, A. A., Baughman, E. J., Maatta, S., Zhao, J., et al. (2006). Wounding stimulates the accumulation of glycerolipids containing oxophytodienoic acid and dinor-oxophytodienoic acid in Arabidopsis leaves. *Plant Physiology, 142*(1), 28–39. https://doi.org/10.1104/pp.106.082115.

Camara, B., Bardat, F., Dogbo, O., Brangeon, J., & Monéger, R. (1983). Terpenoid metabolism in plastids: Isolation and biochemical characteristics of *Capsicum annuum* chromoplasts. *Plant Physiology, 73*(1), 94–99. https://doi.org/10.1104/pp.73.1.94.

Cline, K., & Keegstra, K. (1983). Galactosyltransferases involved in galactolipid biosynthesis are located in the outer membrane of pea chloroplast envelopes. *Plant Physiology, 71*(2), 366–372. https://doi.org/10.1104/pp.71.2.366.

Covés, J., Joyard, J., & Douce, R. (1988). Lipid requirement and kinetic studies of solubilized UDP-galactose:Diacylglycerol galactosyltransferase activity from spinach chloroplast envelope membranes. *Proceedings of the National Academy of Sciences of the United States of America, 85*(14), 4966–4970. https://doi.org/10.1073/pnas.85.14.4966.

Demé, B., Cataye, C., Block, M. A., Maréchal, E., & Jouhet, J. (2014). Contribution of galactoglycerolipids to the 3-dimensional architecture of thylakoids. *FASEB Journal, 28*(8), 3373–3383. https://doi.org/10.1096/fj.13-247395.

Dolch, L. J., Lupette, J., Tourcier, G., Bedhomme, M., Collin, S., Magneschi, L., et al. (2017). Nitric oxide mediates nitrite-sensing and acclimation and triggers a remodeling of lipids. *Plant Physiology, 175*(3), 1407–1423. https://doi.org/10.1104/pp.17.01042.

Dolch, L. J., Rak, C., Perin, G., Tourcier, G., Broughton, R., Leterrier, M., et al. (2017). A palmitic acid elongase affects eicosapentaenoic acid and plastidial monogalactosyldiacylglycerol levels in Nannochloropsis. *Plant Physiology, 173*(1), 742–759. https://doi.org/10.1104/pp.16.01420.

Dörmann, P., Balbo, I., & Benning, C. (1999). *Arabidopsis* galactolipid biosynthesis and lipid trafficking mediated by DGD1. *Science, 284*(5423), 2181–2184. https://doi.org/10.1126/science.284.5423.2181.

Dörmann, P., & Benning, C. (2002). Galactolipids rule in seed plants. *Trends in Plant Science,* 7(3), 112–118. https://doi.org/10.1016/s1360-1385(01)02216-6.

Dörmann, P., Hoffmann-Benning, S., Balbo, I., & Benning, C. (1995). Isolation and characterization of an Arabidopsis mutant deficient in the thylakoid lipid digalactosyl diacylglycerol. *Plant Cell,* 7(11), 1801–1810. https://doi.org/10.1105/tpc.7.11.1801.

Douce, R. (1974). Site of biosynthesis of galactolipids in spinach chloroplasts. *Science, 183*, 852–853.

Douce, R., & Joyard, J. (1980). *Lipids: Structure and function*. New York: Academic Press.

Dubots, E., Audry, M., Yamaryo, Y., Bastien, O., Ohta, H., Breton, C., et al. (2010). Activation of the chloroplast monogalactosyldiacylglycerol synthase MGD1 by phosphatidic acid and phosphatidylglycerol. *Journal of Biological Chemistry, 285*(9), 6003–6011. https://doi.org/10.1074/jbc.M109.071928.

Elkahoui, S., Smaoui, A., Zarrouk, M., Ghrir, R., & Limam, F. (2004). Salt-induced lipid changes in *Catharanthus roseus* cultured cell suspensions. *Phytochemistry, 65*(13), 1911–1917. https://doi.org/10.1016/j.phytochem.2004.06.021.

Feige, G. B., Heinz, E., Wrage, K., Cochems, N., & Ponzelar, E. (1980). *Discovery of a new glyceroglycolipid in blue-green algae and its role in galactolipid biosynthesis*. Amsterdam: Elsevier Biomedical Press.

Fourrier, N., Bédard, J., Lopez-Juez, E., Barbrook, A., Bowyer, J., Jarvis, P., et al. (2008). A role for *SENSITIVE TO FREEZING2* in protecting chloroplasts against freeze-induced damage in Arabidopsis. *Plant Journal, 55*(5), 734–745. https://doi.org/10.1111/j.1365-313X.2008.03549.x.

Froehlich, J. E., Benning, C., & Dörmann, P. (2001). The digalactosyldiacylglycerol (DGDG) synthase DGD1 is inserted into the outer envelope membrane of chloroplasts in a manner independent of the general import pathway and does not depend on direct interaction with monogalactosyldiacylglycerol synthase for DGDG biosynthesis. *Journal of Biological Chemistry, 276*(34), 31806–31812. https://doi.org/10.1074/jbc.M104652200.

Fujii, S., Kobayashi, K., Nakamura, Y., & Wada, H. (2014). Inducible knockdown of *MONOGALACTOSYLDIACYLGLYCEROL SYNTHASE1* reveals roles of galactolipids in organelle differentiation in Arabidopsis cotyledons. *Plant Physiology, 166*(3), 1436–1449. https://doi.org/10.1104/pp.114.250050.

Fujii, S., Nagata, N., Masuda, T., Wada, H., & Kobayashi, K. (2019). Galactolipids are essential for internal membrane transformation during etioplast-to-chloroplast differentiation. *Plant and Cell Physiology, 60*(6), 1224–1238. https://doi.org/10.1093/pcp/pcz041.

Fujii, S., Wada, H., & Kobayashi, K. (2019). Role of Galactolipids in plastid differentiation before and after light exposure. *Plants (Basel), 8*(10), 357. https://doi.org/10.3390/plants8100357.

Garab, G. (2016). Self-assembly and structural-functional flexibility of oxygenic photosynthetic machineries: Personal perspectives. *Photosynthetic Research, 127*(1), 131–150. https://doi.org/10.1007/s11120-015-0192-z.

Garab, G., Ughy, B., & Goss, R. (2016). Role of MGDG and non-bilayer lipid phases in the structure and dynamics of chloroplast thylakoid membranes. *Sub-Cellular Biochemistry, 86*, 127–157. https://doi.org/10.1007/978-3-319-25979-6_6.

Gasulla, F., García-Plazaola, J. I., López-Pozo, M., & Fernández-Marín, B. (2019). Evolution, biosynthesis and protective roles of oligogalactolipids: Key molecules for terrestrial photosynthesis? *Environmental and Experimental Botany, 164*, 135–148. https://doi.org/10.1016/j.envexpbot.2019.05.003.

Gaude, N., Tippmann, H., Flemetakis, E., Katinakis, P., Udvardi, M., & Dörmann, P. (2004). The galactolipid digalactosyldiacylglycerol accumulates in the peribacteroid membrane of nitrogen-fixing nodules of soybean and lotus. *Journal of Biological Chemistry, 279*(33), 34624–34630. https://doi.org/10.1074/jbc.M404098200.

Gu, X., Cao, L., Wu, X., Li, Y., Hu, Q., & Han, D. (2021). A lipid bodies-associated galactosyl hydrolase is involved in triacylglycerol biosynthesis and galactolipid turnover in the unicellular green alga *Chlamydomonas reinhardtii*. *Plants (Basel), 10*(4). https://doi.org/10.3390/plants10040675.

Guskov, A., Kern, J., Gabdulkhakov, A., Broser, M., Zouni, A., & Saenger, W. (2009). Cyanobacterial photosystem II at 2.9-a resolution and the role of quinones, lipids, channels and chloride. *Nature Structural & Molecular Biology, 16*(3), 334–342. https://doi.org/10.1038/nsmb.1559.

Härtel, H., & Benning, C. (2000). Can digalactosyldiacylglycerol substitute for phosphatidylcholine upon phosphate deprivation in leaves and roots of *Arabidopsis*? *Biochemical Society Transactions, 28*(6), 729–732.

Härtel, H., Dormann, P., & Benning, C. (2000). DGD1-independent biosynthesis of extraplastidic galactolipids after phosphate deprivation in *Arabidopsis*. *Proceedings of the National Academy of Sciences of the United States of America, 97*(19), 10649–10654. https://doi.org/10.1073/pnas.180320497.

Härtel, H., Essigmann, B., Lokstein, H., Hoffmann-Benning, S., Peters-Kottig, M., & Benning, C. (1998). The phospholipid-deficient *pho1* mutant of *Arabidopsis thaliana* is affected in the organization, but not in the light acclimation, of the thylakoid membrane. *Biochimica et Biophysica Acta, 1415*(1), 205–218. https://doi.org/10.1016/s0005-2736(98)00197-7.

Hartley, S. E., Eschen, R., Horwood, J. M., Gange, A. C., & Hill, E. M. (2015). Infection by a foliar endophyte elicits novel arabidopside-based plant defence reactions in its host, *Cirsium arvense*. *New Phytologist, 205*(2), 816–827. https://doi.org/10.1111/nph.13067.

Harwood, J. L. (1984). Effects of the environment on the acyl lipids of algae and higher plants. In P. A. Siegenthaler, & W. Eichenberger (Eds.), *Structure, function and metabolism of plant lipids* (pp. 543–550). Amsterdam: Elsevier.

Heemskerk, J. W., Bögemann, G., Helsper, J. P., & Wintermans, J. F. (1988). Synthesis of mono- and digalactosyldiacylglycerol in isolated spinach chloroplasts. *Plant Physiology, 86*(3), 971–977. https://doi.org/10.1104/pp.86.3.971.

Heemskerk, J. W., Bögemann, G., & Wintermans, J. F. G. M. (1983). Turnover of galactolipids incorporated into chloroplast envelopes: An assay for galactolipid: Galactolipid galactosyltransferase. *Biochimica et Biophysica Acta (BBA) - Lipids and Lipid Metabolism, 754*(2), 181–189. https://doi.org/10.1016/0005-2760(83)90160-1.

Harwood, J. L. (1998). Membrane lipids in algae. In P. A. Siegenthaler, & N. Murata (Eds.), *Lipids in photosynthesis: Structure, function and genetics* (pp. 53–64). Dordrech: Springer.

Heemskerk, J., Jacobs, F., Scheijen, M., Helsper, J., & Wintermans, J. (1987). Characterization of galactosyltransferases in spinach chloroplast envelopes. *Biochimica et Biophysica Acta, 918*, 189–203.

Heinz, E., & Tulloch, A. P. (1969). Reinvestigation of the structure of acyl galactosyl diglyceride from spinach leaves. *Hoppe-Seylers Zeitschrift für Physiologische Chemie, 350*(4), 493–498. https://doi.org/10.1515/bchm2.1969.350.1.493.

Higashi, Y., Okazaki, Y., Takano, K., Myouga, F., Shinozaki, K., Knoch, E., et al. (2018). *HEAT INDUCIBLE LIPASE1* remodels chloroplastic monogalactosyldiacylglycerol by liberating α-linolenic acid in Arabidopsis leaves under heat stress. *Plant Cell, 30*(8), 1887–1905. https://doi.org/10.1105/tpc.18.00347.

Hisamatsu, Y., Goto, N., Hasegawa, K., & Shigemori, H. (2003). Arabidopsides a and B, two new oxylipins from *Arabidopsis thaliana*. *Tetrahedron Letters, 44*(29), 5553–5556.

Hisamatsu, Y., Goto, N., Sekiguchi, M., Hasegawa, K., & Shigemori, H. (2005). Oxylipins arabidopsides C and D from Arabidopsis thaliana. *Journal of Natural Products, 68*(4), 600–603. https://doi.org/10.1021/np0495938.

Hölzl, G., & Dörmann, P. (2007). Structure and function of glycoglycerolipids in plants and bacteria. *Progress in Lipid Research, 46*(5), 225–243. https://doi.org/10.1016/j.plipres.2007.05.001.

Hölzl, G., Witt, S., Gaude, N., Melzer, M., Schöttler, M. A., & Dörmann, P. (2009). The role of diglycosyl lipids in photosynthesis and membrane lipid homeostasis in Arabidopsis. *Plant Physiology, 150*(3), 1147–1159. https://doi.org/10.1104/pp.109.139758.

Hölzl, G., Witt, S., Kelly, A. A., Zähringer, U., Warnecke, D., Dörmann, P., et al. (2006). Functional differences between galactolipids and glucolipids revealed in photosynthesis of higher plants. *Proceedings of the National Academy of Sciences of the United States of America, 103*(19), 7512–7517. https://doi.org/10.1073/pnas.0600525103.

Hori, K., Maruyama, F., Fujisawa, T., Togashi, T., Yamamoto, N., Seo, M., et al. (2014). *Klebsormidium flaccidum* genome reveals primary factors for plant terrestrial adaptation. *Nature Communications, 5*, 3978. https://doi.org/10.1038/ncomms4978.

Hori, K., Nobusawa, T., Watanabe, T., Madoka, Y., Suzuki, H., Shibata, D., et al. (2016). Tangled evolutionary processes with commonality and diversity in plastidial glycolipid synthesis in photosynthetic organisms. *Biochimica et Biophysica Acta, 1861*(9 Pt B), 1294–1308. https://doi.org/10.1016/j.bbalip.2016.04.015.

Iwai, M., Hori, K., Sasaki-Sekimoto, Y., Shimojima, M., & Ohta, H. (2015). Manipulation of oil synthesis in *Nannochloropsis* strain NIES-2145 with a phosphorus starvation-inducible promoter from *Chlamydomonas reinhardtii. Frontiers in Microbiology, 6*, 912. https://doi.org/10.3389/fmicb.2015.00912.

Iwai, M., Yamada-Oshima, Y., Asami, K., Kanamori, T., Yuasa, H., Shimojima, M., et al. (2021). Recycling of the major thylakoid lipid MGDG and its role in lipid homeostasis in *Chlamydomonas reinhardtii. Plant Physiology*. https://doi.org/10.1093/plphys/kiab340.

Jordan, P., Fromme, P., Witt, H. T., Klukas, O., Saenger, W., & Krauß, N. (2001). Three-dimensional structure of cyanobacterial photosystem I at 2.5 Å resolution. *Nature, 411*(6840), 909–917. https://doi.org/10.1038/35082000.

Jouhet, J., Maréchal, E., Baldan, B., Bligny, R., Joyard, J., & Block, M. A. (2004). Phosphate deprivation induces transfer of DGDG galactolipid from chloroplast to mitochondria. *Journal of Cell Biology, 167*(5), 863–874. https://doi.org/10.1083/jcb.200407022.

Jouhet, J., Maréchal, E., & Block, M. A. (2007). Glycerolipid transfer for the building of membranes in plant cells. *Progress in Lipid Research, 46*(1), 37–55. https://doi.org/10.1016/j.plipres.2006.06.002.

Kalisch, B., Dörmann, P., & Hölzl, G. (2016). DGDG and glycolipids in plants and algae. *Sub-Cellular Biochemistry, 86*, 51–83. https://doi.org/10.1007/978-3-319-25979-6_3.

Kansy, M., Wilhelm, C., & Goss, R. (2014). Influence of thylakoid membrane lipids on the structure and function of the plant photosystem II core complex. *Planta, 240*(4), 781–796. https://doi.org/10.1007/s00425-014-2130-2.

Kelly, A. A., & Dörmann, P. (2002). *DGD2*, an *Arabidopsis* gene encoding a UDP-galactose-dependent digalactosyldiacylglycerol synthase is expressed during growth under phosphate-limiting conditions. *Journal of Biological Chemistry, 277*(2), 1166–1173. https://doi.org/10.1074/jbc.M110066200.

Kelly, A. A., Froehlich, J. E., & Dörmann, P. (2003). Disruption of the two digalactosyldiacylglycerol synthase genes *DGD1* and *DGD2* in Arabidopsis reveals the existence of an additional enzyme of galactolipid synthesis. *Plant Cell, 15*(11), 2694–2706. https://doi.org/10.1105/tpc.016675.

Kelly, A. A., Kalisch, B., Hölzl, G., Schulze, S., Thiele, J., Melzer, M., et al. (2016). Synthesis and transfer of galactolipids in the chloroplast envelope membranes of *Arabidopsis thaliana. Proceedings of the National Academy of Sciences of the United States of America, 113*(38), 10714–10719. https://doi.org/10.1073/pnas.1609184113.

Klaus, D., Härtel, H., Fitzpatrick, L. M., Froehlich, J. E., Hubert, J., Benning, C., et al. (2002). Digalactosyldiacylglycerol synthesis in chloroplasts of the Arabidopsis *dgd1* mutant. *Plant Physiology, 128*(3), 885–895. https://doi.org/10.1104/pp.010780.

Kleinig, H., & Liedvogel, B. (1978). Fatty acid synthesis by isolated chromoplasts from the daffodil. [^{14}C]acetate incorporation and distribution of labelled acids. *European Journal of Biochemistry, 83*(2), 499–505. https://doi.org/10.1111/j.1432-1033.1978.tb12116.x.

Kobayashi, K. (2016). Role of membrane glycerolipids in photosynthesis, thylakoid biogenesis and chloroplast development. *Journal of Plant Research*, *129*(4), 565–580. https://doi.org/10.1007/s10265-016-0827-y.

Kobayashi, K., Awai, K., Nakamura, M., Nagatani, A., Masuda, T., & Ohta, H. (2009). Type-B monogalactosyldiacylglycerol synthases are involved in phosphate starvation-induced lipid remodeling, and are crucial for low-phosphate adaptation. *Plant Journal*, *57*(2), 322–331. https://doi.org/10.1111/j.1365-313X.2008.03692.x.

Kobayashi, K., Awai, K., Takamiya, K., & Ohta, H. (2004). Arabidopsis type B monogalactosyldiacylglycerol synthase genes are expressed during pollen tube growth and induced by phosphate starvation. *Plant Physiology*, *134*(2), 640–648. https://doi.org/10.1104/pp.103.032656.

Kobayashi, K., Baba, S., Obayashi, T., Sato, M., Toyooka, K., Keränen, M., et al. (2012). Regulation of root greening by light and auxin/cytokinin signaling in *Arabidopsis*. *Plant Cell*, *24*(3), 1081–1095. https://doi.org/10.1105/tpc.111.092254.

Kobayashi, K., Fujii, S., Sasaki, D., Baba, S., Ohta, H., Masuda, T., et al. (2014). Transcriptional regulation of thylakoid galactolipid biosynthesis coordinated with chlorophyll biosynthesis during the development of chloroplasts in *Arabidopsis*. *Frontiers in Plant Science*, *5*, 272. https://doi.org/10.3389/fpls.2014.00272.

Kobayashi, K., Masuda, T., Takamiya, K., & Ohta, H. (2006). Membrane lipid alteration during phosphate starvation is regulated by phosphate signaling and auxin/cytokinin cross-talk. *Plant Journal*, *47*(2), 238–248. https://doi.org/10.1111/j.1365-313X.2006.02778.x.

Kobayashi, K., Nakamura, Y., & Ohta, H. (2009). Type a and type B monogalactosyldiacylglycerol synthases are spatially and functionally separated in the plastids of higher plants. *Plant Physiology and Biochemistry*, *47*(6), 518–525. https://doi.org/10.1016/j.plaphy.2008.12.012.

Kobayashi, K., Narise, T., Sonoike, K., Hashimoto, H., Sato, N., Kondo, M., et al. (2013). Role of galactolipid biosynthesis in coordinated development of photosynthetic complexes and thylakoid membranes during chloroplast biogenesis in Arabidopsis. *Plant Journal*, *73*(2), 250–261. https://doi.org/10.1111/tpj.12028.

Kobayashi, K., & Wada, H. (2016). Role of lipids in chloroplast biogenesis. *Sub-Cellular Biochemistry*, *86*, 103–125. https://doi.org/10.1007/978-3-319-25979-6_5.

Li, X., Moellering, E. R., Liu, B., Johnny, C., Fedewa, M., Sears, B. B., et al. (2012). A galactoglycerolipid lipase is required for triacylglycerol accumulation and survival following nitrogen deprivation in *Chlamydomonas reinhardtii*. *Plant Cell*, *24*(11), 4670–4686. https://doi.org/10.1105/tpc.112.105106.

Li, C., Wang, Y., Liu, L., Hu, Y., Zhang, F., Mergen, S., et al. (2011). A rice plastidial nucleotide sugar epimerase is involved in galactolipid biosynthesis and improves photosynthetic efficiency. *PLoS Genetics*, 7(7). https://doi.org/10.1371/journal.pgen.1002196, e1002196.

Li, H. M., & Yu, C. W. (2018). Chloroplast Galactolipids: The link between photosynthesis, chloroplast shape, Jasmonates, phosphate starvation and freezing tolerance. *Plant and Cell Physiology*, *59*(6), 1128–1134. https://doi.org/10.1093/pcp/pcy088.

Liu, C., Liu, Y., Wang, S., Ke, Q., Yin, L., Deng, X., et al. (2020). Arabidopsis *mgd* mutants with reduced monogalactosyldiacylglycerol contents are hypersensitive to aluminium stress. *Ecotoxicology and Environmental Safety*, *203*, 110999. https://doi.org/10.1016/j.ecoenv.2020.110999.

Liu, Z., Yan, H., Wang, K., Kuang, T., Zhang, J., Gui, L., et al. (2004). Crystal structure of spinach major light-harvesting complex at 2.72 a resolution. *Nature*, *428*(6980), 287–292. https://doi.org/10.1038/nature02373.

Loll, B., Kern, J., Saenger, W., Zouni, A., & Biesiadka, J. (2005). Towards complete cofactor arrangement in the 3.0 a resolution structure of photosystem II. *Nature*, *438*(7070), 1040–1044. https://doi.org/10.1038/nature04224.

Lyu, J., Gao, R., & Guo, Z. (2021). Galactosyldiacylglycerols: From a photosynthesis-associated apparatus to structure-defined in vitro assembling. *Journal of Agricultural and Food Chemistry*, *69*(32), 8910–8928. https://doi.org/10.1021/acs.jafc.1c00204.

Maida, E., & Awai, K. (2016). Digalactosyldiacylglycerol is essential in Synechococcus elongatus PCC 7942, but its function does not depend on its biosynthetic pathway. *Biochimica et Biophysica Acta*, *1861*(9 Pt B), 1309–1314. https://doi.org/10.1016/j.bbalip.2016.03.011.

Maréchal, E., Block, M. A., Joyard, J., & Douce, R. (1991). Purification of UDP-galactose: 1,2-diacylglycerol galactosyltransferase from spinach chloroplast envelope membrane. *C R Acad Sci Paris*, *313*, 521–528.

Maréchal, E., Block, M. A., Joyard, J., & Douce, R. (1994a). Comparison of the kinetic properties of MGDG synthase in mixed micelles and in envelope membranes from spinach chloroplast. *FEBS Letters*, *352*(3), 307–310. https://doi.org/10.1016/0014-5793(94)00978-3.

Maréchal, E., Block, M. A., Joyard, J., & Douce, R. (1994b). Kinetic properties of monogalactosyldiacylglycerol synthase from spinach chloroplast envelope membranes. *Journal of Biological Chemistry*, *269*(8), 5788–5798.

Maréchal, E., Miège, C., Block, M. A., Douce, R., & Joyard, J. (1995). The catalytic site of monogalactosyldiacylglycerol synthase from spinach chloroplast envelope membranes. Biochemical analysis of the structure and of the metal content. *Journal of Biological Chemistry*, *270*(11), 5714–5722. https://doi.org/10.1074/jbc.270.11.5714.

Masuda, S., Harada, J., Yokono, M., Yuzawa, Y., Shimojima, M., Murofushi, K., et al. (2011). A monogalactosyldiacylglycerol synthase found in the green sulfur bacterium *Chlorobaculum tepidum* reveals important roles for galactolipids in photosynthesis. *Plant Cell*, *23*(7), 2644–2658. https://doi.org/10.1105/tpc.111.085357.

Miège, C., Maréchal, E., Shimojima, M., Awai, K., Block, M. A., Ohta, H., et al. (1999). Biochemical and topological properties of type a MGDG synthase, a spinach chloroplast envelope enzyme catalyzing the synthesis of both prokaryotic and eukaryotic MGDG. *European Journal of Biochemistry*, *265*(3), 990–1001. https://doi.org/10.1046/j.1432-1327.1999.00801.x.

Moellering, E. R., & Benning, C. (2011). Galactoglycerolipid metabolism under stress: A time for remodeling. *Trends in Plant Science*, *16*(2), 98–107. https://doi.org/10.1016/j.tplants.2010.11.004.

Moellering, E. R., Muthan, B., & Benning, C. (2010). Freezing tolerance in plants requires lipid remodeling at the outer chloroplast membrane. *Science*, *330*(6001), 226–228. https://doi.org/10.1126/science.1191803.

Murakami, H., Nobusawa, T., Hori, K., Shimojima, M., & Ohta, H. (2018). Betaine lipid is crucial for adapting to low temperature and phosphate deficiency in *Nannochloropsis*. *Plant Physiology*, *177*(1), 181–193. https://doi.org/10.1104/pp.17.01573.

Murakawa, M., Ohta, H., & Shimojima, M. (2019). Lipid remodeling under acidic conditions and its interplay with low pi stress in Arabidopsis. *Plant Molecular Biology*, *101*(1–2), 81–93. https://doi.org/10.1007/s11103-019-00891-1.

Murakawa, M., Shimojima, M., Shimomura, Y., Kobayashi, K., Awai, K., & Ohta, H. (2014). Monogalactosyldiacylglycerol synthesis in the outer envelope membrane of chloroplasts is required for enhanced growth under sucrose supplementation. *Frontiers in Plant Science*, *5*, 280. https://doi.org/10.3389/fpls.2014.00280.

Murphy, D. J. (1982). The importance of non-planar bilayer regions in photosynthetic membranes and their stabilisation by galactolipids. *FEBS Letters*, *150*(1), 19–26. https://doi.org/10.1016/0014-5793(82)81297-0.

Myhre, D. V. (1968). Glycolipids of soft wheat flour. I. Isolation and characterization of 1-O-(6-O-acyl-β-D-galactopyranosyl)-2,3-di-O-acyl-D-glyceritols and phytosteryl 6-O-acyl-β-D-glucopyranosides. *Canadian Journal of Chemistry*, *46*, 3071–3077.
Najine, F., Marzouk, B., & Cherif, A. (1995). Sodium chloride effect on the evolution of fatty acid composition in developing rape seedlings. In J. C. Kader, & P. Mazliak (Eds.), *Plant Lipid Metabolism* (pp. 435–437). Netherlands: Kluwer Academic Publishers.
Nakamura, Y. (2013). Galactolipid biosynthesis in flowers. *Botanical Studies*, *54*(1), 29. https://doi.org/10.1186/1999-3110-54-29.
Nakamura, Y., Arimitsu, H., Yamaryo, Y., Awai, K., Masuda, T., Shimada, H., et al. (2003). Digalactosyldiacylglycerol is a major glycolipid in floral organs of Petunia hybrida. *Lipids*, *38*(10), 1107–1112. https://doi.org/10.1007/s11745-006-1166-x.
Nakamura, Y., Kobayashi, K., & Ohta, H. (2009). Activation of galactolipid biosynthesis in development of pistils and pollen tubes. *Plant Physiology and Biochemistry*, *47*(6), 535–539. https://doi.org/10.1016/j.plaphy.2008.12.018.
Nilsson, A. K., Johansson, O. N., Fahlberg, P., Kommuri, M., Töpel, M., Bodin, L. J., et al. (2015). Acylated monogalactosyl diacylglycerol: Prevalence in the plant kingdom and identification of an enzyme catalyzing galactolipid head group acylation in *Arabidopsis thaliana*. *Plant Journal*, *84*(6), 1152–1166. https://doi.org/10.1111/tpj.13072.
Ohashi, T., Ito, Y., Okada, M., & Sakagami, Y. (2005). Isolation and stomatal opening activity of two oxylipins from *Ipomoea tricolor*. *Bioorganic and Medicinal Chemistry Letters*, *15*(2), 263–265. https://doi.org/10.1016/j.bmcl.2004.10.088.
Ohta, H., Shimojima, M., Arai, T., Masuda, T., Shioi, Y., & Takamiya, K. (1995). UDP-galactose: Diacylglycerol galactosyltransferase in cucumber seedlings: Purification of the enzyme and the activation by phosphatidic acid. *Plant Lipid Metabolism*, 152–155.
Ohta, H., Shimojima, M., Ookata, K., Masuda, T., Shioi, Y., & Takamiya, K. (1995). A close relationship between increases in galactosyltransferase activity and the accumulation of galactolipids during plastid development in cucumber seedlings. *Plant and Cell Physiology*, *36*(6), 1115–1120. https://doi.org/10.1093/oxfordjournals.pcp.a078855.
Omata, T., & Murata, N. (1986). Glucolipid synthesis activities in cytoplasmic and thylakoid membranes from the cyanobacterium *Anacystis nidulans*. *Plant and Cell Physiology*, *27*, 485–490.
Pedersen, M., Linnanto, J., Frigaard, N. U., Nielsen, N. C., & Miller, M. (2010). A model of the protein-pigment baseplate complex in chlorosomes of photosynthetic green bacteria. *Photosynthesis Research*, *104*(2–3), 233–243. https://doi.org/10.1007/s11120-009-9519-y.
Petroutsos, D., Amiar, S., Abida, H., Dolch, L. J., Bastien, O., Rébeillé, F., et al. (2014). Evolution of galactoglycerolipid biosynthetic pathways- -from cyanobacteria to primary plastids and from primary to secondary plastids. *Progress in Lipid Research*, *54*, 68–85. https://doi.org/10.1016/j.plipres.2014.02.001.
Pribil, M., Labs, M., & Leister, D. (2014). Structure and dynamics of thylakoids in land plants. *Journal of Experimental Botany*, *65*(8), 1955–1972. https://doi.org/10.1093/jxb/eru090.
Qi, Y., Yamauchi, Y., Ling, J., Kawano, N., Li, D., & Tanaka, K. (2004). Cloning of a putative monogalactosyldiacylglycerol synthase gene from rice (*Oryza sativa* L.) plants and its expression in response to submergence and other stresses. *Planta*, *219*(3), 450–458. https://doi.org/10.1007/s00425-004-1245-2.
Qin, X., Suga, M., Kuang, T., & Shen, J. R. (2015). Photosynthesis. Structural basis for energy transfer pathways in the plant PSI-LHCI supercomplex. *Science*, *348*(6238), 989–995. https://doi.org/10.1126/science.aab0214.
Reyes-Prieto, A., Weber, A. P., & Bhattacharya, D. (2007). The origin and establishment of the plastid in algae and plants. *Annual Review of Genetics*, *41*, 147–168. https://doi.org/10.1146/annurev.genet.41.110306.130134.

Rocha, J., Audry, M., Pesce, G., Chazalet, V., Block, M. A., Maréchal, E., et al. (2013). Revisiting the expression and purification of MGD1, the major galactolipid synthase in Arabidopsis to establish a novel standard for biochemical and structural studies. *Biochimie, 95*(4), 700–708. https://doi.org/10.1016/j.biochi.2012.11.011.

Rocha, J., Nitenberg, M., Girard-Egrot, A., Jouhet, J., Maréchal, E., Block, M. A., et al. (2018). Do Galactolipid synthases play a key role in the biogenesis of chloroplast membranes of higher plants? *Frontiers in Plant Science, 9*, 126. https://doi.org/10.3389/fpls.2018.00126.

Rocha, J., Sarkis, J., Thomas, A., Pitou, L., Radzimanowski, J., Audry, M., et al. (2016). Structural insights and membrane binding properties of MGD1, the major galactolipid synthase in plants. *Plant Journal, 85*(5), 622–633. https://doi.org/10.1111/tpj.13129.

Roston, R. L., Wang, K., Kuhn, L. A., & Benning, C. (2014). Structural determinants allowing transferase activity in SENSITIVE TO FREEZING 2, classified as a family I glycosyl hydrolase. *Journal of Biological Chemistry, 289*(38), 26089–26106. https://doi.org/10.1074/jbc.M114.576694.

Roughan, P. G., & Slack, C. R. (1982). Cellular Organization of Glycerolipid Metabolism. *Annual Review of Plant Physiology, 33*(1), 97–132. https://doi.org/10.1146/annurev.pp.33.060182.000525.

Sakaki, T., Kondo, N., & Yamada, M. (1990a). Free fatty acids regulate two galactosyltransferases in chloroplast envelope membranes isolated from spinach leaves. *Plant Physiology, 94*(2), 781–787. https://doi.org/10.1104/pp.94.2.781.

Sakaki, T., Kondo, N., & Yamada, M. (1990b). Pathway for the synthesis of triacylglycerols from monogalactosyldiacylglycerols in ozone-fumigated spinach leaves. *Plant Physiology, 94*(2), 773–780. https://doi.org/10.1104/pp.94.2.773.

Sakurai, I., Mizusawa, N., Wada, H., & Sato, N. (2007). Digalactosyldiacylglycerol is required for stabilization of the oxygen-evolving complex in photosystem II. *Plant Physiology, 145*(4), 1361–1370. https://doi.org/10.1104/pp.107.106781.

Sakurai, I., Shen, J. R., Leng, J., Ohashi, S., Kobayashi, M., & Wada, H. (2006). Lipids in oxygen-evolving photosystem II complexes of cyanobacteria and higher plants. *Journal of Biochemistry, 140*(2), 201–209. https://doi.org/10.1093/jb/mvj141.

Sarkis, J., Rocha, J., Maniti, O., Jouhet, J., Vié, V., Block, M. A., et al. (2014). The influence of lipids on MGD1 membrane binding highlights novel mechanisms for galactolipid biosynthesis regulation in chloroplasts. *FASEB Journal, 28*(7), 3114–3123. https://doi.org/10.1096/fj.14-250415.

Sato, N., & Awai, K. (2016). Diversity in biosynthetic pathways of galactolipids in the light of endosymbiotic origin of chloroplasts. *Frontiers in Plant Science, 7*, 117. https://doi.org/10.3389/fpls.2016.00117.

Sato, N., & Murata, N. (1982a). Lipid biosynthesis in the blue-green alga (cyanobacterium), *Anabaena variabilis*. III. UDP-glucose: Diacylglycerol glucosyltransferase activity *in vitro*. *Plant and Cell Physiology, 23*, 1115–1120.

Sato, N., & Murata, N. (1982b). Lipid biosynthesis in the blue-green alga, *Anabaena variabilis*. I. Lipid classes. *Biochimica et Biophysica Acta (BBA), 710*, 271–278.

Sato, N., & Murata, N. (1982c). Lipid biosynthesis in the blue-green alga, *Anabaena variabilis*. II. Fatty acids and lipid molecular species. *Biochimica et Biophysica Acta (BBA), 710*, 279–289.

Schaller, S., Latowski, D., Jemioła-Rzemińska, M., Dawood, A., Wilhelm, C., Strzałka, K., et al. (2011). Regulation of LHCII aggregation by different thylakoid membrane lipids. *Biochimica et Biophysica Acta, 1807*(3), 326–335. https://doi.org/10.1016/j.bbabio.2010.12.017.

Schaller, S., Latowski, D., Jemioła-Rzemińska, M., Wilhelm, C., Strzałka, K., & Goss, R. (2010). The main thylakoid membrane lipid monogalactosyldiacylglycerol (MGDG) promotes the de-epoxidation of violaxanthin associated with the light-harvesting complex of photosystem II (LHCII). *Biochimica et Biophysica Acta, 1797*(3), 414–424. https://doi.org/10.1016/j.bbabio.2009.12.011.

Shimojima, M., & Ohta, H. (2011). Critical regulation of galactolipid synthesis controls membrane differentiation and remodeling in distinct plant organs and following environmental changes. *Progress in Lipid Research, 50*(3), 258–266. https://doi.org/10.1016/j.plipres.2011.03.001.

Shimojima, M., Ohta, H., Iwamatsu, A., Masuda, T., Shioi, Y., & Takamiya, K. (1997). Cloning of the gene for monogalactosyldiacylglycerol synthase and its evolutionary origin. *Proceedings of the National Academy of Sciences of the United States of America, 94*(1), 333–337. https://doi.org/10.1073/pnas.94.1.333.

Simionato, D., Block, M. A., La Rocca, N., Jouhet, J., Maréchal, E., Finazzi, G., et al. (2013). The response of *Nannochloropsis gaditana* to nitrogen starvation includes *de novo* biosynthesis of triacylglycerols, a decrease of chloroplast galactolipids, and reorganization of the photosynthetic apparatus. *Eukaryotic Cell, 12*(5), 665–676. https://doi.org/10.1128/ec.00363-12.

Slabas, T. (1997). Galactolipid biosynthesis genes and endosymbiosis. *Trends in Plant Science, 2*, 161–162.

Song, Y., Zoong Lwe, Z. S., Wickramasinghe, P., & Welti, R. (2021). Head-group acylation of chloroplast membrane lipids. *Molecules, 26*(5). https://doi.org/10.3390/molecules26051273.

Szpryngiel, S., Ge, C., Iakovleva, I., Georgiev, A., Lind, J., Wieslander, A., et al. (2011). Lipid interacting regions in phosphate stress glycosyltransferase atDGD2 from *Arabidopsis thaliana. Biochemistry, 50*(21), 4451–4466. https://doi.org/10.1021/bi200162f.

Szpryngiel, S., & Mäler, L. (2016). Insights into the membrane interacting properties of the C-terminal domain of the monotopic glycosyltransferase DGD2 in *Arabidopsis thaliana. Biochemistry, 55*(49), 6776–6786. https://doi.org/10.1021/acs.biochem.6b00559.

Tajima, N., Sato, S., Maruyama, F., Kurokawa, K., Ohta, H., Tabata, S., et al. (2014). Analysis of the complete plastid genome of the unicellular red alga *Porphyridium purpureum. Journal of Plant Research, 127*(3), 389–397. https://doi.org/10.1007/s10265-014-0627-1.

Teucher, T., & Heinz, E. (1991). Purification of UDP-galactose: Diacylglycerol galactosyltransferase from chloroplast envelopes of spinach (*Spinacia oleracea* L.). *Planta, 184*(3), 319–326. https://doi.org/10.1007/bf00195332.

Thorlby, G., Fourrier, N., & Warren, G. (2004). The *SENSITIVE TO FREEZING2* gene, required for FREEZING tolerance in *Arabidopsis thaliana*, encodes a β-glucosidase. *Plant Cell, 16*(8), 2192–2203. https://doi.org/10.1105/tpc.104.024018.

Tjellström, H., Andersson, M. X., Larsson, K. E., & Sandelius, A. S. (2008). Membrane phospholipids as a phosphate reserve: The dynamic nature of phospholipid-to-digalactosyl diacylglycerol exchange in higher plants. *Plant, Cell & Environment, 31*(10), 1388–1398. https://doi.org/10.1111/j.1365-3040.2008.01851.x.

Torres-Franklin, M. L., Gigon, A., de Melo, D. F., Zuily-Fodil, Y., & Pham-Thi, A. T. (2007). Drought stress and rehydration affect the balance between MGDG and DGDG synthesis in cowpea leaves. *Physiologia Plantarum, 131*(2), 201–210. https://doi.org/10.1111/j.1399-3054.2007.00943.x.

Umena, Y., Kawakami, K., Shen, J. R., & Kamiya, N. (2011). Crystal structure of oxygen-evolving photosystem II at a resolution of 1.9 Å. *Nature, 473*(7345), 55–60. https://doi.org/10.1038/nature09913.

van Besouw, A., & Wintermans, J. F. (1978). Galactolipid formation in chloroplast envelopes. I. Evidence for two mechanisms in galactosylation. *Biochimica et Biophysica Acta, 529*(1), 44–53. https://doi.org/10.1016/0005-2760(78)90102-9.

Vojta, L., Soll, J., & Bölter, B. (2007). Protein transport in chloroplasts - targeting to the intermembrane space. *The FEBS Journal, 274*(19), 5043–5054. https://doi.org/10.1111/j.1742-4658.2007.06023.x.

Wada, H., & Murata, N. (1989). *Synechocystis* PCC6803 defective in desaturation of fatty acids. *Plant and Cell Physiology, 30*, 971–978.

Wang, K., Hersh, H. L., & Benning, C. (2016). SENSITIVE TO FREEZING2 aides in resilience to salt and drought in FREEZING-sensitive tomato. *Plant Physiology, 172*(3), 1432–1442. https://doi.org/10.1104/pp.16.01183.

Wang, S., Uddin, M. I., Tanaka, K., Yin, L., Shi, Z., Qi, Y., et al. (2014). Maintenance of chloroplast structure and function by overexpression of the rice *MONOGALACTOSYLDIACYLGLYCEROL SYNTHASE* gene leads to enhanced salt tolerance in tobacco. *Plant Physiology, 165*(3), 1144–1155. https://doi.org/10.1104/pp.114.238899.

Webb, M. S., & Green, B. R. (1991). Biochemical and biophysical properties of thylakoid acyl lipids. *Biochimica et Biophysica Acta (BBA) - Bioenergetics, 1060*(2), 133–158. https://doi.org/10.1016/S0005-2728(09)91002-7.

Xu, C., Fan, J., Froehlich, J. E., Awai, K., & Benning, C. (2005). Mutation of the TGD1 chloroplast envelope protein affects phosphatidate metabolism in *Arabidopsis*. *Plant Cell, 17*(11), 3094–3110. https://doi.org/10.1105/tpc.105.035592.

Xu, C., Moellering, E. R., Fan, J., & Benning, C. (2008). Mutation of a mitochondrial outer membrane protein affects chloroplast lipid biosynthesis. *Plant Journal, 54*(1), 163–175. https://doi.org/10.1111/j.1365-313X.2008.03417.x.

Yamaryo, Y., Kanai, D., Awai, K., Shimojima, M., Masuda, T., Shimada, H., et al. (2003). Light and cytokinin play a co-operative role in MGDG synthesis in greening cucumber cotyledons. *Plant and Cell Physiology, 44*(8), 844–855. https://doi.org/10.1093/pcp/pcg110.

Yamaryo, Y., Motohashi, K., Takamiya, K., Hisabori, T., & Ohta, H. (2006). In vitro reconstitution of monogalactosyldiacylglycerol (MGDG) synthase regulation by thioredoxin. *FEBS Letters, 580*, 4086–4090.

Yu, C. W., Lin, Y. T., & Li, H. M. (2020). Increased ratio of galactolipid MGDG: DGDG induces jasmonic acid overproduction and changes chloroplast shape. *The New Phytologist, 228*(4), 1327–1335. https://doi.org/10.1111/nph.16766.

Yuzawa, Y., Nishihara, H., Haraguchi, T., Masuda, S., Shimojima, M., Shimoyama, A., et al. (2012). Phylogeny of galactolipid synthase homologs together with their enzymatic analyses revealed a possible origin and divergence time for photosynthetic membrane biogenesis. *DNA Research, 19*(1), 91–102. https://doi.org/10.1093/dnares/dsr044.

Yuzawa, Y., Shimojima, M., Sato, R., Mizusawa, N., Ikeda, K., Suzuki, M., et al. (2014). Cyanobacterial monogalactosyldiacylglycerol-synthesis pathway is involved in normal unsaturation of galactolipids and low-temperature adaptation of Synechocystis sp. PCC 6803. *Biochimica et Biophysica Acta, 1841*(4), 475–483. https://doi.org/10.1016/j.bbalip.2013.12.007.

Zábranská, M., Vrkoslav, V., Sobotníková, J., & Cvačka, J. (2012). Analysis of plant galactolipids by reversed-phase high-performance liquid chromatography/mass spectrometry with accurate mass measurement. *Chemistry and Physics of Lipids, 165*(5), 601–607. https://doi.org/10.1016/j.chemphyslip.2012.03.002.

Zarrouk, M., Seqqat-Dakhma, W., & Cherif, A. (1995). Salt stress effect on polar lipid metabolism of olive leaves. In J. C. Kader, & P. Mazliak (Eds.), *Plant Lipid Metabolism* (pp. 429–431). Netherlands: Kluwer Academic Publishers.

Zepke, H. D., Heinz, E., Radunz, A., Linscheid, M., & Pesch, R. (2004). Combination and positional distribution of fatty acids in lipids from blue-green algae. *Archives of Microbiology, 119*, 157–162.

Zhang, M., Deng, X., Yin, L., Qi, L., Wang, X., Wang, S., et al. (2016). Regulation of galactolipid biosynthesis by overexpression of the rice *MGD* gene contributes to enhanced aluminum tolerance in tobacco. *Frontiers in Plant Science*, 7, 337. https://doi.org/10.3389/fpls.2016.00337.

Zhang, R., Hu, H., Wang, Y., Hu, Z., Ren, S., Li, J., et al. (2020). A novel rice *fragile culm 24* mutant encodes a UDP-glucose epimerase that affects cell wall properties and photosynthesis. *Journal of Experimental Botany*, *71*(10), 2956–2969. https://doi.org/10.1093/jxb/eraa044.

Zhou, F., Liu, S., Hu, Z., Kuang, T., Paulsen, H., & Yang, C. (2009). Effect of monogalactosyldiacylglycerol on the interaction between photosystem II core complex and its antenna complexes in liposomes of thylakoid lipids. *Photosynthesis Research*, *99*(3), 185–193. https://doi.org/10.1007/s11120-008-9388-9.

CHAPTER THREE

Plastoglobules: A hub of lipid metabolism in the chloroplast

Venkatasalam Shanmugabalaji*, Wayne Zita, Joy Collombat, and Felix Kessler
Laboratoire de Physiologie Végétale, Université de Neuchâtel, Neuchâtel, Switzerland
*Corresponding author: e-mail address: shanmugabalaji.venkatasalam@unine.ch

Contents

Abstract

Chloroplasts in plant and algae are the home of photosynthesis and many essential metabolic pathways. The photosynthetic light reactions occur at an extensive membrane system, the thylakoids. The main constituents of thylakoid membranes are the galactolipids, in particular monogalactosyldiacylglycerol (MGDG) and digalactosyldiacylglycerol (DGDG). Apart from these, thylakoid membranes contain a number of other lipid molecules that function as electron transporters (plastoquinone) or as lipid antioxidants (tocopherols) and that are free to diffuse in the membrane system. Others such as phylloquinone, carotenoids and chlorophylls are mostly associated with protein components of the photosystems that reside in the thylakoids. The thylakoids are extremely dynamic and rapidly respond to environmental stresses. Plastoglobules (PG), lipoprotein

Advances in Botanical Research, Volume 101
ISSN 0065-2296
https://doi.org/10.1016/bs.abr.2021.09.002

particles or lipid droplets associated with thylakoids, play a central role in these responses. PG are thought to allocate stored electron carriers as well as lipid antioxidants to the membranes. PG also contribute to the enzymatic capacity for antioxidant and electron carrier synthesis and metabolism. During senescence and under nitrogen deprivation PG actively participate in membrane catabolism. PG become supersized while absorbing triacylglycerols and fatty acid phytol esters resulting from galactolipid and chlorophyll degradation. This chapter aims at giving a broad overview of the role of PG in chloroplast lipid metabolism.

1. Introduction

1.1 Discovery and history

The advent of electron microscopy as a tool in cellular biology in the late 1950s and 1960s allowed a first glimpse inside the plant cell and its subcellular structures. Among other features within the chloroplast the electron micrographs revealed the presence of globular structures. Osmium tetroxide stained these structures intensely which hence were called osmiophile globules. Due to their presence in many plastid types the name was later changed to plastoglobules (abbreviated PG) (Greenwood, Leech, & Williams, 1963; Lichtenthaler, 1968a; Spurr & Harris, 1968). These early studies carried out on multiple species revealed a diversity of osmiophile globules which varied in size (30 nm to several μm) and number under various environmental (for instance drought) or developmental conditions (for instance in older and younger leaves or fruit) (Lichtenthaler, 1968a; Thomson & Platt, 1973). Supersized PG, several μm in diameter, were readily observed in old and senescent leaves (Gaude, Bréhélin, Tischendorf, Kessler, & Dörmann, 2007; Tevini & Steinmuller, 1985). It was also observed that numerous PG were contained in the prolamellar body of etiolated, dark grown seedlings. When exposed to light the thylakoid system rapidly emerged and the number of PG diminished during the greening process. The authors noted this correlation and suggested that PG may contribute to the lipids of the developing photosynthetic membrane system (Lichtenthaler & Sprey, 1966).

It was quickly recognized that the globules were lipidic in nature and of low density, a property which allowed their isolation by flotation centrifugation. Isolated globules were subjected to biochemical analyses that revealed the presence of relatively large quantities of neutral lipids (mostly triacylglycerol (TAG), plastoquinone/plastoquinol (PQ/PQH_2)

Table 1 Lipids in PG from different plastid types.

Plastids	Enriched lipids	References
Chloroplast	Galactolipids DAG (diacylglycerol) TAG (triacylglycerol) α-Tocopherol (Vitamin E) Phylloquinone (Vitamin K1) Plastoquinone/ol (PQ/PQH2) Fatty acid phytyl esters	Steinmüller and Tevini (1985) Leggett Bailey and Whyborn (1963) Lichtenthaler (1968b) Lichtenthaler and Peveling (1966) Lippold et al. (2012)
Chromoplast	Triacylglycerol (TAG) Carotenoids Carotenoid esters Phytol α-Tocopherol (Vitamin E) Galactolipids	Hansmann and Sitte (1982) Deruère et al. (1994) Arango and Heise (1997)
Gerontoplast	Fatty acid phytyl esters Triacylglycerol (TAG)	Gaude et al. (2007) Steinmüller and Tevini (1985) Lippold et al. (2012)

and tocopherol) as well as smaller quantities of polar galactolipids and traces of protein (Table 1) (Gaude et al., 2007; Greenwood et al., 1963; Leggett Bailey & Whyborn, 1963; Lichtenthaler & Peveling, 1966; Tevini & Steinmuller, 1985). A model emerged in which a monolayer of polar galactolipids surrounded a lipophilic core (Fig. 1). The trace amounts of protein were not accounted for as individual proteins could not be identified at the time. The variations in size and number of PG in chloroplast and their correlation with the state of thylakoid membrane system suggested a dynamic relation between the two systems. The increase in PG size correlated with the disassembly of the thylakoid system during chloroplast senescence suggesting that PG function as a storage site for thylakoid membrane catabolites. In the absence of further evidence, PG were thought to function as passive lipid storage sites.

This model started to evolve only recently with the discovery of the PG protein components (Table 2) (Deruère et al., 1994; Kessler et al., 1999; Vidi et al., 2006; Ytterberg et al., 2006). Ensuing studies revealed that PG contained a set of enzymatic activities and participated in a multitude of lipid metabolic pathways associated with PG and thylakoids. It has now been

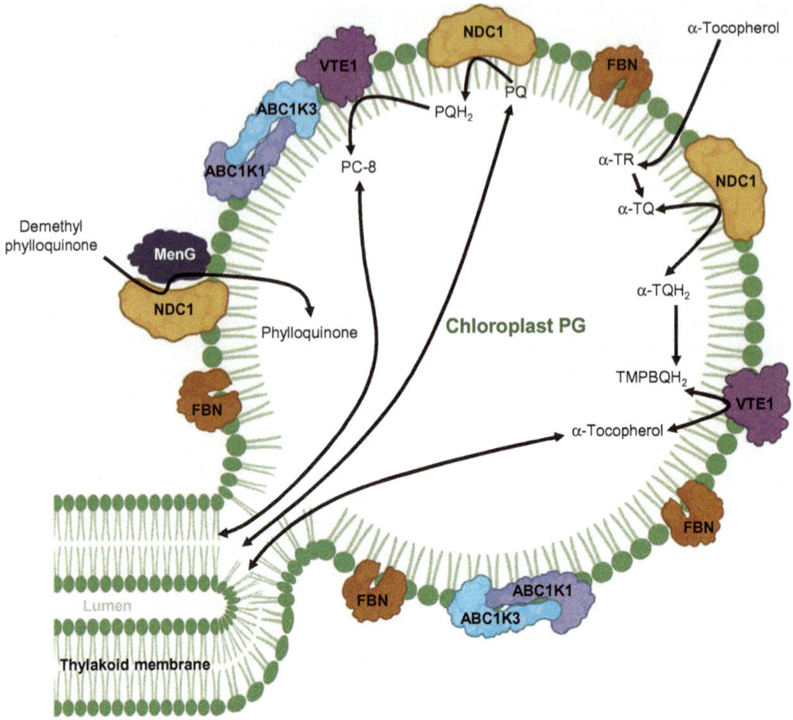

Fig. 1 Prenyl lipid metabolism in the chloroplast plastoglobule. Tocopherol cyclase (VTE1) catalyzes the biosynthesis of plastochromanol (PC8) from plastoquinol (PQH2) resulting from the prior reduction of PQ by type II NAD(P)H dehydrogenase C1 (NDC1). In addition, NDC1 derived trimethylbenzoquinol ($TMPBQH_2$) is converted to α-tocopherol by VTE1. ACTIVITY OF BC1-LIKE 1/3 KINASE (ABC1K1/ABC1K3) have been implicated in the regulation of VTE1 activity and may do so by directly phosphorylating VTE1. NDC1 reduces demethylphylloquinone promoting methylation by MenG resulting in phylloquinone synthesis. Abundant fibrillins (FBNs) are thought to have mainly structural roles. The tocopherol repair and recycling involve NDC1 and VTE1, specifically to convert α-tocopherolquinone (α-TQ) to α-tocopherol (α-T) in PG. Abbreviations: α-TQH2, α-tocopherolquinol; α-TR, α-tocopheroxyl radical

shown that PG even harbor regulatory functions (van Wijk & Kessler, 2017; Vidi et al., 2006; Ytterberg et al., 2006). To present these concepts, the present chapter will give an overview of the current state of knowledge on PG and their protein components. While most of the known functions pertain to prenyl lipids, attention will also be given to the role of PG in galactolipid metabolism.

Table 2 PG core proteins based on proteome studies in *Arabidopsis thaliana* and *Capsicum annuum*.

Protein name	Accession	Functions
Fibrillins		
Fibrillin 1a (FBN1a)	AT4G04020	Structural protein
Fibrillin 1b (FBN1b)	AT4G22240	
Fibrillin 2 (FBN2)	AT2G35490	
Fibrillin 4 (FBN4)	AT3G23400	
Fibrillin 7a (FBN7a)	AT3G58010	
Fibrillin 7b (FBN7b)	AT2G42130	
Fibrillin 8 (FBN8)	AT2G46910	
Enzymes		
Tocopherol cyclase VTE1	AT4G32770	Tocopherol metabolism
NAD(P)H dehydrogenase C1 NDC1	AT5G08740	Phylloquinone and PQ metabolism
Phytyl ester synthase PES1	AT1G54570	TAG and FAPEs biosynthesis
Phytyl ester synthase PES2	AT3G26840	
Regulatory kinases		
Activity of bc1-like kinase ABC1K1	AT4G31390	Metabolic regulation and PQ homeostasis
Activity of bc1-like kinase ABC1K3	AT1G79600	Metabolic regulation and PQ homeostasis
Activity of bc1-like kinase ABC1K5	AT1G71810	
Activity of bc1-like kinaseABC1K6	AT3G24190	
Activity of bc1-like kinase ABC1K7	AT3G07700	Metabolic regulation
Activity of bc1-like kinase ABC1K9	AT5G05200	
Other proteins		
SOUL heme binding protein 4 (SOUL 4)	AT3G10130	Heme metabolism

Continued

Table 2 PG core proteins based on proteome studies in *Arabidopsis thaliana* and *Capsicum annuum*.—cont'd

Protein name	Accession	Functions
UbiE methyltransferase-related	AT1G78140	
UbiE methyltransferase-related	AT2G41040	
Aldo/keto reductase	AT1G06690	
Plastoglobular protein 18 PG18	AT4G13200	Thylakoid biogenesis
PLAT/LH2–1	AT4G39730	
PGM48	AT3G27110	Protein homeostasis
Flavin reductase-related 1	AT1G32220	
Flavin reductase-related 2	AT2G34460	
DUF1350	AT3G43540	
alpha/beta hydrolase family protein	AT1G73750	
Esterase 1	AT5G41120	
Chromoplast PG carotenoid enzymes		
Carotenoid cleavage dioxygenase 4 CCD4	AT4G19170	Carotenoid metabolism
ζ-Carotene desaturase ZDS	gi|12643508 (Capsicum annuum)	Carotenoid biosynthesis
Lycopene β-cyclase LCY-β	gi|12643508 (Capsicum annuum)	
β-Carotene β-hydroxylase CrtR-β	gi|12643508 (Capsicum annuum)	

2. Metabolites and proteins

2.1 The role of plastoglobules in thylakoid (prenyl) lipid metabolism

After their discovery by electron microscopy it quickly became clear that PG contain large and variable quantities of lipids that were in one way or the other linked to thylakoid function (Greenwood et al., 1963; Tevini &

Steinmuller, 1985). PG being lipid droplets, they contain neutral, hydrophobic lipids while the surface consists of a membrane lipid monolayer. Indeed, plastoquinone (large quantities) together with the tocopherols and phylloquinone (both in moderate quantities) were identified as major components (Leggett Bailey & Whyborn, 1963; Lichtenthaler, 1968b; Lichtenthaler & Peveling, 1966; Steinmüller & Tevini, 1985) (Table 1). All three belong to the prenyl quinone family and in part share biosynthetic pathways that are integrated at the PG (Figs. 1–3). It is important to note that all three fulfill important functions at the thylakoid membrane and do not appear to carry out any specifically within PG. Plastoquinone is a membrane mobile electron carrier transferring electrons from Photosystem II to the cytochrome b6f complex (Tikhonov, 2013; Van Eerden et al., 2017). It also "moonlights" as an anti-oxidant, the unsaturated prenyl tail able to react with and quench ROS whereby it is consumed (Apel & Hirt, 2004; Ksas et al., 2018). Tocopherol is considered a membrane anti-oxidant quenching ROS via chromanol ring opening (Krieger-Liszkay & Trebst, 2006; Munné-Bosch, 2005; Neely et al., 1988). The tocopherol oxidation product tocopherol quinone is known to be recycled to tocopherol (Kobayashi & DellaPenna, 2008), the process taking place at least in part at the PG level (Eugeni Piller et al., 2014) (Fig. 1). Phylloquinone is an electron carrier functioning within Photosystem I (Malkin, 1986; Santabarbara & Casazza, 2019; Schoeder & Lockau, 1986). Functions of free phylloquinone within the thylakoid membrane have to our knowledge not been described.

Other neutral lipids in PG are triacylglycerols that are present in large quantities and mostly derived from thylakoid galactolipid hydrolysis (Table 1). Free fatty acids were also detected and likely to be galactolipid hydrolysis products as well (Gaude et al., 2007; Tevini & Steinmuller, 1985). In addition, PG of senescent chloroplasts accumulate large quantities of fatty acid phytyl esters (Fig. 4) (Gaude et al., 2007). Phytol is a product of chlorophyll breakdown and thought to be membrane toxic due to detergent-like properties (Krauß & Vetter, 2018). Esterification with free fatty acid detoxifies phytol and results in very hydrophobic fatty acid phytol esters that can be sequestered from the thylakoid membrane and deposited in PG (Hörtensteiner & Kräutler, 2011; Ischebeck et al., 2006; Krauß & Vetter, 2018) (Table 1). Membrane lipids are present in minor quantities as they only constitute the membrane lipid monolayer at surface of the PG. They include monogalactosyl- and digalactosyldiacylglycerol (MGDG and DGDG) as well sulfoquinovosyldiacylglycerol which are the major membrane lipid components of the thylakoid membrane (Boudière et al., 2014). Phospholipids have also been detected at trace levels.

Electron microscopic images indicate that the membrane lipid monolayer surrounding the PG is contiguous with the outer membrane lipid layer of the thylakoid membrane (Austin et al., 2006; Kessler et al., 1999). This arrangement would allow for a lipid conduit from the PG interior to the thylakoid membrane. Such a conduit has been postulated to explain the lipid exchange processes that must occur between the thylakoid membrane and PG (Havaux, 2020; Kirchhoff, 2019). However, such conduits have not been functionally demonstrated so far.

Interestingly, PG composition may vary with plastid type. Senescent chloroplasts, also known as gerontoplasts, accumulate very large amounts of TAG and fatty acid phytyl esters that result from thylakoid membrane catabolism and chlorophyll breakdown (Fig. 4; Table 1) (Gaude et al., 2007; Tevini & Steinmuller, 1985). This accumulation is directly linked to PG supersizing that can easily be observed by electron microscopy (Barton, 1966). During chloroplast-to-chromoplast transition, thylakoid membranes are dismantled and large PG appear (Egea et al., 2010; Spurr & Harris, 1968). In red pepper large PG transiently appear and are progressively replaced by carotenoid fibrils that are structurally related to PG (Deruère et al., 1994). The PG and fibrils contain large quantities of carotenoids such as the linear lycopene in tomato and/or carotenoid esters that may constitute up to 90% of carotenoids in red pepper (Berry et al., 2019; Deruère et al., 1994) (Table 1). It is interesting to note here that the PG in chromoplasts recruit enzymes of the carotenoid biosynthesis pathway. Presumably, this enhances sequestration of these hydrophobic molecules within a hydrophobic metabolic sink (i.e., PG and/or fibrils).

2.2 The plastoglobule proteome

The first PG protein to be identified was called fibrillin due to its association with carotenoid fibrils (as well as PG) in red pepper fruit (Deruère et al., 1994). Since then, a series of analyses have determined a proteome of around 30 *bona fide* PG proteins (Table 2) that may vary between plastid types and environmental conditions. Rather surprisingly, around a third of the proteins turned out be enzymes with known or predicted roles in lipid metabolism (Eugeni Piller et al., 2011; Vidi et al., 2006). Several members of the family of atypical ABC1K kinases were also present in the proteome suggesting that PG also had regulatory functions (Fig. 1) (Lundquist et al., 2013; Martinis et al., 2014; Pralon et al., 2019, 2020). Finally, members of the fibrillin (FBN) family constituted the largest portion of the mass of the PG proteome and have been attributed roles in PG structure and maintenance (Fig. 1) (Kessler et al., 1999; Pozueta-Romero et al., 1997; Rey et al., 2000; Simkin et al., 2007; Singh et al., 2012). Interestingly, the PG proteins partitioned to various degrees between thylakoid membranes, stroma and PG, but there are few physical explanations for this. But, the distribution of the fibrillin proteins between the compartments correlated with their isoelectric points and hydrophobicity. Fibrillins that associated with PG tended to have a lower isoelectric point and be more hydrophobic (Lundquist et al., 2012). While the precise functions of many PG components remain unknown, a genome-wide coexpression network of PG genes attributed PG proteins to four distinct "modules" (1–4) that were particularly enriched for senescence, carotenoid metabolism, redox regulation/"photoacclimation" and plastid biogenesis (Lundquist et al., 2012). In the following paragraphs, we will discuss notable PG proteins and their functions.

2.2.1 The versatile but elusive fibrillin family

Quantitatively, the largest portion of the PG proteome is made up of members of the fibrillin/FBN family, sometimes also called plastoglobulins (PG) (Kessler et al., 1999; Lundquist et al., 2012). FBNs originated from a cyanobacterial chloroplast progenitor and are part of the green cut (Lohscheider & Río Bártulos, 2016; Merchant et al., 2007). The first fibrillin was discovered in red pepper (*Capsicum annuum*) through its association with carotenoid-sequestering fibrils (Deruère et al., 1994). Red pepper fibrillin protein allowed to reconstitute carotenoid fibrils from the fibril lipid components. This finding was taken as indication for a structural function of fibrillin (Deruère et al., 1994). Additional support for this notion comes from the ability of FBN1a and -1b to dimerize *in vivo*. This may allow formation of a stabilizing FBN network around PG (Gámez-Arjona et al., 2014). Overexpression of FBN in tobacco led to increased numbers and clustering of PG suggesting that FBN may form an interface at the surface of PG and prevent their coalescence (Rey et al., 2000). FBNs share limited sequence homology with lipocalins, small proteins involved in the binding and transport of small hydrophobic compounds. The presence of a lipocalin signature suggested that some fibrillins may directly participate in lipid trafficking between PG and membrane compartments (Flower et al., 2000; Singh & McNellis, 2011).

The Arabidopsis genome encodes 14 FBNs, 7 of which (FBN1a, -1b, -2, -4, -7a, -7b, and -8) are considered PG core proteins (Vidi et al., 2006; Ytterberg et al., 2006). The other FBNs are associated with thylakoid membranes (FBN3a, -3b, -6, and -9 and FBN-like) or present in the chloroplast stroma (FBN5) (Kim et al., 2015). The loss-of-function mutants for FBN1a, -1b, -2, and -4 in Arabidopsis as well as other species suggested the (direct or indirect) involvement of these FBNs in responses to biotic stresses, heat stress, and excess light (Gillet et al., 1998; Langenkämper et al., 2001; Li et al., 2015; Singh et al., 2012; Youssef et al., 2010). Arabidopsis FBN5 is not a PG component but a soluble stromal protein. FBN5 is required for the synthesis of solanesyl diphosphate, an essential step in the PQ-9 biosynthetic pathway (Kim et al., 2017, 2015; Otsubo et al., 2018). FBN5 interacts directly with SPS1 and -2 (solanesyl diphosphate synthase 1 and -2) to stimulate the elongation of the prenyl chain resulting in the essential solanesyl moiety of PQ-9 (Kim et al., 2015) (Fig. 2). FBN4 has also been linked to PQ-9 metabolism. The PG of an apple FBN4

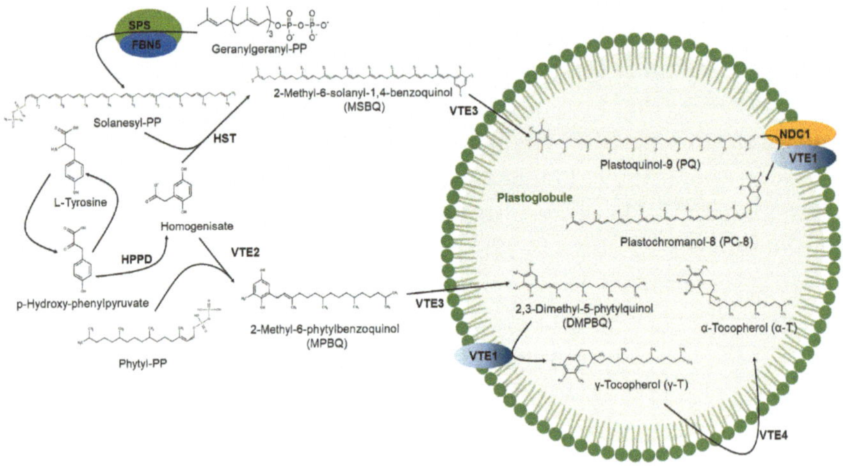

Fig. 2 Plastochromanol and tocopherol biosynthetic pathways in chloroplasts. The tyrosine aminotransferase (TAT) converts the precursor L-tyrosine derived from the shikimate pathway into p-hydroxy-phenylpyruvate. The p-hydroxyphenyl-pyruvate dioxygenase (HPPD) converts p-hydroxy-phenylpyruvate into homogentisate. Homogentisate phytyltransferase (VTE2) adds the phytyl-pp tail to homogentisate resulting in 2,3-dimethyl-6-phytyl-1,4-benzoquinone (MPBQ). Alternatively, homogentisate solanesyltransferase (HST) adds a solanesyl-tail to homogentisate resulting in 2-methylmethyl-6-solanyl-1,4-benzoquinol (MSBQ). Solanesyl-PP is derived from geranylgeranyl-PP elongation catalyzed by solanesyl diphosphate synthases 1 and 2 (SPS1 and 2) in conjunction with fibrillin 5 (FBN5). MPBQ methyltransferase (VTE3) methylates MPBQ resulting in DMPBQ as well as MSBQ resulting in plastoquinol-9 (PQ-9). Tocopherol cyclase (VTE1), localized at the surface of the plastoglobule (PG; shown as green lipid droplet) catalyzes the cyclization of DMPBQ generating γ-Tocopherol (γ-T). Also, at the PG, NDC1 reduces PQ-9 to PQH_2-9 prior to conversion into plastochromanol-8 (PC-8) by VTE1. Tocopherol O-methyltransferase (VTE4) methylates γ-Tocopherol (γ-T) resulting in α-Tocopherol (α-T).

knockdown mutant had strongly reduced levels of PQ-9 while overall concentrations were comparable to WT suggesting that FBN4 participates in partitioning and/or trafficking of PQ-9 between PG and thylakoid membrane (Singh et al., 2012). The current knowledge suggests that FBNs are highly versatile proteins contributing to PG structure as well as several metabolic pathways and metabolite trafficking. However, the functional mechanisms of the fibrillins remain elusive.

2.2.2 Members of the ABC1K family contribute to control the prenyl lipid metabolism and photosynthetic electron flow

Members of ABC1 (Activity of BC1 complex)-like kinase family (ABC1K) constitute the second most abundant family of proteins in PG (Lundquist et al., 2012). The founding member of the ABC1K family [also called UbiB or ABC1 domain–containing kinases (ADCKs)] is UbiB from *E. coli*, which is required for the aerobic biosynthesis of ubiquinone (Poon et al., 2000). The yeast (*Saccharomyces cerevisiae*) mitochondrial ABC1K, also known as coenzyme Q biosynthesis 8 (ScCOQ8), and the mammalian homologs ADCK3 and -4 are required for ubiquinone synthesis in the mitochondria of their respective organisms, possibly through phosphorylation-dependent assembly of a large biosynthetic multi-enzyme complex (Bousquet et al., 1991; Do et al., 2001; Lagier-Tourenne et al., 2008; Mollet et al., 2008). Due to the lack of ubiquinone, the *coq8* mutation compromises the mitochondrial bc1 complex and aerobic metabolic activity. Defects of the kind are linked to hereditary disease in humans (Gerards et al., 2010).

Arabidopsis has 17 ABC1Ks, 8 of which are likely localized in mitochondria and 9 in plastids. Six of nine of the plastid ABC1Ks are localized in PG (ABC1K1, -3, -5, -6, -7,-9) (Lundquist et al., 2012). Notably, the ABC1K family represents the majority of known kinases in plastids and mitochondria; however, relatively little is known about this family in plants. Functional studies of PG-localized ABC1K1, -3 and -7, have been initiated in Arabidopsis, and the role of ABC1K3 has been addressed in rice (Li et al., 2015). PG-localized ABC1K7 is involved in cadmium tolerance, oxidative stress response, iron distribution, lipid metabolism and in crosstalk between abscisic acid and ROS signaling (Manara et al., 2014; Manara et al., 2015). The Chlamydomonas homolog of ABC1K6 (EYE3) located in the eyespot, is involved in the biogenesis of pigment granules (Boyd et al., 2011). Still, the modes of action of the ABC1K family members, for instance phosphorylation of proteins or metabolites, are quite unclear.

Apart from its association with PG, ABC1K1 was identified in at least two genetic screens. In a genetic screen to identify mutants defective in seedling growth under continuous red light, ABC1K1 was identified as BDR1 (bleaching and dwarf under red light 1) (Yang et al., 2016). Under continuous red light *abc1k1/bdr1* exhibited an albino and dwarf phenotype and, while most photosynthetic proteins accumulated normally, the Photosystem II core protein D1 was diminished. To identify additional components required for ABC1K1/BDR1 function, a genetic screen to find suppressors of *abc1k1-2* was carried out and resulted in the isolation of *rbd1* (repressor of *bdr1-2*) that turned out to be identical with ABC1K3 (Huang et al., 2015). It is thought that ABC1K3 in the *abc1k1* background interferes with red light signaling acting downstream of PhyB and HY5 and affects expression of PIF genes. This is supported by the observation that ABC1K3 overexpression produced a phenotype similar to that of the *abc1k1* mutant (Huang et al., 2015).

In a second screen for mutants with reduced non-photochemical quenching (NPQ) of chlorophyll fluorescence (Shikanai et al., 1999), PGR6 (Proton Gradient Regulation 6) was identified and found to be identical with ABC1K1. *abc1k1* has a photosynthetic phenotype characterized by strong photoinhibition and reduced NPQ after high light exposure (Martinis et al., 2014; Pralon et al., 2019). Prenyl lipid metabolism was affected in *abc1k1*: Tocopherol concentrations did not increase under high light stress and carotenoid concentrations were diminished even under moderate light conditions (Martinis et al., 2014; Martinis et al., 2013). After high light treatment, sugar metabolism was perturbed and carbon-allocation shifted from starch to soluble sugars. In addition to the genetic interaction between ABC1K1 and its closest homolog, ABC1K3, the two atypical kinases interact physically (Fig. 1). (Lundquist et al., 2013). Moreover, additional reports demonstrated that ABC1K1 and -3 together are involved in PQ-9 homeostasis (maintenance of sufficient PQ-9 in the photoactive thylakoid membrane pool) (Pralon et al., 2019; Pralon et al., 2020). In the *abc1k1* mutant, the photoactive PQ-9 pool in the thylakoid membranes was selectively depleted under moderate high light stress. The depleted photoactive PQ-9 pool phenotype of *abc1k1*is rescued in the *abc1k1/abc1k3* double mutant. Thus, ABC1K1 and ABC1K3 appear to regulate the distribution of PQ-9 between the plastoglobule and thylakoid membrane pools.

The perturbed plastoquinone homeostasis in the *abc1k1* mutant may explain the "proton gradient regulation"/PGR phenotype that is characterized by decreased NPQ: concomitantly with its role as an electron carrier, plastoquinone transports protons from the stromal side of the thylakoid membrane to the lumenal side generating a ΔpH. An increasing ΔpH serves as a proxy for excess light and induces NPQ, to protect the photosynthetic apparatus. In *abc1k1*, plastoquinone levels in the electron chain are diminished under high light. These insufficient amounts of plastoquinone in *abc1k1* would limit not only electron transport but also the ΔpH across the thylakoid membrane. Therefore, induction of non-photochemical quenching, would be decreased, resulting in the observed PGR phenotype. Thus, ABC1K1 may be a very direct regulator of the proton gradient (Pralon et al., 2019).

It remains unclear how the photosynthetic PGR phenotype relates to the cellular signaling phenotype also observed for ABC1K1 and -K3. It will be important to reconcile the two phenomena in the future. One possibility is that these two atypical kinases transmit signals emanating from the photosynthetic electron transport chain to the nucleus thereby affecting gene expression.

3. Role of PG proteins in the context of thylakoid lipid metabolism

3.1 NDC1 a new enzyme in phylloquinone biosynthesis

Phylloquinone mainly acts as an electron carrier in Photosystem I in the thylakoid membrane (Brettel et al., 1986). But it has been demonstrated that about one-third of phylloquinone is present in PG (Lohmann et al., 2006; Spicher & Kessler, 2015). This suggests phylloquinone trafficking between PG and the thylakoid membrane. The MenG enzyme, a predicted methylase, catalyzes the final methylation step of the biosynthetic pathway from 2-phytyl-1,4-naphthoquinone (demethylphylloquinone) to phylloquinone (Fig. 3). The AtmenG knockout mutant lacks phylloquinone but accumulates the precursor demethylphylloquinone instead. The lack of phylloquinone in *atmenG* has rather minor consequences resulting in reduced levels of Photosystem I and consequently a slight decrease of photosynthetic efficiency upon high light (HL) treatment (Eugeni Piller et al., 2011; Lohmann et al., 2006). Thus, it appears that the precursor of phylloquinone can functionally replace phylloquinone, at least under standard conditions. Surprisingly, the *ndc1* mutant in Arabidopsis mimics the *atmenG* phenotype and has identical phylloquinone and demethylphylloquinone profiles (Eugeni Piller et al., 2011).

In Arabidopsis, NDC1 is one of seven NDH-2 (Type II NAD(P)H-dependent dehydrogenases). NDC1 appears to be localized mostly in chloroplast PG (Fig. 3) whereas the remaining homologs have been localized in mitochondria. However, NDC1 was initially also found in mitochondria

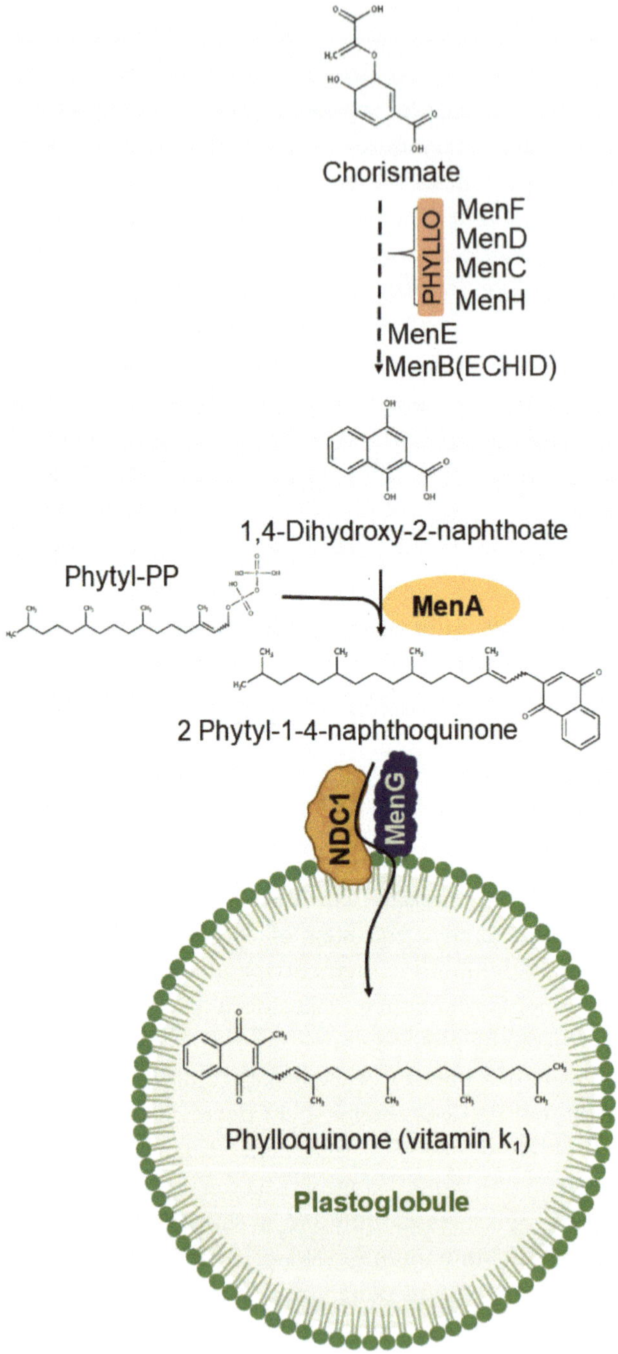

Fig. 3 Phylloquinone biosynthetic pathway in plants. PHYLLO, a fusion protein of menaquinone synthesis enzymes (MenF, D, H, C) followed by MenE and MenB convert chorismate into 1,4-dihydroxydihydroxy-2-naphthoate (DHNA). DHNA phytyltransferase (MenA) converts DHNA to 2-phytyl-1,4-naphthoquinone (demethylphylloquinone). Next, the demethylphylloquinone is reduced by type II NAD(P)H dehydrogenase C1 (NDC1) resulting in demethylphylloquinol. Depending on the prior reduction of demethylphylloquinone by NDC1, MenG methylates demethylphylloquinol to complete the synthesis of phylloquinone.

(Michalecka et al., 2003), and dual localization in mitochondria and chloroplasts cannot be excluded. NDC1 is evolutionarily conserved from cyanobacteria. In Chlamydomonas NDA2 NAD(P)H:quinone oxidoreductase, the homolog of NDC1, is involved in cyclic electron flow and chlororespiration directly transferring electrons from NAD(P)H to plastoquinone (Desplats et al., 2009). NDC1 in cyanobacteria is part of the MEN operon and its deletion also leads to loss of phylloquinone (Fatihi et al., 2015). The mechanism of NDC1 in the methylation step of phylloquinone synthesis has been solved: The NDC1-dependent reduction of demethylphylloquinone intermediate is essential for the subsequent methylation by MenG. NDC1 is therefore considered a *bona fide* enzyme of the phylloquinone biosynthesis pathway (Fig. 3). (Eugeni Piller et al., 2011; Fatihi et al., 2015). Interestingly, Gross et al. (2006) also demonstrated that the PHYLLO metabolon (genetic fusion of MenF, MenD, MenC, and MenH) fused to a fluorescent marker resulted in a plastoglobuli-like fluorescence pattern inside chloroplasts (Gross et al., 2006). The fluorescence pattern suggests that the PHYLLO metabolon may also be recruited to PG coupling the upstream portion of pathway with NDC1 and AtMenG activity (Gross et al., 2006).

3.2 NDC1 regulates the PQ-9 redox state and contributes to plastochromanol biosynthesis and tocopherol recycling

It has been demonstrated that recombinant NDC1 reduces decyl plastoquinone (a PQ-9 analog) in a NAD(P)H dependent fashion (Eugeni Piller et al., 2011). However, NDC1 in Arabidopsis is not directly implicated in either linear or cyclic electron flow in Arabidopsis. In chloroplasts, plastoquinone is distributed between the thylakoids and PG, constituting photoactive and non-photoactive pools, respectively. The non-photoactive pool may communicate with the photoactive plastoquinone pool to maintain adequate PQ-9 levels in the photoactive pool ("plastoquinone homeostasis") as seen in the earlier paragraphs on the ABC1K1 and -3 kinases. Total PQ-9 in *ndc1* was significantly more oxidized than in the wild-type while electron transport rates and other photosynthetic parameters were not affected. This suggested that NDC1 selectively reduces the photoinactive PQ-9 pool inside PG (Eugeni Piller et al., 2011).

In an interesting twist, the *ndc1* mutant accumulates less plastochromanol-8 (PC-8) than WT (Eugeni Piller et al., 2011). PC-8 is the product of tocopherol cyclase/VTE1 activity on reduced PQ-9 (PQH_2-9). VTE1 preferentially cyclizes reduced substrates, in this case PQH_2-9 contained in PG (Fig. 2)

(Woggon, 1999). In the absence of *ndc1* PQ-9 in PG is more oxidized impeding the action of VTE1 (Fig. 2) and hence the biosynthesis of PC-8.

Under stress conditions, Photosystem II generates reactive oxygen species to which the thylakoid membranes are exposed. Tocopherols are one of the primary antioxidants able to quench ROS and protect the membrane lipids from oxidation (Munné-Bosch & Alegre, 2002). Scavenging of lipid peroxy radicals leads to chromanol ring opening in α-tocopherol and α-tocoquinone production (Muñoz & Munné-Bosch, 2019). α-tocoquinone can be "repaired" by a sequence of enzymatic steps culminating in chromanol ring formation by VTE1. This step occurs preferentially in the reduced substrate α-tocopherol quinol (α-TQH2) (Eugeni Piller et al., 2014; Kobayashi & DellaPenna, 2008). In the *ndc1* mutant, α-tocopherol quinol is produced less efficiently leading to diminished α-tocopherol repair and increased concentrations of α-tocoquinone (Eugeni Piller et al., 2014).

3.3 The tocopherol cyclase VTE1 (VITAMIN E DEFICIENT 1), the key enzyme in tocochromanol biosynthesis and metabolism

α-tocopherol is the primary tocochromanol protecting thylakoid membranes from lipid peroxidation under stress (Fryer, 1992; Liu et al., 2008; Munné-Bosch & Alegre, 2002). Most of the enzymes of the tocopherol biosynthesis pathway, including tocopherol cyclase (VTE1), were originally localized at the inner envelope membrane of the chloroplast by activity measurements in isolated membrane fractions (Soll et al., 1985). Tocopherol cyclase introduces the chromanol ring in DMPBQ, converting it to γ-tocopherol (Fig. 2). Subsequently, VTE4 methylates γ-tocopherol resulting in α-tocopherol. Surprisingly, given the earlier results, VTE1 was discovered in the PG proteome (Vidi et al., 2006; Ytterberg et al., 2006). The majority of tocopherol cyclase VTE1 has been localized at PG using a variety of independent techniques: VTE1-fluorescence tagging, immunoelectron microscopy, immunoelectron tomography and Western blotting of isolated chloroplast membrane fractions. Moreover, DMPBQ, the substrate of VTE1, accumulated in the PG of the *vte1* mutant, providing supporting evidence for the localization of VTE1 at PG (Austin et al., 2006; Vidi et al., 2006; Ytterberg et al., 2006). The *vte1* mutant largely resembled the wild type suggesting that DMPBQ or other tocochromanols such as PC-8 can substitute for tocopherol as a functional antioxidant.

Together with plastochromanol-8 (PC-8), tocopherol protects seeds from oxidation and is required for seed longevity (Mène-Saffrané et al., 2010). PC-8 is derived from plastoquinol (PQH_2-9) by tocopherol cyclase/VTE1 activity (Fig. 2). PC-8 has the longest solanesyl side chain of plastoquinones but the chromanol-containing headgroup is identical to that of γ-tocopherol. The VTE1 overexpression lines resulted in a large increase of PC-8 that mainly accumulated in PG. It therefore appeared likely that VTE1 localized at PG converted PQH_2-9 to PC-8 (Zbierzak et al., 2009). Notably, the *ndc1* mutant accumulated less PC-8 likely due to the preference of VTE1 for reduced PQH_2-9 over PQ-9 (Eugeni Piller et al., 2012).

4. Role of PG in thylakoid membrane remodeling under stress

Various abiotic environmental stresses (oxidative stress, high light intensity, nitrate starvation, drought) and plastid developmental transitions (chloroplast-to-chromoplast, chloroplast-to-gerontoplast) are associated with remodeling or even wholesale dismantling of thylakoid membranes and with increased accumulation of lipid antioxidants to cope with such conditions. Typically, PG number and size increase due to lipid deposition in PG resulting from thylakoid membrane dismantling during chloroplast-to-gerontoplast transition or under nitrogen deprivation (Besagni & Kessler, 2013; Domínguez & Cejudo, 2021; Gaude et al., 2007; Hörtensteiner & Kräutler, 2011). This process may result in supersized PG of several micrometers in diameter due to triacylglycerol and fatty acid phytyl ester accumulation. These products result from the combined simultaneous breakdown of galactolipids and chlorophyll in the presence of PG-localized enzymes (Fig. 4) (Zechmann, 2019).

Phytol is released from chlorophyll by pheophytin pheophorbide hydrolase (PPH) (Hörtensteiner & Kräutler, 2011; Schelbert et al., 2009). However, the membrane-toxic phytol does not accumulate. Instead, phytol and free fatty acids, which are hydrolysis products of the thylakoid galactolipids, are converted to fatty acid phytyl esters (FAPE) and triacylglycerol (TAG) in PG by phytyl ester synthases 1 and 2 (PES1 and PES2) (Fig. 4) (Krauß & Vetter, 2018; Lippold et al., 2012). Gerontoplast PG transiently recruit LIPOXYGENASE 3 and 4 (LOX3 and LOX4) that peroxidate 16:3 and 18:3 fatty acids (Lundquist et al., 2012). Together, allene oxide cyclase (AOC) and allene oxide synthase (AOS) convert oxidized 18:3 fatty acid into 12-oxo phytodienoic acid (OPDA) (Fig. 4), a precursor for jasmonic acid synthesis in the peroxisome (Acosta & Farmer, 2010). The initial proteome studies found AOS in PG (Vidi et al., 2006; Ytterberg et al., 2006). A recent study demonstrated strongly increased levels of AOS in PG

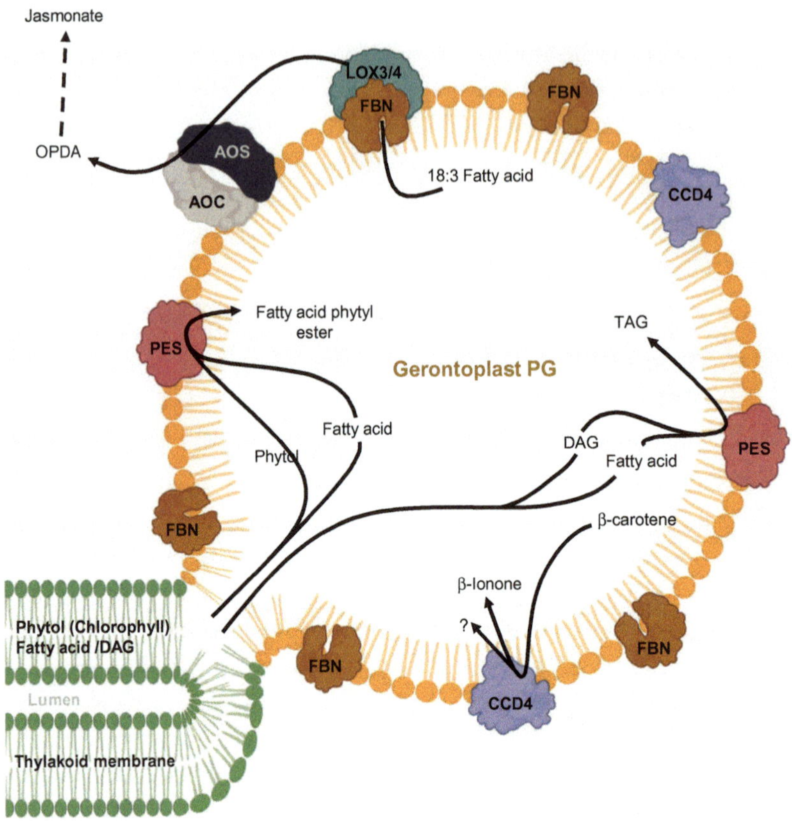

Fig. 4 Lipid metabolism in the gerontoplast plastoglobule. The senescence process is characterized by the transformation of a chloroplast to a gerontoplast and the dismantling of the thylakoid membrane, which comprises galactolipid hydrolysis and chlorophyll breakdown. Phytyl ester synthases (PES) catalyze the synthesis of fatty acid phytyl esters from free fatty acids (from galactolipid hydrolysis) and phytol (from chlorophyll break down). In parallel, phytyl ester synthases contribute to triacylglycerol (TAG) synthesis from free fatty acids and diacylglycerol (DAG). The gerontoplast PG-localized LIPOXYGENASE (LOX) oxidizes unsaturated fatty acid and may be recruited by a fibrillin (FBN). In addition, allene oxide cyclase (AOC) and allene oxide synthase (AOS, also in PG) are essential enzymes to produce 12-oxo-phytodienoic acid (OPDA), a jasmonate precursor. The carotenoid cleavage dioxygenase 4 (CCD4) cleaves β-carotene to release β-ionone and other volatile compounds.

after high light treatment (Espinoza-Corral et al., 2021). Senescence, nitrogen deprivation and high light treatment are useful conditions for studying dynamic behavior of PG morphology and metabolism (Gaude et al., 2007; Zechmann, 2019).

4.1 PES1 and PES2 are involved in phytol and fatty acid metabolism

PES1 and -2 belong to the esterase/lipase/thioesterase (ELT) acyl transferase family. PES1 and -2 are induced during leaf senescence and under nitrogen deprivation. Under nitrogen deprivation, the *pes1pes2* double mutant accumulated only residual levels of phytyl esters and far more free phytol than the wild type (Lippold et al., 2012). The residual phytyl esters may be explained by the presence of additional ELT homologs. PES1 and PES2 converted phytol and fatty acids into FAPE in the heterologous yeast expression system thereby demonstrating phytyl ester synthase activity. Senescence and chlorophyll catabolism was slightly delayed in the *pes1pes2* double mutant emphasizing the physiological importance of the two enzymes in plant senescence (Lippold et al., 2012).

PES1 and PES2 proteins have two distinct domains, a hydrolase domain at the N-terminus and an acyltransferase domain at the C-terminus of the protein. Thus, the PES1 and PES2 proteins potentially harbor two enzymatic functions, namely that of an esterase as well as that of an acyl transferase. In addition, PES1 and PES2 recruit a range of substrates including acyl-CoA species as well as the free fatty acids resulting from galactolipid breakdown. In addition, MGDG can directly serve as an acyl donor for PES1 and PES2 facilitating the release of fatty acids as well as TAG and FAPE biosynthesis (Fig. 4). Acyltransferases typically synthesize triacylglycerol (TAG) from acyl-CoA and diacylglycerol. In a yeast mutant lacking acyltransferase genes and deficient in TAG, expression of Arabidopsis PES1 and PES2 genes rescued TAG accumulation. The results validated that PES1 and PES2 act as acyltransferases and may use DAG as a substrate. Also, the *pes1pes2* double mutant reduces TAG accumulation by about 30% in leaves under nitrogen deficiency (Lippold et al., 2012). These results demonstrate that PES1 and PES2 are important DAG acyltransferases contributing to DAG to TAG conversion under stress. Overall, PES1 and PES2 in PG play a central role in dismantling the photosynthetic membrane during senescence and abiotic stresses.

5. PG act as the metabolic platform for carotenoid metabolism

The chromoplast is an organelle of the plastid family that develops from the chloroplast during fruit ripening (Camara et al., 1995; Egea et al., 2010). This begins with the dismantling of the photosynthetic thylakoid membrane that is accompanied by the tightly controlled degradation of chlorophyll as well as yellow-colored carotenoids, processes visible as degreening (Egea et al., 2010; Hörtensteiner & Kräutler, 2011). Simultaneously, the PG within chromoplasts enlarge and begin to accumulate large amounts of red-colored carotenoids and other prenyl lipids that may be synthesized *de novo* or stem from the mobilization of the thylakoid membranes (Fig. 5). Chromoplast PG may have tubular/fibrillar (example: red pepper) or globular structure (example: tomato) (Nogueira et al., 2013; Pozueta-Romero et al., 1997; Simkin et al., 2007). *In vitro* reconstitution of fibrils was achieved from a mixture of membrane lipids, carotenoids and fibrillin (Deruère et al., 1994). However, the PG of chromoplasts, as already

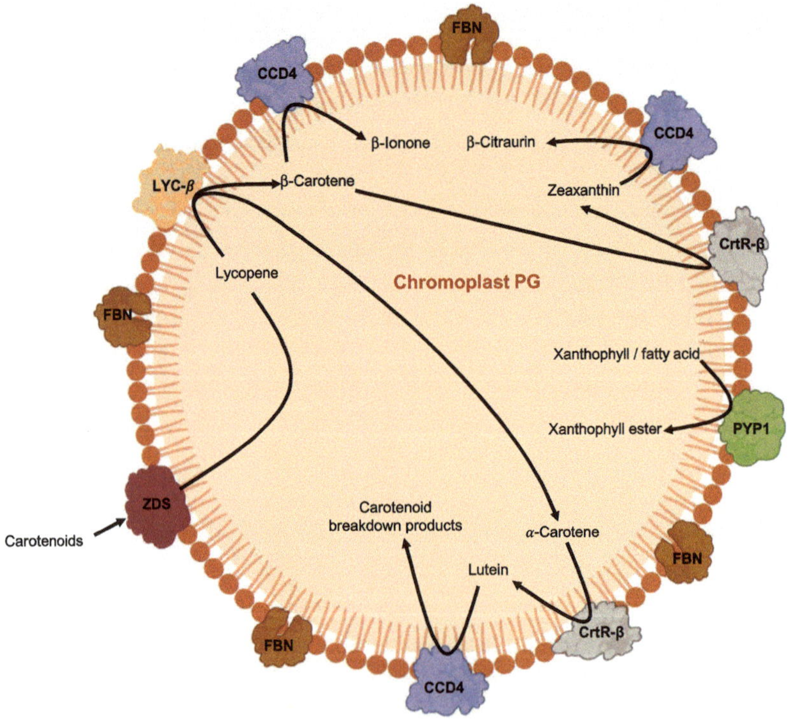

Fig. 5 Carotenoid metabolism in the chromoplast plastoglobule. Chromoplast plastoglobules recruit specific enzymes such as lycopene β-cyclase (LYC-β), zeta-carotene desaturase (ZDS) and β-carotene β-hydroxylase (CrtR-β) to promote carotenoid biosynthesis. Fibrillins (FBNs) may contribute to carotenoid sequestration in PG. PALE YELLOW PETAL 1 (PYP1) catalyzes carotenoid ester synthesis in tomato petal chromoplasts. The carotenoid cleavage dioxygenase 4 (CCD4) cleaves carotenoids, releasing volatile compounds, contributing to aroma and fruit quality in tomatoes.

seen for chloroplast PG, are more than just storage sites and play a central role in the biosynthesis and metabolism of carotenoids.

The proteome of pepper fruit PG has been determined and provided very interesting insight. Chromoplast PG are highly enriched in several FBNs (Fig. 5), particularly FBN1a and -1b, as well as several carotenoid biosynthetic enzymes: ZDS (ζ-carotene desaturase) involved in the synthesis of all-trans-lycopene, a red pigment in tomato, and LYC-β (lycopene beta cyclase) involved in the cyclization reactions that generate carotene (Fig. 5). The β-carotene β-hydroxylases (CrtR-β) introduce hydroxyl groups to carotene, resulting in lutein and zeaxanthin (Fig. 5). These enzymes lack transmembrane domains and are therefore thought to associate with the chromoplast PG surface. The clustering of these enzymes at PG and associated membranes probably facilitates substrate channeling and accumulation of carotenoids inside PG and fibrils (Lundquist et al., 2012). PALE YELLOW PETAL 1 (PYP1) in tomato shares sequence homology with the PES enzymes and is responsible for the production of carotenoid esters, the most abundant form of carotenoids in chromoplasts (Fig. 5) (Ariizumi et al., 2014). Chloroplast and chromoplast PG contain an additional carotenoid metabolic enzyme, carotenoid cleavage dioxygenase 4 (CCD4). In various plant species CCD4s are involved in carotenoid cleavage resulting in various apocarotenoids, especially the volatile β-ionone (Fig. 5) (Varghese et al., 2021). From researches on chromoplasts, it emerges that PG are key determinants of fruit pigmentation, functioning as a site of carotenoid bio synthesis and a sink for carotenoid accumulation. By the action of CCD4, PG may also contribute to the volatile bouquet of tomato fruit.

5.1 Role of CCD4 in chromoplasts and crop quality

CCD4 plays an important role in a surprising number of crops. In addition to its presence in the Arabidopsis PG proteome (Lundquist et al., 2012; Rottet et al., 2016), CCD4 was identified as a major negative regulator of seed carotenoid accumulation in a genome wide association (GWAS) study. *ccd4* mutants had increased β-carotene content upon seed desiccation (Gonzalez-Jorge et al., 2013). Petals of Chrysanthemum CCD4 knock down plant had yellow rather than white petals, demonstrating that CCD4 cleaves carotenoids in chromoplasts and contributes to the accumulation of colorless carotenoids (Ohmiya et al., 2006). The red pigment β-citraurin in mandarin fruit is the CCD4 cleavage product of β-cryptoxanthin and zeaxanthin (Fig. 5) (Rodrigo et al., 2013). In *Brassica napus*, yellow petals contained higher carotenoid levels than white petals due to the disrupted expression of the CCD4 gene (Zhang et al., 2015). Down-regulation of the CCD4 gene in potato tubers led to accumulation of two- to fivefold higher carotenoid content than in wild type tubers (Campbell et al., 2010). In peaches, the expression of CCD4 is repressed in yellow flesh (Redhaven) cultivars as compared to white flesh (Redhaven Bianca) cultivars. Yellow flesh fruits accumulated approximately 10-fold more carotenoids than white flesh (Brandi et al., 2011). CCD4, highly expressed in the stigmata of *Crocus sativus* (saffron), contributes to color, flavor and aroma of the desiccated stigmata by producing apocarotenoids such as safranal, picrocosin, crocetin and β-ionone (Fig. 5). During *Crocus sativus* stigmata development PG size and number increase probably due to the accumulation of CCD4 cleavage products. Engineered CsCCD4 in bacteria preferentially cleaved β-carotene into β-ionone and β-cyclocitral (Frusciante et al., 2014; Rubio-Moraga et al., 2014).

Little is known whether and how CCD4 is regulated. PGM48, a plastoglobule-localized metalloprotease promotes senescence. PGM48 may do so in part by CCD4 degradation which is thought to accelerate senescence through the plant hormone abscisic acid and reduction of apocarotenoid signals (Bhuiyan et al., 2016).

6. Conclusions

In recent years plastoglobules have gone from poorly characterized osmiophile lipid-storing globules to important players in chloroplast lipid metabolism. Nevertheless, many mysteries remain. For one, it is not known how metabolites are trafficked between PG and the associated thylakoid membrane network. Contact sites between the two systems exist and may permit lipid exchange (Austin et al., 2006) but the implication of any proteins (fibrillins have been suggested) remains unknown. It is likely that the trafficking of different lipid species would be separate and highly regulated processes, as has been suggested for the allocation of PQ-9 to the thylakoid membrane (Pralon et al., 2019; Pralon et al., 2020). Early on, it had been hypothesized that plastoglobules contribute to thylakoid membrane formation during early plant development by the conversion of triacylglycerols to membrane lipids. The research tools currently available should allow to investigate this fascinating possibility. It will be of great interest in the future to address these and other open questions and their relevance for photosynthesis and metabolism.

A number of known PG core components have not been characterized so far. These include methyltransferases, reductases and others with various predicted functional domains and enzymatic activities (Lundquist et al., 2012). This remaining wealth of uncharacterized components promises the discovery of further, currently unknown metabolic functions in PG. It will be exciting to see what these will be.

PG have potential biotechnological and agricultural applications. Vitamin E and K enhancement in various crops can easily be imagined. Modification of tomato pigmentation and volatile production can also be envisaged by modifying carotenoid biosynthetic and metabolic pathways that are present in PG. Targeting of recombinant proteins, such as vaccine antigens, has been suggested and attempted in the tobacco experimental system (Shanmugabalaji et al., 2013). Of course, many other applications can also be imagined.

Clearly, research on PG has so far just scratched the surface and many exciting discoveries lie ahead of us.

Acknowledgments

This work was supported by grants from the Swiss National Science Foundation (31003A_156998 and 31003A _176191) and by the University of Neuchâtel.

References

Acosta, I. F., & Farmer, E. E. (2010). Jasmonates. *Arabidopsis Book*, *8*, e0129.

Apel, K., & Hirt, H. (2004). Reactive oxygen species: Metabolism, oxidative stress, and signal transduction. *Annual Review of Plant Biology*, *55*, 373–399.

Arango, Y., & Heise, K. P. (1997). Tocopherol synthesis from homogentisate in Capsicum anuum L. (yellow pepper) chromoplast membranes: Evidence for tocopherol cyclase. *The Biochemical Journal*, *3*, 531–533.

Ariizumi, T., Kishimoto, S., Kakami, R., Maoka, T., Hirakawa, H., Suzuki, Y., et al. (2014). Identification of the carotenoid modifying gene PALE YELLOW PETAL 1 as an essential factor in xanthophyll esterification and yellow flower pigmentation in tomato (Solanum lycopersicum). *The Plant Journal: For Cell and Molecular Biology*, *79*, 453–465.

Austin, J. R., II, Frost, E., Vidi, P.-A., Kessler, F., & Staehelin, L. A. (2006). Plastoglobules are lipoprotein subcompartments of the chloroplast that are permanently coupled to thylakoid membranes and contain biosynthetic enzymes. *The Plant Cell*, *18*, 1693–1703.

Barton, R. (1966). Fine structure of mesophyll cells in senescing leaves of Phaseolus. *Planta*, *71*, 314–325.

Berry, H. M., Rickett, D. V., Baxter, C. J., Enfissi, E. M. A., & Fraser, P. D. (2019). Carotenoid biosynthesis and sequestration in red chilli pepper fruit and its impact on colour intensity traits. *Journal of Experimental Botany*, *70*, 2637–2650.

Besagni, C., & Kessler, F. (2013). A mechanism implicating plastoglobules in thylakoid disassembly during senescence and nitrogen starvation. *Planta*, *237*, 463–470.

Bhuiyan, N. H., Friso, G., Rowland, E., Majsec, K., & van Wijk, K. J. (2016). The plastoglobule-localized metallopeptidase PGM48 is a positive regulator of senescence in *Arabidopsis thaliana*. *Plant Cell*, *28*, 3020–3037.

Boudière, L., Michaud, M., Petroutsos, D., Rébeillé, F., Falconet, D., Bastien, O., et al. (2014). Glycerolipids in photosynthesis: Composition, synthesis and trafficking. *Biochimica et Biophysica Acta (BBA)—Bioenergetics*, *1837*, 470–480.

Bousquet, I., Dujardin, G., & Slonimski, P. P. (1991). ABC1, a novel yeast nuclear gene has a dual function in mitochondria: It suppresses a cytochrome b mRNA translation defect and is essential for the electron transfer in the bc 1 complex. *The EMBO Journal*, *10*, 2023–2031.

Boyd, J. S., Mittelmeier, T. M., Lamb, M. R., & Dieckmann, C. L. (2011). Thioredoxin-family protein EYE2 and Ser/Thr kinase EYE3 play interdependent roles in eyespot assembly. *Molecular Biology of the Cell*, *22*, 1421–1429.

Brandi, F., Bar, E., Mourgues, F., Horváth, G., Turcsi, E., Giuliano, G., et al. (2011). Study of 'Redhaven' peach and its white-fleshed mutant suggests a key role of CCD4 carotenoid dioxygenase in carotenoid and norisoprenoid volatile metabolism. *BMC Plant Biology*, *11*, 24.

Brettel, K., Sétif, P., & Mathis, P. (1986). Flash-induced absorption changes in photosystem I at low temperature: Evidence that the electron acceptor A1 is vitamin K1. *FEBS Letters*, *203*, 220–224.

Camara, B., Hugueney, P., Bouvier, F., Kuntz, M., & Monéger, R. (1995). Biochemistry and molecular biology of chromoplast development. *International Review of Cytology*, *163*, 175–247.

Campbell, R., Ducreux, L. J., Morris, W. L., Morris, J. A., Suttle, J. C., Ramsay, G., et al. (2010). The metabolic and developmental roles of carotenoid cleavage dioxygenase4 from potato. *Plant Physiology*, *154*, 656–664.

Deruère, J., Römer, S., d'Harlingue, A., Backhaus, R. A., Kuntz, M., & Camara, B. (1994). Fibril assembly and carotenoid overaccumulation in chromoplasts: A model for supramolecular lipoprotein structures. *The Plant Cell*, *6*, 119–133.

Desplats, C., Mus, F., Cuiné, S., Billon, E., Cournac, L., & Peltier, G. (2009). Characterization of Nda2, a plastoquinone-reducing type II NAD(P)H dehydrogenase in chlamydomonas chloroplasts. *Journal of Biological Chemistry*, *284*, 4148–4157.

Do, T. Q., Hsu, A. Y., Jonassen, T., Lee, P. T., & Clarke, C. F. (2001). A defect in coenzyme Q biosynthesis is responsible for the respiratory deficiency in saccharomyces cerevisiae abc1 mutants. *Journal of Biological Chemistry*, *276*, 18161–18168.

Domínguez, F., & Cejudo, F. J. (2021). Chloroplast dismantling in leaf senescence. *The Journal of Experimental Botany*.

Egea, I., Barsan, C., Bian, W., Purgatto, E., Latché, A., Chervin, C., et al. (2010). Chromoplast differentiation: Current status and perspectives. *Plant and Cell Physiology*, *51*, 1601–1611.

Espinoza-Corral, R., Schwenkert, S., & Lundquist, P. K. J. T. P. J. (2021). Molecular changes of *Arabidopsis thaliana* plastoglobules facilitate thylakoid membrane remodeling under high light stress. *The Plant Journal*.

Eugeni Piller, L., Abraham, M., Dörmann, P., Kessler, F., & Besagni, C. (2012). Plastid lipid droplets at the crossroads of prenylquinone metabolism. *Journal of Experimental Botany*, *63*, 1609–1618.

Eugeni Piller, L., Besagni, C., Ksas, B., Rumeau, D., Bréhélin, C., Glauser, G., et al. (2011). Chloroplast lipid droplet type II NAD(P)H quinone oxidoreductase is essential for prenylquinone metabolism and vitamin K_1 accumulation. *Proceedings of the National Academy of Sciences*, *108*, 14354–14359.

Eugeni Piller, L., Glauser, G., Kessler, F., & Besagni, C. (2014). Role of plastoglobules in metabolite repair in the tocopherol redox cycle. *Frontiers in Plant Science*, *5*, 298.

Fatihi, A., Latimer, S., Schmollinger, S., Block, A., Dussault, P. H., Vermaas, W. F., et al. (2015). A dedicated type II NADPH dehydrogenase performs the penultimate step in the biosynthesis of vitamin K1 in synechocystis and arabidopsis. *The Plant cell, 27*, 1730–1741.

Flower, D. R., North, A. C., & Sansom, C. E. (2000). The lipocalin protein family: Structural and sequence overview. *Biochim Biophys Acta, 1482*, 9–24.

Frusciante, S., Diretto, G., Bruno, M., Ferrante, P., Pietrella, M., Prado-Cabrero, A., et al. (2014). Novel carotenoid cleavage dioxygenase catalyzes the first dedicated step in saffron crocin biosynthesis. *Proceedings of the National Academy of Sciences of the United States of America, 111*, 12246–12251.

Fryer, M. (1992). The antioxidant effects of thylakoid vitamin E (α-tocopherol). *Plant, Cell & Environment, 15*, 381–392.

Gámez-Arjona, F. M., de la Concepción, J. C., Raynaud, S., & Mérida, Á. (2014). Arabidopsis thaliana plastoglobule-associated fibrillin 1a interacts with fibrillin 1b in vivo. *FEBS Letters, 588*, 2800–2804.

Gaude, N., Bréhélin, C., Tischendorf, G., Kessler, F., & Dörmann, P. (2007). Nitrogen deficiency in Arabidopsis affects galactolipid composition and gene expression and results in accumulation of fatty acid phytyl esters. *The Plant Journal, 49*, 729–739.

Gerards, M., van den Bosch, B., Calis, C., Schoonderwoerd, K., van Engelen, K., Tijssen, M., et al. (2010). Nonsense mutations in CABC1/ADCK3 cause progressive cerebellar ataxia and atrophy. *Mitochondrion, 10*, 510–515.

Gillet, B., Beyly, A., Peltier, G., & Rey, P. (1998). Molecular characterization of CDSP 34, a chloroplastic protein induced by water deficit in Solanum tuberosum L. plants, and regulation of CDSP 34 expression by ABA and high illumination. *The Plant Journal, 16*, 257–262.

Gonzalez-Jorge, S., Ha, S. H., Magallanes-Lundback, M., Gilliland, L. U., Zhou, A., Lipka, A. E., et al. (2013). Carotenoid cleavage dioxygenase4 is a negative regulator of β-carotene content in Arabidopsis seeds. *Plant Cell, 25*, 4812–4826.

Greenwood, A. D., Leech, R. M., & Williams, J. P. (1963). The osmiophilic globules of chloroplasts: I. Osmiophilic globules as a normal component of chloroplasts and their isolation and composition in Vicia faba L. *Biochimica et Biophysica Acta, 78*, 148–162.

Gross, J., Cho, W. K., Lezhneva, L., Falk, J., Krupinska, K., Shinozaki, K., et al. (2006). A plant locus essential for phylloquinone (vitamin K1) biosynthesis originated from a fusion of four eubacterial genes. *Journal of Biological Chemistry, 281*, 17189–17196.

Hansmann, P., & Sitte, P. (1982). Composition and molecular structure of chromoplast globules of Viola tricolor. *Plant Cell Reports, 1*, 111–114.

Havaux, M. (2020). Plastoquinone in and beyond photosynthesis. *Trends in Plant Science, 25*, 1252–1265.

Hörtensteiner, S., & Kräutler, B. (2011). Chlorophyll breakdown in higher plants. *Biochimica et Biophysica Acta (BBA)—Bioenergetics, 1807*, 977–988.

Huang, H., Yang, M., Su, Y.e., Qu, L., & Deng, X. W. (2015). Arabidopsis atypical kinases ABC1K1 and ABC1K3 act oppositely to cope with photodamage under red light. *Molecular Plant, 8*, 1122–1124.

Ischebeck, T., Zbierzak, A. M., Kanwischer, M., & Dormann, P. (2006). A salvage pathway for phytol metabolism in Arabidopsis. *Journal of Biological Chemistry, 281*, 2470–2477.

Kessler, F., Schnell, D., & Blobel, G. (1999). Identification of proteins associated with plastoglobules isolated from pea (Pisum sativum L.) chloroplasts. *Planta, 208*, 107–113.

Kim, E. H., Lee, Y., & Kim, H. U. (2015). Fibrillin 5 is essential for plastoquinone-9 biosynthesis by binding to solanesyl diphosphate synthases in arabidopsis. *Plant Cell, 27*, 2956–2971.

Kim, E. H., Lee, D. W., Lee, K. R., Jung, S. J., Jeon, J. S., & Kim, H. U. (2017). Conserved function of fibrillin5 in the plastoquinone-9 biosynthetic pathway in arabidopsis and rice. *Frontiers in Plant Science, 8*, 1197.

Kirchhoff, H. (2019). Chloroplast ultrastructure in plants. *The New Phytologist, 223*, 565–574.

Kobayashi, N., & DellaPenna, D. (2008). Tocopherol metabolism, oxidation and recycling under high light stress in Arabidopsis. *The Plant Journal, 55*, 607–618.

Krauß, S., & Vetter, W. (2018). Phytol and phytyl fatty acid esters: Occurrence, concentrations, and relevance. *European Journal of Lipid Science and Technology, 120*, 1700387.

Krieger-Liszkay, A., & Trebst, A. (2006). Tocopherol is the scavenger of singlet oxygen produced by the triplet states of chlorophyll in the PSII reaction centre. *Journal of Experimental Botany, 57*, 1677–1684.

Ksas, B., Légeret, B., Ferretti, U., Chevalier, A., Pospíšil, P., Alric, J., et al. (2018). The plastoquinone pool outside the thylakoid membrane serves in plant photoprotection as a reservoir of singlet oxygen scavengers. *Plant, Cell & Environment, 41*, 2277–2287.

Lagier-Tourenne, C., Tazir, M., López, L. C., Quinzii, C. M., Assoum, M., Drouot, N., et al. (2008). ADCK3, an ancestral kinase, is mutated in a form of recessive ataxia associated with coenzyme Q10 deficiency. *The American Journal of Human Genetics, 82*, 661–672.

Langenkämper, G., Manac'h, N., Broin, M., Cuiné, S., Becuwe, N., Kuntz, M., et al. (2001). Accumulation of plastid lipid-associated proteins (fibrillin/CDSP34) upon oxidative stress, ageing and biotic stress in Solanaceae and in response to drought in other species. *Journal of Experimental Botany, 52*, 1545–1554.

Leggett Bailey, J., & Whyborn, A. G. (1963). The osmiophilic globules of chloroplasts II. Globules of the spinach-beet chloroplast. *Biochimica et Biophysica Acta, 78*, 163–174.

Li, T., Jiang, J., Zhang, S., Shu, H., Wang, Y., Lai, J., et al. (2015). OsAGSW1, an ABC1-like kinase gene, is involved in the regulation of grain size and weight in rice. *The Journal of Experimental Botany, 66*, 5691–5701.

Lichtenthaler, H. J. E. (1968a). Plastoglobuli and fine structure of plastids. *Endeavor, 27*, 144.

Lichtenthaler, H. K. (1968b). Distribution and relative concentrations of lipophilic plastid quinones in green plants. *Planta, 81*, 140–152.

Lichtenthaler, H. K., & Peveling, E. (1966). Plastoglobuli in different types of plastids from Allium cepa L. *Planta, 72*, 1–13.

Lichtenthaler, H. K., & Sprey, B. (1966). Über die osmiophilen globulären Lipideinschlüsse der Chloroplasten. *Zeitschrift für Naturforschung B, 21*, 690–697.

Lippold, F., vom Dorp, K., Abraham, M., Hölzl, G., Wewer, V., Yilmaz, J. L., et al. (2012). Fatty acid phytyl ester synthesis in chloroplasts of arabidopsis. *The Plant Cell, 24*, 2001–2014.

Liu, X., Hua, X., Guo, J., Qi, D., Wang, L., Liu, Z., et al. (2008). Enhanced tolerance to drought stress in transgenic tobacco plants overexpressing VTE1 for increased tocopherol production from *Arabidopsis thaliana*. *Biotechnology Letters, 30*, 1275–1280.

Lohmann, A., Schöttler, M. A., Bréhélin, C., Kessler, F., Bock, R., Cahoon, E. B., et al. (2006). Deficiency in phylloquinone (vitamin K1) methylation affects prenyl quinone distribution, photosystem I abundance, and anthocyanin accumulation in the arabidopsis atmenG mutant. *Journal of Biological Chemistry, 281*, 40461–40472.

Lohscheider, J. N., & Río Bártulos, C. (2016). Plastoglobules in algae: A comprehensive comparative study of the presence of major structural and functional components in complex plastids. *Marine Genomics, 28*, 127–136.

Lundquist, P. K., Poliakov, A., Bhuiyan, N. H., Zybailov, B., Sun, Q., & van Wijk, K. J. (2012). The functional network of the arabidopsis plastoglobule proteome based on quantitative proteomics and genome-wide coexpression analysis. *Plant Physiology, 158*, 1172–1192.

Lundquist, P. K., Poliakov, A., Giacomelli, L., Friso, G., Appel, M., McQuinn, R. P., et al. (2013). Loss of plastoglobule kinases ABC1K1 and ABC1K3 causes conditional degreening, modified prenyl-lipids, and recruitment of the jasmonic acid pathway. *The Plant Cell, 25*, 1818–1839.

Malkin, R. (1986). On the function of two vitamin K1 molecules in the PS I electron acceptor complex. *FEBS Letters*, *208*, 343–346.
Manara, A., DalCorso, G., Guzzo, F., & Furini, A. (2015). Loss of the atypical kinases ABC1K7 and ABC1K8 changes the lipid composition of the chloroplast membrane. *Plant and Cell Physiology*, *56*, 1193–1204.
Manara, A., DalCorso, G., Leister, D., Jahns, P., Baldan, B., & Furini, A. (2014). AtSIA1 AND AtOSA1: Two Abc1 proteins involved in oxidative stress responses and iron distribution within chloroplasts. *The New phytologist*, *201*, 452–465.
Martinis, J., Glauser, G., Valimareanu, S., & Kessler, F. (2013). A chloroplast ABC1-like kinase regulates vitamin E metabolism in arabidopsis. *Plant Physiology*, *162*, 652–662.
Martinis, J., Glauser, G., Valimareanu, S., Stettler, M., Zeeman, S. C., Yamamoto, H., et al. (2014). ABC1K1/PGR6 kinase: A regulatory link between photosynthetic activity and chloroplast metabolism. *The Plant Journal*, *77*, 269–283.
Mène-Saffrané, L., Jones, A. D., & DellaPenna, D. (2010). Plastochromanol-8 and tocopherols are essential lipid-soluble antioxidants during seed desiccation and quiescence in Arabidopsis. *Proceedings of the National Academy of Sciences of the United States of America*, *107*, 17815–17820.
Merchant, S. S., Prochnik, S. E., Vallon, O., Harris, E. H., Karpowicz, S. J., Witman, G. B., et al. (2007). The Chlamydomonas genome reveals the evolution of key animal and plant functions. *Science (New York, NY)*, *318*, 245–250.
Michalecka, A. M., Svensson, A. S., Johansson, F. I., Agius, S. C., Johanson, U., Brennicke, A., et al. (2003). Arabidopsis genes encoding mitochondrial type II NAD(P)H dehydrogenases have different evolutionary origin and show distinct responses to light. *Plant Physiology*, *133*, 642–652.
Mollet, J., Delahodde, A., Serre, V., Chretien, D., Schlemmer, D., Lombes, A., et al. (2008). CABC1 gene mutations cause ubiquinone deficiency with cerebellar ataxia and seizures. *The American Journal of Human Genetics*, *82*, 623–630.
Munné-Bosch, S. (2005). The role of α-tocopherol in plant stress tolerance. *Journal of Plant Physiology*, *162*, 743–748.
Munné-Bosch, S., & Alegre, L. (2002). The function of tocopherols and tocotrienols in plants. *Critical Reviews in Plant Sciences*, *21*, 31–57.
Muñoz, P., & Munné-Bosch, S. (2019). Vitamin E in plants: Biosynthesis, transport, and function. *Trends in Plant Science*, *24*, 1040–1051.
Neely, W. C., Martin, J. M., & Barker, S. A. (1988). Products and relative reaction rates of the oxidation of tocopherols with singlet molecular oxygen. *Photochemistry and Photobiology*, *48*, 423–428.
Nogueira, M., Mora, L., Enfissi, E. M., Bramley, P. M., & Fraser, P. D. (2013). Subchromoplast sequestration of carotenoids affects regulatory mechanisms in tomato lines expressing different carotenoid gene combinations. *Plant Cell*, *25*, 4560–4579.
Ohmiya, A., Kishimoto, S., Aida, R., Yoshioka, S., & Sumitomo, K. (2006). Carotenoid cleavage dioxygenase (CmCCD4a) contributes to white color formation in chrysanthemum petals. *Plant Physiology*, *142*, 1193–1201.
Otsubo, M., Ikoma, C., Ueda, M., Ishii, Y., & Tamura, N. (2018). Functional role of fibrillin 5 in acclimation to photooxidative stress. *Plant & Cell Physiology*, *59*, 1670–1682.
Poon, W. W., Davis, D. E., Ha, H. T., Jonassen, T., Rather, P. N., & Clarke, C. F. (2000). Identification of *Escherichia coli*, a gene required for the first monooxygenase step in ubiquinone biosynthesis. *Journal of Bacteriology*, *182*, 5139–5146.
Pozueta-Romero, J., Rafia, F., Houlné, G., Cheniclet, C., Carde, J. P., Schantz, M. L., et al. (1997). A ubiquitous plant housekeeping gene, PAP, encodes a major protein component of bell pepper chromoplasts. *Plant Physiology*, *115*, 1185–1194.

Pralon, T., Collombat, J., Pipitone, R., Ksas, B., Shanmugabalaji, V., Havaux, M., et al. (2020). Mutation of the atypical kinase ABC1K3 partially rescues the PROTON GRADIENT REGULATION 6 phenotype in *Arabidopsis thaliana*. *Frontiers in Plant Science, 11*.

Pralon, T., Shanmugabalaji, V., Longoni, P., Glauser, G., Ksas, B., Collombat, J., et al. (2019). Plastoquinone homoeostasis by Arabidopsis proton gradient regulation 6 is essential for photosynthetic efficiency. *Communications Biology, 2*, 220.

Rey, P., Gillet, B., Römer, S., Eymery, F., Massimino, J., Peltier, G., et al. (2000). Over-expression of a pepper plastid lipid-associated protein in tobacco leads to changes in plastid ultrastructure and plant development upon stress. *The Plant Journal, 21*, 483–494.

Rodrigo, M. J., Alquézar, B., Alós, E., Medina, V., Carmona, L., Bruno, M., et al. (2013). A novel carotenoid cleavage activity involved in the biosynthesis of Citrus fruit-specific apocarotenoid pigments. *The Journal of Experimental Botany, 64*, 4461–4478.

Rottet, S., Devillers, J., Glauser, G., Douet, V., Besagni, C., & Kessler, F. (2016). Identification of plastoglobules as a site of carotenoid cleavage. *Frontiers in Plant Science*, 7, 1855.

Rubio-Moraga, A., Rambla, J. L., Fernández-de-Carmen, A., Trapero-Mozos, A., Ahrazem, O., Orzáez, D., et al. (2014). New target carotenoids for CCD4 enzymes are revealed with the characterization of a novel stress-induced carotenoid cleavage dioxygenase gene from Crocus sativus. *Plant Molecular Biology, 86*, 555–569.

Santabarbara, S., & Casazza, A. P. (2019). Kinetics and energetics of phylloquinone reduction in photosystem i: insight from modeling of the site directed mutants. *Frontiers in Plant Science, 10*.

Schelbert, S., Aubry, S., Burla, B., Agne, B., Kessler, F., Krupinska, K., et al. (2009). Pheophytin pheophorbide hydrolase (pheophytinase) is involved in chlorophyll breakdown during leaf senescence in Arabidopsis. *Plant Cell, 21*, 767–785.

Schoeder, H.-U., & Lockau, W. (1986). Phylloquinone copurifies with the large subunit of photosystem I. *FEBS Letters, 199*, 23–27.

Shanmugabalaji, V., Besagni, C., Piller, L. E., Douet, V., Ruf, S., Bock, R., et al. (2013). Dual targeting of a mature plastoglobulin/fibrillin fusion protein to chloroplast plastoglobules and thylakoids in transplastomic tobacco plants. *Plant Molecular Biology, 81*, 13–25.

Shikanai, T., Munekage, Y., Shimizu, K., Endo, T., & Hashimoto, T. (1999). Identification and characterization of arabidopsis mutants with reduced quenching of chlorophyll fluorescence. *Plant and Cell Physiology, 40*, 1134–1142.

Simkin, A. J., Gaffé, J., Alcaraz, J. P., Carde, J. P., Bramley, P. M., Fraser, P. D., et al. (2007). Fibrillin influence on plastid ultrastructure and pigment content in tomato fruit. *Phytochemistry, 68*, 1545–1556.

Singh, D. K., Laremore, T. N., Smith, P. B., Maximova, S. N., & McNellis, T. W. (2012). Knockdown of FIBRILLIN4 gene expression in apple decreases plastoglobule plastoquinone content. *PLos One*, 7, e47547.

Singh, D. K., & McNellis, T. W. (2011). Fibrillin protein function: The tip of the iceberg? *Trends in Plant Science, 16*, 432–441.

Soll, J., Schultz, G., Joyard, J., Douce, R., & Block, M. A. (1985). Localization and synthesis of prenylquinones in isolated outer and inner envelope membranes from spinach chloroplasts. *Archives of Biochemistry and Biophysics, 238*, 290–299.

Spicher, L., & Kessler, F. (2015). Unexpected roles of plastoglobules (plastid lipid droplets) in vitamin K1 and E metabolism. *Current Opinion in Plant Biology, 25*, 123–129.

Spurr, A. R., & Harris, W. M. (1968). Ultrastructure of chloroplasts and chromoplasts in capsicum annuum I. Thylakoid membrane changes during fruit ripening. *American Journal of Botany, 55*, 1210–1224.

Steinmüller, D., & Tevini, M. (1985). Composition and function of plastoglobuli: I. Isolation and purification from chloroplasts and chromoplasts. *Planta, 163*, 201–207.
Tevini, M., & Steinmuller, D. (1985). Composition and function of plastoglobuli: II. Lipid composition of leaves and plastoglobuli during beech leaf senescence. *Planta, 163*, 91–96.
Thomson, W. W., & Platt, K. (1973). Plastid ultrastructure in the barrel cactus, echinocactus acanthodes. *New Phytologist, 72*, 791–797.
Tikhonov, A. N. (2013). pH-Dependent regulation of electron transport and ATP synthesis in chloroplasts. *Photosynthesis Research, 116*, 511–534.
Van Eerden, F. J., Melo, M. N., Frederix, P., Periole, X., & Marrink, S. J. (2017). Exchange pathways of plastoquinone and plastoquinol in the photosystem II complex. *Nature Communications, 8*, 15214.
van Wijk, K. J., & Kessler, F. (2017). Plastoglobuli: Plastid microcompartments with integrated functions in metabolism, plastid developmental transitions, and environmental adaptation. *Annual Review of Plant Biology, 68*, 253–289.
Varghese, R., Udhaya Kumar, S., Doss, C. G. P., & Ramamoorthy, S. (2021). Unraveling the versatility of CCD4: Metabolic engineering, transcriptomic and computational approaches. *Plant Science: an International Journal of Experimental Plant Biology, 310*, 110991.
Vidi, P. A., Kanwischer, M., Baginsky, S., Austin, J. R., Csucs, G., Dormann, P., et al. (2006). Tocopherol cyclase (VTE1) localization and vitamin E accumulation in chloroplast plastoglobule lipoprotein particles. *Journal of Biological Chemistry, 281*, 11225–11234.
Woggon, W. D. (1999). Improving the yield of alpha-tocopherol from natural sources-chemistry versus molecular biology. *Angewandte Chemie (International ed in English), 38*, 2715–2716.
Yang, M., Huang, H., Zhang, C., Wang, Z., Su, Y., Zhu, P., et al. (2016). Arabidopsis atypical kinase ABC1K1 is involved in red light-mediated development. *Plant Cell Reports, 35*, 1213–1220.
Youssef, A., Laizet, Y., Block, M. A., Maréchal, E., Alcaraz, J. P., Larson, T. R., et al. (2010). Plant lipid-associated fibrillin proteins condition jasmonate production under photosynthetic stress. *The Plant Journal, 61*, 436–445.
Ytterberg, A. J., Peltier, J. B., & van Wijk, K. J. (2006). Protein profiling of plastoglobules in chloroplasts and chromoplasts. A surprising site for differential accumulation of metabolic enzymes. *Plant Physiology, 140*, 984–997.
Zbierzak, A. M., Kanwischer, M., Wille, C., Vidi, P. A., Giavalisco, P., Lohmann, A., et al. (2009). Intersection of the tocopherol and plastoquinol metabolic pathways at the plastoglobule. *The Biochemical Journal, 425*, 389–399.
Zechmann, B. (2019). Ultrastructure of plastids serves as reliable abiotic and biotic stress marker. *PLoS One, 14*, e0214811.
Zhang, B., Liu, C., Wang, Y., Yao, X., Wang, F., Wu, J., et al. (2015). Disruption of a CAROTENOID CLEAVAGE DIOXYGENASE 4 gene converts flower colour from white to yellow in Brassica species. *The New Phytologist, 206*, 1513–1526.

CHAPTER FOUR

Non-vesicular glycerolipids transport in plant cells

Sébastien Leterme and Morgane Michaud*
Université Grenoble-Alpes, CNRS, INRAE, CEA, IRIG, LPCV, Grenoble, France
*Corresponding author: e-mail address: morgane.michaud@cea.fr

Contents

Abstract

In Eukaryotic cells, organelles are delineated by one or several membranes that are composed of a specific assembly of lipids and proteins ensuring the identity of each cell compartment and regulating their biogenesis and function. Lipids regulate many aspects of cell biology including vesicular trafficking, protein targeting, programmed cell death or signaling. They also play key roles in response to stress, which is particularly important for organisms such as plants that constantly cope with a multitude of environmental variations. Membranes are mainly composed of glycerolipids, sterols and sphingolipids. Lipid precursors are primarily synthesized in plastids and in the endoplasmic reticulum. Then, they have to be specifically directed to other cell compartments where they can be modified and transported to other organelles to regulate membrane

Advances in Botanical Research, Volume 101
ISSN 0065-2296
https://doi.org/10.1016/bs.abr.2021.07.001

biogenesis and function. Thus, lipid transfer is a key process involved in the maintenance of membrane homeostasis, in particular in stress conditions. Lipid exchanges between organelles can rely on vesicular or non-vesicular transport. Whereas vesicular transport concerns mainly organelles connected to the endomembrane system, such as the Golgi, vacuoles or the plasma membrane, non-vesicular lipid trafficking concerns virtually all organelles. This lipid transfer mechanism is thought to occur at membrane contact sites, corresponding to specific cell structures formed by the close apposition of two membranes at a distance ≤30 nm. In this chapter, we will summarize our current knowledge about the mechanisms involved in non-vesicular lipid transport in plant cells, with a particular focus on glycerolipids, the main components of organelle membranes.

1. Introduction

1.1 Glycerolipid synthesis in plant cells

Glycerolipids are composed of a glycerol backbone esterified with two fatty acids (FAs) in position *sn-1* and *sn-2* (Fig. 1A). FAs are defined by the length of their carbon chain and the number of double bonds (or unsaturations). In *Arabidopsis thaliana*, glycerolipid acyl-chains are mainly composed of 16 or 18 carbons (C16, C18) with up to three unsaturations (C16:0 to C16:3 and C18:0 to C18:3) (Li-Beisson et al., 2013). The nature of the polar head associated to the position *sn-3* will define the class of the glycerolipid molecule. Three main classes are found in plants: (1) the phospholipids, containing a phosphate (Pi) moiety, (2) the galactoglycerolipids, containing one or two galactose residues for the monogalactosyldiacylglycerol (MGDG) or the digalactosyldiacylglycerol (DGDG), respectively, and (3) the sulfolipid sulfoquinovosyldiacylglycerol (SQDG), containing a sulfoquinovosyl group in its polar head (Fig. 1B). Galactoglycerolipids and SQDG are the main constituents of plastid membranes, where they play a major structural and functional role in photosynthesis, whereas phospholipids are the main glycerolipids found in extra-plastidial membranes. Enzymes involved in the synthesis of galactoglycerolipids, SQDG and of plastidial phosphatidylglycerol (PG) are localized in plastid envelopes (Boudière et al., 2014; Li-Beisson et al., 2013). Enzymes involved in phospholipids synthesis are mainly located in the endoplasmic reticulum (ER) and some are also found in mitochondria or in the endomembrane system (Li-Beisson et al., 2013; Michaud, Prinz, & Jouhet, 2017). In plants, FA biosynthesis occurs in the plastid stroma and gives rise to C16:0-acyl carrier protein (ACP) and C18:1-ACP (Li, Kabbage, Liu, & Dickman, 2016; Li, Xu,

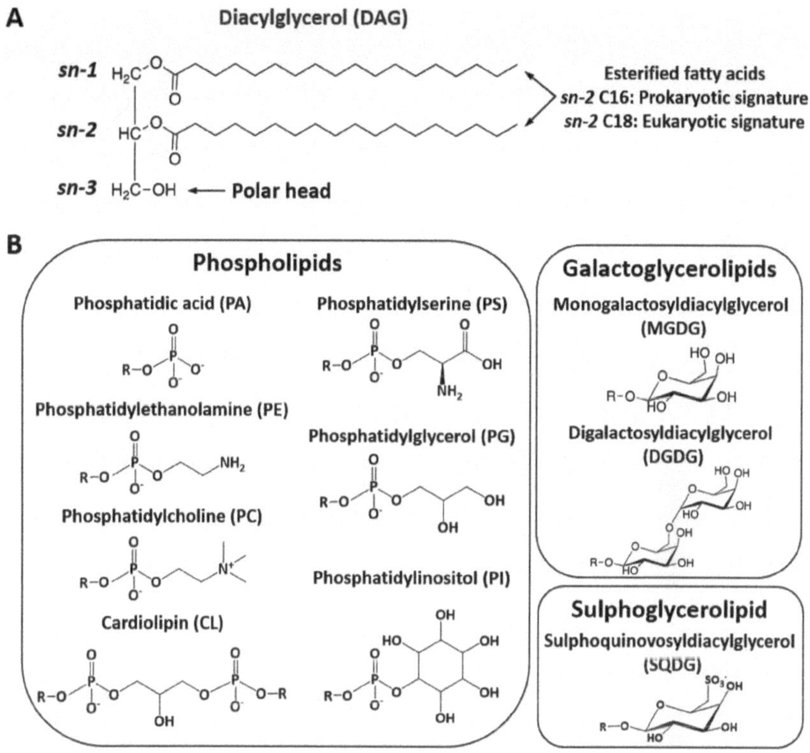

Fig. 1 Glycerolipid families found in plant cells. (A) Glycerolipids are composed of a glycerol backbone esterified with two fatty acids (FA) in *sn-1* and *sn-2* positions. In *Arabidopsis thaliana*, the main FAs found in glycerolipids are 16 (C16) or 18 (C18) carbons-long and have up to three unsaturations (C16:0 to C16:3 and C18:0 to C18:3). In *A. thaliana*, the length of the carbon chain in *sn-2* position defines the origin of the glycerolipid backbone: a C16 is a signature of a backbone synthesized by the plastid by the Prokaryotic pathway while a C18 is a signature of a backbone synthesized in the endoplasmic reticulum by the Eukaryotic pathway. (B) Three main glycerolipid families are found in plants depending on the nature of the polar head located in *sn-3* position of the glycerol backbone: (1) the phospholipids, all presenting a phosphate group in their polar head, (2) the galactoglycerolipids, having a polar head containing one or two galactose residues, and (3) the sulfoglycerolipid, with a sugar-based polar head containing a negatively charged sulfate group

Li-Beisson, & Philippar, 2016). These acyl-chains can then be directly used to feed the synthesis of plastidial lipids by the so-called Prokaryotic pathway or they can be exported in the cytosol and integrated into glycerolipids in the ER by the so-called Eukaryotic pathway (LaBrant, Barnes, & Roston, 2018; Li-Beisson et al., 2013). These two pathways lead to the biogenesis of phosphatidic acid (PA) and diacylglycerol (DAG), two precursor molecules that will be used to synthesize the whole set of plant glycerolipids (Fig. 1). Importantly, after their synthesis, glycerolipid molecules can be subjected to a wide variety of modifications, such as acyl-chain remodeling or polar head modifications, combined to their transport to different cell compartments or translocation from one membrane leaflet to another. The vast variety of pathways involved in these processes highlights the complexity of maintaining membrane lipid homeostasis.

1.2 Non-vesicular glycerolipid transport in plant cells

1.2.1 Where does non-vesicular glycerolipid transport occur in plant cells?

Lipid transport can be mediated by vesicular or non-vesicular pathways. The vesicular pathway involves the transport of lipids, as well as proteins and soluble molecules, in vesicles that will emerge from a donor organelle, be transported through the cytosol and finally fuse with the acceptor organelle to deliver their content. This trafficking route concerns organelles connected to the endomembrane system, such as the ER, the vacuole, the Golgi or the plasma membrane (PM) (Cevher-Keskin, 2020). In plants, vesicular transport also occurs in plastid stroma where this process is involved in the synthesis of thylakoids (Lindquist & Aronsson, 2018). Non-vesicular lipid transport corresponds to the direct transfer of lipid molecules from one membrane to another. This process is involved in the biogenesis of virtually all organelles, but those not connected to the vesicular transport pathway, like mitochondria or plastids, are completely dependent on this lipid transport route to ensure their biogenesis and function (Bayer, Sparkes, Vanneste, & Rosado, 2017; Hurlock, Roston, Wang, & Benning, 2014; LaBrant et al., 2018; Michaud et al., 2017). Non-vesicular lipid transport is thought to occur at membrane contact sites (MCSs), corresponding to particular cell structures formed by the close apposition of two membranes at a distance $\leq$30 nm (Michaud & Jouhet, 2019; Petit, Immel, Lins, & Bayer, 2019; Scorrano et al., 2019). The short distance between the two membranes at MCSs is supposed to decrease the soluble interface that amphiphilic lipid molecules have to cross from the donor to the acceptor membranes. Three types of proteins can be found at MCSs: (1) the tethers, which are proteins able to bridge and maintain the two membranes in close apposition, (2) the effectors, corresponding to proteins providing functionality to the MCSs, and (3) proteins that share both functions (Bohnert & Schuldiner, 2018; Prinz, Toulmay, & Balla, 2020; Scorrano et al., 2019). Beside lipid transport, MCSs play essential roles in many cellular processes including ion exchanges, programmed cell death, organelle division, and appear now as cellular hubs for the regulation of organelle biogenesis and function and cell signaling (Michaud & Jouhet, 2019; Prinz et al., 2020; Scorrano et al., 2019). Contact sites between many organelles have been observed in plants, but we are just starting to understand their role and importance. Plant MCSs are highly dynamic and regulated during development and in response to many biotic and abiotic stresses (Liu & Li, 2019; Mathur, 2020; Michaud & Jouhet, 2019). In this chapter, we will mainly focus on their role in plant lipid transport. For recent reviews about plant MCSs, see Liu and Li (2019), Michaud and Jouhet (2019), Mueller-Schuessele and Michaud (2018), Perez-Sancho et al. (2015), and Petit et al. (2019).

In plants, lipid transport could virtually occur between all organelles. However, the currently available information about the transported molecules and the mechanisms involved are scarce. By analogy to yeast and animal systems, it can be suspected that huge lipid transport between the ER and mitochondria and the ER and the PM occurs, but direct proofs of such transport and the nature of the transported lipids are still missing (Michaud et al., 2017; Perez-Sancho et al., 2016; Petrungaro & Kornmann, 2019; Saheki & De Camilli, 2017a). The most investigated non-vesicular lipid trafficking route in plants corresponds to the ER to chloroplast transport of lipid precursors to synthesize plastid lipids. As described above, plastids have the ability to synthesize *de novo* lipid precursors by the Prokaryotic pathway and to import precursors synthesized in the ER from the Eukaryotic pathway. Due to the specificity of the plastidial and ER *sn-2* acyltransferases, glycerolipids arising from both pathways have an acyl-chain signature at the *sn-2* position corresponding to a C16 or a C18 for the glycerolipids arising from the Prokaryotic or the Eukaryotic pathways, respectively (Fig. 1A) (Hurlock et al., 2014; LaBrant et al., 2018). In *A. thaliana*, around 50% of plastidial lipids come from the Eukaryotic pathway whereas in other plants, like grasses, they almost all arise from this pathway (Browse, Warwick, Somerville, & Slack, 1986; Mongrand, Bessoule, Cabantous, & Cassagne, 1998). Thus, an extensive transfer of lipids from the ER to plastids is required to sustain chloroplast biogenesis. The identity of the transported precursor molecule is still under debate, but it is now widely accepted that different precursors, such as phosphatidylcholine (PC), lyso-PC (*i.e.*, a PC molecule with one acyl-chain), PA or DAG, could be imported into plastids by several co-existing pathways.

The particularity of plant organisms is their ability to adapt to a wide variety of stresses including pathogen attacks, nutrient starvation, extreme temperatures, high light exposure or drought. The modification of plant lipid composition is a widespread phenomenon involved in cell adaptation to stress (Liu et al., 2019; Liu & Li, 2019; Michaud & Jouhet, 2019; Moellering & Benning, 2011). The modification of the plant lipidome can play a role in signal transduction by the generation of lipid signaling molecules such as PA, lyso-lipids, free FAs or phosphatidylinositide phosphates (PIPs) (Hou, Ufer, & Bartels, 2016). It can also play a direct role in the adaptation of plant cells to environmental changes, for instance by modifying membrane fluidity during cold or heat stress or by triggering energy storage by the accumulation of triacylglycerols (TAGs) into lipid droplets (LDs) in response to nitrogen starvation (Liu et al., 2019; Meï et al., 2015; Upchurch, 2008). The best-studied lipid remodeling process occurs in response to phosphate (Pi) starvation, a frequent nutrient stress encountered by plants in nature. Phospholipids retain up to one third of the intracellular Pi content and constitute an important Pi reserve (Poirier, Thoma, Somerville, & Schiefelbein, 1991). During Pi starvation, plants partially degrade their phospholipids to remobilize intracellular Pi (Plaxton & Tran, 2011; Poirier et al., 1991). To maintain membrane integrity, the degraded phospholipids are replaced by DGDG (Hartel, Dormann, & Benning, 2001; Jouhet, Maréchal, Bligny, Joyard, & Block, 2003; Tjellström, Andersson, Larsson, & Sandelius, 2008). Since DGDG is exclusively synthetized in plastids, a massive DGDG export from this organelle to extra-plastidial membranes like vacuole, mitochondria and PM takes place (Andersson, Larsson, Tjellström, Liljenberg, & Sandelius, 2005; Jouhet et al., 2004). The transfer of DGDG from plastids to mitochondria is thought to occur by non-vesicular routes at MCSs between the two organelles as their number significantly increases in response to Pi starvation (Jouhet et al., 2004). However, the mechanisms involved in DGDG transport from plastids to vacuoles and PM (*i.e.*, vesicular and/or non-vesicular) are still unknown.

1.2.2 How does non-vesicular glycerolipid transport occur in plant cells?

The non-vesicular transport of lipids can be divided in three steps: (1) the desorption of the lipid from the donor membrane, (2) its transport through the cytosol and (3) its insertion into the acceptor membrane. The limiting step is the desorption of the lipid from the donor membrane because of the high-energy barrier linked to the extrusion of amphipathic molecules from their hydrophobic environment (Somerharju, 2015). Different mechanisms

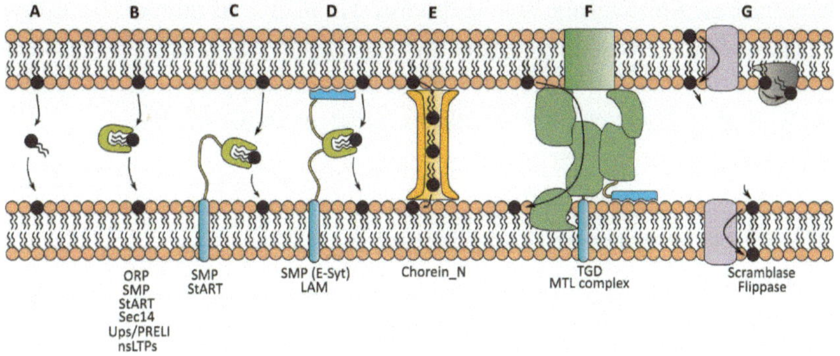

Fig. 2 Mechanisms involved in non-vesicular glycerolipid transport at membrane contact sites. (A) Spontaneous transfer of glycerolipids between membranes. (B–E) Lipid transport mediated by lipid transfer proteins (LTPs). LTPs have a soluble domain with a hydrophobic cavity accommodating lipid molecule(s). They can be soluble (B), membrane bound (C) or anchored at two membranes (D), to shuttle lipids between membranes, or they can adopt a bridge-like structure forming a hydrophobic tunnel for lipid channeling between membranes (E). (F) Lipid transport can be mediated by multi-subunits complexes crossing two or several membranes. (G) Some membrane protein can promote non-vesicular lipid transport by favoring lipid extrusion or insertion from the donor or the acceptor membranes, respectively. The different proteins and complexes using each mechanisms that will be discussed in this chapter are indicated.

have been proposed to mediate non-vesicular lipid transport in cells. Spontaneous transport of glycerolipids can occur (Fig. 2A), but this process is likely irrelevant in the context of cell division as the half time of spontaneous PC transport *in vitro* ranges from several hours to days (Somerharju, 2015). Obviously, many factors can influence the rate of spontaneous transport of glycerolipids, including the composition of the donor membrane and the hydrophobicity of the transported molecules (length and unsaturation level of the acyl-chains). For instance, the *in vitro* spontaneous transfer rate of lyso-lipids significantly decreases compared to glycerolipids, ranging from seconds to minutes (Somerharju, 2015). It has also been proposed that MCSs might form microdomains with a specific lipid and protein composition favoring non-vesicular lipid transport (Gaigg, Simbeni, Hrastnik, Paltauf, & Daum, 1995; Pichler et al., 2001; Somerharju, 2015), but this remains to be determined. *In vivo*, many proteins, such as lipid transfer proteins (LTPs) (Fig. 2B–E), as well as multi-subunit complexes that can cross at least two membranes (Fig. 2F), are involved in non-vesicular lipid transport. In addition, some membrane proteins can act in concert with LTPs to promote membrane lipid desorption or insertion (Fig. 2G). In this chapter, these mechanisms will be discussed in plants. As our current knowledge is scarce, we will rely on the lipid transfer processes described in other organisms to draw a more global picture of the putative mechanisms involved in non-vesicular glycerolipid transport in plant cells.

2. Lipid transport mediated by LTPs

LTPs can be divided into different families depending on their structure and mechanism of lipid transport, but they all share common features: (1) they possess a soluble domain with a hydrophobic cavity accommodating one or several lipid molecules, favoring the transport of lipids through soluble compartments, and (2) they are able to significantly increase the rate of glycerolipid transfer between membranes (Reinisch & Prinz, 2021; Somerharju, 2015; Wong, Gatta, & Levine, 2019). Some LTPs have also been proposed to directly facilitate membrane lipid desorption or to associate to membrane proteins favoring lipid extrusion (Lipp, Ikhlef, Milanini, & Drin, 2020; Reinisch & Prinz, 2021; Wong et al., 2019). Two main mechanisms have been proposed for LTPs: (1) the shuttle mechanism, where the lipid-binding domain of LTPs is able to diffuse through the soluble compartment to shuttle lipids from one membrane to another (Fig. 2B–D) and (2) the bridge-like mechanism in which LTPs are able to form hydrophobic tunnels contacting simultaneously two membranes between which lipids will be transported (Fig. 2E) (Reinisch & Prinz, 2021; Wong et al., 2019). LTPs can be soluble and transiently associated to membranes *via* protein-lipid or protein-protein interactions (Fig. 2B) or anchored to membranes by transmembrane (TM) domains (Fig. 2C). Indeed, as further described in this chapter, the lipid-binding domain of LTPs is often associated to additional domains, the most frequent being the pleckstrin-homology (PH) domain, which specifically binds to PIPs, and the C2 domains that can bind a wider range of membrane lipids. These additional domains allow LTPs targeting to particular organelles or MCSs (Reinisch & Prinz, 2021; Wong et al., 2019). Furthermore, some LTPs are able to simultaneously bind two membranes and are able to act as lipid transporters and membrane tethers at MCSs (Fig. 2D). In plants, a few LTPs have been characterized until now but they are involved in a wide variety of cellular processes beside lipid transport including lipid signaling, plant resistance to pathogens, adaptation to many abiotic stresses or regulation of gene expression (Table 1).

Table 1 Roles of the different lipid transfer proteins (LTPs) characterized until now in plants.

Roles	Characterized LTP proteins in plants
Lipid transfer between membranes	Syt1, Syt3, At1g73200, ORP3A, AtSfh1, AtPITP11, Ssh2, NbSec14
Membrane tethering	Syt1, Syt 3, Syt5, Syt7, At1g73200
Signaling	AtSfh1, Ssh1, AtPIPT7, MtN5
Vesicular trafficking	Syt2, AtPATL1, AtPATL2, AtPITP7
Plant development	PiORP3, AtSfh1, GhLTPG1
Abiotic stress response	Syt1, Syt3, AtPATL1, Ssh1, AsE246, MtN5
Biotic stress response	Syt1, Syt5, Syt7, VAD1, BAGP1, ERD2, ORPs, AtPATL3, AtPATL6, NbSec14
Regulation of gene expression	Syt7, GL2, HD-StART, GmOSBP

A detailed description of the role of these LTPs can be found in this chapter.

2.1 The SMP family

The SMP (synaptotagmin-like mitochondrial-lipid binding protein) domain is a lipid binding/transport domain belonging to the TULIPs (tubular lipid binding proteins) superfamily of LTPs (Kopec, Alva, & Lupas, 2010; Wong & Levine, 2017). SMP domain-containing proteins are evolutionarily conserved among Eukaryotes (Craxton, 2004; Lee & Hong, 2006). This family has been mostly studied in yeast and mammals and shown to be located at various MCSs: (1) at ER-PM MCSs, for the Extended synaptotagmins (*E*-Syts), also called Tricalbins in yeast (Perez-Sancho et al., 2015; Saheki & De Camilli, 2017b), (2) at ER-mitochondria MCSs, for some ERMES (ER-mitochondria encounter structure) complex subunits in yeast and the PDZD8 protein in mammals (Hirabayashi et al., 2017; Kornmann et al., 2009), (3) at ER-Golgi MCSs for Nvj2 in yeast (Liu, Choudhary, Toulmay, & Prinz, 2017) and (4) at the ER-endosomes MCSs for PDZD8 (Elbaz-Alon et al., 2020; Shirane et al., 2020). The SMP proteins are mostly predicted to be anchored to membranes by one TM segment linked to the SMP domain by a flexible linker (Lee & Hong, 2006; Saheki & De Camilli, 2017b). Other domains can also be associated, as exemplified by C2 domains located in *E*-Syts allowing the binding of the protein to a second membrane enriched in PIPs (see below). Cristal structures of the SMP domains show that they fold into a β-barrel structure oligomerizing to form hydrophobic tunnels that can accommodate phospholipid molecules (Jeong, Park, Jun, & Lee, 2017; Saheki & De Camilli, 2017b; Schauder et al., 2014). SMP domains are able to form homodimers or heteromultimers (AhYoung et al., 2015; Schauder et al., 2014). Analyses of lipids bound to purified proteins as well as *in vitro* transfer assays support the role of SMP proteins in phospholipid transport between membranes at MCSs, independently of the polar head, suggesting that SMP proteins are non-specific glycerolipid transporters (AhYoung et al., 2015; Jeong et al., 2017; Saheki et al., 2016). The flexible linkers located between the TM and the SMP domain of these proteins allow the movement of the lipid binding domain in the cytosol to shuttle lipids from one membrane to another in an ATP-independent manner (Fig. 2D) (Jeong et al., 2017; Schauder et al., 2014). By searching in the literature and in domain databases, 13 proteins containing a SMP domain can be identified in *A. thaliana* (Table 2) (Ishikawa et al., 2020; Perez-Sancho et al., 2016). Seven show a domain organization similar to *E*-Syts and two to Nvj2 proteins.

2.1.1 E-Syts

E-Syts contain a TM domain anchoring the protein in the ER, and one or several C2 domains located in the cytosol (Perez-Sancho et al., 2016; Saheki & De Camilli, 2017b). C2 domains are able to bind PIPs-enriched PM in a Ca^{2+}-dependent manner, allowing the localization of E-Syts at ER-PM MCSs (Fig. 2D). At this junction, E-Syts play a dual role as tethers and lipid transporters (Jeyasimman & Saheki, 2020; Saheki et al., 2016). Whereas yeast and human genomes contain three E-Syt genes (Reinisch & De Camilli, 2016), this family is expanded in plants, with seven members found in *A. thaliana* genome (Table 2) (Ishikawa et al., 2020). Among this family, Syt1, Syt3, Syt5 and Syt7 (also called AtCLB1) are enriched at ER-PM MCSs and are ubiquitously expressed at a high level (de Silva, Laska, Brown, Sederoff, & Khodakovskaya, 2011; Ishikawa et al., 2020; Lee et al., 2020). Intriguingly, Syt7, which binds ceramides, was also observed in the nuclear envelope of root tip cells. Syt7 associates to AtTASH1 promoter to negatively regulate the expression of this gene involved in the synthesis of triterpenoids required for root development (de Silva et al., 2011). The regulation of the localization of this membrane protein to either ER-PM MCSs or the nucleus remains to be investigated. Syt2 is highly expressed in pollen and localized in Golgi and PM (Wang et al., 2015; Zhang et al., 2011). Syt2 is involved in non-conventional secretion pathways and plays a role in pollen development and pollen tube

Table 2 Genes of the synaptotagmin-like mitochondrial-lipid binding protein (SMP) family present in the *A. thaliana* genome.

AGI	Name	Sub-family	Localization	Domain organization	References
At2g20990	SYT1/SYTA	E-Syt	ER/ER-PM MCSs	TM-SMP-C2-C2	Perez-Sancho et al. (2015)
At1g20080	SYT2/SYTB		Golgi/PM	TM-SMP-C2-C2	Wang et al. (2015) and Zhang et al. (2011)
At5g04220	SYT3/SYTC		ER-PM MCSs	TM-SMP-C2-C2	Ruiz-Lopez et al. (2020)
At5g11100	SYT4/SYTD		ER (prediction)	TM-SMP-C2-C2	
At1g05500	SYT5/SYTE		ER/ER-PM MCSs	TM-SMP-C2-C2	Ishikawa, Fukao, and Shimada (2020) and Lee et al. (2020)
At3g18370	SYT6/SYTF		ER (prediction)	TM-SMP-C2-C2-C2-CC	
At3g61050	Syt7/AtCLB1		ER/ER-PM MCSs/nucleus	TM-SMP-C2-CC	Ishikawa et al. (2020)
At3g60950			Mitochondria (prediction)	SMP-C2-DNaseI-like	
At3g61030			Mitochondria (prediction)	SMP-C2-DNaseI-like	
At1g73200			ER-Golgi MCSs (yeast complementation)	TM-PH-SMP	Liu et al. (2017)
At1g17820			ER-Golgi MCSs (by homology to At1g73200)	TM-PH-SMP	
At1g50260	NTMC2T5.1		Plastids (prediction)	TM-SMP-TM-TM-C2-TM	
At3g19830	NTMC2T5.2		Plastids (prediction)	TM-SMP-TM-TM-C2-TM	

Protein localization were determined experimentally or the consensus localizations from SUBA and ARAMEMNON websites (prediction) are indicated (Hooper, Castleden, Tanz, Aryamanesh, & Millar, 2017; Schwacke et al., 2003). The domain organization found for each protein is also indicated. CC: coiled-coil domain, PH: pleckstrin homology domain, TM: transmembrane domain.

elongation (Wang et al., 2015; Zhang et al., 2011). These results show that, in contrast to yeast and mammals, plant *E*-Syts have a broader subcellular localization and could be involved in the regulation of a wider range of cellular processes besides ER-PM tethering and lipid transport, including gene expression regulation.

The most characterized E-Syt in plants is Syt1 in *A. thaliana*. Syt1 is anchored in the ER by its N-terminal TM segment and is able to bind negatively charged lipids in the PM *via* the two C2 domains protruding in the cytosol (Fig. 3A) (Lee et al., 2020; Perez-Sancho et al., 2015; Ruiz-Lopez et al., 2020; Schapire et al., 2008; Yamazaki, Kawamura, Minami, & Uemura, 2008; Yamazaki, Takata, Uemura, & Kawamura, 2010). Whether the PM binding is Ca^{2+} dependent or not is still a matter of debate (Perez-Sancho et al., 2015; Ruiz-Lopez et al., 2020; Schapire et al., 2008). The localization of Syt1 to ER-PM MCSs also requires VST (VAP-related suppressor of TMM) proteins, which are peripheral PM proteins implicated in stomatal development (Ho, Paciorek, Abrash, & Bergmann, 2016). Syt1 is involved in many aspects of plant development and biotic and abiotic stress response. Indeed, ER-PM proximity is highly regulated during plant development and in response to different stresses exerting huge constraints on the PM, such as high salinity, osmotic stress, mechanical stress or freezing (Ho et al., 2016; Lee et al., 2019, 2020; McFarlane et al., 2017; Perez-Sancho et al., 2015). In such situations, increasing of ER-PM MCSs is thought to stabilize the PM and to favor its recovery after damages. In accordance with the key role of Syt1 in these processes: (1) Syt1 is overexpressed during cold acclimation or mechanical stress, (2) *syt1* mutants are sensitive to freezing, high NaCl concentration, drought, osmotic or mechanical stress and (3) a decrease of PM integrity is observed in *syt1* mutants (Perez-Sancho et al., 2015; Ruiz-Lopez et al., 2020; Schapire et al., 2008; Yamazaki et al., 2008). Syt1 is also located at ER-PM MCSs close to plasmodesmata (PD) entries. PDs are channels crossing the plant cell wall, ensuring a continuity of PM, ER and cytosol between plant cells and allowing cell-to-cell communication (Petit et al., 2019). In PDs, ER diameter is reduced and ER-PM MCSs mediated by Syt1, as well as by Syt5 and 7, are suspected to play a role in ER constriction to enter and cross PDs (Ishikawa et al., 2020). In contrast to what is observed in the case of abiotic stress, *syt1* mutants exhibit a higher resistance to fungi and viral infection, notably by reducing pathogen entry into cells or limiting cell-to-cell transmission of viral particles through PDs (Ishikawa et al., 2020; Kim et al., 2016; Levy, Zheng, & Lazarowitz, 2015). Syt1 is able to interact with

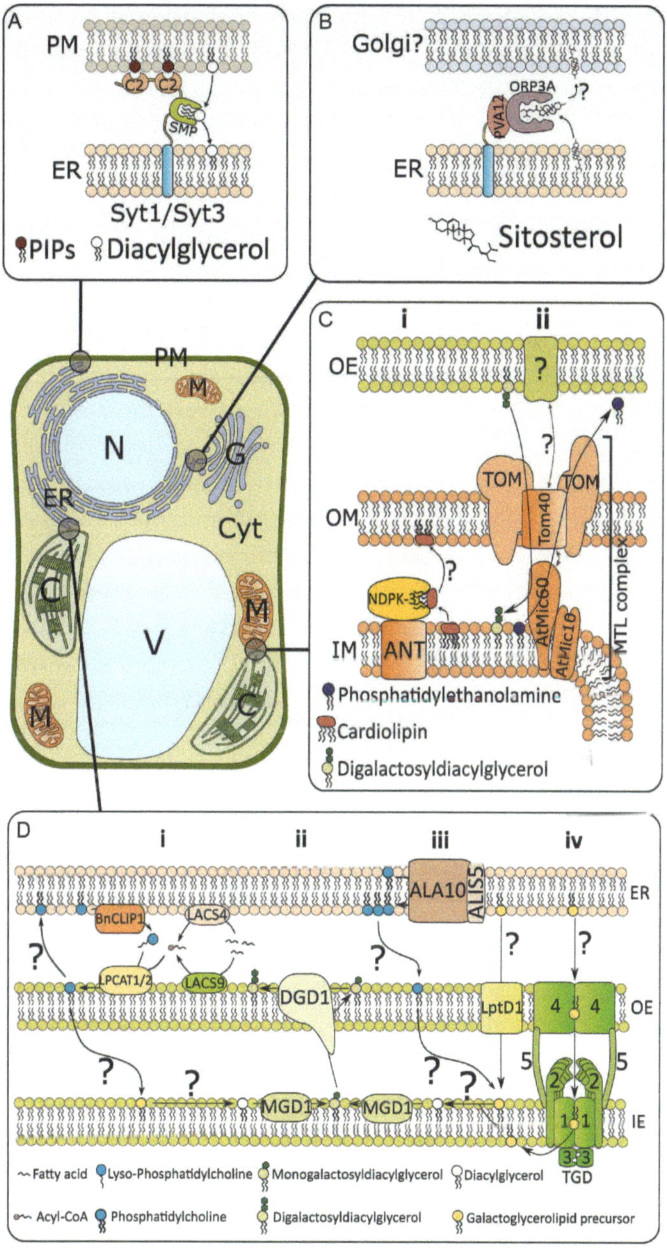

Fig. 3 Proteins involved in non-vesicular glycerolipid transport at different membrane contact sites (MCSs) in plant cells. (A) Transport of diacylglycerol (DAG) by Syt1/3 at endoplasmic reticulum (ER)-plasma membrane (PM) MCSs. Syt1/3 are anchored to the ER by a transmembrane domain and to the PM by the binding of the C2 domains *(Continued)*

three other *E*-Syts, Syt3, Syt5 and Syt7, suggesting it could also form heteromultimeric complexes at ER-PM MCSs, maybe to promote lipid transfer as in mammals (Ishikawa et al., 2020; Lee et al., 2020; Ruiz-Lopez et al., 2020). Recently, it has been shown that Syt3 has a redundant function with Syt1. Both proteins are involved in ER-PM MCSs formation and are more abundant in these sites during cold stress (Ruiz-Lopez et al., 2020). Interestingly, lipidomic analyses of the double *syt1/3* mutant under cold stress show no changes in the whole leaf lipidome but a specific accumulation of DAG and PC in PM fraction, suggesting that DAG clearance from PM is impaired in the double mutant (Ruiz-Lopez et al., 2020). DAG accumulation in PM could be responsible for the cold hypersensitivity of the *syt1/3* mutant by altering PM integrity. In addition, Syt1 co-purifies with DAG (Ruiz-Lopez et al., 2020), further supporting a role of Syt1, and likely Syt3, in DAG transport at ER-PM MCSs during cold stress (Fig. 3A). This first study confirms that, like in yeast and metazoans, *E*-Syts are involved in both ER-PM tethering and lipid transport, opening important perspectives about the role of these proteins in PM biogenesis and response to stress. Interestingly, five mRNA splice variants were detected for Syt3

Fig. 3—Cont'd to phosphatidylinositides (PIPs). DAG transport is mediated by the synaptotagmin-like mitochondrial-lipid binding protein (SMP) domain. (B) Putative transport of sitosterol by ORP3A at ER-golgi MCS. ORP3A interacts in the ER with the PVA12 VAP protein. (C) (i) Putative transport of cardiolipin at mitochondrial (M) outer (OM)-inner (IM) membrane MCSs mediated by NDPK-3. NDPK-3 interacts in the IM with the ANT translocator. (ii) Digalactosyldiacylglycerol (DGDG) and phosphatidylethanolamine transport at OM-IM MCSs mediated by the mitochondrial transmembrane lipoprotein (MTL) complex during phosphate (Pi) starvation. In this complex, AtMic60 interacts with Tom40 and plays a key role in the lipid transport process. Because of the presence of chloroplast (C) proteins in the MTL complex, it could also be located at chloroplasts-mitochondria MCSs to mediate DGDG import during Pi starvation. (D) Lipid transport at ER-chloroplasts MCSs and outer (OE)-inner (IE) envelopes MCSs. (i) Putative lipid transport mediated by phosphatidylcholine (PC) remodeling at ER-chloroplasts MCSs. BnCLIP1 lipase could form lyso-PC from PC that could be acylated in the OE by lyso-PC acyltransferases (LPCATs) using a local pool of acyl-CoA generated by the long-chain acyl-CoA synthetases LACS4 and LACS9. (ii) The DGDG synthase DGD1 might play a role in OE-IE lipid transport thanks to its N-terminal extension. (iii) The ALA10 flippase could promote PC transport from the ER to the chloroplasts by promoting enrichment of 18:2-PC in the cytosolic leaflet of the ER. (iv) Import of lipids precursors in chloroplasts IE by the TGD complex and the LptD1 protein. TDG1, 2 and 3 form an ABC transporter in the IE whereas TGD5 bridges the IE and OE *via* its interaction with TGD1/2/3 and TGD4. LptD1 is a homolog of TGD4 and might be involved in galactoglycerolipid synthesis during Pi starvation.

(Ruiz-Lopez et al., 2020), increasing the putative number of Syts present in plant cells and the functions they can perform.

2.1.2 Other SMP-domain containing proteins

The localization and role of the six other *A. thaliana* SMP domain-containing proteins are currently unknown. At1g73200 and At1g17820 have a similar domain organization than yeast Nvj2 with a TM domain separated of the SMP domain by a PH domain (Table 2). Nvj2 is involved in the formation of ER-Golgi contacts and ceramide transport to Golgi to alleviate the ER cytotoxicity linked to ceramide accumulation (Liu et al., 2017). At1g73200 is able to restore ceramide transport and ER-Golgi tethering in the yeast *Δnvj2* mutant, suggesting that At1g73200 and At1g17820 might be Nvj2 orthologs (Liu et al., 2017). The At1g50260 and At3g19830 proteins present an unusual domain organization with four predicted TM domains flanking a SMP and a C2 domain (Table 2). This organization seems to be found only in plants (Lee & Hong, 2006). The At3g60950 and At3g61030 proteins are the only *A. thaliana* SMP proteins that do not contain a TM domain (Table 2). The SMP domain is associated to a C2 and an endonuclease/exonuclease/phosphatase super family domain that was not found associated to SMP domain in other organisms. No putative orthologs of the ERMES complex subunits or PDZD8, involved in ER-mitochondria and ER-endosomes lipid transport in yeast and mammals, respectively, were found in *A. thaliana* genome.

2.1.3 The StARkin super-family

The StARkin (relatives (kin) of steroidogenic acute regulatory protein (StAR)) is a large superfamily of LTPs that gather families sharing a common lipid binding/transfer domain folding despite a low sequence conservation (Hanada, 2018; Wong & Levine, 2016). The StARkin lipid binding/transfer domain is composed of a curved β-sheet containing seven strands and of α-helices that form a hydrophobic cavity accommodating amphiphilic molecules (Wong & Levine, 2016). StARkin super family includes the StARTs (StAR related transfers), the LAMs (lipid transfer at membrane contact sites), the PITPs (phosphoinositol (PI) transfer proteins), the Ups/PRELI and the Bet-V1 protein family. Bet-V1 (also called PR10) proteins are plant-specific. They are able to bind a wide variety of amphipathic molecules and are involved in the synthesis of many compounds such as flavonoids, alkaloids, sporopollenin or polyphenolics, for instance in response to pathogen infection. As their main substrates are not glycerolipids, this family will not be further described in this chapter (for a recent review about Bet_V1 proteins, see Morris, Caldo, Liang, & Facchini, 2021). PITPs with a StARkin domain play a role in the transport of PI, PC, PA or sphingolipids in different cell compartments and are involved in many cellular processes such as PIPs signaling, vesicular trafficking or sphingolipid synthesis. This family is mainly present in metazoans but is not conserved either in plants or in yeasts and thus will not be further discussed in this chapter (for recent reviews, see Grabon, Bankaitis, & McDermott, 2019; Selitrennik & Lev, 2016).

2.1.4 The LAM family

LAM proteins are lipid transporters located at MCSs and conserved throughout the Eukaryotic lineage. All the LAM proteins studied until now are sterol transporters (Khafif, Cottret, Balague, & Raffaele, 2014; Wong & Levine, 2016). These proteins have a conserved organization with (1) a GRAM (glucosyltransferases, Rab-like GTPase activators and myotubularins) domain corresponding to a PH-like domain involved in protein/protein or protein/lipid interactions, (2) one or two VASt (VAD1 analog of StAR-related lipid transfer) domains, involved in lipid binding and transport and (3) a C-terminal TM domain (Hanada, 2018; Wong & Levine, 2016). The three human LAMs, as well as the yeast Lam1–4 proteins are located at ER-PM MCSs, whereas Lam5 and Lam6 are detected at multiple MCSs in yeast (Elbaz-Alon et al., 2015; Gatta et al., 2015; Murley et al., 2015; Sandhu et al., 2018). They are anchored in the ER by their TM domain but the domains and mechanisms involved in their association with a second organelle remain elusive. Lam6 associates to mitochondria and vacuole through interactions with the protein import receptors Tom70/71 and Vac8, respectively (Murley et al., 2015). *In vitro*, the VASt domains of Lam2/4 is able to bind phosphatidylserine (PS)-containing liposomes. The transport efficiency of the VASt domain of GRAMD1 proteins is enhanced by the addition of anionic lipids (PS or PIPs) in the acceptor liposome (Horenkamp, Valverde, Nunnari, & Reinisch, 2018; Tong, Manik, & Im, 2018), suggesting that the anchoring to a second membrane could also be achieved by interactions of LAMs with lipids. Structural analyses of the VASt domain of several LAM proteins show it is monomeric and folds into a hydrophobic pocket accommodating one sterol molecule (Gatta, Sauerwein, Zhuravleva, Levine, & Matthews, 2018; Horenkamp et al., 2018). *In vitro*, the VASt domain of most of the LAMs is able to bind and transport sterols between membranes in an ATP-independent manner (Gatta et al., 2015, 2018; Horenkamp et al., 2018; Murley et al., 2015). This transport activity was also confirmed *in vivo*, where LAM proteins are able to shuttle sterols at ER-PM or ER-vacuole junctions (Gatta et al., 2015; Murley et al., 2017; Sandhu et al., 2018). In yeast, Lam2/4 and Lam6 trigger the formation of sterol-enriched microdomains in the PM and in the vacuole, respectively, playing a role in signaling during stress response (Murley et al., 2015, 2017).

A. thaliana genome contains four putative LAM proteins (Khafif et al., 2014; Wong & Levine, 2016). Besides VAD1, plant LAMs have an unusual domain organization compared to yeast and mammals with the presence of additional C2 domains and two TM domains located in the middle of the protein for At1g03370 and At5g50170 (Table 3). The C2 domains might be involved in protein binding to membranes enriched in negatively-charged lipids, such as the PM. Few data are available concerning plant LAMs. VAD1-RFP fusion does not show a punctate pattern and seems to be localized in different cell compartments, including, PM, Golgi, tonoplast, ER, cytosol or nucleus when overexpressed (Khafif et al., 2017). This localization pattern might be linked to the RFP fused in C-terminal, close to the TM domain, which might alter membrane targeting of VAD1. VAD1 is a negative regulator of the programmed cell death hypersensitive response (HR) occurring in response to pathogen invasion to avoid their propagation

Table 3 Genes of the lipid transfer at membrane contact sites (LAM) family present in the *A. thaliana* genome.

AGI	Name	Localization	Domain organization	References
At1g02120	VAD1	PM, ER, golgi, cytosol, tonoplast, nucleus	GRAM-VASt-TM-CC	Khafif, Balague, Huard-Chauveau, and Roby (2017)
At1g03370		Cytosol (prediction)	C2-VASt-TM-TM-C2-GRAM-VaST	
At3g59660	BAGP1	PM (prediction)	TM-C2-GRAM-VASt	
At5g50170		Cytosol (prediction)	C2-VASt-TM-TM-C2-GRAM-VaST	

Protein localization were determined experimentally or the consensus localizations from SUBA and ARAMEMNON websites (prediction) are indicated (Hooper et al., 2017; Schwacke et al., 2003). The domain organization found for each protein is also indicated. CC: coiled-coil domain, GRAM: glucosyltransferases, Rab-like GTPase activators and myotubularins domain, TM: transmembrane domain, VASt: VAD1 analog of StAR-related lipid transfer domain.

in plant tissues (Khafif et al., 2017; Lorrain et al., 2004). Because of its role in cell death regulation, VAD1 is involved in pathogen resistance and the *vad1* mutant shows an enhanced resistance to pathogen attack linked to constitutive cell death activation (Khafif et al., 2017; Lorrain et al., 2004). Interestingly, the role of VAD1 in HR, as well as its nuclear localization, are dependent on the VASt, but not on the GRAM domain, suggesting that VAD1 localization and function might be regulated by lipid binding to the VASt domain (Khafif et al., 2017). The second LAM protein studied in *A. thaliana* is BAGP1, which also plays a role in biotic stress response by promoting basal immunity against the pathogenic fungus *Botrytis cinerea*. In response to pathogen infection, BAGP1 interacts with the protease APCB1 to promote the cleavage of BAG6 (Bcl2-associated athanogene), an autophagy activator inducing basal immunity against fungi (Li, Kabbage, et al., 2016; Li, Xu, et al., 2016). However, the localization of BAGP1, how it regulates BAG6 cleavage and whether lipids are involved in the regulation of its function are still unknown. Overall, LAM proteins seem to play important roles in biotic stress response in plants, but despite their homology with yeast and mammalian LAMs, their localization and their ability to bind and/or transport lipids remain to be determined.

2.1.5 The Ups/PRELI family

Proteins of the Ups/PRELI family are soluble proteins located in the intermembrane space (IMS) of mitochondria acting in heterodimers with the molecular chaperone Mdm35/TRIAP1 to shuttle lipids from the outer (OM) to the inner (IM) membrane. Three Ups/PRELI and one Mdm35/TRIAP1 proteins exist in yeast (Ups1–3, Mdm35) and mammals (PRELID1, PRELID3a, PRELID3b, TRIAP1). The Ups1/PRELID1 and Ups2/PRELID3b proteins regulate mitochondrial CL and PE metabolism, respectively, by promoting the transport of lipid substrates, PA and PS, respectively, from the OM to the IM where cardiolipin (CL) and phosphatidylethanolamine (PE) synthesizing enzymes are located (Aaltonen et al., 2016; Connerth et al., 2012; Miyata, Watanabe, Tamura, Endo, & Kuge, 2016; Osman et al., 2009). Mdm35/TRIAP1 acts as a chaperone facilitating import and stability of Ups/PRELI proteins in the IMS (Potting et al., 2013; Potting, Wilmes, Engmann, Osman, & Langer, 2010). Mdm35/TRIAP1 dissociates from Ups/PRELI upon binding to membrane enriched in anionic lipids, such as CL, allowing lipid releasing and protein recycling (Aaltonen et al., 2016; Connerth et al., 2012; Watanabe, Tamura, Kawano, & Endo, 2015). The Ups/PRELI proteins fold is characteristic of the StARkin domain superfamily with a seven-stranded β-sheet decorated with three flexible helices forming a hydrophobic cavity (Miliara et al., 2019; Watanabe et al., 2015). Interestingly, in plants, only one Ups/PRELI (At5g13070) and one Mdm35/TRIAP1 (At4g33100) homologs are found (Wideman & Muñoz-Gómez, 2016). At5g13070 protein is uncharacterized in *A. thaliana* but is predicted to be located in mitochondria IMS (Hooper et al., 2017; Schwacke et al., 2003). Sequence analysis suggests that At5g13070 is closer to the Ups2 than the Ups1 family, and might transport PS because of the absence of a conserved lysine residue (K58), present in Ups1 family, which impaired PS transport (Miliara et al., 2019; Wideman & Muñoz-Gómez, 2016). *In vitro*, yeast Ups2 can also transport PA, but *in vivo*, no decrease of CL is observed in *Δups2* showing that Ups2 does not play an essential role in PA transport (Miyata et al., 2016; Osman et al., 2009). Thus, the plant Ups/PRELI protein might be involved in the transport of both PS and PA and/or alternative pathways might exist to transport PA for CL synthesis in the IM.

2.1.6 The StART family

StART domain-containing proteins are conserved throughout evolution (Ponting & Aravind, 1999). StART family is expanded in higher plants with 35 and 29 members in *A. thaliana* and in rice, respectively, whereas 15 members are found in humans and only one in yeast (Clark, 2020; Schrick, Nguyen, Karlowski, & Mayer, 2004). StART proteins are divided into subfamilies according to their domain organization and lipid binding specificity (Clark, 2020; Schrick et al., 2004). They are able to bind different ligands such as sterols, ceramides or phospholipids (Clark, 2020). As only members of the StARD2 subfamily have been shown to transport phospholipids (Kang, Wei, & Cohen, 2010), we will focus on this subfamily (for reviews about StART proteins, see Clark, 2012, 2020). In humans, there are three phospholipid-binding StART proteins (StARD2, StARD7 and StARD10) also called PC-TPs (PC transporter proteins), which are small proteins (around 200–300 aa) only composed of a StART domain (Clark, 2020; Schrick et al., 2004). StARD2 binds and transports PC between liposomes *in vitro* (Wirtz, 1991). It folds into a nine-stranded curved β-sheet with four helices forming a binding pocket accommodating one PC molecule (Roderick et al., 2002). This protein is localized in the cytosol and the nucleus and has been shown to regulate lipid droplet (LD) size in rat liver as well as the interplay between glucose and lipid metabolism (Clark, 2020;

Kang et al., 2010; Kanno et al., 2007; Nicholls, Hornick, & Cohen, 2017). In the nucleus, StARD2 also interacts with the Homeodomain (HD) transcription factor Pax3, involved in tissues and organs development (Buckingham & Relaix, 2007), and stimulates its activity (Kang et al., 2010; Kanno et al., 2007). Thus, it seems that StARD2 is more likely a PC-sensor regulating different cellular pathways than a PC transporter between membranes. In contrast, StARD7 is involved in PC transport from the mitochondrial OM to the IM in mammals (Horibata et al., 2016; Saita et al., 2018). *Stard7* KO or KD cell lines present a variety of mitochondrial defects including a reduced respiration rate and ATP level, mtDNA damages and defects in mitochondrial cristae structure, highlighting the importance of this PC-transporter for mitochondria biogenesis and functions (Flores-Martín, Reyna, Ridano, Panzetta-Dutari, & Genti-Raimondi, 2016; Horibata et al., 2016; Saita et al., 2018; Yang et al., 2017). Two isoforms, StARD7-I, with a 75aa N-terminal mitochondrial targeting sequence, and StARD7-II, co-exist in cells and are able to bind PC and ceramides, but only PC is transported between liposomes (Bockelmann et al., 2018; Horibata & Sugimoto, 2010; Saita et al., 2018). StARD10 is able to bind and transport PC and PE *in vitro* and its transport activity is inhibited by phosphorylation, showing that lipid transfer and membrane binding can be regulated by post-translational modifications (Olayioye et al., 2005, 2007). In lung alveolar type II epithelial cells, StARD10 interacts with the ER lyso-PC acyltransferase 1 (LPCAT1) and promotes lipid transport on lamellar bodies (LB) where lipids are stored until secretion in the lung surfactant (Goss, Hunt, & Postle, 2013; Lin, Ikegami, Moon, Naren, & Shannon, 2015).

Six homologs of the PC-PT family can be found in *A. thaliana* (Table 4). They are longer than their yeast and human counterparts (around 400aa) and predicted to possess one or several TM domains (Schrick et al., 2004). TM domain-containing StART proteins are also found in rice and chickpea (Satheesh et al., 2016; Schrick et al., 2004). They are expressed at different stages of plant development and expression of some of them are induced or repressed in response to biotic or abiotic stress (Satheesh et al., 2016). Information about the function, localization and lipid binding properties of these proteins is scarce. 3D homology modeling of the At3g13062 shows that the StART domain overlaps with the crystal structure of StARD2 despite a low primary sequence similarity (Schrick et al., 2004). None of them is predicted to be targeted in mitochondria, questioning the existence of a StARD7 homolog in plants. Concerning the other StART domain-containing proteins, five members in *A. thaliana* have in C-terminal a

Table 4 Genes of the StAR related transfers (StART) family present in the *A. thaliana* genome.

AGI	Name	Localization	Domain organization	References
At1g05230		Nucleus (prediction)	HD-ZLZ-StART	
At1g17920		Nucleus (prediction)		
At1g34650		Nucleus (prediction)		
At1g73360	HDG11	Nucleus		Khosla et al. (2014)
At1g79840	GL2	Nucleus		Schrick et al. (2014)
At2g32370		Nucleus (prediction)		
At3g03260		Nucleus (prediction)		
At3g61150		Nucleus (prediction)		
At4g00730	ANL2, AHDP	Nucleus (prediction)		
At4g04890	PDF2	Nucleus (prediction)		
At4g17710		Nucleus (prediction)		
At4g21750	ATML1	Nucleus (prediction)		
At4g25530	FWA	Nucleus (prediction)		
At5g17320		Nucleus (prediction)		
At5g46880		Nucleus (prediction)		
At5g52170		Nucleus (prediction)		

Continued

Table 4 Genes of the StAR related transfers (StART) family present in the *A. thaliana* genome.—cont'd

AGI	Name	Localization	Domain organization	References
At1g30490	PHV-ATHB9	Nucleus (prediction)	HD-ZIP-StART	
At1g52170		Nucleus (prediction)		
At2g34710	PHB, ATHB-14	Nucleus		Lee et al. (2006)
At4g32880	ATHB-8	Nucleus		Lee et al. (2006)
At5g60690	REV, IFL1	Nucleus		Lee et al. (2006)
At2g28320		Cyto/nucleus (prediction)	PH-StART-DUF1336	
At3g54800		Cyto/Mito/ soluble (prediction)		
At4g19040	EDR2	ER, PM, endosomes		Vorwerk et al. (2007)
At5g45560		Mito (prediction)		
At5g35180		Cyto/plaste/ nucleus (prediction)	StART-DUF1336	
At1g55960	Putatives PC-PT	PM (prediction)	TM-StART	
At1g64720		Vacuole (prediction)	TM-StART	
At3g13062		PM (prediction)	TM-StART	
At3g23080		PM (prediction)	TM-StART-TM	
At4g14500		PM (prediction)	TM-StART-TM	
At5g54170		PM (prediction)	TM-StART-TM	
At4g26920		PM (prediction)	TM-StART	
At5g07260		Cyto/Mito (prediction)	StART	
At5g49800		Nucleus/ Chloroplast (prediction)	StART	

Protein localization were determined experimentally or the consensus localizations from SUBA and ARAMEMNON websites (prediction) are indicated (Hooper et al., 2017; Schwacke et al., 2003). T-he domain organization found for each protein is also indicated. HD: homeodomain, PH: pleckstrin homology domain, TM: transmembrane domain, ZIP: leucine zipper, ZLZ: leucine zipper-loop-zipper.

domain of unknown function (DUF1336), which is plant-specific (Table 4) (Schrick et al., 2004). One member, EDR2, is a negative regulator of cell death during pathogen attack and binds to PI-4-monophosphate (PI(4)P) *via* its PH domain (Vorwerk et al., 2007), but the role of PI(4)P binding or of the StART domain has not been further investigated. It has to be noted that plants possess a unique subfamily of transcription factors composed of a StART domain associated to an homeodomain (HD) (21 members in *A. thaliana*, Table 4) (Schrick et al., 2004). In accordance with their putative role in the regulation of gene expression, they are predicted to be located in the nucleus and this localization was confirmed for some of them (Khosla et al., 2014; Lee et al., 2006; Schrick et al., 2014). Mutant analyses highlighted the importance of these transcription factors in plant development (Schrick et al., 2014). However, the role of the StART domains in the regulation of plant development, as well as their ligands, are still poorly understood. The StART domain of Glabra 2 (GL2), is required for the function of this transcription factor but not for its nuclear localization (Schrick et al., 2014). Interestingly, the StART domain of the mouse cholesterol transporter StAR can functionally replace GL2 StART domain and a mutation affecting StAR cholesterol binding site prevents functional complementation, indicating that sterol binding to GL2 is required for its transcriptional activator function (Schrick et al., 2014). In the same study, the authors analyzed the metabolites associated to the StART domain of three *A. thaliana* proteins (GL2, PDF2 and At3g13062) expressed and purified from yeast cells. Most of the detected metabolites are lipids. GL2 and PDF2 are associated to similar metabolites including lyso-PC, PC, DAG or sphingolipids whereas the putative PC-PT At3g13062 was associated to DAG and TAG (Schrick et al., 2014). However, because the study was performed in yeast, which contains only one sort of sterol (ergosterol) and missed some plant-specific lipids such as galactoglycerolipids, the lipid-binding ability of these *A. thaliana* StART domains might be misevaluated. Because of the association of a StARkin with a HD domain, regulation of gene expression by these plant-specific transcription factors might involve lipid binding.

2.2 ORP

ORPs (Oxysterol-binding protein (OSBP) related proteins), also called Osh (Oxysterol-binding protein homolog), are a widespread family of LTPs that regulate sterol and phospholipid homeostasis (Lipp et al., 2020; Moser von

Filseck & Drin, 2016; Pietrangelo & Ridgway, 2018; Santos, Girik, & Nunes-Hasler, 2020). Twelve and seven homologs are found in human and yeast genomes, respectively, and most of them are located at MCSs and are able to bind and transport sterols and PIPs between different organelles (Moser von Filseck & Drin, 2016; Pietrangelo & Ridgway, 2018; Schulz & Prinz, 2007). Some of them, Osh6/7 in yeast and their homologs in human ORP5/8, are also able to bind and transport PS and will be further described below (Lipp et al., 2020; Santos et al., 2020). ORP proteins all contain an ORD (oxysterol-binding related domain), which folds into a hydrophobic β-barrel accommodating one lipid molecule with a flexible N-terminal lid that regulates the opening/closing of the tunnel entry (Dong, Wang, Ye, & Zhang, 2020; Im, Raychaudhuri, Prinz, & Hurley, 2005; Maeda et al., 2013). The ORD domain can be associated to other domains regulating membrane binding and protein sub-cellular localization, such as a PH domain, a FFAT (Phenylalanine-Phenylalanine Acidic Track) motif targeting proteins to ER, or a TM domain (Pietrangelo & Ridgway, 2018; Schulz & Prinz, 2007). Interestingly, some of them (ORP5/8 and OSBP in human and Osh4/6/7 in yeast) use an original counter-exchange mechanism to transport lipids from one membrane to another against their concentration gradient (Chung et al., 2015; de Saint-Jean et al., 2011; Lipp et al., 2020; Moser von Filseck et al., 2015; Moser von Filseck & Drin, 2016). Osh6/7 and ORP5/8 are located at ER-PM MCSs where they play a role in PS transport to PM, PS being enriched in the cytosolic leaflet. Those proteins are able to bind PS or PI(4)P, which are abundant in the PM, by their ORD domain in a mutually exclusive manner (Maeda et al., 2013; Moser von Filseck et al., 2015). *In vivo*, Osh6/7 and ORP5/8 extract PS from its site of synthesis, the ER, to shuttle and release it into the PM where PI(4)P is extracted and shuttled back to the ER (Maeda et al., 2013; Moser von Filseck et al., 2015). The hydrolysis of PI(4)P by the ER phosphatase Sac1 is suspected to provide sufficient energy for PS transport and to maintain low PI(4)P content in the ER compared to PM, favoring directional PS transport (Lipp et al., 2020; Moser von Filseck & Drin, 2016). Interestingly, ORP5/8 are also located at ER-mitochondria and ER-LD MCSs. ORP5 interacts with LD *via* an amphipathic helix in the ORP domain and has been shown to regulate LD PS and PI(4)P content and LD size (Du et al., 2020). At ER-mitochondria MCSs, ORP5 and ORP8 interact with each other and with PTPIP51, an OM protein involved in mitochondria-ER tethering *via* its interaction with the ER protein VAP-B (Galmes et al., 2016; Stoica et al., 2014). ORP5/8 are involved in PS transport at ER-mitochondria to fulfill PE synthesis in the IM by PSD1 (Galmes et al., 2016; Rochin et al., 2020). Interestingly, most of the ORP5 proteins located at ER-mitochondria MCSs are close to cristae junctions (CJ), which are OM-IM MCSs maintained by the IM MICOS complex (Mukherjee, Ghosh, & Meinecke, 2021; Rochin et al., 2020), suggesting that ORP5/8 might link PS transport from ER to the mitochondrial IM. As PIPs are barely detected in mitochondria, the transport of PS might be directed by a concentration gradient of PS rather than by a counter-exchange mechanism. In addition to its role in PS transport to mitochondria and PM, ORP5 is also involved in cholesterol transfer from low density lipoproteins (LDL) to late endosomes/lysosomes in Niemann Pick C (NPC) fibroblasts (Du et al., 2011) and regulates Ca^{2+} accumulation in the mitochondria and in the PM microdomains called caveolae (Pulli et al., 2018; Rochin et al., 2020). This example shows the versatility of functions and localizations associated to ORP proteins.

A. thaliana and rice genomes encode 12 and six ORP domain-containing proteins, respectively (Skirpan et al., 2006; Umate, 2011; Ye, Ji, Guo, & Jiang, 2020). *A. thaliana* proteins cluster in four groups and five proteins contain a PH domain and a FFAT motif that might be involved in their localization to ER and endomembrane compartments whereas the other members only have an ORD domain (Table 5) (Ye et al., 2020). Interestingly, when phylogeny analyses are performed with yeast and mammalian proteins, no *A. thaliana* proteins group with the PS transfer proteins ORP5/8 and Osh6/7 (Skirpan et al., 2006; Ye et al., 2020), suggesting that ORP proteins might not transport PS in plants or that this transport is mediated by a more distantly related family of ORPs. All ORPs are expressed in different tissues and developmental stages, ORP3 group being the most highly expressed (Skirpan et al., 2006; Umate, 2011). Most of the ORPs genes are induced by biotic and/or abiotic stress as well as in response to toxic compounds such as metals (Umate, 2011). In accordance with this, GmOSBP is highly induced during high salt stress and senescence and its PH domain targets GFP protein to the nucleus in onion epidermal cells, suggesting the protein might be involved in gene expression regulation in soybean (Li et al., 2008). In petunia, PiORP1 is highly expressed in pollen tubes where it interacts with the receptor-like kinase PRK1, a protein essential for pollen tube development, leading to PiORP1 phosphorylation (Skirpan et al., 2006). PiORP1-GFP localizes in the cytosol and forms PH domain-dependent punctates in the PM, suggesting a localization at PM MCSs. However, the lipid binding specificity of PiORP1 and its role in pollen tube development, maybe to provide lipids to PM for elongation, is

Table 5 Genes of the OSBP related protein (ORP) family present in the *A. thaliana* genome.

AGI	Name	Localization	Domain organization	References
At2g31020	ORP1A	Cyto/Chloro (prediction)	PH-ORD	
At2g31030	ORP1B	Cyto (prediction)	ORD	
At4g08180	ORP1C	Nucleus/Chloro (prediction)	PH-ORD	
At1g13170	ORP1D	Nucleus/Chloro (prediction)	PH-ORD	
At4g22540	ORP2A	Nucleus (prediction)	PH-ORD	
At4g12460	ORP2B	Nucleus (prediction)	PH-ORD	
At5g02100	ORP3A	ER	ORD	Saravanan et al. (2009)
At3g09300	ORP3B	Cyto (prediction)	ORD	
At5g59420	ORP3C	Cyto (prediction)	ORD	
At4g25860	ORP4A	Cyto (prediction)	ORD	
At4g25850	ORP4B	Cyto (prediction)	ORD	
At5g57240	ORP4C	Cyto (prediction)	ORD	

Protein localization were determined experimentally or the consensus localizations from SUBA and ARAMEMNON websites (prediction) are indicated (Hooper et al., 2017; Schwacke et al., 2003). The domain organization found for each protein is also indicated. ORD: oxysterol-binding related domain, PH: pleckstrin homology domain.

still unknown. In *A. thaliana*, ORP3A is localized in the ER *via* a WFDE motif required for its interaction with the ER VAP protein PVA12 (Saravanan et al., 2009). When this interaction is disrupted, ORP3A localizes to the Golgi compartment. *In vitro*, ORP3A ORD domain binds sitosterols but PVA12 interaction as well as the ER localization of ORP3A are independent of its ability to bind lipids (Saravanan et al., 2009). All these results suggest that ORP3A could shuttle sitosterol from ER to Golgi (Fig. 3B). Plants ORP proteins could also be involved in the replication of some viruses that require specific lipids or membrane compartments for replication (Barajas et al., 2014; Nagy, 2015; Rodriguez-Peña, Mounadi, & Garcia-Ruiz, 2021). For instance, the bushy stunt virus (TBSV), a plant (+)RNA virus belonging to the Tombusvirus family, induces the formation of specific compartments arising from peroxisome membrane invagination, called spherules. Within spherules, the assembly of the virus replicase complex (VRC) and the replication process occur, protected from degradation (Nagy, Barajas, & Pogany, 2012; Rodriguez-Peña et al., 2021). The TBSV protein p33 is involved in the recruitment of the VRC to peroxisome membrane. Using a yeast model system to investigate the mechanism of TBSV replication, it has been shown that sterols are redistributed from the PM to the virus replication site and that sterols are essential for efficient replication of the virus (Barajas et al., 2014). Interestingly, p33 interacts with six *A. thaliana* ORPs (ORP3A-C and ORP4A-C) as well as with five VAP proteins and ORP3A and PVA12 have been shown to co-localize with p33 close to VRCs in *Nicotiana benthamiana* (Barajas et al., 2014). In both yeast and plant systems, the presence of a replicative TBSV virus induces ER-peroxisomes MCSs formation close to spherules, p33 being located in both peroxisomes and ER membranes (Barajas et al., 2014). This suggests that p33 induces peroxisome-ER MCSs formation and recruits ORP proteins to promote sterol transfer from ER to peroxisomes and to stimulate TBSV replication and spherule formation.

2.3 VPS13-like protein family/Chorein_N domain-containing proteins

VPS13-like proteins are highly conserved from yeast to human. This protein family is composed of two main protein types: the huge (>3000 aa) vacuolar protein sorting 13 (VPS13) and the autophagy-related 2 (ATG2) proteins. Among all the functional domains found in VPS13-like proteins, the N-terminal Chorein_N domain and the adjacent VPS13_N domain, both involved in lipid binding, are the best conserved (De et al., 2017; Pfisterer et al., 2014). In VPS13 proteins, these domains are usually followed by a long region containing repeated coiled regions and the characteristic VPS13 adaptor binding (VAB) domain. Their C-terminus is typically composed of an APT1-homology domain, an ATG-homology domain (ATG_C) and finally a C-terminal PH domain. Most of these domains are involved in the subcellular targeting of VPS13 proteins (Dziurdzik & Conibear, 2021). VPS13-like proteins associate to membranes and MCSs and are implicated in a large variety of cellular processes like sporulation and *trans*-Golgi network homotypic fusion in yeast (De et al., 2017; Park & Neiman, 2012; Rabitsch et al., 2001) or regulation of autophagy, cytoskeletal architecture and cell survival in humans (Föller et al., 2012; Muñoz-Braceras, Calvo, & Escalante, 2015). The *Saccharomyces cerevisiae* genome contains only one VPS13 protein, which localizes to nuclear-vacuolar junctions, mitochondria-endosome and vacuole-mitochondria MCSs depending on the cell physiological condition (Bean et al., 2018). Four homologs have been identified in human which also localize to different MCSs (for recent review, see Dziurdzik & Conibear, 2021). Yeast and human genomes encode one and two ATG2 protein, respectively, that are localized to ER-autophagosome MCSs (Graef, Friedman, Graham, Babu, & Nunnari, 2013; Suzuki, Akioka, Kondo-Kakuta, Yamamoto, & Ohsumi, 2013) and are part of the autophagosome forming machinery (Nakatogawa, 2020). Both VPS13 and ATG2 proteins play a major role in lipid transport. *In vitro*, the N-terminal parts of *Chaetomium thermophilum* VPS13 and human ATG2A are able to bind at least 10 glycerolipids and to transport them between liposomes, strongly suggesting they are non-specific glycerolipid transporters (Kumar et al., 2018; Valverde et al., 2019). Structural analyses show that the N-terminal extremity containing the Chorein_N domain forms a scoop-like hydrophobic cavity that could presumably accommodate glycerolipids and that is extended by a long hydrophobic groove with a rod-like β-strand structure (Kumar et al., 2018; Li, Lees, Lusk, & Reinisch, 2020; Osawa et al., 2019; Valverde et al., 2019). Due to their localization at MCSs, their lipid-friendly elongated groove structure and their size exceeding 20 nm (De et al., 2017), it has been broadly hypothesized that VPS13 proteins form bridges between membranes at MCSs to mediate non-specific bulk glycerolipid transport by a bridge-like mechanism (Fig. 2E) (Dziurdzik & Conibear, 2021; Kumar et al., 2018; Li et al., 2020; Reinisch & Prinz, 2021).

Nothing is known about VPS13-like plant homologs. In *A. thaliana*, five Chorein_N domain-containing proteins have been identified, three of which have already been previously identified (Table 6) (Michaud et al., 2017; Velayos-Baeza, Vettori, Copley, Dobson-Stone, & Monaco, 2004). Three unknown proteins (At4g17140, At1g48090 and At5g24740) harbor the typical VPS13 protein conserved domains, especially the Chorein_N domain and the VPS13_N domain present at their N-terminal extremity. Given the high degree of homology found with other VPS13 proteins, we propose to rename At4g17140, At1g48090 and At5g24740, respectively, AtVPS13A, AtVPS13B and AtVPS13C. Currently, nothing is known about AtVPS13A-C cellular localization. Predictions suggest a localization in the Golgi for all three proteins. Surprisingly, AtVPS13A and AtVPS13B have an additional VPS62-homology domain of unknown function in the middle of the protein. AtVPS13A has an additional predicted PH domain in the repeating coiled region and AtVPS13B harbors a C2 domain just before the VAB domain while AtVPS13C lacks the C-terminal PH domain

Table 6 Genes of the VPS13-like/Chorein_N family present in the *A. thaliana* genome.

AGI	Name	Localization	Domain organization
At4g17140	AtVPS13A	Golgi (prediction)	Chorein–VPS13_N–RCR–PH–RCR–VPS62–VAB–APT1–ATG_C–PH
At1g48090	AtVPS13B	Golgi (prediction)	Chorein–VPS13_N–RCR–VPS62–C2–VAB–APT1–ATG_C–PH
At5g24740	AtVPS13C, SHBY	Golgi (prediction)	Chorein–VPS13_N–RCR–RCR–VAB–APT1–ATG_C
At3g19190	ATG2/ PEUP1	Nucleus/Golgi (prediction)	Chorein–VPS13_N–ATG2_CAD–ATG_C
At3g20720		Mitochondria/ Cytosol (prediction)	Chorein–VPS13_N

Protein localization corresponds to the consensus localizations from SUBA and ARAMEMNON websites (prediction) (Hooper et al., 2017; Schwacke et al., 2003). The domain organization found for each protein is also indicated. APT1: Aberrant pollen transmission 1, PH: pleckstrin homology domain, RCR: Repeated coiled region, VAB: Vps13 adaptor binding.

(Table 6). The conservation of the domains involved in VPS13 protein targeting in yeast and mammals suggest that plant VPS13 proteins are also associated to multiple cell compatments, including MCSs. In addition, the presence of slight differences in domain organization and the presence of additional domains, like C2 or VPS62-like domains, preclude a higher complexity in the pattern of VPS13 protein localization and their regulation in plant cells. Interestingly, AtVPS13A/At4g17140 was identified in the mitochondrial OM proteome of *A. thaliana* (Duncan et al., 2011). Even though the GFP fusions of the 200 N-terminal and C-terminal amino acid residues failed to give a consistent localization, this study gives a first hint for the potential mitochondrial localization of AtVPS13A. AtVPS13A was also found in the *A. thaliana* mitochondrial complexome and migrated at a molecular weight of 650 kDa on BN PAGE (Senkler et al., 2017). A FFAT motif localized in the N-terminal region of AtVPS13A was also identified, suggesting an association with the ER membrane (Slee & Levine, 2019). In light of these preliminary results, AtVPS13A-C seem to be good candidates as novel VPS13-family LTPs in plants and require further investigation. At3g19190 corresponds to a plant ATG2, also known as PEUP1. Its domain organization and its cellular functions are similar to yeast and mammalian ATG2 proteins (Chen, Dong, & Wang, 2021; Ding, Zhang, & Otegui, 2018).

2.4 The Sec14-superfamily

The Sec14 domain is widely conserved in Eukaryotes and the Sec14-superfamily is expanded in higher Eukaryotes where more than 20 members can be found in mammals and plants whereas yeast contains only six Sec14 proteins (Bankaitis, Mousley, & Schaaf, 2010). In lower Eukaryotes, proteins of this family usually contain only one Sec14 domain whereas in higher Eukaryotes, the Sec14 domain is mainly associated to other domains including the GOLD (Golgi dynamics) domain, the PRELI domain or the plant-specific nodulin domain (Bankaitis et al., 2010). Because of their high number and huge diversity in domain organization, the Sec14 proteins are involved in a wide range of cellular processes including membrane trafficking, lipid metabolism and signaling pathways by regulating PIPs levels and the activity of different GTPases (Bankaitis et al., 2010). Sec14 domains fold in an approximatively 280 aa globular structure with two lobes, formed by an assembly of α-helices and one β-sheet (for a review, see Bankaitis et al., 2010). The lipid-binding pocket is composed of a β-sheet defining the pocket floor and of α-helices forming the wall, the number of β-strands and the orientation and number of α-helices varying among the Sec14 proteins. A helical flexible gate regulates the entry of the cavity with an "open" conformation in absence of lipids and a "closed" conformation when a lipid molecule is inserted in the binding pocket (Bankaitis et al., 2010). Sec14 proteins are able to bind and transport PI, as well as a second ligand that can be diverse hydrophobic compounds like PC, PS, PE, sterols, squalene or α-tocopherol (Li et al., 2000; Schaaf et al., 2008; Tripathi et al., 2019). The role and mechanisms of action of Sec14 PIPTs have been mostly studied in yeast where they are involved in PIPs metabolism and signaling, vesicular trafficking, lipid droplet biogenesis or PS synthesis (Lipp et al., 2020; Ren et al., 2014; Schaaf et al., 2008). It appears that Sec14 proteins are likely involved in regulatory processes rather than in interorganellar bulk lipid transport. The most studied protein is yeast Sec14, that regulates PIPs metabolism and vesicular trafficking at the *trans*-Golgi network (TGN) (for reviews, see Bankaitis et al., 2010; Grabon, Khan, & Bankaitis, 2015; Lipp et al., 2020). Sec14 binds and transports PC and PI in an ATP-independent manner and regulates PI(4)P level in the Golgi by transporting PI in this organelle and promoting its phosphorylation by the PI(4) kinase Pik1 (Bankaitis, Aitken, Cleves, & Dowhan, 1990; Lipp et al., 2020). The delivery of PI to the Golgi is coupled to the exchange with one PC molecule (Schaaf et al., 2008). From structural analyses of PC- or PI-bound Sec14 and Sfh1, the closest Sec14 homolog in yeast, it was possible to establish PC- and PI-binding barcodes, corresponding to aa residues establishing specific interactions with PI or PC polar head (Lipp et al., 2020; Schaaf et al., 2008). The main role of Sec14 is not to transport PI *per se*, but to render PI accessible to Pik1 for phosphorylation which otherwise happens with very low efficiency when PI is buried in a lipid bilayer (Lipp et al., 2020; Schaaf et al., 2008). This mechanism is called the "presentation model" in the literature (Kf de Campos & Schaaf, 2017; Lipp et al., 2020). It has to be noticed that efficient PI presentation to Pik1 requires PC binding, explaining why PC transfer activity is required for regulation of PIPs signaling by Sec14 and why PC metabolism is coupled to this signaling pathway (Grabon et al., 2015; Kf de Campos & Schaaf, 2017; Schaaf et al., 2008). *A. thaliana* and rice genomes encode for 32 and 27 Sec14 domain-containing proteins, respectively (Table 7) (Huang, Ghosh, & Bankaitis, 2016). They are classified into three different families depending on domain organization and sequence similarities.

2.4.1 The Sec14-nodulin family

Fourteen proteins (AtSfh1–14) in *A. thaliana* present a Sec14 domain fused to a nodulin domain only found in plants (Huang, Ghosh, & Bankaitis, 2016; Montag, Hornbergs, Ivanov, & Bauer, 2020; Vincent et al., 2005). The Sec14-nodulin organization is found in angiosperms and some gymnosperms but is absent from more primitive plants such as moss and fern (Huang, Ghosh, & Bankaitis, 2016; Huang, Ghosh, Tripathi, et al., et al., 2016; Montag et al., 2020). Most of these proteins are highly expressed in tissues where extensive polarized membrane growth is required, such as in pollens or roots (Ghosh et al., 2015; Vincent et al., 2005). It is predicted to adopt a α-helical coiled-coil structure composed of three antiparallel α-helices (Ghosh et al., 2015). 3D modeling of AtSfh1 Sec14 domain shows a similar fold as for the yeast Sec14 protein and highlights a conservation of the yeast PI- and PC-binding barcodes among the Sec14-nodulin proteins (Huang, Ghosh, & Bankaitis, 2016; Huang, Ghosh, Tripathi, et al., 2016). The ability of AtSfh1 and other Sec14-nodulin proteins to transport both PI and PC was confirmed by complementation of the yeast *sec14* mutant combined with *in vitro* transfer assays (Böhme et al., 2004; Huang, Ghosh, & Bankaitis, 2016; Huang, Ghosh, Tripathi, et al., 2016; Vincent et al., 2005). These results suggest that this family might also regulate PIPs signaling by stimulating PIPs kinase activity. The occurrence of this mechanism for plant Sec14-nodulin proteins was further supported by

Table 7 Genes of the Sec14 super-family present in the *A. thaliana* genome.

AGI	Name	Sub-family	Localization	Domain organization	References
At4g34580	AtSFH1/COW1	Sec14-nodulin	PM/cytosol	Sec14-nodulin	Böhme et al. (2004) and Vincent et al. (2005)
At4g39180	AtSFH2		Cytosol (Prediction)	Sec14	
At2g21540	AtSFH3		Cytosol (Prediction)	Sec14-nodulin	
At1g19650	AtSFH4		Cytosol (Prediction)	Sec14-nodulin	
At1g75370	AtSFH5		Cytosol (Prediction)	Sec14-nodulin	
At4g39170	AtSFH6		Cytosol (Prediction)	Sec14-nodulin	
At2g16380	AtSFH7		Cytosol (Prediction)	Sec14-nodulin	
At2g21520	AtSFH8		Cytosol (Prediction)	Sec14-nodulin	
At3g24840	AtSFH9		Cytosol (Prediction)	Sec14-nodulin	
At2g18180	AtSFH10		Cytosol (Prediction)	Sec14-nodulin	
At5g47510	AtSFH11		Cytosol (Prediction)	Sec14	
At4g36490	AtSFH12		PM (Prediction)	Sec14-nodulin	
At1g55690	AtSFH13		Nucleus (Prediction)	Sec14-nodulin	
At5g56160	AtSFH14		Cytosol (Prediction)	Sec14	
At1g72150	AtPATL1	Patellin	Cytosol/Cell plate/PM	Sec14-GOLD	Peterman, Sequeira, Samia, and Lunde (2006), Tejos et al. (2018), and Wu et al. (2017)
At1g22530	AtPATL2		Cytosol/Cell plate/PM	Sec14-GOLD	Suzuki et al. (2016), Tejos et al. (2018), and Wu et al. (2017)

At1g72160	AtPATL3		Cell periphery/PM/cytosol/Cell plate	Sec14-GOLD	Peiro et al. (2014), Tejos et al. (2018), and Wu et al. (2017)
At1g30690	AtPATL4		Cytosol/Cell plate/PM	Sec14-GOLD	Tejos et al. (2018) and Wu et al. (2017)
At4g09160	AtPATL5		PM	Sec14-GOLD	Tejos et al. (2018)
At3g51670	AtPATL6		Cell periphery/PM/cytosol/Cell plate	Sec14-GOLD	Peiro et al. (2014), Tejos et al. (2018), and Wu et al. (2017)
At4g08690	AtPITP1	Stand-alone Sec14	Cytosol (Prediction)	Sec14	
At1g22180	AtPITP2		Cytosol (Prediction)	Sec14	
At4g36640	AtPITP3		Cytosol (Prediction)	Sec14	
At1g75170	AtPITP4		Cytosol (Prediction)	Sec14	
At1g05370	AtPITP5		Cytosol (Prediction)	Sec14	
At3g22410	AtPITP6		Cytosol (Prediction)	Sec14	
At5g63060	AtPITP7/CPSFL1		Plastids	Sec14	Hertle et al. (2020)
At1g01630	AtPITP8		Cytosol (Prediction)	Sec14	
At1g14820	AtPITP9		Cytosol (Prediction)	Sec14	
At3g46450	AtPITP10		Cytosol (Prediction)	Sec14	
At1g55840	AtPITP11		Cytosol (Prediction)	Sec14	
At5g47730	AtPITP12		Cytosol (Prediction)	Sec14	

Protein localization were determined experimentally or the consensus localizations from SUBA and ARAMEMNON websites (prediction) are indicated (Hooper et al., 2017; Schwacke et al., 2003). The domain organization found for each protein is also indicated. GOLD: Golgi dynamic.

functional analysis of AtSfh1, a protein mainly expressed during and involved in root hair elongation (Böhme et al., 2004; Vincent et al., 2005). During root hair elongation, a huge polarized transport of vesicles to the root tip elongation zone is required and involves the maintenance of a concentration gradient of PI(4,5)P_2 in the PM along the elongating root hair. AtSfh1 associates specifically to PI(4,5)P_2 in PM *via* a polybasic stretch located at the C-terminal of the nodulin domain (Ghosh et al., 2015; Vincent et al., 2005). The binding of AtSfh1 to PM and its oligomerization *via* the nodulin domain were suggested to promote the formation of microdomains enriched in PI(4,5)P_2 and to maintain the PI(4,5)P_2 gradient (Ghosh et al., 2015; Huang, Ghosh, & Bankaitis, 2016). In addition, AtSfh1 might also regulate the level of PI(4,5)P_2 by stimulating PI(4) kinase activity and PI(4)P synthesis, which can be subsequently used as a substrate for PI(4,5)P_2 synthesis. Because the Sec14 domain of several Sec14-nodulin proteins are able to complement yeast phenotype and to transport both PI and PC, it is likely that the regulation of PIPs metabolism by the presentation model is a general mechanism of action of Sec14-nodulin proteins (Huang, Ghosh, & Bankaitis, 2016; Huang, Ghosh, Tripathi, et al., 2016; Kf de Campos & Schaaf, 2017).

2.4.2 The Patellin family

Patellins are characterized by the association of the Sec14 domain to a GOLD domain and are present in plants as well as in metazoans. In plants, Patellins are found in angiosperms and gymnosperms and *A. thaliana* genome encodes six Patellin genes (AtPATL1–6) (Table 7) (Peterman et al., 2006; Zhou et al., 2019). Patellins have a variable N-terminal extension differing in both length and aa composition (Peterman et al., 2006; Peterman, Ohol, McReynolds, & Luna, 2004). They are expressed in many tissues during plant development, in particular in cells with high mitotic activity, and are responsive to biotic and abiotic stress (Hua-Ying et al., 2019; Peterman et al., 2004; Tejos et al., 2018). Patellins do not contain TM domains and are found to be cytosolic and associated to membranes, in particular to PM (Peterman et al., 2004; Suzuki et al., 2016; Tejos et al., 2018; Wu et al., 2017). In dividing cells, AtPATLs are also localized to the cell plate, suggesting a role during cytokinesis (Peterman et al., 2004; Tejos et al., 2018). The PI-binding barcode, but not the PC-binding barcode, is conserved among the *A. thaliana* Patellins, suggesting they can transport PI and probably other yet unidentified substrates other than PC (Huang, Ghosh, & Bankaitis, 2016; Huang, Ghosh, Tripathi, et al., 2016; Peterman et al., 2004). All Patellins studied until now are able to bind several PIPs *in vitro* (Montag et al., 2020; Peterman et al., 2004; Suzuki et al., 2016). However, the association of AtPATL2 and AtPATL3 with the PM is mediated by protein-protein interactions *via* the GOLD domain and not by PIPs binding (Suzuki et al., 2016; Wu et al., 2017). Information concerning the roles of Patellins during plant development and in response to stress is scarce. AtPATL1 is involved in cold and salt stress resistance (Chu et al., 2018; Zhou et al., 2018) whereas AtPATL3 and 6 are involved in plant resistance to Alfalfa mosaic virus (AMV) by interacting with the AMV movement protein, impairing viral particle targeting to PDs and cell-to-cell movement (Peiro et al., 2014). A role in membrane trafficking during cytokinesis, salt stress response, vacuole biogenesis and autophagy is also suspected. AtPATLs are recruited to cell plate during cytokinesis, but their role in the formation or maturation of such structure is unknown (Peterman et al., 2004; Tejos et al., 2018). AtPATL1 is overexpressed in response to high salts and interacts with the PM Na^+ transporter SOS1, reducing Na^+ transport and likely promoting salt stress tolerance (Zhou et al., 2018). Finally, AtPATL1 and 2 interact with AMSH3, a deubiquitinase enzyme involved in vesicular trafficking, vacuole biogenesis and autophogosome degradation (Isono et al., 2010; Katsiarimpa et al., 2013). Because of their ability to bind a wide range of PIPs, roles in cell signaling rather than bulk lipid transport are suspected (Zhou et al., 2019). Unfortunately, the molecular mechanisms involved in the regulation of Patellin function and whether their putative lipid transfer activity is required to sustain this function is still unknown.

2.4.3 Stand-alone Sec14

A. thaliana contains 12 stand-alone Sec14 proteins that can be divided into five clades that might define different biochemical properties and biological functions (Table 7) (Huang, Ghosh, & Bankaitis, 2016). A few proteins of this huge family were characterized and they seem to play a role in chloroplast development as well as in biotic and abiotic stress response (Hertle et al., 2020; Kearns et al., 1998; Kiba et al., 2014). Most of these proteins in plants were first identified by their ability to complement the yeast *sec14* mutant. This includes AtPITP11 (Jouannic et al., 1998), soybean Ssh1 and Ssh2 (Kearns et al., 1998) and *N. benthamiana* NbSec14 (Kiba et al., 2012). AtPITP11 is mainly expressed in leaves and flowers and transport PI *in vitro* (Jouannic et al., 1998), but no further information about its subcellular localization and function are available. Soybean Ssh1 is mainly expressed in leaves and roots whereas Ssh2 shows the highest expression level in developing seeds (Kearns et al., 1998). Both proteins are membrane associated but only Ssh2 is able to transport PI, but not PC, *in vitro*, whereas Ssh1 was not able transport PI and PC in the tested condition, suggesting it has no lipid transport activity or transports other unidentified molecules (Kearns et al., 1998). Both proteins bind PI(4,5)P_2 and Ssh1 also associates to PI(3,5)P_2 *in vitro* whereas Ssh2 seems to associate to PS and PA (Kearns et al., 1998). In response to hyperosmotic stress, a rapid increase of Ssh1 phosphorylation mediated by the salt-induced SPK1 and SPK2 kinases was observed, decreasing Ssh1 affinity for membranes (Monks, Aghoram, Courtney, DeWald, & Dewey, 2001). As Ssh1 increases *in vitro* PI(3) kinase and PI(4) kinase activities, it was proposed that Ssh1 regulates PIPs signaling in response to osmotic stress in soybean *via* modulation of PI kinases activity (Monks et al., 2001). NbSec14, a stand-alone Sec14 protein able to transfer PI and PC *in vitro*, is involved in lipid signaling during plant immune response to the bacteria *Ralstonia solanacearum* (Kiba et al., 2012, 2014). NbSec14 is upregulated during infection and NbSec14 siRNA plants show a higher susceptibility to *R. solanacearum*. In these lines, no increase in PA and DAG levels, two key lipids involved in plant immune response signaling, was observed upon infection, in accordance with a decreased DAG kinase, phospholipase D and phospholipase C activities in total plant extracts (Kiba et al., 2014). More recently, the first identified chloroplastic Sec14 protein, AtPITP7 (also called CPSFL1), was shown to be involved in chloroplast development and photoautotropic growth (Hertle et al., 2020). AtPITP7 accumulates in plastid stroma vesicles and is involved in vesicular trafficking required for thylakoid biogenesis. *In vitro*, AtPITP7 binds PA, SQDG and PIPs but is only able to extract and transfer PI(4)P between liposomes (Hertle et al., 2020). Thus, this protein might regulate lipid signaling between the inner envelope of plastids and thylakoids.

2.5 nsLTPs

The non-specific LTP (nsLTP) is a land plant-specific class of LTPs that can bind a wide variety of substrates *in vitro*. They are encoded by large gene families, with 79 and 77 members in *A. thaliana* and rice, respectively (Salminen, Blomqvist, & Edqvist, 2016). NsLTPs are small basic proteins (7–10 kDa) characterized by the presence of eight conserved Cys residues. They fold into a compact globular structure composed of four α-helices with a C-terminal unstructured tail (for detailed reviews, see Carvalho & Gomes, 2007; Salminen et al., 2016). NsLTPs possess an N-terminal peptide sequence that targets proteins to the secretory pathway, most of them being then secreted in the apoplast space. They are soluble but some of them can bind membranes and liposomes *in vitro* while others can be post-translationally associated to a glycosylphosphatidylinositol (GPI) anchor. NsLTPs are able to bind a wide variety of substrates, like phospholipids, galactoglycerolipids, free FAs, acyl-CoA, lyso-lipids or prostaglandin (Kader, 1996; Salminen et al., 2016). Proteins from this family are mainly involved in (1) cell wall synthesis, (2) pollen and seed development, (3) embryogenesis, (4) defense again pathogens, (5) long-distance signaling and (6) stress response (Carvalho & Gomes, 2007; Edqvist, Blomqvist, Nieuwland, & Salminen, 2018; Kader, 1996; Salminen et al., 2016). Most of nsLTPs are likely not involved in the intracellular transfer of lipids dedicated to membrane biogenesis. However, such a role might be supported by two nsLTPs, AsE246 and MtN5, in legumes during nodulation in response to nitrogen starvation and by one nsLTP, GhLTPG1, involved in cotton fiber elongation (Deng et al., 2016; Lei et al., 2014; Pii et al., 2009).

During nitrogen starvation, legume roots are able to establish symbiotic relationships with rhizobial bacteria or arbuscular mycorrhizal fungi that will provide nitrogen to the plant in exchange of carbon (Downie, 2014; Udvardi & Poole, 2013; Verma, 1996). This symbiotic process occurs in special structures formed in roots of infected plants called nodules where bacteria are enclosed in a special cell compartment, the symbiosome, in which they will divide and differentiate into nitrogen-fixing bacteroids. Symbiosome membrane lipids arise from the host plant cell and a tremendous amount of lipid synthesis and transfer is required for nodule formation (Harrison & Ivanov, 2017; Verma, 1996). Symbiosome membranes formation relies on Golgi-derived vesicle trafficking (Harrison & Ivanov, 2017). They are composed of phospholipids (PC, PE, PI and PG) and DGDG (Gaude et al., 2004; Lei et al., 2014), suggesting that chloroplast and likely non-vesicular lipid trafficking might be involved in the biogenesis of these membranes. In line with this, AsE246, a nsLTP specifically expressed during nodulation and located to the symbiosome membrane, is required for nodule formation and symbiosomes development in *Astragalus sinicus* infected by the rhizobacteria *Mesorhizobium huakuii* (Chou, 2006; Lei et al., 2014). AsE246 is able to bind different phospholipids and galactoglycerolipids (Lei et al., 2014). However, whether AsE246 promotes lipid transport from cell organelles to symbiosome membranes remains to be defined. Another *Medicago truncatula* nsLTP, MtN5, plays a role in the regulation of root nodulation during *Sinorhizobium meliloti* symbiosis (Pii et al., 2009; Pii, Molesini,

Masiero, & Pandolfini, 2012; Pii, Molesini, & Pandolfini, 2013). MtN5 is expressed early during nodule formation and is required for optimal bacterial infection and nodule invasion. *In vitro*, MtN5 binds to some lyso-PC species (Pii et al., 2009). Lipidomic analyses of roots overexpressing MtN5 highlighted a global increase in lipid content, in particular of DGDG, MGDG, PE and glucosyl-ceramide. This increase is concomitant with the modification of the expression of genes involved in lipid metabolism and lipid transport (Santi et al., 2017). These results suggest that MtN5 might regulate lipid synthesis and/or transport during nodulation. Another role in the regulation of nodulation by lipid signalization has been proposed as (1) MtN5 is able to bind lyso-PC *in vitro* (Pii et al., 2009), a lipid signal involved in arbuscular mycorrhizal symbiosis (Drissner et al., 2007), and (2) overexpression of MtN5 triggers deregulation of the expression of genes involved in the synthesis of lipid signaling molecules such as PIPs kinases, phospholipases C or Sec14-like proteins (Santi et al., 2017). All these data suggest that non-vesicular lipid transport could also be involved in membrane biogenesis during plant-microorganism symbiosis in response to nitrogen starvation. This might be particularly true for DGDG, as this plastidial lipid has never been shown to be transported *via* vesicular routes.

Cotton fiber synthesis starts from ovule epidermal cells and is divided in four steps: (1) fiber initiation, (2) cell elongation, (3) secondary wall deposition and (4) maturation (Kim & Triplett, 2001). The cell elongation step requires a huge amount of lipids to sustain membrane synthesis in growing cells (Wanjie, Welti, Moreau, & Chapman, 2005). The GPI-anchored nsLTP GhLTPG1 is highly expressed in elongating fibers and is able to specifically bind and transport PI-monophosphates (Deng et al., 2016). GhLTPG1 harbors a peptide signal and a GPI anchor, both required for its localization to the PM (Deng et al., 2016). Silencing of GhLTPG1 slightly reduces cotton fiber length and the expression of cotton-fiber elongation genes. A small decrease in polar lipids, PI in particular, was also observed (Deng et al., 2016), suggesting that GhLTPG1 might be involved in lipid transport to sustain fiber elongation.

3. Lipid transport mediated by other proteins

3.1 NDPK-D

Nucleoside diphosphate kinases (NDPKs) are a family of proteins conserved throughout evolution with a primary role in the exchange of γ-phosphate between nucleosides-tri-phosphate and nucleoside-di-phosphate (Boissan,

Schlattner, & Lacombe, 2018). They act as symmetrical homo-hexamers and are present in different cell compartments where they can play a role in many cellular processes (Boissan et al., 2018; Janin et al., 2000). Interestingly, one well studied homolog in mammals, NDPK-D, is located in mitochondria IMS and is known to bind membranes and transfer CL (Epand, Schlattner, Wallimann, Lacombe, & Epand, 2007; Schlattner et al., 2013; Tokarska-Schlattner et al., 2008). NDPK-D membrane binding is mediated by a cationic motif R89-R90-K91 located at the surface of the hexamer, which associates with negatively charged lipids, in particular CL (Tokarska-Schlattner et al., 2008). The R90 residue, specific to the mitochondrial NDPK-D ortholog in humans, is essential for CL and membrane binding (Tokarska-Schlattner et al., 2008). As one CL binding site is present per monomer, NDPK-D hexamers are able to tether liposomes containing CL *in vitro* and OM-IM *in vivo* (Epand et al., 2007; Kagan et al., 2016; Tokarska-Schlattner et al., 2008). Membrane binding inhibits NDPK-D phosphotransferase activity and promotes CL transfer from IM to OM, inducing apoptosis and mitophagy and linking NDPK-D lipid transport function to CL signaling (Kagan et al., 2016; Schlattner et al., 2013, 2018)

Two NDPK-D homologs are found in *A. thaliana* (NDPK3a and NDPK3b), NDPK3b being expressed at a very low level (Table 8) (Hammargren et al., 2007). Both proteins have been detected in mitochondrial proteome (Heazlewood et al., 2004) and NDPK3a-GFP fusion was found to be located in mitochondria (Carrie et al., 2009) or in mitochondria and plastids (Hammargren et al., 2007; Luzarowski et al., 2017). Interestingly, NDPK3a was also found in proteomic analysis of the MTL complex, a mitochondrial complex involved in lipid transport (see below) (Michaud et al., 2016). However, Western blot analysis suggests that this protein probably co-migrates with the MTL complex on native gel rather than being a real MTL complex component. In addition, the R90 residue required for membrane binding of human NDPK-D is not conserved in *A. thaliana*, suggesting that NDPK3 could not bind membrane lipids (Fig. 4). However, in pea, NDPK3 is strongly associated with the intermembrane space (IMS) leaflet of the IM and interacts with the ANT carrier (Knorpp, Johansson, & Baird, 2003), suggesting that membrane binding could be mediated by protein-protein interactions in plants (Fig. 3C(i)). Further analyses are required to understand if NDPK3a is a component of the MTL complex and if it transports CL between mitochondrial OM and IM.

Table 8 Diverse genes playing a role in *A. thaliana* lipid transport.

AGI	Name	Localization	References
At4g11010	NDPK3a	Mitochondria/ plastids	Carrie et al. (2009), Hammargren, Sundström, Johansson, Bergman, and Knorpp (2007), and Luzarowski et al. (2017)
At4g23900	NDPK3b	Mitochondria (prediction)	
At3g25610	ALA10	PM, ER, ER-plastids MCSs, ER-PM MCSs	Botella et al. (2016) and Poulsen et al. (2015)
At1g19800	TGD1	Chloroplast inner envelope	Xu, Fan, Froehlich, Awai, and Benning (2005)
At3g20320	TGD2	Chloroplast inner envelope	Awai, Xu, Tamot, and Benning (2006)
At1g65410	TGD3	Stroma	Lu, Xu, Awai, Jones, and Benning (2007)
At3g06960	TGD4	Chloroplast outer envelope	Wang, Xu, and Benning (2012)
At1g27695	TGD5	Chloroplast envelopes	Fan, Zhai, Yan, and Xu (2015)
At2g44640	atLPTD1	Chloroplast outer envelope	Hsueh, Ehmann, Flinner, Ladig, and Schleiff (2017)
At4g39690	AtMic60	Mitochondrial inner membrane	Michaud et al. (2016)
At1g22520	AtMic10	Mitochondria (Prediction)	
At3g20000	AtTom40	Mitochondrial outer membrane	Duncan et al. (2011)

Protein localization were determined experimentally or the consensus localizations from SUBA and ARAMEMNON websites (prediction) are indicated (Hooper et al., 2017; Schwacke et al., 2003).

Fig. 4 Sequence alignment of the mitochondrial NDPK enzymes from human (HsNDPK-D) and *Arabidopsis thaliana* (AtNDPK3a, b) showing the non-conservation in plants of the RRK motif involved in cardiolipin binding in humans.

3.2 Enzymes involved in lipid metabolism

In yeast and mammals, several enzymes involved in lipid synthesis are enriched at MCSs, in particular at ER-mitochondria MCSs and at ER-PM MCSs (Gaigg et al., 1995; Pichler et al., 2001; Vance, 1990). In plants, enzymes involved in lipid metabolism have also been detected in ER-PM and in ER-plastids MCSs (Andersson, Goksor, & Sandelius, 2007; Larsson, Kjellberg, Tjellström, & Sandelius, 2007). In addition, during Pi starvation, the phospholipase D PLDζ2 is enriched at vacuole-mitochondria and vacuole-chloroplasts MCSs, whereas an even distribution along the vacuole membrane is observed when *A. thaliana* plants are grown in Pi-sufficient conditions (Yamaryo, Dubots, Albrieux, Baldan, & Block, 2008). An increasing number of evidences suggests that such localization is important to provide an efficient transport of lipids between membranes. This is the case for instance in yeast for the transport of PS at ER-mitochondria MCSs, where the localization of PS synthase at this junction increases the efficiency of PS transport to mitochondria (Kannan, Lahiri, Liu, Choudhary, & Prinz, 2017). Several hypotheses can explain why the localization of lipid-synthesizing enzymes at MCSs is important for non-vesicular lipid transfer: (1) local synthesis can create microdomains with special lipid composition facilitating lipid extrusion and transfer, (2) the local enrichment of a lipid at MCSs can favor its transport by energy-independent LTPs following its concentration gradient, (3) the modification of a lipid at MCSs can promote the formation of lipid species more competent for non-vesicular transport (*i.e.*, less hydrophobic), and/or (4) the coupling between synthesis and transfer can be important to rapidly regulate organelle biogenesis during cell division or in response to stress. In addition, several examples have demonstrated that enzymes can also work *in trans* at MCSs, meaning that they can modify a lipid substrate located in the juxtaposed membrane. This is the case for the phosphatidylserine decarboxylase PSD1 in mitochondria IM that can decarboxylate OM PS to produce PE at OM-IM MCSs (Aaltonen et al., 2016). This mechanism is involved in the regulation of lipid composition at MCSs but is not involved in the lipid transport process *per se*.

3.2.1 PC remodeling at ER-plastids MCSs

Lyso-PC can be spontaneously transferred between membranes because of its low hydrophobicity. In plants, lyso-PC synthesis at ER-plastids MCSs might be one of the mechanisms involved in the transport of lipid precursors from ER to chloroplast to fulfill galactoglycerolipid synthesis *via* the Eukaryotic pathway. This idea was supported by pulse chase experiments and by the presence of several enzymes involved in the synthesis of lyso-PC from PC and in its reacylation at ER-chloroplasts MCSs or in chloroplasts OE (Fig. 3D(i)) (Bessoule, Testet, & Cassagne, 1995; Jessen, Roth, Wiermer, & Fulda, 2015; Mongrand, Cassagne, & Bessoule, 2000; Tan et al., 2011). In *Brassica napus*, the lipase BnCLIP1 is located at ER-chloroplasts MCSs, where it might be involved in lyso-lipid synthesis from ER phospholipids (Tan et al., 2011). Andersson et al. show that cytosolic lipases can generate lyso-PC molecules which are subsequently used for galactoglycerolipid synthesis (Andersson, Kjellberg, & Sandelius, 2004). In *A. thaliana*, the long-chain acyl-CoA synthetases LACS4 and LASC9 are localized in the ER and the chloroplast OE, respectively, and double *lacs4lacs9* mutants show a decrease in the content of galactoglycerolipids arising from the Eukaryotic pathway (Jessen et al., 2015). Thus, these LACSs have been proposed to play a role in the creation of a local pool of acyl-CoA at ER-chloroplasts MCSs that might be used by LPCAT to promote PC synthesis in the chloroplast envelope (Jessen et al., 2015). In addition, LPCAT activity has been detected in the chloroplast envelope of several plant species (Bessoule et al., 1995; Karki, Johnson, & Bates, 2019). However, the analyses of *lpcat1lpcat2* mutants in *A. thaliana*, the two main enzymes responsible for LPCAT activity, demonstrated that LPCAT activity, and likely lyso-PC transport, is not essential for chloroplastic lipid synthesis *via* the Eukaryotic pathway (Karki et al., 2019). Interestingly, pulse chase labelling experiments performed in this work support a model in which the LPCAT1/2 enzymes are involved in the incorporation of newly synthesized chloroplastic FAs into PC by lyso-PC acylation. PC is then translocated back to the ER where it can also be synthesized by the *de novo* pathway from glycerol-3-phosphate and acyl-CoA, leading to two distinct pools of PC in the ER (Karki et al., 2019). Thus, this current model suggests that LPCAT1/2 are more likely involved in chloroplast to ER PC transport than in the Eukaryotic pathway of galactoglycerolipid synthesis. Transport of PC *via* the lyso-PC intermediate could also occur at ER-PM MCSs in plants where LPCAT activity has been detected in pea (Larsson et al., 2007).

3.2.2 DGD1

MGDG synthesis starts by the addition of a galactose residue to a DAG backbone. Bulk MGDG synthesis takes place at the plastid IE and is catalyzed by MGD1, while MGD2/3 operate at the plastid OE in non-photosynthetic tissues or during specific stress conditions, typically Pi starvation (Awai et al., 2001; Kobayashi et al., 2009). DGDG synthesis requires the addition of another galactose residue moiety to MGDG, a reaction catalyzed by DGD1 and DGD2 at the plastid OE (Fig. 3D(ii)) (Kelly, Froehlich, & Dörmann, 2003). Hence, DGDG synthesis requires the transport of MGDG from the IE to the OE. The N-terminal region of DGD1, NDGD1, has been shown to promote lipid transport between the plastid IE and OE (Kelly et al., 2016). NDGD1 is necessary for DGDG synthesis in the OE. Furthermore, NDGD1 binds PA and enables fusion of PA-containing liposomes *in vitro* (Kelly et al., 2016). Thus, it has been suggested that NDGD1 could facilitate lipid transport between both chloroplast membranes by triggering membrane-hemifusion (Kelly et al., 2016). Whether DGD1 acts as a tether or as a direct lipid transport protein remains to be determined.

3.3 Enzymes flipping lipids between membrane leaflets

Several recent studies suggest that enzymes involved in lipid translocation between membrane leaflets act as key players in non-vesicular lipid transport between membranes, notably by acting in concert with LTPs. Flippases can flip lipids against their concentration gradient in an ATP-dependent manner to create asymmetric membranes while scramblases can translocate lipids between the two leaflets following their concentration gradient in an ATP-independent manner, thus equilibrating lipids between membrane leaflets (Maxfield & Menon, 2016). The putative roles of these proteins in non-vesicular lipid exchanges might be (1) to ensure the selectivity and directionality of transport, (2) to facilitate the loading of lipids by LTPs, and/or (3) to facilitate the insertion of the transported lipid in the acceptor membrane (Dziurdzik & Conibear, 2021; Reinisch & Prinz, 2021). A good example of this mechanism is found in Gram-negative bacteria: lipopolysaccharide (LPS) is synthetized on the cytosolic leaflet of the IM and is subsequently transported to the extracellular leaflet of the OM (Sperandeo, Martorana, & Polissi, 2017). LPS is first flipped across the IM by the flippase MsbA, then extracted from the IM by the LptB2FGC ABC transporter complex, ensuring the directionality of the transport. LPS transport through the periplasm is facilitated by LptACDFG proteins that form a hollow rod-like structure composed of hydrophobic β-jellyrolls reminiscent of the VPS13 hydrophobic groove (Thélot, Orlando, Li, & Liao, 2020). The insertion of LPS in the OM and its translocation to the outer leaflet is ensured by the LptDE translocon (Sperandeo et al., 2017). A similar mechanism was proposed for lipid transport at ER-autophagosomes MCSs in yeast where the

LTP Atg2 associates with the autophagosome-located Atg9 scramblase and with the ER-located scramblases TMEM41B and VMP1 (Matoba et al., 2020; Noda, 2021; Orii, Tsuji, Ogasawara, & Fujimoto, 2021). Other LTPs, such as the human VPS13A or the yeast Osh6 protein have also been shown to associate with lipid scramblases (D'Ambrosio et al., 2020; Park & Neiman, 2020), suggesting further that lipid-flipping enzymes could play an active role in non-vesicular lipid transfer in association with LTPs.

In plants, lipid scramblases are also conserved but their role in lipid transport has never been investigated. However, in *A. thaliana*, a P4-ATPase flippase, ALA10, has been shown to regulate plastid MGDG synthesis in leaves (Table 8 and Fig. 3D(iii)) (Botella et al., 2016). ALA10 is located in the ER where it can interact with the ALA-interacting subunits (ALIS) ALIS1 or ALIS5. This association regulates ALA10 subcellular localization: in the ER close to PM when ALA10 is co-expressed with ALIS1 or in the ER close to chloroplasts when co-expressed ALIS5 (Botella et al., 2016). More recently, ALA10 localization was also shown to be impacted by its level of ubiquitination by the E3-Ubiquitin ligase PUB11, showing the complexity of the mechanisms regulating P4-ATPase localization in cells (López-Marqués, Davis, Harper, & Palmgren, 2021; Salvaing et al., 2020). Overexpression of ALA10 increases the ratio MGDG/PC in leaves and the 18:2 content in chloroplastic PC, suggesting it can stimulate MGDG synthesis and enrich 18:2-PC species in chloroplasts (Botella et al., 2016). ALA10 interacts with the ER FA desaturase FAD2, involved in 18:1 acyl chain desaturation to 18:2, but not with FAD3, involved in 18:2 acyl chain desaturation to 18:3 (Botella et al., 2016). Thus, it was proposed that ALA10, by interacting with FAD2, translocates PC species enriched in 18:2 from the lumenal to the cytosolic leaflet of the ER, promoting a local enrichment of this lipid and favoring its transport by an unknown mechanism to chloroplasts. To explain the higher MGDG/PC ratio in ALA10 overexpressing plants, the authors suggested that the translocated PC could be a substrate for phospholipases D to generate PA, a lipid known to activate MGD1 activity (Botella et al., 2016; Dubots et al., 2012). This example illustrates that lipid flipping across membrane leaflets can also be involved in inter-organellar lipid transfer in plants.

4. Lipid transport mediated by protein complexes

4.1 The TGD complex

An important molecular effector implicated in ER to chloroplast lipid transport is the TGD (Trigalactosyldiacylglycerol) complex composed of at least five subunits: TGD1–5 (Table 8 and Fig. 3D(iv)). Plants deficient in any of the TGD proteins show substantial trigalactosyldiacylglycerol (TGDG) accumulation (hence their name) (Xu, Fan, Riekhof, Froehlich, & Benning, 2003), reduction in growth and photosynthetic capacity (Li, Gao, Benning, & Sharkey, 2012; Xu, Fan, Cornish, & Benning, 2008), and compromised embryogenesis (Xu et al., 2005). Importantly, they show reduced amounts of C18 acyl chains in *sn*-2 position of chloroplast lipids, consistent with an Eukaryotic lipid import defect into chloroplasts (Xu et al., 2003, 2008).

TGD1, TGD2 and TGD3 form a non-canonical ABC transporter system at the chloroplast IE. It probably derives from a Gram-negative bacterial ABC import system that was present in cyanobacteria prior to endosymbiosis (Roston, Gao, Murcha, Whelan, & Benning, 2012). Bacterial ABC importers are divided into two structural categories but show conserved features comprising (1) two TM domains composed of multiple transmembrane helices forming a permease to allow substrate passage, (2) two nucleotide binding domains specialized in ATP hydrolysis that provide energy for substrate transport and (3) a usually periplasmic substrate binding domain (SBD) that captures and delivers the substrate to the transporter subunits (Beis, 2015; Ford & Beis, 2019; Hollenstein, Dawson, & Locher, 2007; Wilkens, 2015). TGD1 structure prediction is reminiscent of a half-permease unit of a bacterial ABC importer harboring six transmembrane helices and is therefore thought to be the permease subunit of the TGD1–3 complex (Xu et al., 2003, 2005). Consistent with its putative role, TGD1 is an integral membrane protein localized in the chloroplast IE (Xu et al., 2005). TGD2 is the SBD-containing subunit of the TGD1–3 complex. In contrast to bacterial SBD subunits, TGD2 is anchored to the chloroplast IE by a N-terminal TM domain while the C-terminal part of the protein is located in the intermembrane space and specifically binds PA (Awai et al., 2006; Lu & Benning, 2009; Roston, Gao, Xu, & Benning, 2011). A mycobacterial cell entry (MCE) domain is juxtaposed to the TM domain (Awai et al., 2006). Interestingly, bacterial MCE proteins are known to assemble into hexamers forming an elongated hydrophobic channel ideal for lipid transport between the bacterial OM and IM (Ekiert et al., 2017). A 25 amino acid sequence partially overlapping the MCE domain is sufficient to mediate PA binding, but the complete TGD2 C-terminus still provides better binding affinity suggesting the existence of multiple binding determinants (Lu & Benning, 2009). TGD3 is a small ATPase that provides the energy for the TGD-mediated lipid transport (Lu et al., 2007). TGD3 is a stromal protein that is associated with the chloroplast IE, consistent with its bacterial ABC importer counterparts (Fig. 3D(iv)) (Lu et al., 2007).

TGD1–3 form a highly stable complex of about 500 kDa presumably composed of two TGD1 permease subunits, two TGD3 ATPase subunits and between 8 and 12 TGD2 SBD subunits (Roston et al., 2011, 2012). The large number of TGD2 proteins is surprising as it diverges from the canonical bacterial ABC importer architecture that usually contains only one SBD subunit. Why the TGD1–3 complex contains multiple SBD subunits is still unclear. Since TGD2 binds its putative substrate, PA, in a cooperative manner (Lu & Benning, 2009), it has been suggested that an increased amount of SBDs could help to sense and bind PA even at extremely low concentrations (Roston et al., 2012). Alternatively, it has been proposed that the assembly of multiple SBDs could concentrate PA, a non-bilayer forming lipid, and thus facilitate lipid extraction from the donor membrane (LaBrant et al., 2018). Consistent with the latter hypothesis, TGD2 has been found to disrupt membranes *in vitro* (Roston et al., 2011). Furthermore, the MCE domain contained in TGD2 suggests that its oligomerization might form a lipid transport channel like MlaD in the bacterial non-canonical ABC lipid transporter (Chi et al., 2020; Ekiert et al., 2017; Thong et al., 2016).

The TGD complex also harbors two additional proteins: TGD4 and TGD5. TGD4 is a β-barrel protein that forms homodimers in the chloroplast OE (Fig. 3D(iv)) (Wang et al., 2012; Wang, Anderson, & Benning, 2013; Xu et al., 2008). TGD4 is distantly related to the lipid A transporter D (LptD) subunit that transports LPS from the bacterial IM to the bacterial OM (see above). Even though TGD4 was initially thought to be localized in the ER, further investigations indicate a localization in the chloroplast OE (Wang et al., 2012; Xu et al., 2008). Thus, it has been proposed that TGD4 could be present at ER-chloroplast MCSs, thus providing a direct lipid transport route from the ER to the chloroplast (Wang & Benning, 2012). TGD4 binds PA with two cytosol-facing N-terminal domains that are both necessary and sufficient for lipid import, supporting a lipid transport function within the TGD complex (Wang et al., 2012, 2013). Another β-barrel protein, AtLPTD1, homologous to LptD like TGD4, is present in the chloroplast OE (Fig. 3D(iv)) (Hsueh et al., 2017). Mutants depleted in AtLPTD1 show phenotypic traits similar to *tgd* mutants, including TGDG accumulation (Hsueh et al., 2017; LaBrant et al., 2018). AtLPTD1 is expressed during plant adaptation to Pi starvation and is presumably implicated in DAG import into chloroplasts (Hsueh et al., 2017). It is presently not known if AtLPTD1 is part of the TGD complex. TGD5 is a small glycine-rich protein interacting with the TGD1–3 complex and TGD4, strongly suggesting that it bridges the TGD complex subunits located in both chloroplast envelope membranes (Fan et al., 2015). A recent study showed that *A. thaliana* TGD5 mutants have a strong increase in plastid stromules (Itoh et al., 2021). Since stromules are known to promote ER-chloroplast contact sites (Schattat, Barton, & Mathur, 2011), it is tempting to hypothesize that stromule formation is the response to a lipid import defect caused by the lack of TGD5 subunit.

The current picture of the TGD complex suggests that ER-synthesized lipids in the chloroplast are directly channeled through the chloroplast OE and IE by TGD4 and TGD1–3, respectively (Fig. 3D(iv)). A big gap of knowledge still concerns the nature of the transported lipid (LaBrant et al., 2018; Michaud & Jouhet, 2019). Since TGD2 and TGD4 have been shown to bind PA (Awai et al., 2006; Wang et al., 2012), it is tempting to speculate that PA is the transported lipid. However, a mathematical model mimicking chloroplast lipid synthesis *via* the Eukaryotic pathway suggests the import of DAG rather than PA (Maréchal & Bastien, 2014). Thus, PA could have a regulatory function of TGD, as it is the case for MGD1 for instance (Dubots et al., 2012). Remaining unanswered questions concern the connection between the TGD complex and the ER and its putative involvement in lipid precursor import from the ER to the OE.

4.2 The MTL complex

The MTL complex is a complex involved in mitochondrial lipid transport, notably in response to Pi starvation (Michaud, 2018; Michaud et al., 2016). The MTL complex is a huge lipid-containing entity with roughly 200 proteins and migrating at about 800kDa on native gel (Michaud et al., 2016). Proteomic analysis revealed that the MTL complex contains mainly mitochondrial proteins and also several proteins located in non-mitochondrial compartments, such as the plastid OE, suggesting a possible location at mitochondrial MCSs. Interestingly, all the components of the translocase of the outer membrane (TOM) complex and AtMic60 are present in the MTL complex (Table 8 and Fig. 3C(ii)). AtMic60 is a confirmed functional homolog of yeast Mic60, a core component of the mitochondrial contact site cristae organizing system (MICOS) complex (Pfanner et al., 2014).

MICOS plays a central role in mitochondrial membrane architecture, dynamics of protein import, lipid homeostasis and lipid trafficking (Michaud et al., 2017; Mukherjee et al., 2021; Rampelt, Zerbes, van der Laan, & Pfanner, 2017). In yeast and mammals, Mic60, a core component located in the IM, interacts with multiple partners in the mitochondrial OM, *i.e.*, the TOM complex, SAM50, VDAC and DNAJC11, thus tightly tethering the mitochondrial OM and IM (Mukherjee et al., 2021). AtMic60 interacts with AtTom40 and a decrease in AtTom40 content was observed in the MTL complex of *atmic60* mutants, suggesting that AtMic60 also modulates OM-IM tethering (Michaud et al., 2016). Furthermore, the abundance of AtMic60-AtTom40 in the MTL complex increases in Pi deprived conditions, suggesting a role in lipid remodeling (Michaud et al., 2016). Lipid analysis of the MTL complex in *atmic60* KO lines revealed that DGDG accumulation in the MTL complex is compromised whereas PE accumulates during Pi starvation, suggesting that PE export and DGDG import are impaired (Fig. 3C(ii)). A role in intra-mitochondrial lipid trafficking was further supported by mitochondrial lipid analyses showing a decreased DGDG and increased PE levels in *atmic60* compared to WT during Pi starvation (Michaud et al., 2016). AtMic60 also destabilizes liposomes *in vitro*, suggesting that it could promote lipid desorption from membranes prior to transport (Michaud et al., 2016). Even if AtMic60 and AtTom40 were identified as key players, the exact lipid transport mechanism exerted by the MTL complex as a whole, as well as the role of its partners, remains to be unraveled. Since the MTL complex also encloses several extra-mitochondrial proteins, notably chloroplast OE proteins of unknown function, it is tempting to speculate that the MTL complex might form a structure connecting mitochondrial membranes to other cell compartments such as chloroplasts, promoting direct lipid transport between both compartments (Michaud et al., 2016, 2017). Such a hypothetical structure is reminiscent of the ER–mitochondria organizing network (ERMIONE) that has been suggested to organize ER-mitochondria MCSs functions in yeast and mammals and to be involved in ER to IM lipid transport (Wideman & Muñoz-Gómez, 2016).

5. Conclusion and perspectives

The general mechanisms described in mammals and yeast for non-vesicular glycerolipid transport are conserved in plants. For instance, most of the LTP families present in yeast and mammals are also found in plants, even if only a few of them have been characterized in terms of subcellular localization, lipid transport specificities and cellular functions. Peculiarities have to be highlighted concerning non-vesicular lipid transport in plants:

- Plants cope with a wide variety of stresses where lipid signaling and membrane remodeling play a crucial role in the adaptation process (Hou et al., 2016; Liu et al., 2019; Michaud & Jouhet, 2019). Thus, specific lipid signaling and trafficking systems might have been developed during evolution to empower the plant response to these particular situations. In accordance with this, many LTP families are expanded in plants and most of them are highly responsive to biotic and abiotic stresses. In addition, most of the plant LTPs characterized until now have a confirmed role in plant stress susceptibility (Table 1). It also has to be noticed that many *A. thaliana* LTP genes are predicted to be transcribed into several mRNA isoforms in the TAIR database, increasing the putative number of lipid transporter that could co-exist in a cell.

- Plant have a chloroplast which biogenesis requires extensive exchanges of lipids with the ER. Thus, specific lipid transport machineries, such as the TGD complex, have evolved during evolution to ensure chloroplast membranes biogenesis and functions. However, the mechanisms involved in ER-plastids MCSs formation and in the transport of lipids between both organelles remain poorly characterized. In addition, chloroplast lipids can be major contributors for the biogenesis of extra-plastidial membranes in stress conditions, such as Pi starvation where DGDG is exported to mitochondria, PM and vacuoles, or nitrogen starvation in legumes where symbiosome membrane formation requires DGDG. Beside the poorly understood mechanisms involved in those transports, the question of whether chloroplast glycerolipids could also be involved in extra-plastidial membrane synthesis in normal growth conditions remains open.
- Finally, the synthesis of cell wall structures in plants, such as cutin, wax or sporopollenin, requires the secretion to the extracellular space of hydrophobic components. To ensure this function, a specific family of LTPs, the nsLTPs, appeared in plants during evolution. In this huge family, some of them are also able to bind glycerolipids and are involved in membrane biogenesis during stress, highlighting that plants have evolved specific pathways for non-vesicular lipid transport.

In summary, the understanding of the processes involved in non-vesicular lipid transfer in plants is in its infancy. Many studies still have to be performed to understand how this process occurs and how it is regulated in response to stress.

Fundings

This work is supported by the French National Research Agency in the framework of the "investissement d'avenir" program (ANR-15-IDEX-O2) and of the ANRJCJC MiCoSLiT (ANR-19-C13–0013). It also received funding from GRAL a program from the Chemistry Biology Health (CBH) Graduate School of University Grenoble Alpes (ANR-17-EURE-0003).

References

Aaltonen, M. J., Friedman, J. R., Osman, C., Salin, B., di Rago, J., Nunnari, J., et al. (2016). MICOS and phospholipid transfer by Ups2–Mdm35 organize membrane lipid synthesis in mitochondria. *The Journal of Cell Biology*, *213*(5), 525–534.

AhYoung, A. P., Jiang, J., Zhang, J., Khoi Dang, X., Loo, J. A., Zhou, Z. H., et al. (2015). Conserved SMP domains of the ERMES complex bind phospholipids and mediate tether assembly. *Proceedings of the National Academy of Sciences of the United States of America*, *112*(25), E3179–E3188. https://doi.org/10.1073/pnas.1422363112.

Andersson, M. X., Goksor, M., & Sandelius, A. S. (2007). Optical manipulation reveals strong attracting forces at membrane contact sites between endoplasmic reticulum and chloroplasts. *Journal of Biological Chemistry*, *282*(2), 1170–1174. https://doi.org/10.1074/Jbc.M608124200.

Andersson, M. X., Kjellberg, J. M., & Sandelius, A. S. (2004). The involvement of cytosolic lipases in converting phosphatidyl choline to substrate for galactolipid synthesis in the chloroplast envelope. *Biochimica et Biophysica Acta*, *1684*(1–3), 46–53. https://doi.org/10.1016/j.bbalip.2004.06.003.

Andersson, M. X., Larsson, K. E., Tjellström, H., Liljenberg, C., & Sandelius, A. S. (2005). Phosphate-limited oat. *Journal of Biological Chemistry*, *280*(30), 27578–27586. https://doi.org/10.1074/jbc.M503273200.

Awai, K., Marechal, E., Block, M. A., Brun, D., Masuda, T., Shimada, H., et al. (2001). Two types of MGDG synthase genes, found widely in both 16:3 and 18:3 plants, differentially mediate galactolipid syntheses in photosynthetic and nonphotosynthetic tissues in Arabidopsis thaliana. *Proceedings of the National Academy of Sciences of the United States of America*, *98*(19), 10960–10965. https://doi.org/10.1073/pnas.181331498.

Awai, K., Xu, C., Tamot, B., & Benning, C. (2006). A phosphatidic acid-binding protein of the chloroplast inner envelope membrane involved in lipid trafficking. *Proceedings of the National Academy of Sciences of the United States of America*, *103*(28), 10817–10822. https://doi.org/10.1073/pnas.0602754103.

Bankaitis, V. A., Aitken, J. R., Cleves, A. E., & Dowhan, W. (1990). An essential role for a phospholipid transfer protein in yeast Golgi function. *Nature*, *347*(6293), 561–562. https://doi.org/10.1038/347561a0.

Bankaitis, V. A., Mousley, C. J., & Schaaf, G. (2010). The Sec14 superfamily and mechanisms for crosstalk between lipid metabolism and lipid signaling. *Trends in Biochemical Sciences*, *35*(3), 150–160. https://doi.org/10.1016/j.tibs.2009.10.008.

Barajas, D., Xu, K., de Castro Martin, I. F., Sasvari, Z., Brandizzi, F., Risco, C., et al. (2014). Co-opted oxysterol-binding ORP and VAP proteins channel sterols to RNA virus replication sites via membrane contact sites. *PLoS Pathogens*, *10*(10), e1004388. https://doi.org/10.1371/journal.ppat.1004388.

Bayer, E. M., Sparkes, I., Vanneste, S., & Rosado, A. (2017). From shaping organelles to signalling platforms: The emerging functions of plant ER-PM contact sites. *Current Opinion in Plant Biology*, *40*, 89–96. https://doi.org/10.1016/j.pbi.2017.08.006.

Bean, B. D. M., Dziurdzik, S. K., Kolehmainen, K. L., Fowler, C. M. S., Kwong, W. K., Grad, L. I., et al. (2018). Competitive organelle-specific adaptors recruit Vps13 to membrane contact sites. *The Journal of Cell Biology*, *217*(10), 3593–3607. https://doi.org/10.1083/jcb.201804111.

Beis, K. (2015). Structural basis for the mechanism of ABC transporters. *Biochemical Society Transactions*, *43*(5), 889–893. https://doi.org/10.1042/BST20150047.

Bessoule, J. J., Testet, E., & Cassagne, C. (1995). Synthesis of phosphatidylcholine in the chloroplast envelope after import of lysophosphatidylcholine from endoplasmic reticulum membranes. *European Journal of Biochemistry*, *228*(2), 490–497.

Bockelmann, S., Mina, J. G. M., Korneev, S., Hassan, D. G., Müller, D., Hilderink, A., et al. (2018). A search for ceramide binding proteins using bifunctional lipid analogs yields

CERT-related protein StarD7. *Journal of Lipid Research*, *59*(3), 515–530. https://doi.org/10.1194/jlr.M082354.

Böhme, K., Li, Y., Charlot, F., Grierson, C., Marrocco, K., Okada, K., et al. (2004). The Arabidopsis COW1 gene encodes a phosphatidylinositol transfer protein essential for root hair tip growth. *The Plant Journal: For Cell and Molecular Biology*, *40*(5), 686–698. https://doi.org/10.1111/j.1365-313X.2004.02245.x.

Bohnert, M., & Schuldiner, M. (2018). Stepping outside the comfort zone of membrane contact site research. *Nature Reviews. Molecular Cell Biology*, *19*, 483–484. https://doi.org/10.1038/s41580-018-0022-1.

Boissan, M., Schlattner, U., & Lacombe, M.-L. (2018). The NDPK/NME superfamily: State of the art. *Laboratory Investigation*, *98*(2), 164–174. https://doi.org/10.1038/labinvest.2017.137.

Botella, C., Sautron, E., Boudiere, L., Michaud, M., Dubots, E., Yamaryo-Botté, Y., et al. (2016). ALA10, a phospholipid flippase, controls FAD2/FAD3 desaturation of phosphatidylcholine in the ER and affects chloroplast lipid composition in Arabidopsis thaliana. *Plant Physiology*, *170*(3), 1300–1314. https://doi.org/10.1104/pp.15.01557.

Boudière, L., Michaud, M., Petroutsos, D., Rébeillé, F., Falconet, D., Bastien, O., et al. (2014). Glycerolipids in photosynthesis: Composition, synthesis and trafficking. *Biochimica et Biophysica Acta (BBA)—Bioenergetics*, *1837*(4), 470–480. https://doi.org/10.1016/j.bbabio.2013.09.007.

Browse, J., Warwick, N., Somerville, C. R., & Slack, C. R. (1986). Fluxes through the prokaryotic and eukaryotic pathways of lipid synthesis in the '16:3' plant Arabidopsis thaliana. *The Biochemical Journal*, *235*(1), 25–31. https://doi.org/10.1042/bj2350025.

Buckingham, M., & Relaix, F. (2007). The role of Pax genes in the development of tissues and organs: Pax3 and Pax7 regulate muscle progenitor cell functions. *Annual Review of Cell and Developmental Biology*, *23*, 645–673. https://doi.org/10.1146/annurev.cellbio.23.090506.123438.

Carrie, C., Kühn, K., Murcha, M. W., Duncan, O., Small, I. D., O'Toole, N., et al. (2009). Approaches to defining dual-targeted proteins in Arabidopsis. *The Plant Journal*, *57*(6), 1128–1139. https://doi.org/10.1111/j.1365-313X.2008.03745.x.

Carvalho, A.d. O., & Gomes, V. M. (2007). Role of plant lipid transfer proteins in plant cell physiology-a concise review. *Peptides*, *28*(5), 1144–1153. https://doi.org/10.1016/j.peptides.2007.03.004.

Cevher-Keskin, B. (2020). Endomembrane trafficking in plants. In T. Yonar (Ed.), *Electrodialysis* IntechOpen. https://doi.org/10.5772/intechopen.91642.

Chen, H., Dong, J., & Wang, T. (2021). Autophagy in plant abiotic stress management. *International Journal of Molecular Sciences*, *22*(8), 4075. https://doi.org/10.3390/ijms22084075.

Chi, X., Fan, Q., Zhang, Y., Liang, K., Wan, L., Zhou, Q., et al. (2020). Structural mechanism of phospholipids translocation by MlaFEDB complex. *Cell Research*, *30*(12), 1127–1135. https://doi.org/10.1038/s41422-020-00404-6.

Chou, M.-X. (2006). Thirteen nodule-specific or nodule-enhanced genes encoding products homologous to cysteine cluster proteins or plant lipid transfer proteins are identified in Astragalus sinicus L. by suppressive subtractive hybridization. *Journal of Experimental Botany*, *57*(11), 2673–2685. https://doi.org/10.1093/jxb/erl030.

Chu, M., Li, J., Zhang, J., Shen, S., Li, C., Gao, Y., et al. (2018). AtCaM4 interacts with a Sec14-like protein, PATL1, to regulate freezing tolerance in Arabidopsis in a CBF-independent manner. *Journal of Experimental Botany*, *69*(21), 5241–5253. https://doi.org/10.1093/jxb/ery278.

Chung, J., Torta, F., Masai, K., Lucast, L., Czapla, H., Tanner, L. B., et al. (2015). PI4P/phosphatidylserine countertransport at ORP5- and ORP8-mediated ER-plasma membrane contacts. *Science*, *349*(6246), 428–432. https://doi.org/10.1126/science.aab1370.

Clark, B. J. (2012). The mammalian START domain protein family in lipid transport in health and disease. *Journal of Endocrinology*, *212*(3), 257–275. https://doi.org/10.1530/JOE-11-0313.

Clark, B. J. (2020). The START-domain proteins in intracellular lipid transport and beyond. *Molecular and Cellular Endocrinology*, *504*, 110704. https://doi.org/10.1016/j.mce.2020.110704.

Connerth, M., Tatsuta, T., Haag, M., Klecker, T., Westermann, B., & Langer, T. (2012). Intramitochondrial transport of phosphatidic acid in yeast by a lipid transfer protein. *Science*, *338*(6108), 815–818. https://doi.org/10.1126/science.1225625.

Craxton, M. (2004). Synaptotagmin gene content of the sequenced genomes. *BMC Genomics*, *5*(1), 43. https://doi.org/10.1186/1471-2164-5-43.

D'Ambrosio, J. M., Albanèse, V., Lipp, N.-F., Fleuriot, L., Debayle, D., Drin, G., et al. (2020). Osh6 requires Ist2 for localization to ER–PM contacts and efficient phosphatidylserine transport in budding yeast. *Journal of Cell Science*, *133*(11), jcs243733. https://doi.org/10.1242/jcs.243733.

De, M., Oleskie, A. N., Ayyash, M., Dutta, S., Mancour, L., Abazeed, M. E., et al. (2017). The Vps13p–Cdc31p complex is directly required for TGN late endosome transport and TGN homotypic fusion. *Journal of Cell Biology*, *216*(2), 425–439. https://doi.org/10.1083/jcb.201606078.

de Saint-Jean, M., Delfosse, V., Douguet, D., Chicanne, G., Payrastre, B., Bourguet, W., et al. (2011). Osh4p exchanges sterols for phosphatidylinositol 4-phosphate between lipid bilayers. *Journal of Cell Biology*, *195*(6), 965–978. https://doi.org/10.1083/jcb.201104062.

de Silva, K., Laska, B., Brown, C., Sederoff, H. W., & Khodakovskaya, M. (2011). Arabidopsis thaliana calcium-dependent lipid-binding protein (AtCLB): A novel repressor of abiotic stress response. *Journal of Experimental Botany*, *62*(8), 2679–2689. https://doi.org/10.1093/jxb/erq468.

Deng, T., Yao, H., Wang, J., Wang, J., Xue, H., & Zuo, K. (2016). GhLTPG1, a cotton GPI-anchored lipid transfer protein, regulates the transport of phosphatidylinositol monophosphates and cotton fiber elongation. *Scientific Reports*, *6*, 26829. https://doi.org/10.1038/srep26829.

Ding, X., Zhang, X., & Otegui, M. S. (2018). Plant autophagy: New flavors on the menu. *Current Opinion in Plant Biology*, *46*, 113–121. https://doi.org/10.1016/j.pbi.2018.09.004.

Dong, X., Wang, Z., Ye, S., & Zhang, R. (2020). The crystal structure of ORP3 reveals the conservative PI4P binding pattern. *Biochemical and Biophysical Research Communications*, *529*(4), 1005–1010. https://doi.org/10.1016/j.bbrc.2020.06.090.

Downie, J. A. (2014). Legume nodulation. *Current Biology*, *24*(5), R184–R190. https://doi.org/10.1016/j.cub.2014.01.028.

Drissner, D., Kunze, G., Callewaert, N., Gehrig, P., Tamasloukht, M., Boller, T., et al. (2007). Lyso-phosphatidylcholine is a signal in the arbuscular mycorrhizal symbiosis. *Science*, *318*(5848), 265–268. https://doi.org/10.1126/science.1146487.

Du, X., Kumar, J., Ferguson, C., Schulz, T. A., Ong, Y. S., Hong, W., et al. (2011). A role for oxysterol-binding protein-related protein 5 in endosomal cholesterol trafficking. *The Journal of Cell Biology*, *192*(1), 121–135. https://doi.org/10.1083/jcb.201004142.

Du, X., Zhou, L., Aw, Y. C., Mak, H. Y., Xu, Y., Rae, J., et al. (2020). ORP5 localizes to ER–lipid droplet contacts and regulates the level of PI(4)P on lipid droplets. *Journal of Cell Biology*, *219*(1), e201905162. https://doi.org/10.1083/jcb.201905162.

Dubots, E., Botté, C., Boudière, L., Yamaryo-Botté, Y., Jouhet, J., Maréchal, E., et al. (2012). Role of phosphatidic acid in plant galactolipid synthesis. *Biochimie*, *94*(1), 86–93. https://doi.org/10.1016/j.biochi.2011.03.012.

Duncan, O., Taylor, N. L., Carrie, C., Eubel, H., Kubiszewski-Jakubiak, S., Zhang, B., et al. (2011). Multiple lines of evidence localize signaling, morphology, and lipid biosynthesis machinery to the mitochondrial outer membrane of arabidopsis. *Plant Physiology*, *157*(3), 1093–1113. https://doi.org/10.1104/pp.111.183160.

Dziurdzik, S. K., & Conibear, E. (2021). The Vps13 family of lipid transporters and its role at membrane contact sites. *International Journal of Molecular Sciences*, *22*, 2905.

Edqvist, J., Blomqvist, K., Nieuwland, J., & Salminen, T. A. (2018). Plant lipid transfer proteins: Are we finally closing in on the roles of these enigmatic proteins? *Journal of Lipid Research*, *59*(8), 1374–1382. https://doi.org/10.1194/jlr.R083139.

Ekiert, D. C., Bhabha, G., Isom, G. L., Greenan, G., Ovchinnikov, S., Henderson, I. R., et al. (2017). Architectures of lipid transport systems for the bacterial outer membrane. *Cell*, *169*(2), 273–285.e17. https://doi.org/10.1016/j.cell.2017.03.019.

Elbaz-Alon, Y., Eisenberg-Bord, M., Shinder, V., Stiller, S. B., Shimoni, E., Wiedemann, N., et al. (2015). Lam6 regulates the extent of contacts between organelles. *Cell Reports*, *12*(1), 7–14. https://doi.org/10.1016/j.celrep.2015.06.022.

Elbaz-Alon, Y., Guo, Y., Segev, N., Harel, M., Quinnell, D. E., Geiger, T., et al. (2020). PDZD8 interacts with Protrudin and Rab7 at ER-late endosome membrane contact sites associated with mitochondria. *Nature Communications*, *11*(1), 3645. https://doi.org/10.1038/s41467-020-17451-7.

Epand, R. F., Schlattner, U., Wallimann, T., Lacombe, M.-L., & Epand, R. M. (2007). Novel lipid transfer property of two mitochondrial proteins that bridge the inner and outer membranes. *Biophysical Journal*, *92*(1), 126–137. https://doi.org/10.1529/biophysj.106.092353.

Fan, J., Zhai, Z., Yan, C., & Xu, C. (2015). Arabidopsis TRIGALACTOSYLDIACYLGLYCEROL5 interacts with TGD1, TGD2, and TGD4 to facilitate lipid transfer from the endoplasmic reticulum to plastids. *The Plant Cell*, *27*(10), 2941–2955. tpc.15.00394 https://doi.org/10.1105/tpc.15.00394.

Flores-Martín, J., Reyna, L., Ridano, M. E., Panzetta-Dutari, G. M., & Genti-Raimondi, S. (2016). Suppression of StarD7 promotes endoplasmic reticulum stress and induces ROS production. *Free Radical Biology & Medicine*, *99*, 286–295. https://doi.org/10.1016/j.freeradbiomed.2016.08.023.

Föller, M., Hermann, A., Gu, S., Alesutan, I., Qadri, S. M., Borst, O., et al. (2012). Chorein-sensitive polymerization of cortical actin and suicidal cell death in chorea-acanthocytosis. *The FASEB Journal*, *26*(4), 1526–1534. https://doi.org/10.1096/fj.11-198317.

Ford, R. C., & Beis, K. (2019). Learning the ABCs one at a time: Structure and mechanism of ABC transporters. *Biochemical Society Transactions*, *47*(1), 23–36. https://doi.org/10.1042/BST20180147.

Gaigg, B., Simbeni, R., Hrastnik, C., Paltauf, F., & Daum, G. (1995). Characterization of a microsomal subfraction associated with mitochondria of the yeast, Saccharomyces cerevisiae. Involvement in synthesis and import of phospholipids into mitochondria. *Biochimica et Biophysica Acta*, *1234*(2), 214–220. https://doi.org/10.1016/0005-2736(94)00287-y.

Galmes, R., Houcine, A., Vliet, A. R., Agostinis, P., Jackson, C. L., & Giordano, F. (2016). ORP5/ORP8 localize to endoplasmic reticulum–mitochondria contacts and are involved in mitochondrial function. *EMBO Reports*, *17*(6), 800–810. https://doi.org/10.15252/embr.201541108.

Gatta, A. T., Sauerwein, A. C., Zhuravleva, A., Levine, T. P., & Matthews, S. (2018). Structural insights into a StART-like domain in Lam4 and its interaction with sterol ligands. *Biochemical and Biophysical Research Communications*, *495*(3), 2270–2274. https://doi.org/10.1016/j.bbrc.2017.12.109.

Gatta, A. T., Wong, L. H., Sere, Y. Y., Calderón-Noreña, D. M., Cockcroft, S., Menon, A. K., et al. (2015). A new family of StART domain proteins at membrane contact sites has a role in ER-PM sterol transport. *eLife*, *4*, e07253. https://doi.org/10.7554/eLife.07253.

Gaude, N., Tippmann, H., Flemetakis, E., Katinakis, P., Udvardi, M., & Dörmann, P. (2004). The galactolipid digalactosyldiacylglycerol accumulates in the peribacteroid membrane of nitrogen-fixing nodules of soybean and lotus. *Journal of Biological Chemistry*, *279*(33), 34624–34630. https://doi.org/10.1074/jbc.M404098200.

Ghosh, R., de Campos, M. K. F., Huang, J., Huh, S. K., Orlowski, A., Yang, Y., et al. (2015). Sec14-nodulin proteins and the patterning of phosphoinositide landmarks for developmental control of membrane morphogenesis. *Molecular Biology of the Cell*, *26*(9), 1764–1781. https://doi.org/10.1091/mbc.E14-10-1475.

Goss, V., Hunt, A. N., & Postle, A. D. (2013). Regulation of lung surfactant phospholipid synthesis and metabolism. *Biochimica et Biophysica Acta*, *1831*(2), 448–458. https://doi.org/10.1016/j.bbalip.2012.11.009.

Grabon, A., Bankaitis, V. A., & McDermott, M. I. (2019). The interface between phosphatidylinositol transfer protein function and phosphoinositide signaling in higher eukaryotes. *Journal of Lipid Research*, *60*(2), 242–268. https://doi.org/10.1194/jlr.R089730.

Grabon, A., Khan, D., & Bankaitis, V. A. (2015). Phosphatidylinositol transfer proteins and instructive regulation of lipid kinase biology. *Biochimica et Biophysica Acta (BBA)—Molecular and Cell Biology of Lipids*, *1851*(6), 724–735. https://doi.org/10.1016/j.bbalip.2014.12.011.

Graef, M., Friedman, J. R., Graham, C., Babu, M., & Nunnari, J. (2013). ER exit sites are physical and functional core autophagosome biogenesis components. *Molecular Biology of the Cell*, *24*(18), 2918–2931. https://doi.org/10.1091/mbc.e13-07-0381.

Hammargren, J., Sundström, J., Johansson, M., Bergman, P., & Knorpp, C. (2007). On the phylogeny, expression and targeting of plant nucleoside diphosphate kinases. *Physiologia Plantarum*, *129*(1), 79–89. https://doi.org/10.1111/j.1399-3054.2006.00794.x.

Hanada, K. (2018). Lipid transfer proteins rectify inter-organelle flux and accurately deliver lipids at membrane contact sites. *Journal of Lipid Research*, *59*(8), 1341–1366. https://doi.org/10.1194/jlr.R085324.

Harrison, M. J., & Ivanov, S. (2017). Exocytosis for endosymbiosis: Membrane trafficking pathways for development of symbiotic membrane compartments. *Current Opinion in Plant Biology*, *38*, 101–108. https://doi.org/10.1016/j.pbi.2017.04.019.

Hartel, H., Dormann, P., & Benning, C. (2001). Galactolipids not associated with the photosynthetic apparatus in phosphate-deprived plants. *Journal of Photochemistry and Photobiology B: Biology*, *61*, 46–51.

Heazlewood, J. L., Tonti-Filippini, J. S., Gout, A. M., Day, D. A., Whelan, J., & Millar, A. H. (2004). Experimental analysis of the Arabidopsis mitochondrial proteome highlights signaling and regulatory components, provides assessment of targeting prediction programs, and indicates plant-specific mitochondrial proteins. *The Plant Cell*, *16*(1), 241–256. https://doi.org/10.1105/tpc.016055.

Hertle, A. P., García-Cerdán, J. G., Armbruster, U., Shih, R., Lee, J. J., Wong, W., et al. (2020). A Sec14 domain protein is required for photoautotrophic growth and chloroplast vesicle formation in Arabidopsis thaliana. *Proceedings of the National Academy of Sciences of the United States of America*, *117*(16), 9101–9111. https://doi.org/10.1073/pnas.1916946117.

Hirabayashi, Y., Kwon, S. K., Paek, H., Pernice, W. M., Paul, M. A., Lee, J., et al. (2017). ER-mitochondria tethering by PDZD8 regulates Ca(2+) dynamics in mammalian neurons. *Science*, *358*(6363), 623–630. https://doi.org/10.1126/science.aan6009.

Ho, C. M., Paciorek, T., Abrash, E., & Bergmann, D. C. (2016). Modulators of stomatal lineage signal transduction alter membrane contact sites and reveal specialization among ERECTA kinases. *Developmental Cell*, *38*(4), 345–357. https://doi.org/10.1016/j.devcel.2016.07.016.

Hollenstein, K., Dawson, R. J., & Locher, K. P. (2007). Structure and mechanism of ABC transporter proteins. *Current Opinion in Structural Biology*, *17*(4), 412–418. https://doi.org/10.1016/j.sbi.2007.07.003.

Hooper, C. M., Castleden, I. R., Tanz, S. K., Aryamanesh, N., & Millar, A. H. (2017). SUBA4: The interactive data analysis centre for Arabidopsis subcellular protein locations. *Nucleic Acids Research*, *45*(D1), D1064–D1074. https://doi.org/10.1093/nar/gkw1041.

Horenkamp, F. A., Valverde, D. P., Nunnari, J., & Reinisch, K. M. (2018). Molecular basis for sterol transport by StART-like lipid transfer domains. *The EMBO Journal*, *37*(6). https://doi.org/10.15252/embj.201798002.

Horibata, Y., Ando, H., Zhang, P., Vergnes, L., Aoyama, C., Itoh, M., et al. (2016). StarD7 protein deficiency adversely affects the phosphatidylcholine composition, respiratory activity, and cristae structure of mitochondria. *The Journal of Biological Chemistry*, *291*(48), 24880–24891. https://doi.org/10.1074/jbc.M116.736793.

Horibata, Y., & Sugimoto, H. (2010). StarD7 mediates the intracellular trafficking of phosphatidylcholine to mitochondria. *Journal of Biological Chemistry*, *285*(10), 7358–7365. https://doi.org/10.1074/jbc.M109.056960.

Hou, Q., Ufer, G., & Bartels, D. (2016). Lipid signalling in plant responses to abiotic stress: Lipid signalling in plant responses to abiotic stress. *Plant, Cell & Environment*, *39*(5), 1029–1048. https://doi.org/10.1111/pce.12666.

Hsueh, Y.-C., Ehmann, C., Flinner, N., Ladig, R., & Schleiff, E. (2017). The plastid outer membrane localized LPTD1 is important for glycerolipid remodelling under phosphate starvation: LptD orthologues in *A. thaliana*. *Plant, Cell & Environment*, *40*(8), 1643–1657. https://doi.org/10.1111/pce.12973.

Huang, J., Ghosh, R., & Bankaitis, V. A. (2016). Sec14-like phosphatidylinositol transfer proteins and the biological landscape of phosphoinositide signaling in plants. *Biochimica et Biophysica Acta*, *1861*(9 Pt. B), 1352–1364. https://doi.org/10.1016/j.bbalip.2016.03.027.

Huang, J., Ghosh, R., Tripathi, A., Lönnfors, M., Somerharju, P., & Bankaitis, V. A. (2016). Two-ligand priming mechanism for potentiated phosphoinositide synthesis is an evolutionarily conserved feature of Sec14-like phosphatidylinositol and phosphatidylcholine exchange proteins. *Molecular Biology of the Cell*, *27*(14), 2317–2330. https://doi.org/10.1091/mbc.E16-04-0221.

Hua-Ying, M., Wen-Ju, W., Wei-Hua, S., Ya-Chun, S., Feng, L., Cong-Na, L., et al. (2019). Genome-wide identification, phylogeny, and expression analysis of Sec14-like PITP gene family in sugarcane. *Plant Cell Reports*, *38*(5), 637–655. https://doi.org/10.1007/s00299-019-02394-1.

Hurlock, A. K., Roston, R. L., Wang, K., & Benning, C. (2014). Lipid trafficking in plant cells. *Traffic*, *15*(9), 915–932. https://doi.org/10.1111/tra.12187.

Im, Y. J., Raychaudhuri, S., Prinz, W. A., & Hurley, J. H. (2005). Structural mechanism for sterol sensing and transport by OSBP-related proteins. *Nature*, *437*(7055), 154–158. https://doi.org/10.1038/nature03923.

Ishikawa, K., Tamura, K., Fukao, Y., & Shimada, T. (2020). Structural and functional relationships between plasmodesmata and plant endoplasmic reticulum–plasma membrane contact sites consisting of three synaptotagmins. *New Phytologist*, *226*(3), 798–808. https://doi.org/10.1111/nph.16391.

Isono, E., Katsiarimpa, A., Müller, I. K., Anzenberger, F., Stierhof, Y.-D., Geldner, N., et al. (2010). The deubiquitinating enzyme AMSH3 is required for intracellular trafficking and vacuole biogenesis in Arabidopsis thaliana. *The Plant Cell*, *22*(6), 1826–1837. https://doi.org/10.1105/tpc.110.075952.

Itoh, R. D., Nakajima, K. P., Sasaki, S., Ishikawa, H., Kazama, Y., Abe, T., et al. (2021). TGD5 is required for normal morphogenesis of non-mesophyll plastids, but not mesophyll chloroplasts, in *Arabidopsis*. *The Plant Journal*, *107*, 237–255. https://doi.org/10.1111/tpj.15287.

Janin, J., Dumas, C., Moréra, S., Xu, Y., Meyer, P., Chiadmi, M., et al. (2000). Three-dimensional structure of nucleoside diphosphate kinase. *Journal of Bioenergetics and Biomembranes*, *32*(3), 215–225. https://doi.org/10.1023/a:1005528811303.

Jeong, H., Park, J., Jun, Y., & Lee, C. (2017). Crystal structures of Mmm1 and Mdm12-Mmm1 reveal mechanistic insight into phospholipid trafficking at ER-mitochondria contact sites. *Proceedings of the National Academy of Sciences of the United States of America*, *114*(45), E9502–E9511. https://doi.org/10.1073/pnas.1715592114.

Jessen, D., Roth, C., Wiermer, M., & Fulda, M. (2015). Two activities of long-chain acyl-coenzyme A synthetase are involved in lipid trafficking between the endoplasmic reticulum and the plastid in Arabidopsis. *Plant Physiology*, *167*(2), 351–366. https://doi.org/10.1104/pp.114.250365.

Jeyasimman, D., & Saheki, Y. (2020). SMP domain proteins in membrane lipid dynamics. *Biochimica et Biophysica Acta (BBA)—Molecular and Cell Biology of Lipids*, *1865*(1), 158447. https://doi.org/10.1016/j.bbalip.2019.04.007.

Jouannic, N., Lepetit, M., Vergnolle, C., Cantrel, C., Gardies, A. M., Kader, J. C., et al. (1998). Isolation of a cDNA from Arabidopsis thaliana that complements the sec14 mutant of yeast. *European Journal of Biochemistry*, *258*(2), 402–410. https://doi.org/10.1046/j.1432-1327.1998.2580402.x.

Jouhet, J., Maréchal, E., Baldan, B., Bligny, R., Joyard, J., & Block, M. A. (2004). Phosphate deprivation induces transfer of DGDG galactolipid from chloroplast to mitochondria. *Journal of Cell Biology*, *167*(5), 863–874. https://doi.org/10.1083/jcb.200407022.

Jouhet, J., Maréchal, E., Bligny, R., Joyard, J., & Block, M. A. (2003). Transient increase of phosphatidylcholine in plant cells in response to phosphate deprivation. *FEBS Letters*, *544*(1–3), 63–68. https://doi.org/10.1016/S0014-5793(03)00477-0.

Kader, J.-C. (1996). Lipid-transfer proteins in plants. *Annual Review of Plant Physiology and Plant Molecular Biology*, *47*(1), 627–654. https://doi.org/10.1146/annurev.arplant.47.1.627.

Kagan, V. E., Jiang, J., Huang, Z., Tyurina, Y. Y., Desbourdes, C., Cottet-Rousselle, C., et al. (2016). NDPK-D (NM23-H4)-mediated externalization of cardiolipin enables elimination of depolarized mitochondria by mitophagy. *Cell Death and Differentiation*, *23*(7), 1140–1151. https://doi.org/10.1038/cdd.2015.160.

Kang, H. W., Wei, J., & Cohen, D. E. (2010). PC-TP/StARD2: Of membranes and metabolism. *Trends in Endocrinology and Metabolism: TEM*, *21*(7), 449–456. https://doi.org/10.1016/j.tem.2010.02.001.

Kannan, M., Lahiri, S., Liu, L.-K., Choudhary, V., & Prinz, W. A. (2017). Phosphatidylserine synthesis at membrane contact sites promotes its transport out of the ER. *Journal of Lipid Research*, *58*(3), 553–562. https://doi.org/10.1194/jlr.M072959.

Kanno, K., Wu, M. K., Agate, D. S., Fanelli, B. J., Wagle, N., Scapa, E. F., et al. (2007). Interacting proteins dictate function of the minimal START domain phosphatidylcholine transfer protein/StarD2. *Journal of Biological Chemistry*, *282*(42), 30728–30736. https://doi.org/10.1074/jbc.M703745200.

Karki, N., Johnson, B. S., & Bates, P. D. (2019). Metabolically distinct pools of phosphatidylcholine are involved in trafficking of fatty acids out of and into the chloroplast for membrane production. *The Plant Cell*, *31*, 2768–2788. https://doi.org/10.1105/tpc.19.00121.

Katsiarimpa, A., Kalinowska, K., Anzenberger, F., Weis, C., Ostertag, M., Tsutsumi, C., et al. (2013). The deubiquitinating enzyme AMSH1 and the ESCRT-III subunit VPS2.1 are required for autophagic degradation in *Arabidopsis*. *The Plant Cell*, *25*(6), 2236–2252. https://doi.org/10.1105/tpc.113.113399.

Kearns, M. A., Monks, D. E., Fang, M., Rivas, M. P., Courtney, P. D., Chen, J., et al. (1998). Novel developmentally regulated phosphoinositide binding proteins from soybean whose expression bypasses the requirement for an essential phosphatidylinositol transfer protein in yeast. *The EMBO Journal*, *17*(14), 4004–4017. https://doi.org/10.1093/emboj/17.14.4004.

Kelly, A. A., Froehlich, J. E., & Dörmann, P. (2003). Disruption of the two digalactosyldiacylglycerol synthase genes *DGD1* and *DGD2* in Arabidopsis reveals the existence of an additional enzyme of galactolipid synthesis. *The Plant Cell*, *15*(11), 2694–2706. https://doi.org/10.1105/tpc.016675.

Kelly, A. A., Kalisch, B., Hölzl, G., Schulze, S., Thiele, J., Melzer, M., et al. (2016). Synthesis and transfer of galactolipids in the chloroplast envelope membranes of *Arabidopsis thaliana*. *Proceedings of the National Academy of Sciences of the United States of America*, *113*(38), 10714–10719. https://doi.org/10.1073/pnas.1609184113.

Kf de Campos, M., & Schaaf, G. (2017). The regulation of cell polarity by lipid transfer proteins of the SEC14 family. *Current Opinion in Plant Biology*, *40*, 158–168. https://doi.org/10.1016/j.pbi.2017.09.007.

Khafif, M., Balague, C., Huard-Chauveau, C., & Roby, D. (2017). An essential role for the VASt domain of the Arabidopsis VAD1 protein in the regulation of defense and cell death in response to pathogens. *PLoS One*, *12*(7), e0179782. https://doi.org/10.1371/journal.pone.0179782.

Khafif, M., Cottret, L., Balague, C., & Raffaele, S. (2014). Identification and phylogenetic analyses of VASt, an uncharacterized protein domain associated with lipid-binding domains in Eukaryotes. *BMC Bioinformatics*, *15*, 222. https://doi.org/10.1186/1471-2105-15-222.

Khosla, A., Paper, J. M., Boehler, A. P., Bradley, A. M., Neumann, T. R., & Schrick, K. (2014). HD-zip proteins GL2 and HDG11 have redundant functions in Arabidopsis trichomes, and GL2 activates a positive feedback loop via MYB23. *The Plant Cell*, *26*(5), 2184–2200. https://doi.org/10.1105/tpc.113.120360.

Kiba, A., Galis, I., Hojo, Y., Ohnishi, K., Yoshioka, H., & Hikichi, Y. (2014). SEC14 phospholipid transfer protein is involved in lipid signaling-mediated plant immune responses in Nicotiana benthamiana. *PLoS One*, *9*(5), e98150. https://doi.org/10.1371/journal.pone.0098150.

Kiba, A., Nakano, M., Vincent-Pope, P., Takahashi, H., Sawasaki, T., Endo, Y., et al. (2012). A novel Sec14 phospholipid transfer protein from Nicotiana benthamiana is up-regulated in response to Ralstonia solanacearum infection, pathogen associated molecular patterns and effector molecules and involved in plant immunity. *Journal of Plant Physiology*, *169*(10), 1017–1022. https://doi.org/10.1016/j.jplph.2012.04.002.

Kim, H., Kwon, H., Kim, S., Kim, M. K., Botella, M. A., Yun, H. S., et al. (2016). Synaptotagmin 1 negatively controls the two distinct immune secretory pathways to powdery mildew fungi in Arabidopsis. *Plant & Cell Physiology*, *57*(6), 1133–1141. https://doi.org/10.1093/pcp/pcw061.

Kim, H. J., & Triplett, B. A. (2001). Cotton fiber growth in planta and in vitro. Models for plant cell elongation and cell wall biogenesis. *Plant Physiology*, *127*(4), 1361–1366.

Knorpp, C., Johansson, M., & Baird, A.-M. (2003). Plant mitochondrial nucleoside diphosphate kinase is attached to the membrane through interaction with the adenine nucleotide translocator. *FEBS Letters*, *555*(2), 363–366. https://doi.org/10.1016/s0014-5793(03)01288-2.

Kobayashi, K., Awai, K., Nakamura, M., Nagatani, A., Masuda, T., & Ohta, H. (2009). Type-B monogalactosyldiacylglycerol synthases are involved in phosphate starvation-induced lipid remodeling, and are crucial for low-phosphate adaptation. *The Plant Journal*, *57*(2), 322–331. https://doi.org/10.1111/j.1365-313X.2008.03692.x.

Kopec, K. O., Alva, V., & Lupas, A. N. (2010). Homology of SMP domains to the TULIP superfamily of lipid-binding proteins provides a structural basis for lipid exchange between ER and mitochondria. *Bioinformatics*, *26*(16), 1927–1931. https://doi.org/10.1093/bioinformatics/btq326.

Kornmann, B., Currie, E., Collins, S. R., Schuldiner, M., Nunnari, J., Weissman, J. S., et al. (2009). An ER-mitochondria tethering complex revealed by a synthetic biology screen. *Science, 325*(5939), 477–481. https://doi.org/10.1126/science.1175088.

Kumar, N., Leonzino, M., Hancock-Cerutti, W., Horenkamp, F. A., Li, P., Lees, J. A., et al. (2018). VPS13A and VPS13C are lipid transport proteins differentially localized at ER contact sites. *Journal of Cell Biology, 217*(10), 3625–3639. https://doi.org/10.1083/jcb.201807019.

LaBrant, E., Barnes, A. C., & Roston, R. L. (2018). Lipid transport required to make lipids of photosynthetic membranes. *Photosynthesis Research, 138*(3), 345–360. https://doi.org/10.1007/s11120-018-0545-5.

Larsson, K. E., Kjellberg, J. M., Tjellström, H., & Sandelius, A. (2007). LysoPC acyltransferase/PC transacylase activities in plant plasma membrane and plasma membrane-associated endoplasmic reticulum. *BMC Plant Biology*, 7(1), 64. https://doi.org/10.1186/1471-2229-7-64.

Lee, J.-Y., Colinas, J., Wang, J. Y., Mace, D., Ohler, U., & Benfey, P. N. (2006). Transcriptional and posttranscriptional regulation of transcription factor expression in Arabidopsis roots. *Proceedings of the National Academy of Sciences of the United States of America, 103*(15), 6055–6060. https://doi.org/10.1073/pnas.0510607103.

Lee, I., & Hong, W. (2006). Diverse membrane-associated proteins contain a novel SMP domain. *The FASEB Journal, 20*(2), 202–206. https://doi.org/10.1096/fj.05-4581hyp.

Lee, E., Santana, B. V. N., Samuels, E., Benitez-Fuente, F., Corsi, E., Botella, M. A., et al. (2020). Rare earth elements induce cytoskeleton-dependent and PI4P-associated rearrangement of SYT1/SYT5 endoplasmic reticulum–plasma membrane contact site complexes in Arabidopsis. *Journal of Experimental Botany, 71*(14), 3986–3998. https://doi.org/10.1093/jxb/eraa138.

Lee, E., Vanneste, S., Perez-Sancho, J., Benitez-Fuente, F., Strelau, M., Macho, A. P., et al. (2019). Ionic stress enhances ER-PM connectivity via phosphoinositide-associated SYT1 contact site expansion in Arabidopsis. *Proceedings of the National Academy of Sciences of the United States of America, 116*(4), 1420–1429. https://doi.org/10.1073/pnas.1818099116.

Lei, L., Chen, L., Shi, X., Li, Y., Wang, J., Chen, D., et al. (2014). A nodule-specific lipid transfer protein AsE246 participates in transport of plant-synthesized lipids to symbiosome membrane and is essential for nodule organogenesis in chinese milk vetch. *Plant Physiology, 164*(2), 1045–1058. https://doi.org/10.1104/pp.113.232637.

Levy, A., Zheng, J. Y., & Lazarowitz, S. G. (2015). Synaptotagmin SYTA forms ER-plasma membrane junctions that are recruited to plasmodesmata for plant virus movement. *Current Biology, 25*(15), 2018–2025. https://doi.org/10.1016/j.cub.2015.06.015.

Li, Z., Gao, J., Benning, C., & Sharkey, T. D. (2012). Characterization of photosynthesis in Arabidopsis ER-to-plastid lipid trafficking mutants. *Photosynthesis Research, 112*(1), 49–61. https://doi.org/10.1007/s11120-012-9734-9.

Li, D. Y., Inoue, H., Takahashi, M., Kojima, T., Shiraiwa, M., & Takahara, H. (2008). Molecular characterization of a novel salt-inducible gene for an OSBP (oxysterol-binding protein)-homologue from soybean. *Gene, 407*(1–2), 12–20. https://doi.org/10.1016/j.gene.2007.02.029.

Li, Y., Kabbage, M., Liu, W., & Dickman, M. B. (2016). Aspartyl protease-mediated cleavage of BAG6 Is necessary for autophagy and fungal resistance in plants. *Plant Cell, 28*(1), 233–247. https://doi.org/10.1105/tpc.15.00626.

Li, P., Lees, J. A., Lusk, C. P., & Reinisch, K. M. (2020). Cryo-EM reconstruction of a VPS13 fragment reveals a long groove to channel lipids between membranes. *Journal of Cell Biology, 219*(5), e202001161. https://doi.org/10.1083/jcb.202001161.

Li, X., Routt, S. M., Xie, Z., Cui, X., Fang, M., Kearns, M. A., et al. (2000). Identification of a novel family of nonclassic yeast phosphatidylinositol transfer proteins whose function modulates phospholipase D activity and Sec14p-independent cell growth. *Molecular Biology of the Cell*, *11*(6), 1989–2005. https://doi.org/10.1091/mbc.11.6.1989.

Li, N., Xu, C., Li-Beisson, Y., & Philippar, K. (2016). Fatty acid and lipid transport in plant cells. *Trends in Plant Science*, *21*(2), 145–158. https://doi.org/10.1016/j.tplants.2015.10.011.

Li-Beisson, Y., Shorrosh, B., Beisson, F., Andersson, M. X., Arondel, V., Bates, P. D., et al. (2013). Acyl-lipid metabolism. *The Arabidopsis Book*, *11*, e0161. https://doi.org/10.1199/tab.0161.

Lin, S., Ikegami, M., Moon, C., Naren, A. P., & Shannon, J. M. (2015). Lysophosphatidylcholine acyltransferase 1 (LPCAT1) specifically interacts with phospholipid transfer protein StarD10 to facilitate surfactant phospholipid trafficking in alveolar type II cells. *Journal of Biological Chemistry*, *290*(30), 18559–18574. https://doi.org/10.1074/jbc.M115.666701.

Lindquist, E., & Aronsson, H. (2018). Chloroplast vesicle transport. *Photosynthesis Research*, *138*(3), 361–371. https://doi.org/10.1007/s11120-018-0566-0.

Lipp, N.-F., Ikhlef, S., Milanini, J., & Drin, G. (2020). Lipid exchangers: cellular functions and mechanistic links with phosphoinositide metabolism. *Frontiers in Cell and Development Biology*, *8*, 663. https://doi.org/10.3389/fcell.2020.00663.

Liu, L., & Li, J. (2019). Communications between the endoplasmic reticulum and other organelles during abiotic stress response in plants. *Frontiers in Plant Science*, *10*(749), 1–14. https://doi.org/10.3389/fpls.2019.00749.

Liu, L.-K., Choudhary, V., Toulmay, A., & Prinz, W. A. (2017). An inducible ER–Golgi tether facilitates ceramide transport to alleviate lipotoxicity. *Journal of Cell Biology*, *216*(1), 131–147. https://doi.org/10.1083/jcb.201606059.

Liu, X., Ma, D., Zhang, Z., Wang, S., Du, S., Deng, X., et al. (2019). Plant lipid remodeling in response to abiotic stresses. *Environmental and Experimental Botany*, *165*, 174–184. https://doi.org/10.1016/j.envexpbot.2019.06.005.

López-Marqués, R. L., Davis, J. A., Harper, J. F., & Palmgren, M. (2021). Dynamic membranes: The multiple roles of P4 and P5 ATPases. *Plant Physiology*, *185*(3), 619–631. https://doi.org/10.1093/plphys/kiaa065.

Lorrain, S., Lin, B., Auriac, M. C., Kroj, T., Saindrenan, P., Nicole, M., et al. (2004). Vascular associated death1, a novel GRAM domain-containing protein, is a regulator of cell death and defense responses in vascular tissues. *Plant Cell*, *16*(8), 2217–2232. https://doi.org/10.1105/tpc.104.022038.

Lu, B., & Benning, C. (2009). A 25-amino acid sequence of the Arabidopsis TGD2 protein is sufficient for specific binding of phosphatidic acid. *Journal of Biological Chemistry*, *284*(26), 17420–17427. https://doi.org/10.1074/jbc.M109.016014.

Lu, B., Xu, C., Awai, K., Jones, A. D., & Benning, C. (2007). A small ATPase protein of Arabidopsis, TGD3, involved in chloroplast lipid import. *Journal of Biological Chemistry*, *282*(49), 35945–35953. https://doi.org/10.1074/jbc.M704063200.

Luzarowski, M., Kosmacz, M., Sokolowska, E., Jasińska, W., Willmitzer, L., Veyel, D., et al. (2017). Affinity purification with metabolomic and proteomic analysis unravels diverse roles of nucleoside diphosphate kinases. *Journal of Experimental Botany*, *68*(13), 3487–3499. https://doi.org/10.1093/jxb/erx183.

Maeda, K., Anand, K., Chiapparino, A., Kumar, A., Poletto, M., Kaksonen, M., et al. (2013). Interactome map uncovers phosphatidylserine transport by oxysterol-binding proteins. *Nature*, *501*(7466), 257–261. https://doi.org/10.1038/nature12430.

Maréchal, E., & Bastien, O. (2014). Modeling of regulatory loops controlling galactolipid biosynthesis in the inner envelope membrane of chloroplasts. *Journal of Theoretical Biology*, *361*, 1–13. https://doi.org/10.1016/j.jtbi.2014.07.013.

Mathur, J. (2020). Review: Morphology, behaviour and interactions of organelles. *Plant Science*, *301*, 110662. https://doi.org/10.1016/j.plantsci.2020.110662.

Matoba, K., Kotani, T., Tsutsumi, A., Tsuji, T., Mori, T., Noshiro, D., et al. (2020). Atg9 is a lipid scramblase that mediates autophagosomal membrane expansion. *Nature Structural & Molecular Biology*, *27*(12), 1185–1193. https://doi.org/10.1038/s41594-020-00518-w.

Maxfield, F. R., & Menon, A. K. (2016). Intramembrane and intermembrane lipid transport. In *Biochemistry of lipids, lipoproteins and membranes* (pp. 415–436). Elsevier. https://doi.org/10.1016/B978-0-444-63438-2.00014-6.

McFarlane, H. E., Lee, E. K., van Bezouwen, L. S., Ross, B., Rosado, A., & Samuels, A. L. (2017). Multiscale structural analysis of plant ER-PM contact sites. *Plant & Cell Physiology*, *58*(3), 478–484. https://doi.org/10.1093/pcp/pcw224.

Meï, C., Michaud, M., Cussac, M., Albrieux, C., Gros, V., Maréchal, E., et al. (2015). Levels of polyunsaturated fatty acids correlate with growth rate in plant cell cultures. *Scientific Reports*, *5*, 15207. https://doi.org/10.1038/srep15207.

Michaud, M. (2018). Analysis of the MTL supercomplex at contact sites between mitochondria and plastids. *Methods in Molecular Biology*, *1829*, 173–188. https://doi.org/10.1007/978-1-4939-8654-5_12.

Michaud, M., Gros, V., Tardif, M., Brugière, S., Ferro, M., Prinz, W. A., et al. (2016). AtMic60 Is involved in plant mitochondria lipid trafficking and is part of a large complex. *Current Biology*, *26*(5), 627–639. https://doi.org/10.1016/j.cub.2016.01.011.

Michaud, M., & Jouhet, J. (2019). Lipid trafficking at membrane contact sites during plant development and stress response. *Frontiers in Plant Science*, *10*, 2. https://doi.org/10.3389/fpls.2019.00002.

Michaud, M., Prinz, W. A., & Jouhet, J. (2017). Glycerolipid synthesis and lipid trafficking in plant mitochondria. *The FEBS Journal*, *284*(3), 376–390. https://doi.org/10.1111/febs.13812.

Miliara, X., Tatsuta, T., Berry, J.-L., Rouse, S. L., Solak, K., Chorev, D. S., et al. (2019). Structural determinants of lipid specificity within Ups/PRELI lipid transfer proteins. *Nature Communications*, *10*(1), 1130. https://doi.org/10.1038/s41467-019-09089-x.

Miyata, N., Watanabe, Y., Tamura, Y., Endo, T., & Kuge, O. (2016). Phosphatidylserine transport by Ups2-Mdm35 in respiration-active mitochondria. *The Journal of Cell Biology*, *214*(1), 77–88. https://doi.org/10.1083/jcb.201601082.

Moellering, E. R., & Benning, C. (2011). Galactoglycerolipid metabolism under stress: A time for remodeling. *Trends in Plant Science*, *16*(2), 98–107. https://doi.org/10.1016/j.tplants.2010.11.004.

Mongrand, S., Bessoule, J.-J., Cabantous, F., & Cassagne, C. (1998). The C16:3\C18:3 fatty acid balance in photosynthetic tissues from 468 plant species. *Phytochemistry*, *49*(4), 1049–1064. https://doi.org/10.1016/S0031-9422(98)00243-X.

Mongrand, S., Cassagne, C., & Bessoule, J. J. (2000). Import of lyso-phosphatidylcholine into chloroplasts likely at the origin of eukaryotic plastidial lipids. *Plant Physiology*, *122*(3), 845–852. https://doi.org/10.1104/pp.122.3.845.

Monks, D. E., Aghoram, K., Courtney, P. D., DeWald, D. B., & Dewey, R. E. (2001). Hyperosmotic stress induces the rapid phosphorylation of a soybean phosphatidylinositol transfer protein homolog through activation of the protein kinases SPK1 and SPK2. *The Plant Cell*, *13*(5), 1205–1219. https://doi.org/10.1105/tpc.13.5.1205.

Montag, K., Hornbergs, J., Ivanov, R., & Bauer, P. (2020). Phylogenetic analysis of plant multi-domain SEC14-like phosphatidylinositol transfer proteins and structure–function properties of PATELLIN2. *Plant Molecular Biology*, *104*(6), 665–678. https://doi.org/10.1007/s11103-020-01067-y.

Morris, J. S., Caldo, K. M. P., Liang, S., & Facchini, P. J. (2021). PR10/Bet v1-like proteins as novel contributors to plant biochemical diversity. *Chembiochem*, *22*(2), 264–287. https://doi.org/10.1002/cbic.202000354.

Moser von Filseck, J., & Drin, G. (2016). Running up that hill: How to create cellular lipid gradients by lipid counter-flows. *Biochimie*, *130*, 115–121. https://doi.org/10.1016/j.biochi.2016.08.001.

Moser von Filseck, J., opi, A., Delfosse, V., Vanni, S., Jackson, C. L., Bourguet, W., et al. (2015). Phosphatidylserine transport by ORP/Osh proteins is driven by phosphatidylinositol 4-phosphate. *Science*, *349*(6246), 432–436. https://doi.org/10.1126/science.aab1346.

Mueller-Schuessele, S. J., & Michaud, M. (2018). Plastid transient and stable interactions with other cell compartments. *Methods in Molecular Biology*, *1829*, 87–109. https://doi.org/10.1007/978-1-4939-8654-5_6.

Mukherjee, I., Ghosh, M., & Meinecke, M. (2021). MICOS and the mitochondrial inner membrane morphology—When things get out of shape. *FEBS Letters*, *595*(8), 1159–1183.

Muñoz-Braceras, S., Calvo, R., & Escalante, R. (2015). TipC and the chorea-acanthocytosis protein VPS13A regulate autophagy in *Dictyostelium* and human HeLa cells. *Autophagy*, *11*(6), 918–927. https://doi.org/10.1080/15548627.2015.1034413.

Murley, A., Sarsam, R. D., Toulmay, A., Yamada, J., Prinz, W. A., & Nunnari, J. (2015). Ltc1 is an ER-localized sterol transporter and a component of ER–mitochondria and ER–vacuole contacts. *Journal of Cell Biology*, *209*(4), 539–548. https://doi.org/10.1083/jcb.201502033.

Murley, A., Yamada, J., Niles, B. J., Toulmay, A., Prinz, W. A., Powers, T., et al. (2017). Sterol transporters at membrane contact sites regulate TORC1 and TORC2 signaling. *The Journal of Cell Biology*, *216*(9), 2679–2689. https://doi.org/10.1083/jcb.201610032.

Nagy, P. D. (2015). Viral sensing of the subcellular environment regulates the assembly of new viral replicase complexes during the course of infection. *Journal of Virology*, *89*(10), 5196–5199. https://doi.org/10.1128/JVI.02973-14.

Nagy, P. D., Barajas, D., & Pogany, J. (2012). Host factors with regulatory roles in tombusvirus replication. *Current Opinion in Virology*, *2*(6), 691–698. https://doi.org/10.1016/j.coviro.2012.10.004.

Nakatogawa, H. (2020). Mechanisms governing autophagosome biogenesis. *Nature Reviews Molecular Cell Biology*, *21*(8), 439–458. https://doi.org/10.1038/s41580-020-0241-0.

Nicholls, H. T., Hornick, J. L., & Cohen, D. E. (2017). Phosphatidylcholine transfer protein/StarD2 promotes microvesicular steatosis and liver injury in murine experimental steatohepatitis. *American Journal of Physiology. Gastrointestinal and Liver Physiology*, *313*(1), G50–G61. https://doi.org/10.1152/ajpgi.00379.2016.

Noda, N. N. (2021). Atg2 and Atg9: Intermembrane and interleaflet lipid transporters driving autophagy. *Biochimica et Biophysica Acta (BBA)—Molecular and Cell Biology of Lipids*, *1866*(8), 158956. https://doi.org/10.1016/j.bbalip.2021.158956.

Olayioye, M. A., Buchholz, M., Schmid, S., Schoöffler, P., Hoffmann, P., & Pomorski, T. (2007). Phosphorylation of StarD10 on serine 284 by casein kinase II modulates its lipid transfer activity. *Journal of Biological Chemistry*, *282*(31), 22492–22498. https://doi.org/10.1074/jbc.M701990200.

Olayioye, M. A., Vehring, S., Müller, P., Herrmann, A., Schiller, J., Thiele, C., et al. (2005). StarD10, a START domain protein overexpressed in breast cancer, functions as a phospholipid transfer protein. *Journal of Biological Chemistry*, *280*(29), 27436–27442. https://doi.org/10.1074/jbc.M413330200.

Orii, M., Tsuji, T., Ogasawara, Y., & Fujimoto, T. (2021). Transmembrane phospholipid translocation mediated by Atg9 is involved in autophagosome formation. *Journal of Cell Biology*, *220*(3), e202009194. https://doi.org/10.1083/jcb.202009194.

Osawa, T., Kotani, T., Kawaoka, T., Hirata, E., Suzuki, K., Nakatogawa, H., et al. (2019). Atg2 mediates direct lipid transfer between membranes for autophagosome formation. *Nature Structural & Molecular Biology*, *26*(4), 281–288. https://doi.org/10.1038/s41594-019-0203-4.

Osman, C., Haag, M., Potting, C., Rodenfels, J., Dip, P. V., Wieland, F. T., et al. (2009). The genetic interactome of prohibitins: Coordinated control of cardiolipin and phosphatidylethanolamine by conserved regulators in mitochondria. *Journal of Cell Biology, 184*(4), 583–596. https://doi.org/10.1083/jcb.200810189.

Park, J.-S., & Neiman, A. M. (2012). *VPS13* regulates membrane morphogenesis during sporulation in *Saccharomyces cerevisiae*. *Journal of Cell Science, 125*(12), 3004–3011. https://doi.org/10.1242/jcs.105114.

Park, J.-S., & Neiman, A. M. (2020). XK is a partner for VPS13A: A molecular link between Chorea-Acanthocytosis and McLeod Syndrome. *Molecular Biology of the Cell, 31*(22), 2425–2436. https://doi.org/10.1091/mbc.E19-08-0439-T.

Peiro, A., Izquierdo-Garcia, A. C., Sanchez-Navarro, J. A., Pallas, V., Mulet, J. M., & Aparicio, F. (2014). Patellins 3 and 6, two members of the Plant Patellin family, interact with the movement protein of Alfalfa mosaic virus and interfere with viral movement. *Molecular Plant Pathology, 15*(9), 881–891. https://doi.org/10.1111/mpp.12146.

Perez-Sancho, J., Tilsner, J., Samuels, A. L., Botella, M. A., Bayer, E. M., & Rosado, A. (2016). Stitching organelles: Organization and function of specialized membrane contact sites in plants. *Trends in Cell Biology, 26*(9), 705–717. https://doi.org/10.1016/j.tcb.2016.05.007.

Perez-Sancho, J., Vanneste, S., Lee, E., McFarlane, H. E., Esteban Del Valle, A., Valpuesta, V., et al. (2015). The Arabidopsis synaptotagmin1 is enriched in endoplasmic reticulum-plasma membrane contact sites and confers cellular resistance to mechanical stresses. *Plant Physiology, 168*(1), 132–143. https://doi.org/10.1104/pp.15.00260.

Peterman, T. K., Ohol, Y. M., McReynolds, L. J., & Luna, E. J. (2004). Patellin1, a novel Sec14-like protein, localizes to the cell plate and binds phosphoinositides. *Plant Physiology, 136*(2), 3080–3094. discussion 3001-3002 https://doi.org/10.1104/pp.104.045369.

Peterman, T. K., Sequeira, A. S., Samia, J. A., & Lunde, E. E. (2006). Molecular cloning and characterization of patellin1, a novel sec14-related protein, from zucchini (Cucurbita pepo). *Journal of Plant Physiology, 163*(11), 1150–1158. https://doi.org/10.1016/j.jplph.2006.01.009.

Petit, J. D., Immel, F., Lins, L., & Bayer, E. M. (2019). Lipids or proteins: who is leading the dance at membrane contact sites? *Frontiers in Plant Science, 10*, 198. https://doi.org/10.3389/fpls.2019.00198.

Petrungaro, C., & Kornmann, B. (2019). Lipid exchange at ER-mitochondria contact sites: A puzzle falling into place with quite a few pieces missing. *Current Opinion in Cell Biology, 57*, 71–76. https://doi.org/10.1016/j.ceb.2018.11.005.

Pfanner, N., van der Laan, M., Amati, P., Capaldi, R. A., Caudy, A. A., Chacinska, A., et al. (2014). Uniform nomenclature for the mitochondrial contact site and cristae organizing system. *Journal of Cell Biology, 204*(7), 1083–1086. https://doi.org/10.1083/jcb.201401006.

Pfisterer, S. G., Bakula, D., Frickey, T., Cezanne, A., Brigger, D., Tschan, M. P., et al. (2014). Lipid droplet and early autophagosomal membrane targeting of Atg2A and Atg14L in human tumor cells. *Journal of Lipid Research, 55*(7), 1267–1278. https://doi.org/10.1194/jlr.M046359.

Pichler, H., Gaigg, B., Hrastnik, C., Achleitner, G., Kohlwein, S. D., Zellnig, G., et al. (2001). A subfraction of the yeast endoplasmic reticulum associates with the plasma membrane and has a high capacity to synthesize lipids. *European Journal of Biochemistry, 268*(8), 2351–2361. https://doi.org/10.1046/j.1432-1327.2001.02116.x.

Pietrangelo, A., & Ridgway, N. D. (2018). Bridging the molecular and biological functions of the oxysterol-binding protein family. *Cellular and Molecular Life Sciences, 75*(17), 3079–3098. https://doi.org/10.1007/s00018-018-2795-y.

Pii, Y., Astegno, A., Peroni, E., Zaccardelli, M., Pandolfini, T., & Crimi, M. (2009). The Medicago truncatula N5 gene encoding a root-specific lipid transfer protein is required for the symbiotic interaction with Sinorhizobium meliloti. *Molecular Plant-Microbe Interactions, 22*(12), 1577–1587.

Pii, Y., Molesini, B., Masiero, S., & Pandolfini, T. (2012). The non-specific lipid transfer protein N5 of Medicago truncatula is implicated in epidermal stages of rhizobium-host interaction. *BMC Plant Biology*, *12*(233), 1–13.

Pii, Y., Molesini, B., & Pandolfini, T. (2013). The involvement of Medicago truncatula non-specific lipid transfer protein N5 in the control of rhizobial infection. *Plant Signaling & Behavior*, *8*(7), e24836. https://doi.org/10.4161/psb.24836.

Plaxton, W. C., & Tran, H. T. (2011). Metabolic adaptations of phosphate-starved plants. *Plant Physiology*, *156*(3), 1006–1015. https://doi.org/10.1104/pp.111.175281.

Poirier, Y., Thoma, S., Somerville, C., & Schiefelbein, J. (1991). A mutant of Arabidopsis deficient in xylem loading of phosphate. *Plant Physiology*, *97*(7), 1087–1093.

Ponting, C. P., & Aravind, L. (1999). START: A lipid-binding domain in StAR, HD-ZIP and signalling proteins. *Trends in Biochemical Sciences*, *24*(4), 130–132. https://doi.org/10.1016/S0968-0004(99)01362-6.

Potting, C., Tatsuta, T., König, T., Haag, M., Wai, T., Aaltonen, M. J., et al. (2013). TRIAP1/PRELI complexes prevent apoptosis by mediating intramitochondrial transport of phosphatidic acid. *Cell Metabolism*, *18*(2), 287–295. https://doi.org/10.1016/j.cmet.2013.07.008.

Potting, C., Wilmes, C., Engmann, T., Osman, C., & Langer, T. (2010). Regulation of mitochondrial phospholipids by Ups1/PRELI-like proteins depends on proteolysis and Mdm35. *The EMBO Journal*, *29*(17), 2888–2898. https://doi.org/10.1038/emboj.2010.169.

Poulsen, L. R., López-Marqués, R. L., Pedas, P. R., McDowell, S. C., Brown, E., Kunze, R., et al. (2015). A phospholipid uptake system in the model plant Arabidopsis thaliana. *Nature Communications*, *6*(1), 7649. https://doi.org/10.1038/ncomms8649.

Prinz, W. A., Toulmay, A., & Balla, T. (2020). The functional universe of membrane contact sites. *Nature Reviews. Molecular Cell Biology*, *21*, 7–24. https://doi.org/10.1038/s41580-019-0180-9.

Pulli, I., Lassila, T., Pan, G., Yan, D., Olkkonen, V. M., & Törnquist, K. (2018). Oxysterol-binding protein related-proteins (ORPs) 5 and 8 regulate calcium signaling at specific cell compartments. *Cell Calcium*, *72*, 62–69. https://doi.org/10.1016/j.ceca.2018.03.001.

Rabitsch, K. P., Tóth, A., Gálová, M., Schleiffer, A., Schaffner, G., Aigner, E., et al. (2001). A screen for genes required for meiosis and spore formation based on whole-genome expression. *Current Biology*, *11*(13), 1001–1009. https://doi.org/10.1016/S0960-9822(01)00274-3.

Rampelt, H., Zerbes, R. M., van der Laan, M., & Pfanner, N. (2017). Role of the mitochondrial contact site and cristae organizing system in membrane architecture and dynamics. *Biochimica et Biophysica Acta (BBA) - Molecular Cell Research*, *1864*(4), 737–746. https://doi.org/10.1016/j.bbamcr.2016.05.020.

Reinisch, K. M., & De Camilli, P. (2016). SMP-domain proteins at membrane contact sites: Structure and function. *Biochimica et Biophysica Acta (BBA)—Molecular and Cell Biology of Lipids*, *1861*(8), 924–927. https://doi.org/10.1016/j.bbalip.2015.12.003.

Reinisch, K. M., & Prinz, W. A. (2021). Mechanisms of nonvesicular lipid transport. *Journal of Cell Biology*, *220*(3), e202012058. https://doi.org/10.1083/jcb.202012058.

Ren, J., Pei-Chen Lin, C., Pathak, M. C., Temple, B. R. S., Nile, A. H., Mousley, C. J., et al. (2014). A phosphatidylinositol transfer protein integrates phosphoinositide signaling with lipid droplet metabolism to regulate a developmental program of nutrient stress–induced membrane biogenesis. *Molecular Biology of the Cell*, *25*(5), 712–727. https://doi.org/10.1091/mbc.e13-11-0634.

Rochin, L., Sauvanet, C., Jääskeläinen, E., Houcine, A., Kivelä, A. M., Ma, X., et al. (2020). ORP5 transfers phosphatidylserine to mitochondria and regulates mitochondrial calcium uptake at endoplasmic reticulum—Mitochondria contact sites. *Preprint*, *84*. https://doi.org/10.1101/695577.

Roderick, S. L., Chan, W. W., Agate, D. S., Olsen, L. R., Vetting, M. W., Rajashankar, K. R., et al. (2002). Structure of human phosphatidylcholine transfer protein in complex with its ligand. *Nature Structural Biology*, *9*(7), 507–511. https://doi.org/10.1038/nsb812.

Rodriguez-Peña, R., Mounadi, K. E., & Garcia-Ruiz, H. (2021). Changes in subcellular localization of host proteins induced by plant viruses. *Viruses*, *13*(4), 677. https://doi.org/10.3390/v13040677.

Roston, R. L., Gao, J., Murcha, M. W., Whelan, J., & Benning, C. (2012). TGD1, -2, and -3 proteins involved in lipid trafficking form ATP-binding cassette (ABC) transporter with multiple substrate-binding proteins. *Journal of Biological Chemistry*, *287*(25), 21406–21415. https://doi.org/10.1074/jbc.M112.370213.

Roston, R., Gao, J., Xu, C., & Benning, C. (2011). Arabidopsis chloroplast lipid transport protein TGD2 disrupts membranes and is part of a large complex: TGD2 involved in chloroplast lipid transport. *The Plant Journal*, *66*(5), 759–769. https://doi.org/10.1111/j.1365-313X.2011.04536.x.

Ruiz-Lopez, N., Pérez-Sancho, J., del Valle, Haslam, R. P., Vanneste, S., … Botella, M. A. (2020). Synaptotagmins maintain diacylglycerol homeostasis at endoplasmic reticulum-plasma membrane contact sites during abiotic stress [preprint]. *BioRxiv*. https://doi.org/10.1101/2020.07.28.222919.

Saheki, Y., Bian, X., Schauder, C. M., Sawaki, Y., Surma, M. A., Klose, C., et al. (2016). Control of plasma membrane lipid homeostasis by the extended synaptotagmins. *Nature Cell Biology*, *18*(5), 504–515. https://doi.org/10.1038/ncb3339.

Saheki, Y., & De Camilli, P. (2017a). Endoplasmic reticulum-plasma membrane contact sites. *Annual Review of Biochemistry*, *86*, 659–684. https://doi.org/10.1146/annurev-biochem-061516-044932.

Saheki, Y., & De Camilli, P. (2017b). The extended-synaptotagmins. *Biochimica et Biophysica Acta*, *1864*(9), 1490–1493. https://doi.org/10.1016/j.bbamcr.2017.03.013.

Saita, S., Tatsuta, T., Lampe, P. A., Konig, T., Ohba, Y., & Langer, T. (2018). PARL partitions the lipid transfer protein STARD7 between the cytosol and mitochondria. *The EMBO Journal*, *37*(4). https://doi.org/10.15252/embj.201797909.

Salminen, T. A., Blomqvist, K., & Edqvist, J. (2016). Lipid transfer proteins: Classification, nomenclature, structure, and function. *Planta*, *244*(5), 971–997. https://doi.org/10.1007/s00425-016-2585-4.

Salvaing, J., Botella, C., Albrieux, C., Gros, V., Block, M. A., & Jouhet, J. (2020). PUB11-dependent ubiquitination of the phospholipid flippase ALA10 modifies ALA10 localization and affects the pool of linolenic phosphatidylcholine. *Frontiers in Plant Science*, *11*, 1070. https://doi.org/10.3389/fpls.2020.01070.

Sandhu, J., Li, S., Fairall, L., Pfisterer, S. G., Gurnett, J. E., Xiao, X., et al. (2018). Aster proteins facilitate nonvesicular plasma membrane to er cholesterol transport in mammalian cells. *Cell*, *175*(2), 514–529.e20. https://doi.org/10.1016/j.cell.2018.08.033.

Santi, C., Molesini, B., Guzzo, F., Pii, Y., Vitulo, N., & Pandolfini, T. (2017). Genome-wide transcriptional changes and lipid profile modifications induced by medicago truncatula N5 overexpression at an early stage of the symbiotic interaction with sinorhizobium meliloti. *Genes (Basel)*, *8*(12), 396. https://doi.org/10.3390/genes8120396.

Santos, N. C., Girik, V., & Nunes-Hasler, P. (2020). ORP5 and ORP8: Sterol sensors and phospholipid transfer proteins at membrane contact sites? *Biomolecules*, *10*(6), 928. https://doi.org/10.3390/biom10060928.

Saravanan, R. S., Slabaugh, E., Singh, V. R., Lapidus, L. J., Haas, T., & Brandizzi, F. (2009). The targeting of the oxysterol-binding protein ORP3a to the endoplasmic reticulum relies on the plant VAP33 homolog PVA12. *The Plant Journal*, *58*(5), 817–830. https://doi.org/10.1111/j.1365-313X.2009.03815.x.

Satheesh, V., Chidambaranathan, P., Jagannadham, P. T., Kumar, V., Jain, P. K., Chinnusamy, V., et al. (2016). Transmembrane START domain proteins: In silico

identification, characterization and expression analysis under stress conditions in chickpea (Cicer arietinum L.). *Plant Signaling & Behavior, 11*(2), e992698. https://doi.org/10.4161/15592324.2014.992698.
Schaaf, G., Ortlund, E. A., Tyeryar, K. R., Mousley, C. J., Ile, K. E., Garrett, T. A., et al. (2008). Functional anatomy of phospholipid binding and regulation of phosphoinositide homeostasis by proteins of the sec14 superfamily. *Molecular Cell, 29*(2), 191–206. https://doi.org/10.1016/j.molcel.2007.11.026.
Schapire, A. L., Voigt, B., Jasik, J., Rosado, A., Lopez-Cobollo, R., Menzel, D., et al. (2008). Arabidopsis synaptotagmin 1 is required for the maintenance of plasma membrane integrity and cell viability. *Plant Cell, 20*(12), 3374–3388. https://doi.org/10.1105/tpc.108.063859.
Schattat, M., Barton, K., & Mathur, J. (2011). Correlated behavior implicates stromules in increasing the interactive surface between plastids and ER tubules. *Plant Signaling & Behavior, 6*(5), 715–718. https://doi.org/10.4161/psb.6.5.15085.
Schauder, C. M., Wu, X., Saheki, Y., Narayanaswamy, P., Torta, F., Wenk, M. R., et al. (2014). Structure of a lipid-bound extended synaptotagmin indicates a role in lipid transfer. *Nature, 510*(7506), 552–555. https://doi.org/10.1038/nature13269.
Schlattner, U., Tokarska-Schlattner, M., Epand, R. M., Boissan, M., Lacombe, M.-L., & Kagan, V. E. (2018). NME4/nucleoside diphosphate kinase D in cardiolipin signaling and mitophagy. *Laboratory Investigation: A Journal of Technical Methods and Pathology, 98*(2), 228–232. https://doi.org/10.1038/labinvest.2017.113.
Schlattner, U., Tokarska-Schlattner, M., Ramirez, S., Tyurina, Y. Y., Amoscato, A. A., Mohammadyani, D., et al. (2013). Dual function of mitochondrial Nm23-H4 protein in phosphotransfer and intermembrane lipid transfer. *Journal of Biological Chemistry, 288*(1), 111–121. https://doi.org/10.1074/jbc.M112.408633.
Schrick, K., Bruno, M., Khosla, A., Cox, P. N., Marlatt, S. A., Roque, R. A., et al. (2014). Shared functions of plant and mammalian StAR-related lipid transfer (START) domains in modulating transcription factor activity. *BMC Biology, 12*, 70. https://doi.org/10.1186/s12915-014-0070-8.
Schrick, K., Nguyen, D., Karlowski, W. M., & Mayer, K. F. X. (2004). START lipid/sterol-binding domains are amplified in plants and are predominantly associated with homeodomain transcription factors. *Genome Biology, 5*(R41).
Schulz, T. A., & Prinz, W. A. (2007). Sterol transport in yeast and the oxysterol binding protein homologue (OSH) family. *Biochimica et Biophysica Acta (BBA)—Molecular and Cell Biology of Lipids, 1771*(6), 769–780. https://doi.org/10.1016/j.bbalip.2007.03.003.
Schwacke, R., Schneider, A., van der Graaff, E., Fischer, K., Catoni, E., Desimone, M., et al. (2003). ARAMEMNON, a novel database for Arabidopsis integral membrane proteins. *Plant Physiology, 131*(1), 16–26. https://doi.org/10.1104/pp.011577.
Scorrano, L., De Matteis, M. A., Emr, S., Giordano, F., Hajnoczky, G., Kornmann, B., et al. (2019). Coming together to define membrane contact sites. *Nature Communications, 10*(1), 1287. https://doi.org/10.1038/s41467-019-09253-3.
Selitrennik, M., & Lev, S. (2016). The role of phosphatidylinositol-transfer proteins at membrane contact sites. *Biochemical Society Transactions, 44*(2), 419–424. https://doi.org/10.1042/BST20150182.
Senkler, J., Senkler, M., Eubel, H., Hildebrandt, T., Lengwenus, C., Schertl, P., et al. (2017). The mitochondrial complexome of *Arabidopsis thaliana*. *The Plant Journal, 89*(6), 1079–1092. https://doi.org/10.1111/tpj.13448.
Shirane, M., Wada, M., Morita, K., Hayashi, N., Kunimatsu, R., Matsumoto, Y., et al. (2020). Protrudin and PDZD8 contribute to neuronal integrity by promoting lipid extraction required for endosome maturation. *Nature Communications, 11*(1), 4576. https://doi.org/10.1038/s41467-020-18413-9.
Skirpan, A. L., Dowd, P. E., Sijacic, P., Jaworski, C. J., Gilroy, S., & Kao, T. H. (2006). Identification and characterization of PiORP1, a Petunia oxysterol-binding-protein

related protein involved in receptor-kinase mediated signaling in pollen, and analysis of the ORP gene family in Arabidopsis. *Plant Molecular Biology*, *61*(4–5), 553–565. https://doi.org/10.1007/s11103-006-0030-y.

Slee, J. A., & Levine, T. P. (2019). Systematic prediction of FFAT motifs across eukaryote proteomes identifies nucleolar and eisosome proteins with the predicted capacity to form bridges to the endoplasmic reticulum. *Contact*, *2*, 251525641988313. https://doi.org/10.1177/2515256419883136.

Somerharju, P. (2015). Is spontaneous translocation of polar lipids between cellular organelles negligible? *Lipid Insights*, *8s1*, LPI.S31616. https://doi.org/10.4137/LPI.S31616.

Sperandeo, P., Martorana, A. M., & Polissi, A. (2017). The lipopolysaccharide transport (Lpt) machinery: A nonconventional transporter for lipopolysaccharide assembly at the outer membrane of Gram-negative bacteria. *The Journal of Biological Chemistry*, *292*(44), 17981–17990. https://doi.org/10.1074/jbc.R117.802512.

Stoica, R., De Vos, K. J., Paillusson, S., Mueller, S., Sancho, R. M., Lau, K.-F., et al. (2014). ER-mitochondria associations are regulated by the VAPB-PTPIP51 interaction and are disrupted by ALS/FTD-associated TDP-43. *Nature Communications*, *5*, 3996. https://doi.org/10.1038/ncomms4996.

Suzuki, K., Akioka, M., Kondo-Kakuta, C., Yamamoto, H., & Ohsumi, Y. (2013). Fine mapping of autophagy-related proteins during autophagosome formation in *Saccharomyces cerevisiae*. *Journal of Cell Science*, *126*(11), 2534–2544. https://doi.org/10.1242/jcs.122960.

Suzuki, T., Matsushima, C., Nishimura, S., Higashiyama, T., Sasabe, M., & Machida, Y. (2016). Identification of phosphoinositide-binding protein PATELLIN2 as a substrate of Arabidopsis MPK4 MAP kinase during septum formation in cytokinesis. *Plant & Cell Physiology*, *57*(8), 1744–1755. https://doi.org/10.1093/pcp/pcw098.

Tan, X., Wang, Q., Tian, B., Zhang, H., Lu, D., & Zhou, J. (2011). A Brassica napus lipase locates at the membrane contact sites involved in chloroplast development. *PLoS One*, *6*(10), e26831. https://doi.org/10.1371/journal.pone.0026831.

Tejos, R., Rodriguez-Furlán, C., Adamowski, M., Sauer, M., Norambuena, L., & Friml, J. (2018). PATELLINS are regulators of auxin-mediated PIN1 relocation and plant development in Arabidopsis thaliana. *Journal of Cell Science*, *131*(2), jcs204198. https://doi.org/10.1242/jcs.204198.

Thélot, F., Orlando, B. J., Li, Y., & Liao, M. (2020). High-resolution views of lipopolysaccharide translocation driven by ABC transporters MsbA and LptB2FGC. *Current Opinion in Structural Biology*, *63*, 26–33. https://doi.org/10.1016/j.sbi.2020.03.005.

Thong, S., Ercan, B., Torta, F., Fong, Z. Y., Wong, H. Y. A., Wenk, M. R., et al. (2016). Defining key roles for auxiliary proteins in an ABC transporter that maintains bacterial outer membrane lipid asymmetry. *eLife*, *5*, e19042. https://doi.org/10.7554/eLife.19042.

Tjellström, H., Andersson, M. X., Larsson, K. E., & Sandelius, A. S. (2008). Membrane phospholipids as a phosphate reserve: The dynamic nature of phospholipid-to-digalactosyl diacylglycerol exchange in higher plants. *Plant, Cell & Environment*, *31*(10), 1388–1398.

Tokarska-Schlattner, M., Boissan, M., Munier, A., Borot, C., Mailleau, C., Speer, O., et al. (2008). The nucleoside diphosphate kinase D (NM23-H4) binds the inner mitochondrial membrane with high affinity to cardiolipin and couples nucleotide transfer with respiration. *The Journal of Biological Chemistry*, *283*(38), 26198–26207. https://doi.org/10.1074/jbc.M803132200.

Tong, J., Manik, M. K., & Im, Y. J. (2018). Structural basis of sterol recognition and nonvesicular transport by lipid transfer proteins anchored at membrane contact sites. *Proceedings of the National Academy of Sciences of the United States of America*, *115*(5), E856–E865. https://doi.org/10.1073/pnas.1719709115.

Tripathi, A., Martinez, E., Obaidullah, A. J., Lete, M. G., Lönnfors, M., Khan, D., et al. (2019). Functional diversification of the chemical landscapes of yeast Sec14-like phosphatidylinositol transfer protein lipid-binding cavities. *Journal of Biological Chemistry*, *294*(50), 19081–19098. https://doi.org/10.1074/jbc.RA119.011153.

Udvardi, M., & Poole, P. S. (2013). Transport and metabolism in legume-rhizobia symbioses. *Annual Review of Plant Biology*, *64*(1), 781–805. https://doi.org/10.1146/annurev-arplant-050312-120235.

Umate, P. (2011). Oxysterol binding proteins (OSBPs) and their encoding genes in Arabidopsis and rice. *Steroids*, *76*(5), 524–529. https://doi.org/10.1016/j.steroids.2011.01.007.

Upchurch, R. G. (2008). Fatty acid unsaturation, mobilization, and regulation in the response of plants to stress. *Biotechnology Letters*, *30*(6), 967–977. https://doi.org/10.1007/s10529-008-9639-z.

Valverde, D. P., Yu, S., Boggavarapu, V., Kumar, N., Lees, J. A., Walz, T., et al. (2019). ATG2 transports lipids to promote autophagosome biogenesis. *Journal of Cell Biology*, *218*(6), 1787–1798.

Vance, J. E. (1990). Phospholipid synthesis in a membrane fraction associated with mitochondria. *The Journal of Biological Chemistry*, *265*(13), 7248–7256.

Velayos-Baeza, A., Vettori, A., Copley, R. R., Dobson-Stone, C., & Monaco, A. P. (2004). Analysis of the human VPS13 gene family. *Genomics*, *84*(3), 536–549. https://doi.org/10.1016/j.ygeno.2004.04.012.

Verma, D. (1996). Biogenesis of the peribacteriod membrane in root nodules. *Trends in Microbiology*, *4*(9), 364–368. https://doi.org/10.1016/0966-842X(96)10053-6.

Vincent, P., Chua, M., Nogue, F., Fairbrother, A., Mekeel, H., Xu, Y., et al. (2005). A Sec14p-nodulin domain phosphatidylinositol transfer protein polarizes membrane growth of Arabidopsis thaliana root hairs. *Journal of Cell Biology*, *168*(5), 801–812. https://doi.org/10.1083/jcb.200412074.

Vorwerk, S., Schiff, C., Santamaria, M., Koh, S., Nishimura, M., Vogel, J., et al. (2007). EDR2 negatively regulates salicylic acid-based defenses and cell death during powdery mildew infections of Arabidopsis thaliana. *BMC Plant Biology*, 7(1), 35. https://doi.org/10.1186/1471-2229-7-35.

Wang, Z., Anderson, N. S., & Benning, C. (2013). The phosphatidic acid binding site of the arabidopsis trigalactosyldiacylglycerol 4 (TGD4) protein required for lipid import into chloroplasts. *Journal of Biological Chemistry*, *288*(7), 4763–4771. https://doi.org/10.1074/jbc.M112.438986.

Wang, Z., & Benning, C. (2012). Chloroplast lipid synthesis and lipid trafficking through ER–plastid membrane contact sites. *Biochemical Society Transactions*, *40*(2), 457–463. https://doi.org/10.1042/BST20110752.

Wang, H., Han, S., Siao, W., Song, C., Xiang, Y., Wu, X., et al. (2015). Arabidopsis synaptotagmin 2 participates in pollen germination and tube growth and is delivered to plasma membrane via conventional secretion. *Molecular Plant*, *8*(12), 1737–1750. https://doi.org/10.1016/j.molp.2015.09.003.

Wang, Z., Xu, C., & Benning, C. (2012). TGD4 involved in endoplasmic reticulum-to-chloroplast lipid trafficking is a phosphatidic acid binding protein: TGD4 involvement in ER-to-chloroplast trafficking. *The Plant Journal*, *70*(4), 614–623. https://doi.org/10.1111/j.1365-313X.2012.04900.x.

Wanjie, S. W., Welti, R., Moreau, R. A., & Chapman, K. D. (2005). Identification and quantification of glycerolipids in cotton fibers: Reconciliation with metabolic pathway predictions from DNA databases. *Lipids*, *40*(8), 773–785. https://doi.org/10.1007/s11745-005-1439-4.

Watanabe, Y., Tamura, Y., Kawano, S., & Endo, T. (2015). Structural and mechanistic insights into phospholipid transfer by Ups1–Mdm35 in mitochondria. *Nature Communications*, *6*(1), 7922. https://doi.org/10.1038/ncomms8922.

Wideman, J. G., & Muñoz-Gómez, S. A. (2016). The evolution of ERMIONE in mitochondrial biogenesis and lipid homeostasis: An evolutionary view from comparative cell biology. *Biochimica et Biophysica Acta, 1861*(8 Pt. B), 900–912. https://doi.org/10.1016/j.bbalip.2016.01.015.

Wilkens, S. (2015). Structure and mechanism of ABC transporters. *F1000Prime Reports, 7, 14*. https://doi.org/10.12703/P7-14.

Wirtz, K. W. (1991). Phospholipid transfer proteins. *Annual Review of Biochemistry, 60*, 73–99. https://doi.org/10.1146/annurev.bi.60.070191.000445.

Wong, L. H., Gatta, A. T., & Levine, T. P. (2019). Lipid transfer proteins: The lipid commute via shuttles, bridges and tubes. *Nature Reviews Molecular Cell Biology, 20*(2), 85–101. https://doi.org/10.1038/s41580-018-0071-5.

Wong, L. H., & Levine, T. P. (2016). Lipid transfer proteins do their thing anchored at membrane contact sites. But what is their thing? *Biochemical Society Transactions, 44*(2), 517–527. https://doi.org/10.1042/BST20150275.

Wong, L. H., & Levine, T. P. (2017). Tubular lipid binding proteins (TULIPs) growing everywhere. *Biochimica et Biophysica Acta (BBA)—Molecular Cell Research, 1864*(9), 1439–1449. https://doi.org/10.1016/j.bbamcr.2017.05.019.

Wu, C., Tan, L., van Hooren, M., Tan, X., Liu, F., Li, Y., et al. (2017). Arabidopsis EXO70A1 recruits Patellin3 to the cell membrane independent of its role as an exocyst subunit: A subunit-specific role of EXO70A1. *Journal of Integrative Plant Biology, 59*(12), 851–865. https://doi.org/10.1111/jipb.12578.

Xu, C., Fan, J., Cornish, A. J., & Benning, C. (2008). Lipid trafficking between the endoplasmic reticulum and the plastid in *Arabidopsis* requires the extraplastidic TGD4 protein. *The Plant Cell, 20*(8), 2190–2204. https://doi.org/10.1105/tpc.108.061176.

Xu, C., Fan, J., Froehlich, J. E., Awai, K., & Benning, C. (2005). Mutation of the TGD1 chloroplast envelope protein affects phosphatidate metabolism in *Arabidopsis*. *The Plant Cell, 17*(11), 3094–3110. https://doi.org/10.1105/tpc.105.035592.

Xu, C., Fan, J., Riekhof, W., Froehlich, J. E., & Benning, C. (2003). A permease-like protein involved in ER to thylakoid lipid transfer in Arabidopsis. *The EMBO Journal, 22*(10), 2370–2379. https://doi.org/10.1093/emboj/cdg234.

Yamaryo, Y., Dubots, E., Albrieux, C., Baldan, B., & Block, M. A. (2008). Phosphate availability affects the tonoplast localization of PLDζ2, an *Arabidopsis thaliana* phospholipase D. *FEBS Letters, 582*(5), 685–690. https://doi.org/10.1016/j.febslet.2008.01.039.

Yamazaki, T., Kawamura, Y., Minami, A., & Uemura, M. (2008). Calcium-dependent freezing tolerance in Arabidopsis involves membrane resealing via synaptotagmin SYT1. *Plant Cell, 20*(12), 3389–3404. https://doi.org/10.1105/tpc.108.062679.

Yamazaki, T., Takata, N., Uemura, M., & Kawamura, Y. (2010). Arabidopsis synaptotagmin SYT1, a type I signal-anchor protein, requires tandem C2 domains for delivery to the plasma membrane. *The Journal of Biological Chemistry, 285*(30), 23165–23176. https://doi.org/10.1074/jbc.M109.084046.

Yang, L., Na, C.-L., Luo, S., Wu, D., Hogan, S., Huang, T., et al. (2017). The phosphatidylcholine transfer protein Stard7 is required for mitochondrial and epithelial cell homeostasis. *Scientific Reports*, 7, 46416. https://doi.org/10.1038/srep46416.

Ye, H., Ji, C., Guo, R., & Jiang, L. (2020). Membrane contact sites and organelles interaction in plant autophagy. *Frontiers in Plant Science, 11*, 477. https://doi.org/10.3389/fpls.2020.00477.

Zhang, H., Zhang, L., Gao, B., Fan, H., Jin, J., Botella, M. A., et al. (2011). Golgi apparatus-localized synaptotagmin 2 is required for unconventional secretion in arabidopsis. *PLoS One, 6*(11), e26477. https://doi.org/10.1371/journal.pone.0026477.

Zhou, H., Duan, H., Liu, Y., Sun, X., Zhao, J., & Lin, H. (2019). Patellin protein family functions in plant development and stress response. *Journal of Plant Physiology*, *234-235*, 94–97. https://doi.org/10.1016/j.jplph.2019.01.012.

Zhou, H., Wang, C., Tan, T., Cai, J., He, J., & Lin, H. (2018). Patellin1 negatively modulates salt tolerance by regulating PM Na+/H+ antiport activity and cellular redox homeostasis in arabidopsis. *Plant and Cell Physiology*, *59*(8), 1630–1642. https://doi.org/10.1093/pcp/pcy081.

CHAPTER FIVE

Lipid droplets in plants: More than a simple fat storage

Damien Le Moigne, Nolwenn Guéguen, and Juliette Salvaing*
Université Grenoble-Alpes, CEA, INRAE, CNRS, IRIG, LPCV, Grenoble, France
*Corresponding author: e-mail address: juliette.salvaing@cea.fr

Contents

Abstract

Lipid droplets (LDs) are found in all lineages of living organisms, whether prokaryotes or eukaryotes. Their structure is rather simple with a core of hydrophobic molecules, mostly Triacylglycerols and Sterol Esters, surrounded by a monolayer of polar lipids with associated proteins. They are constitutively present in storage organs, such as seeds in plants, or adipose tissue in mammals, while they can form in response to various abiotic or biotic stresses in other plant organs and in microalgae. Their biogenesis has been mostly studied in yeast and animals, yet remains poorly understood in plants. In particular, the central role of the photosynthetic plastid in lipid metabolism raises questions about its role in LD biogenesis. Over the years, many functions of LDs have emerged in addition to the obvious role in carbon and energy storage. While some functions appear broadly distributed among organisms and organs, other appear specific to certain

Advances in Botanical Research, Volume 101
ISSN 0065-2296
https://doi.org/10.1016/bs.abr.2021.07.004

tissues and/or lineages. The present review aims at highlighting specific aspects of LD biogenesis and functions in all photosynthetic organisms, from the historic model plant *Arabidopsis thaliana* to emerging secondary endosymbiont models such as the diatom *Phaeodactylum tricornutum*.

1. Introduction

Lipid Droplets are intracellular lipid inclusions that can be found in very different organisms, and are present in almost all clades composing the tree of life, from Archaea and Bacteria to eukaryotic cells (Lundquist, Shivaiah, & Espinoza-Corral, 2020; Lupette & Maréchal, 2020; Murphy, 2012). These structures have been observed and described for the first time in the late 19th century (Altmann, 1894) and were originally called microsomes or liposomes, but they also have been referred to as lipid bodies, lipid droplets, adiposomes, granules, oleosomes or oil bodies (Murphy, 2012), depending on the research field, or on the considered organisms or tissues. We will adopt the increasingly used term Lipid Droplets (LDs) in the present review. LDs can be categorized in three different classes depending on their subcellular location (Lundquist et al., 2020). "Cytoplasmic LDs" are present in eukaryote's cytoplasms, "Prokaryotic LDs" in prokaryote's cytoplasms and Plastid LDs, also called plastoglobules, in the plastids of photosynthetic eukaryotes. We will restrict ourselves to the study of Cytoplasmic LDs in eukaryotic photosynthetic organisms, while plastoglobules are discussed in chapter "Plastoglobules: A hub of lipid metabolism in the chloroplast" by Shanmugabalaji et al.

Photosynthetic organisms are characterized by the presence of a plastid, which can derive from a primary symbiotic event (Archaeplastida that regroup land plants, green, blue and red algae) or from a secondary endosymbiosis event (Guéguen, Le Moigne, Amato, Salvaing, & Maréchal, 2021; Petroutsos et al., 2014) (Fig. 1). In pluricellular organisms, LDs can be found in various tissues. This great variety is reflected in the composition, biogenesis and main roles of LDs. Many recent reviews (Chapman, Aziz, Dyer, & Mullen, 2019; Guéguen et al., 2021; Ischebeck, Krawczyk, Mullen, Dyer, & Chapman, 2020; Leyland, Boussiba, & Khozin-Goldberg, 2020) have explored the composition and biogenesis pathways of LDs in plants and/or microalgae, of which we give a rapid overview before discussing more extensively the functions of LDs in these organisms.

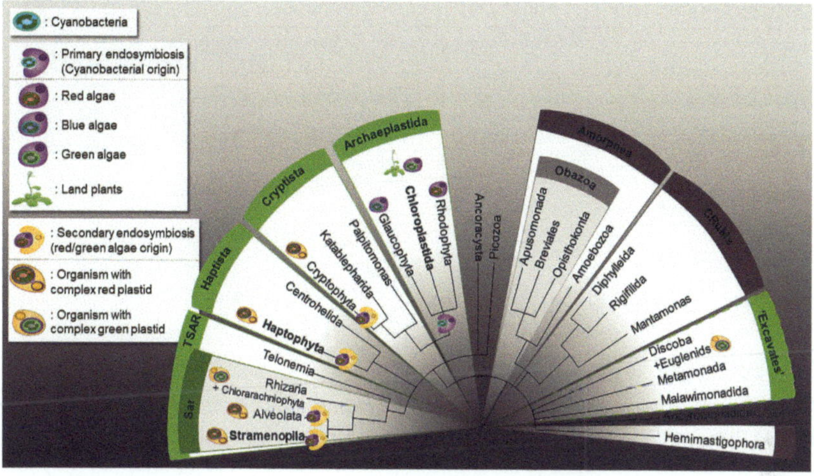

Fig. 1 Photosynthetic organisms in the tree of Life. Different endosymbiosis events leading to the formation of photosynthetic organisms are represented on the eukaryotic tree of life. Archaeplasida result from a single endosymbiotic event, engulfment of a cyanobacteria by a heterotrophic eukaryote. They are composed of blue algae (Glaucophyta), red algae (Rhodophyta) and Chloroplastida. The latter contains both land plants and green algae, such as *Chlamydomonas reinhardtii, Dunaliella* sp. and *Chromochloris zofingiensis*. Several independent secondary endosymbiosis events involving different heterotrophic eukaryotes and either green or red algae have later taken place and given rise to a great variety of photosynthetic organisms, commonly known as microalgae. Studies cited in the review concern mostly Stramenopila including *Nannochloropsis oceanica* and the diatoms *Fistulifera solaris* and *Phaeodactylum tricornutum*, but also Haptophyta such as the coccolithophore *Emiliana huxleyi*. *Tree topology adapted from Burki, F., Roger, A. J., Brown, M. W., and Simpson, A. G. B. (2020). The new tree of eukaryotes. Trends in Ecology & Evolution, 35, 43–55. doi:10.1016/j.tree.2019.08.008.*

2. Lipid droplets composition and biogenesis

LDs contain a hydrophobic core composed of neutral lipids such as Triacylglycerols (TAGs) and Steryl Esters (SEs) and surrounded by a glycerolipid monolayer surface with associated or embedded proteins, which most often play structural or metabolic roles but can have other functions as we will discuss later (Huang, 2018) (Fig. 2).

2.1 Lipid droplet biogenesis

LD biogenesis is dependent on and starts with the core lipids synthesis. TAG synthesis takes place between the two leaflets of the ER membrane

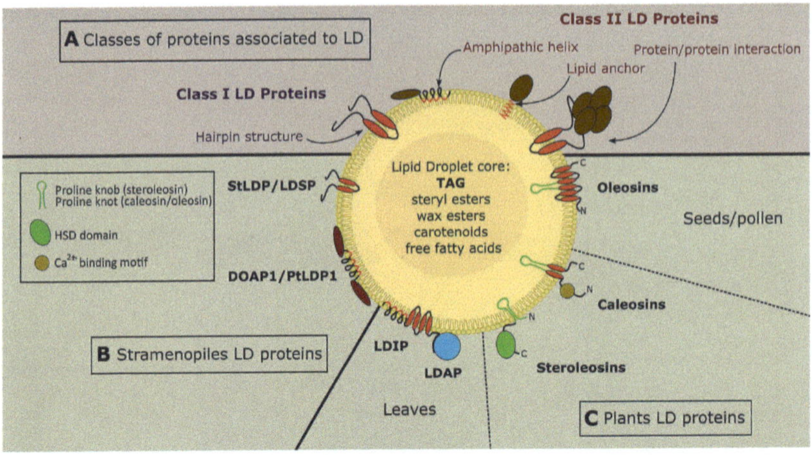

Fig. 2 Structure and content of LDs. The LD core is mainly composed of TAGs with very diverse fatty acid compositions, and steryl esters. Other hydrophobic compounds can be found, such as wax esters in certain land plants, and carotenoids and free fatty acids in some algae. Different classes of proteins can be found at the surface (A) and LD major proteins differ between organisms/tissues (B and C). (A) The different classes of proteins associated to the LD. Class I LD proteins are initially inserted via an hydrophobic hairpin structure in the ER membrane and are later transferred to the LD. Class II proteins are present in the cytosol and can become associated with the LDs. This association can be mediated by an amphipathic helix, or indirect, through protein-protein interaction or addition of a lipid anchor by post-translational modifications. Hydrophobic regions that mediate interaction with the droplet are indicated in red. (B) Major LD proteins in Stramenopiles. Predictions of major lipid droplet proteins in stramenopiles suggest that StLDP (*Phaeodactylum tricornutum*) and LDSP (*Nannochloropsis s.l.*) are class I proteins, with two hydrophobic α-helixes forming a hairpin. On the other hand, DOAP1 (*Fistulifera solaris*) and PtLDP1 (*Phaeodactylum tricornutum*) are class II proteins; they do not contain a clear hydrophobic region but the central region is predicted to form an amphipathic β-barrel. (C) Major LD proteins in plants. Oleosins are the major and the most abundant proteins of seeds and pollen LD, while Caleosins and Steroleosins are less abundant but found in all plant tissues. Oleosins and Caleosins are class I proteins, strongly anchored in the LD through several hydrophobic β-sheets that form hairpin structures. They also contain a central "Proline knot" that penetrates into the LD core and is essential for targeting the Oleosins to the LD. Steroleosins are anchored to the LD by a "Proline knob" that mainly associates with the LD surrounding monolayer. Plant leaf LDs are mainly covered by LDAPs and LDIP. LDIP is a class I protein with four transmembrane domains, while LDAPs are class II proteins that associate to the LD surface through their interactions with LDIP. *Panel (A) adapted from Dhiman, R., Caesar, S., Thiam, A. R., & Schrul, B. (2020). Mechanisms of protein targeting to lipid droplets: a unified cell biological and biophysical perspective.* Seminars in Cell & Developmental Biology, 108, *4–13. https://doi.org/10.1016/j.semcdb.2020.03.004. Panel (B) adapted from Guéguen, N., Le Moigne, D., Amato, A., Salvaing, J., & Maréchal, E. (2021). Lipid droplets in unicellular photosynthetic stramenopiles.* Frontiers in Plant Science, 12, *639276 https://doi.org/10.3389/fpls.2021.639276. Panel (C) adapted from Shao, Q., Liu, X., Su, T., Ma, C., & Wang, P. (2019). New insights into the role of seed oil body proteins in metabolism and plant development.* Frontiers in Plant Science, 10. *1568 https://doi.org/10.3389/fpls.2019.01568 and Chapman, K. D., Aziz, M., Dyer, J. M., & Mullen, R. T. (2019). Mechanisms of lipid droplet biogenesis. The Biochemical Journal, 476(13), 1929–1942. https://doi.org/10.1042/BCJ20180021.*

(Walther, Chung, & Farese, 2017). They can be synthetized *de novo* by the Kennedy pathway, which is conserved across eukaryotes (Chapman & Ohlrogge, 2012). Briefly, fatty acids (FA) synthetized in the plastid are transported to the ER, where they are esterified on glycerol-3-phosphate. Glycerol-3-phosphate acyl transferases (GPATs) catalyze the first esterification on position sn-1 to form lysophosphatidic acid (LPA), then lysophosphatidic acid acyl transferases (LPATs) esterify a second fatty acid on position sn-2 to generate phosphatidic acid (PA). PA's polar head can be further modified to form other phospholipids such as phosphatidylglycerol (PG) or removed by phosphatidic acid phosphatases (PAPs) to generate diacylglycerol (DAG), the latter serving as a platform for most glycerolipids synthesis. DAG can also be generated by recycling of other glycerolipids through different pathways that are presented in Fig. 3. The last reaction to form TAG is then catalyzed by diacylglycerol acyl transferases (DGATs), which esterify a third fatty acid on the sn-3 position of DAG. These three successive acyl esterifications are using acyl-CoAs as donor substrates. Alternatively, TAG can be formed directly by transferring a fatty acid from the sn-2 position of a donor glycerolipid (generally Phosphatidylcholine (PC)) to the sn-3 position of DAG, a reaction catalyzed by phospholipid:diacylglycerol acyl transferases (PDATs) (Fig. 3). Other core lipids include Steryl Esters (SEs) or wax esters (Baud, 2018; Chapman & Ohlrogge, 2012; Ferrer, Altabella, Arro, & Boronat, 2017) and are also synthesized in the ER membrane of plant cells (Ischebeck et al., 2020).

When TAG and/or SE reach a certain concentration, they condense into lenses at specific nucleation sites (Choudhary, Ojha, Golden, & Prinz, 2015; Thiam & Forêt, 2016). LD nucleation is favored at points showing higher membrane curvature, especially in ER tubules, and can be modulated by several proteins, in particular Seipin (Santinho et al., 2020) that can trap TAG and DAG, according to *in silico* molecular simulations (Prasanna et al., 2021; Zoni et al., 2021). In yeast, Seipin can associate with PEX30, a protein involved in peroxisome biogenesis, thus defining common nucleation sites for LDs and peroxisome (Joshi et al., 2018; Wang et al., 2018). In mammals, Multiple C2 domain containing transmembrane protein 2 (MCTP2) plays a role similar to that of PEX30 (Joshi et al., 2018; Joshi, Ragusa, Prinz, & Cohen, 2021). In addition, Seipin can be recruited by LDAF1/Promethin, thus forming a complex that promotes LD formation; LDAF1/Promethin later dissociates from Seipin and covers the surface of LDs (Castro et al., 2019; Chung et al., 2019). Seipin function

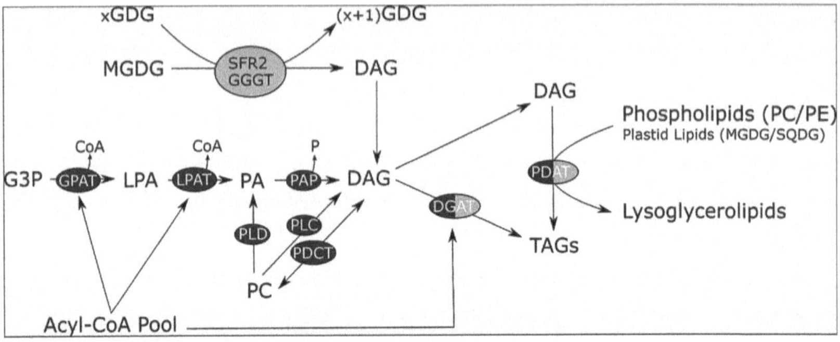

Fig. 3 TAG biosynthesis in photosynthetic organisms. Triacylglycerols (TAGs) can be generated either *de novo* or through the recycling of other glycerolipids. *De novo* synthesis starts with glycerol-3-phosphate (G3P) and leads to diacylglycerol (DAG) formation after esterification of two fatty acids, respectively, by a Glycerol-3-phosphate acyl-transferase (GPAT) and a lysophosphatidic acid acyl-transferase (LPAAT), and then dephosphorylation by a phospatidic acid phosphatase (PAP). DAG can also be generated by the recycling of phospholipids, either directly through the action of phospholipase C (PLC), or indirectly through phospholipase D (PLD) which generates PA. Diacylglycerol cholinephosphotransferase (PDCT) transfers the polar head from PC to DAG, thus leading to an exchange of the fatty acids contained in both molecules. Another recycling route involves the transfer of the galactose polar head of monogalactosyl-diacylglycerol (MGDG) to another galactolipid (xGDG) by a Galactolipid: Galactolipid Galactosyltransferase, thus generating DAG as well as oligogalactolipids ((x+1)GDG)). The last step of TAG synthesis can involve a Diacylglycerol acyl-transferase (DGAT) or a phospholipid:diacylglycerol acyl-transferase (PDAT) that directly transfers a fatty acid from a glycerolipid to DAG thus generating lyso-glycerolipids and TAG. While most PDATs use phospholipids and, in particular, phosphatidylcholine (PC) and phosphatidylethanolamine (PE), the one from *Chlamydomonas reinhardtii* (CrPDAT) can also use plastid lipids such as MGDG and sulfoquinovosyldiacylglycerol (SQDG) (Yoon, Han, Li, Sommerfeld, & Hu, 2012). Extra-plastidial enzymes are represented by black ellipses, and plastidial enzymes by light gray ellipses. DGATs and PDATs can have both localization depending on the considered isoform, organism or physiological state. *Figure based on Ischebeck, T., Krawczyk, H. E., Mullen, R. T., Dyer, J. M., & Chapman, K. D. (2020). Lipid droplets in plants and algae: distribution, formation, turnover and function.* Seminars in Cell & Developmental Biology, 108, *82–93. https://doi.org/10.1016/j.semcdb.2020.02.014 and Higashi, Y., & Saito, K. (2019). Lipidomic studies of membrane glycerolipids in plant leaves under heat stress.* Progress in Lipid Research, 75, *100990. https://doi.org/10.1016/j.plipres.2019.100990.*

in LD biogenesis is conserved in plants (Cai et al., 2015) and algae (Lu et al., 2017), although differences are observed. The three Seipin homologs identified in *Arabidopsis thaliana* have overlapping but not redundant functions (Cai et al., 2015; Taurino et al., 2018). Seipin 2 and 3, but not Seipin 1, can associate with the vesicle-associated membrane protein

(VAMP)-associated protein VAP27, an interaction crucial for LD formation (Greer et al., 2020). LDAF1/Promethin homologs, as well have those of FIT2, another important protein in yeast and animals LD biogenesis (Choudhary et al., 2015), have not been identified thus far in plants. However, some specific plant proteins are involved in LD biogenesis such as Lipid Droplet Associated Proteins (LDAPs) (Gidda et al., 2016) and the associated LDAP-interacting protein (LDIP) (Pyc et al., 2017). A recent study has shown that LDIP interacts with Seipin proteins and that this interaction has functional similarities with the Seipin/LDAF1 interaction (Pyc et al., 2021).

The monolayer surrounding LDs likely reflects the membrane composition of the compartment where LD biogenesis occurs. In plants, as well as in non-photosynthetic organisms, it mostly contains phospholipids (Tauchi-Sato, Ozeki, Houjou, Taguchi, & Fujimoto, 2002; Tzen, Cao, Laurent, Ratnayake, & Huang, 1993). However, other classes of lipids have been identified in microalgae. Upon nitrogen starvation, Sulfoquinovosyldiacylglycerol (SQDG) is found in the LDs of the green alga *Chlamydomonas reinhardtii* (Goold et al., 2016; Tsai et al., 2015) and the diatom *Phaeodactylum tricornutum* (Lupette et al., 2019), along with a betaine lipid, respectively, 1,2-diacylglyceryl-3-O 4′-(N,N,N-trimethyl)-homoserine (DGTS) and 1,2-diacylglyceryl-3-O-2′-(hydroxymethyl)-(N,N,N-trimethyl)-β-alanine (DGTA). Mono- and digalactosylglycerol (MGDG and DGDG) are also found in the LD monolayer of *Chlamydomonas*, especially when algae are submitted to high light stress (Goold et al., 2016). The simultaneous presence of ER lipids (betaine and phospholipids) and plastid lipids (SQDG, MGDG and DGDG) underlines a possible cooperation between the ER and plastid for LD biogenesis in microalgae. Electronic microscopy observations (Flori, Jouneau, Finazzi, Marechal, & Falconet, 2016; Goodson, Roth, Wang, & Goodenough, 2011; Jaussaud et al., 2020; Lupette et al., 2019) comfort this hypothesis, showing tight associations between the plastid and LDs. Interestingly, the role of the chloroplast in LD biogenesis could be conserved in the green tissues of higher plants. Indeed, older studies have shown an association of DGAT with chloroplast envelopes in spinach and *Arabidopsis* leaves (Kaup, Froese, & Thompson, 2002; Martin & Wilson, 1984), while contacts between LDs and chloroplasts were observed by electron microscopy in aging leaves (Brocard et al., 2017).

Altogether, the specificities of the proteins involved in LD biogenesis, as well as the increasingly obvious involvement of plastids, underline the interest in further exploring this process in plants, as well as in microalgae where the diversity of the plastid origins and architectures (cf. Fig. 1) raises even more questions.

2.2 Lipid droplet associated proteins

Proteins are essential components of LDs and can be associated with LDs in different ways. Class I proteins are transferred from the ER to the LD and contain a hydrophobic domain forming a hairpin that anchors the protein in the LD. Class II proteins are transferred to the LD surface from the cytosol. Some contain an amphipathic helix that associates with the LD surface while others interact more loosely with LDs through association either with other LD proteins or following post-translational modifications that add a lipid anchor, such as farnesyl groups (Bersuker & Olzmann, 2017; Dhiman, Caesar, Thiam, & Schrul, 2020; Olzmann & Carvalho, 2019) (Fig. 2A). The most abundant proteins at the surface of LDs are often structural proteins, although they can have other functions as will be exposed later in the review. Land plants have four major families of LD proteins, with different tissue specificities and different associated roles (Huang, 2018) (Fig. 2C). Oleosins (Huang, 1992) are the major LD proteins in plant seeds and pollen, but are absent from leaves LDs, which are instead covered with LDAPs (Gidda et al., 2016; Horn et al., 2013). Oleosins are involved in the regulation of LD size, but also in TAG biosynthesis (Parthibane, Rajakumari, Venkateshwari, Iyappan, & Rajasekharan, 2012). The other two families, Caleosins (Næsted et al., 2000), and Steroleosins (Lin, Tai, Peng, & Tzen, 2002), are found in all plant tissues. Oleosins, caleosins and LDAPs homologs could be identified in green algae (Huang, 2018), but the most abundant LD protein in *Chlamydomonas reinhardtii* is the Major Lipid Droplet Protein (Moellering & Benning, 2010; Nguyen et al., 2011). Finally, the major LD proteins in secondary endosymbionts show little conservation: in the diatom *Phaeodactylum tricornutum,* the most abundant LD protein is Stramenopile Lipid Droplet Protein (StLDP) (Yoneda, Yoshida, Suzuki, & Watanabe, 2016), while it is the Lipid Droplet Surface Protein in other stramenopiles belonging to the *Nannochloropsis* and *Microchloropsis* genera (Vieler, Brubaker, Vick, & Benning, 2012). However, another abundant LD protein in *Phaeodactylum* called PtLDP1 (Lipid Droplet Protein 1) (Wang et al., 2017) is highly similar to the DOAP1 protein (diatom oleosome-associated protein 1) from another diatom, *Fistulifera* sp. (Guéguen et al., 2021; Nojima et al., 2013) (Fig. 2B).

Thus, major LD proteins show specificities depending on the considered organism or tissue. Yet, it only gives a glimpse into the proteomic variety found in LDs, which in turns reflects the many roles of this fascinating compartment. Indeed, while energy storage has been the first identified role of LDs, many more have since emerged.

3. Functions for lipid droplets in normal conditions and abiotic stress

3.1 Lipid droplets as a source of carbon

In plant seeds, TAGs stored in LDs constitute a reservoir of energy and carbon skeletons to sustain germination and post-germination growth, when the plantlet cannot depend on photosynthesis (Graham, 2008; Huang, 1992). While energy storage can rely on other molecules, in particular proteins and starch, TAGs are widely used and can be stored in the embryo itself or in the endosperm (Miray, Kazaz, To, & Baud, 2021). However, we will not distinguish here between the different tissues that compose the seed. LD mobilization can occur in two ways: lipophagy, a specialized form of autophagy, and lipolysis. A recent study has shown that lipophagy plays a role in LD breakdown during seed germination and post-germination growth in *Ricinus communis* (Han et al., 2020). Yet, whether this is specific to certain species or a more general pathway remains to be determined. The most important LD degradation pathway in seeds is lipolysis (Graham, 2008). It starts by the degradation of Oleosins covering seed LDs by the 26S proteasome (Deruyffelaere et al., 2015, 2018; Kretzschmar et al., 2018). Then, the TAG lipases SUGAR-DEPENDENT1 (SDP1) and SUGAR-DEPENDENT1-LIKE (SDP1L) (Eastmond, 2006; Kelly, Quettier, Shaw, & Eastmond, 2011) catalyze the first step of TAG hydrolysis that ultimately leads to the release of glycerol and free fatty acids. The control of seed LD size, which is dependent on Oleosins (Shimada, Shimada, Takahashi, Fukao, & Hara-Nishimura, 2008; Siloto et al., 2006), is crucial to ensure lipolysis efficiency (Shimada et al., 2008). Indeed, smaller LDs have a higher surface/volume ratio that facilitates their degradation. Fatty acids can then be converted to acetyl-CoA by β-oxidation (Goepfert & Poirier, 2007) in specialized peroxisomes called glyoxysomes (Beevers, 1979), and used for gluconeogenesis (Goepfert & Poirier, 2007; Hayashi & Nishimura, 2003; Miray et al., 2021; Quettier, Shaw, & Eastmond, 2008). In most albuminous seeds, the TAGs stored in the endosperm are thus converted to sucrose before being transported to the seedling itself (Miray et al., 2021). Alternatively, fatty acids can be reused to build new membrane lipids.

Similarly to what occurs in seeds, the formation of LDs is essential for pollen development as disruption of TAG synthesis leads to pollen abortion (Zhang, Fan, Taylor, & Ohlrogge, 2010). Studies on olive pollen have shown that LD mobilization provides energy for pollen germination (Zienkiewicz, Zienkiewicz, Rejón, Rodríguez-García, & Castro, 2013). Moreover, impairment of fatty acid transport to peroxisome (Footitt et al., 2007), disruption of the TAG lipase OBL1 (Oil body lipase 1) (Muller & Ischebeck, 2018) as well as enlargement of LDs caused by disruption of the three SEIPINs (Taurino et al., 2018), reduce male fertility. However, the amount of TAGs in LDs is likely insufficient to store enough energy and lipid precursors to sustain the pollen tube growth. Moreover, there is a continuous synthesis of TAGs in this organ, raising the hypothesis that LD essential role lies elsewhere (Ischebeck, 2016; Ischebeck et al., 2020).

In microalgae, LDs accumulate mainly in response to abiotic stress, in particular deprivation of nutrients, which induces growth arrest. While the photosynthetic apparatus and chloroplast lipids eventually get partially degraded, photosynthesis remains active (Abida et al., 2015; Simionato et al., 2013). In this situation, LDs allow the storage of the excess carbon produced, and provide energy and carbon to quickly restart when the conditions get better. In agreement with this, nitrogen resupply following nitrogen starvation leads to a very rapid degradation of TAGs, as well as other storage components such as starch (Mulders, Lamers, Wijffels, & Martens, 2015; Tsai, Uygun, Roston, Shiu, & Benning, 2018).

3.2 Lipid droplets, lipid synthesis and lipid remodeling

The function of LDs in lipid membrane remodeling and homeostasis has been well described in yeast (Graef, 2018; Henne, Goodman, & Hariri, 2020). Indeed, in *Saccharomyces cerevisiae*, the KO of the four genes involved in TAG (the *DGAT* ortholog *Dga1p* and the *PDAT* ortholog *Lro1*), and SE (the *acyl-CoA:cholesterol O-acyltransferase* orthologs *Are1p* and *Are2p*) syntheses severely impairs the yeast ability to respond to stress. Exposure of these mutants to unsaturated free fatty acids (FFA), which are normally taken in and incorporated into TAGs, leads to aberrant membrane formation and ultimately to cell death, suggesting that LD storage of TAGs and SEs acts as a buffer to protect the cell against toxic FFA (Petschnigg et al., 2009). Moreover, growth of these mutants at high temperature in the absence of Inositol leads to severe growth defects and the authors observed a decreased synthesis of membrane phospholipids even in normal growing conditions, suggesting a coordination of membrane lipids and storage lipid syntheses to maintain lipid homeostasis (Gaspar, Hofbauer, Kohlwein, & Henry, 2011). In agreement with this idea, a more recent study has shown that LD formation and subsequent degradation by microautophagy are crucial for adaptation of yeast cells to lipid imbalance (Vevea et al., 2015). The quadruple TAG/SE synthesis mutants also show macroautophagy defects. This may be linked to general ER homeostasis defects or failure to supply the FA necessary for autophagophore formation (Henne et al., 2020; Velázquez, Tatsuta, Ghillebert, Drescher, & Graef, 2016), as it is the case for prospore membrane building during sporulation (Hsu, Chen, Cheng, & Wang, 2017). Finally, a recent study has shown an important function for *Lro1,* the yeast homolog of *PDAT,* in the homeostasis of the nuclear envelop and localization of the nucleolus during the cell cycle or in response to stress (Barbosa et al., 2019).

3.2.1 Lipid droplets, lipid synthesis and remodeling in microalgae

LD proteomes from green microalgae such as *Chlamydomonas reinhardtii* (Goold et al., 2016; Moellering & Benning, 2010; Nguyen et al., 2011; Tsai et al., 2015), as well as from secondary endosymbionts such as *Phaeodactylum tricornutum* (Leyland, Zarka, Didi-Cohen, Boussiba, & Khozin Goldberg, 2020; Lupette et al., 2019) have shown that their LDs contain high amounts of proteins involved in lipid synthesis. Most can be directly related to TAG biosynthesis (such as enzymes involved in FA synthesis, GPATs, LPATs and DGATs, see Fig. 3), while others are involved in different lipid synthesis pathways. Interestingly, the DGTS synthase BTA1 has been identified in the LD proteomes of the green algae *Dunaliella bardawil* (Davidi, Levin, Ben-Dor, & Pick, 2015), *Chromochloris zofingiensis* (Wang, Wei, Mao, & Liu, 2019) and *Chlamydomonas reinhardtii*, in which it is one of the most abundant proteins (Goold et al., 2016; Nguyen et al., 2011; Tsai et al., 2015). It has been proposed that LD-resident BTA1 synthetizes the DGTS that surrounds *Chlamydomonas* LDs (Goold et al., 2016; Nguyen et al., 2011). However, while DGTA is found in the monolayer surrounding *Phaeodactylum* LDs (Lupette et al., 2019) the putative DGTA synthase has not been identified in the corresponding LD proteomes (Leyland, Zarka, et al., 2020; Lupette et al., 2019). This could be explained by differences in experimentation time-courses, as LD proteomes from green algae were determined after 24–48 h of nitrogen starvation while those of *Phaeodactylum* were established after 7 days. However, the different growth rates of these different algae make the comparison difficult. Alternatively, it could reflect metabolic differences between the considered organisms; DGTA is only a minor lipid in *Phaeodactylum*, while DGTS is major in *Chlamydomonas* where it replaces Phosphatidylcholine (PC).

Several proteins involved in sterol biosynthesis have also been identified in many LD proteomes. This is the case after nitrogen starvation in the previously cited green algae (Davidi et al., 2015; Goold et al., 2016; Nguyen et al., 2011; Tsai et al., 2015; Wang et al., 2019) and in the diatom *Phaeodactylum tricornutum* (Leyland, Zarka, et al., 2020), as well as in *Chlamydomonas* after exposure to high light. One of these proteins, the Sterol methyl-transferase 1, has also been identified in the proteome of *Nicotiana benthamiana* pollen tube LDs, along with cycloartenol synthase, another enzyme of sterol metabolism (Kretzschmar et al., 2018). Interestingly, a function of LDs in sterol metabolism has been shown in yeast (Sorger, Athenstaedt, Hrastnik, & Daum, 2004), but whether a similar function exists in algae or plants remains unknown.

Altogether, the presence of proteins involved in different lipid synthesis pathways in several independent LD proteomes raises questions as to potential roles of LDs in coordinating lipid production pathways in the considered organisms/organs.

In addition to proteins involved in lipid synthesis, proteins involved in membrane remodeling and lipid trafficking have also been found in the previously cited proteomes as well as in another diatom, *Fistulifera solaris* (Nonoyama et al., 2019), along with cytoskeleton and intracellular vesicle trafficking proteins. Several studies have pointed out the function of LDs in lipid remodeling. In the absence of stress, TAG synthesis in *Chlamydomonas reinhardtii* is mainly dependent on PDAT. This enzyme (CrPDAT) can not only use phospholipids but also the plastid lipids MGDG, DGDG and SQDG as fatty acid donors (cf. Fig. 2). Moreover, reduction of CrPDAT levels results in a slightly slower growth, suggesting that lipid remodeling by this pathway is important for the cell fitness (Yoon et al., 2012). While this study shows that CrPDAT contribution is less important in stress conditions, degradation of glycerolipids is still important to feed TAGs under such situations. The disruption of the lipase PLASTID GALACTOGLYCEROLIPID DEGRADATION1 (PGD1), which hydrolyses MGDG, reduced TAG accumulation under nitrogen starvation (Li, Moellering, et al., 2012), while other classes of glycerolipids are converted to TAGs under heat stress (Légeret et al., 2016). This recycling of glycerolipids to TAGs is likely conserved as high levels of very long-chain unsaturated fatty acids are found in TAGs in Stramenopiles (reviewed in Guéguen et al., 2021) as well as in the Haptophyte *Tisochrysis lutea* (Huang et al., 2019) following nitrogen or phosphate deprivation. Finally, study of the green microalga *Parachlorella kessleri* under salt stress has shown that LDs accumulate at the cell periphery and likely contribute to remodeling the FA composition of the plasmid membrane lipids to modify its permeability (You, Zhang, Peng, & Miao, 2019).

3.2.2 Lipid droplets and lipid remodeling in the pollen tube

In higher plants, lipid trafficking appears as a major function during pollen formation and later during pollen tube growth. Sporopollenin, a very resistant polymer that covers pollen, is synthetized and deposited from a specific tissue localized in the anther and called the tapetum. Tapetum cells form specific LDs associated with ER membranes, named tapetosomes, as well as specialized plastids filled with plastoglobuli, called elaioplasts. Upon programmed cell death, tapetal cells release the lipids contained in those structures that contribute to the formation of the pollen coat (reviewed by Ischebeck, 2016). Pollen germination and pollen tube growth require LD mobility. Indeed, in olive pollen after germination LDs are no more found in the grain, but many are present in the pollen tube (Zienkiewicz et al., 2010). This change of localization is hindered by the enlarged size of LDs in triple Seipin mutants of *Arabidopsis,* and leads to male fertility defects (Taurino et al., 2018). The continuous TAG synthesis during pollen tube growth is at least partly fed by the recycling of phospholipids. Indeed, mutants deficient for the non-specific phospholipases C2 and C6 (Bose, Ngo, Nguyen, & Nakamura, 2021) as well as for DGAT1 and PDAT1 (Bose et al., 2021; Zhang et al., 2010) (cf. Fig. 2) show reduced TAG synthesis. In his 2016 review, Ischebeck proposed several interesting hypotheses as to the function of LDs in pollen tube, including the involvement of LDs in lipid transport, regulation of lipid synthesis, acyl-editing or adaptation to environmental changes occurring during the pollen tube growth (Ischebeck, 2016). These functions have been known in other organisms for some time, in particular mammals (Murphy, 2012), and may not be restricted to the pollen tube.

3.2.3 Lipid droplets and lipid remodeling in leaves

In higher plants, TAGs represent a very low percentage of glycerolipids in healthy leaves and very few LDs are observed. However, disruption of TAG degradation leads to ectopic accumulation of LDs in the leaves (James et al., 2010), suggesting that TAG are constantly synthetized and degraded in this tissue. Later studies have shown that LD abundance varies with the diurnal cycle, and that LDs are the most abundant at the end of the night phase (Gidda et al., 2016). As FA synthesis is promoted by photosynthesis, it seems unlikely that these TAG are synthetized *de novo* and more likely that they derive from recycling of membrane lipids. Mechanisms explaining TAG function in lipid homeostasis and remodeling in higher plants have been proposed in a recent review (Yu, Zhou, Fan, Shanklin, & Xu, 2021). Indeed, the *Arabidopsis* genome encodes three different DGATs and two PDATs with different substrate specificities in terms of acyl chain length and insaturation levels. Therefore, expression levels of the various DGATs may affect the acyl-CoA pool available for other lipid productions, while expression levels of the PDATs may promote recycling of certain PC species, thus modifying membrane compositions. As TAG storage organelles, LDs could just serve as a transitory sink or help to preserve certain fatty acids that are costly to produce or that can be used in case of rapidly changing conditions. Interestingly, a recent paper has enlightened the important role of autophagy for TAG formation and membrane lipid turnover in mature and senescent leaves, highlighting a possible role of LDs for the degradation of damaged organelles and recycling of their lipid contents (Fan, Yu, & Xu, 2019).

Indeed, LD formation in leaves is mainly triggered by senescence and abiotic stresses. During senescence, the degradation of membrane lipids and subsequent formation of TAGs may allow the redistribution of FA to other tissues. Consistently, the proteome of LDs in senescent leaves contains several cytoskeleton proteins as well as proteins involved in lipid but also protein transports (Brocard et al., 2017). In addition to the formation of LDs, abiotic stresses trigger changes in the fatty acid composition of glycerolipids, in particular of the plastid lipids MGDG and DGDG, and of PC. These changes vary depending on the type of stress, as does TAG composition (reviewed in Higashi & Saito, 2019). Heat, drought and salt stress can induce a very rapid TAG accumulation, in contrast to cold, high light or osmotic stress that require a longer exposure (Mueller, Krause, Mueller, & Fekete, 2015). The response to heat stress has been extensively studied and shows that TAG generation comes from the recycling of other lipids rather than from *de novo* synthesis (Higashi & Saito, 2019). Indeed, transcriptomic data indicate a decrease of fatty acid and chloroplast glycerolipid syntheses (Higashi, Okazaki, Myouga, Shinozaki, & Saito, 2015), while the generated TAGs contain high levels of unsaturated fatty acids, in particular C18:3 and C16:3 (Higashi et al., 2015; Mueller et al., 2015). Analysis of fatty acid desaturation mutants shows that these unsaturated fatty acids are likely recycled from MGDG (Mueller, Unger, Guender, Fekete, & Mueller, 2017), after the initial removal of C18:3 from MGDG by the Heat-Inducible Lipase 1 (HIL1) (Higashi et al., 2018). The analysis of several TAG synthesis mutants has shown that PDAT1, but not DGAT1, is the major actor in TAG biosynthesis upon heat stress (Mueller et al., 2017). Finally, the thermosensitivity observed in *pdat1* and *hil1* mutants enlightens the importance of MGDG remodeling and of TAG synthesis in this process (Higashi et al., 2018; Mueller et al., 2017). MGDG remodeling also appears essential in tolerance to freezing, albeit for different reasons. Indeed, Moellering and collaborators have identified Sensitive-to-Freezing2 (SFR2) as a galactolipid:galactolipid galactosyltransferase that can transfer galactose from MGDG to another galactolipid, thus depleting the MGDG pool and forming oligogalactolipids and DAGs, the latter being converted into TAGs (cf. Fig. 3). The authors hypothesized that this reduction of MGDG and consequent increase in DGDG and other oligogalactolipids serve to stabilize membranes following freezing-induced desiccation (Moellering, Muthan, & Benning, 2010). Consistently, similar changes in galactolipids and TAGs were observed upon drought (Gasulla et al., 2013; Tarazona, Feussner, & Feussner, 2015) and cold (Degenkolbe et al., 2012; Tarazona et al., 2015) stresses, and SFR2 was also associated with resistance to salt and drought stresses in tomato (Wang, Hersh, & Benning, 2016). More recent studies have shown that in contrast to heat stress, DGAT1 is the essential player in TAG formation upon freezing and is important for freezing tolerance (Arisz et al., 2018; Tan et al., 2018).

3.3 Lipid droplets and cell protection

In addition to their role in membrane lipid remodeling and transitory storage of fatty acids during stress conditions, leaf LDs appear essential to cell protection. Indeed, free fatty acids, as well as glycerolipids that cannot assemble into membranes such as PA or DAG, can be toxic for the cell by disorganizing the complex intracellular membrane network (Li et al., 2020;

Petschnigg et al., 2009). Storage into TAG serves as a transient but essential sink before degradation (Fan, Yan, Roston, Shanklin, & Xu, 2014; Fan, Yu, & Xu, 2017).

In addition to TAG and sterol esters, LDs can store other hydrophobic compounds such as carotenoids. In the marine green algae *Dunaliella bardawil* and *salina*, carotenoids are stored in plastoglobuli, while in freshwater green algae such as *Haematococcus pluvialis,* carotenoids are found in cytosolic LDs. In both cases, carotenoid production is induced by reactive oxygen species (ROS) that are generated following stress such as high light or nutrient deprivation, and they can serve both to protect the photosynthetic system against photodamage created by excessive irradiation and to quench the effect of ROS (review Pick, Zarka, Boussiba, & Davidi, 2019). Carotenoids have also been found in *Phaeodactylum* LDs but whether they are stored in cytosolic LDs or plastoglobuli is not clear (Lupette et al., 2019).

Finally, several proteins of the ER associated protein degradation (ERAD) pathway, which serve to remove misfolded proteins and target them for degradation by the 26S-proteasome, have been identified in LD proteomes of the diatom *Phaeodactylum tricornutum* after 7 days of nitrogen deprivation (Leyland, Zarka, et al., 2020; Lupette et al., 2019) as well as in *Arabidopsis* seeds (Deruyffelaere et al., 2018; Kretzschmar et al., 2018), senescent leaves (Fernandez-Santos et al., 2020) and *Nicotiana* pollen (Kretzschmar et al., 2018). Their function in *Arabidopsis* seeds seems to be mainly the degradation of Oleosins and Caleosins preceding the remobilization of oil reserves upon germination (Deruyffelaere et al., 2015, 2018). However, many additional proteins associated with the regulation of protein folding and ubiquitinylation have been identified in LDs from pollen tubes, leaves and *Phaeodactylum*, suggesting a broader function of lipid droplets in protein processing and catabolism.

3.4 Lipid droplets, a reserve to restart

In addition to proteins associated with lipid metabolism or LD structure, which are expected to be present, more unexpected proteins are found in LDs. In many cases, they are assumed to result from contaminations, and, as such, are little discussed in the literature. Yet, there are lines of evidence suggesting that they should be considered with more attention.

Many of the microalgae LD proteomes performed under nutrient starvation conditions include proteins linked to DNA organization (histones), and regulation of protein translation (ribosomal proteins, translation elongation factors) (Davidi et al., 2015; Goold et al., 2016; Leyland, Zarka, et al., 2020; Lupette et al., 2019; Moellering & Benning, 2010; Nguyen et al., 2011; Tsai et al., 2015; You et al., 2019). Interestingly, these are generally not found in higher plant seed proteomes.

Upon nutrient starvation, LDs can include all histones or a subset of histones, including core histones (H2A, H2B, H3 and H4), histone variants, as well as the linker histone H1. While histones are abundant proteins, they are mostly contained in the nucleus and are highly hydrophilic proteins, suggesting that their association with LDs in stressed microalgae is unlikely to result from contamination. The association of histones to LDs has been described in the model fly *Drosophila melanogaster*, where LDs serve as histone storage (Cermelli, Guo, Gross, & Welte, 2006) and regulate histone supply during the early phases of embryonic development (Li, Thiel, et al., 2012). As nutrient starvation severely impairs microalgae growth, it could be speculated that LDs, similarly to what occurs in *Drosophila*, could serve as a histone deposit during stress, when cell growth is severely impaired, to favor quick restart once the conditions become more favorable. The storage of ribosomal proteins and translation elongation factors could serve the same purpose. Ribosomal proteins are among the most abundant proteins in the cell and their presence in LD proteomes has long been considered as contaminations. Yet, similarly to histones, they are not found in plant seeds proteomes. Moreover, proteins associated with translation processes are the most strongly down-regulated upon nitrogen and phosphorus starvation (Hulatt et al., 2020), and storage in LDs could serve to preserve a minimal pool from degradation. In support of this possible role for LDs, neither histones nor translation-associated proteins are found in the LD proteome of *Chlamydomonas* grown under high light, a condition that triggers LD accumulation but not cell growth arrest (Goold et al., 2016).

Finally, some RNA-binding proteins have been identified in several microalgae proteomes (Lupette et al., 2019; Moellering & Benning, 2010; Nguyen et al., 2011; Wang et al., 2019), raising the possibility that LDs could serve to preserve a RNA pool.

3.5 Lipid droplets and brassinosteroid signaling

Together with Oleosins and Caleosins, Steroleosins are one of the main families of proteins associated with LDs in plants (Fig. 2C). They may be associated with the synthesis of brassinosteroids, which are plant hormones involved in different developmental processes as well as resistance to biotic and abiotic stresses (Divi & Krishna, 2009; Singh & Savaldi-Goldstein, 2015). The first two steroleosins have been identified in *Sesamum indicum* (Chen, Tai, Peng, & Tzen, 1998), and the first identified *Arabidopsis* homolog possesses 11β- and 17β-hydroxysteroid dehydrogenase activities and was thus renamed HYDROXYSTEROID DEHYDROGENASE 1 (AtHSD1) (d'Andréa et al., 2007). Initial studies have shown that over-expression of AtHSD1 leads to reduced seed dormancy (Baud et al., 2009; Li, Asami, Wu, Tsang, & Cutler, 2007) as well as to increased growth and resistance to salt stress (Li et al., 2007). How the association of AtHSD1 to LDs modulates its function and contributes to brassinosteroid signaling remain open questions.

4. LD and biotic stress

Interactions between LDs and pathogens have been mostly studied in mammalian cells, where studies highlighted two antagonist roles for LDs: they can either be used by the pathogen or participate in cell and organism defense.

4.1 Hijacking of the host metabolism

The current state of the art suggests that mammalian LDs can participate in the infection process, providing invaders with substrates for survival and/or growth (Bosch et al., 2020). This phenomenon is described as the "hijacking" of the host metabolism. Bacteria such as *Chlamydia trachomatis* (Kumar, Cocchiaro, & Valdivia, 2006), eukaryotic parasites such as *Toxoplasma gondii* and *Neospora caninum* (Hu, Binns, & Reese, 2017), or even viruses such as Hepatitis C Virus (Herker et al., 2010; Miyanari et al., 2007) and Sars-Cov2 (Dias et al., 2020), can induce the accumulation of LDs in the host and use it as a lipid source. During viral infection, this hijacking can go even further. For example, the Hepatitis C Virus (HCV) targets the liver, where LDs are consumed for the production and assembly of new infectious virus particles. In a model proposed by Miyanari et al., the core HCV protein binds to the lipid monolayer surrounding the LD, brings ER membranes close to the LD and recruits virus proteins as well as replication complexes to this microenvironment in order to produce new viral particles (Miyanari et al., 2007).

Interestingly, this hijacking of the host metabolism by pathogens also occurs in viral infection of eukaryotic microalgae. Due to high ecological relevance, several studies have taken interest in the infection of

Haptophytes, and in particular of the coccolithophore *Emiliana Huxleyi*, by large dsDNA viruses. An initial study showed that infection of *Emiliana* by the lytic virus EhV201, belonging to the Phycodnaviridae, leads to a complete rewiring of the host lipidome (Rosenwasser et al., 2014). An increase in *de novo* fatty acid synthesis proved crucial for viral production. Indeed, lipid biosynthesis reoriented to favor the production of specific glycosphingolipids at the expense of host sphingolipids. These virus-derived glycosphingolipids (vGSL) are central structural components of the virus membrane and can trigger programmed cell death of infected cell (Vardi et al., 2009). Finally, sterols appeared depleted in the host, feeding the new virions (Rosenwasser et al., 2014). Malitsky et al. (2016) completed this study and showed that Ehv201 can induce specific TAG accumulation and LD formation by upregulating genes involved in *de novo* TAG synthesis (LPAT, PAP, DGAT2) and downregulating genes involved in TAG recycling from phospholipids (such as PDAT). Virus-induced TAG, characterized by a high enrichment in unsaturated and monounsaturated fatty acids, are later found in high amounts in the virions. All those results led the authors to speculate that virus-induced LDs play a crucial role in virus assembly (Malitsky et al., 2016) (Fig. 1, left panel) as has been described for HCV (Miyanari et al., 2007). Other virus families may share this mechanism. Indeed, a recent study showed that the giant haptophyte-infecting virus, *Prymnesium kappa* virus RF01 (PkV RF01) belonging to the Mimiviridae family, encodes many proteins involved in lipid catabolism, among which a TAG lipase and other key enzymes of the β-oxidation pathway. This suggests that this virus is also capable to break down host LDs for its propagation (Blanc-Mathieu et al., 2021).

4.2 Lipid droplets as defense organelles

While this LD hijacking by pathogens may paint LDs as weak points in the cell, their active role in innate defense have been equally reviewed (Pereira-Dutra, Teixeira, de Costa, & Bozza, 2019). Indeed, in mammals, proteins with anti-pathogen properties such as viperin (Hinson & Cresswell, 2009) and interferon-γ (IFN-γ) inducible guanosine triphosphatase (GTPase) (IGTP; Bougnères et al., 2009; Haldar et al., 2013) are localized in the LDs of infected cells, and similar roles of LDs have been shown in higher plants.

Several studies regarding LDs in *Arabidopsis thaliana* leaves highlight their function in defense against various pathogens (Brocard et al., 2017;

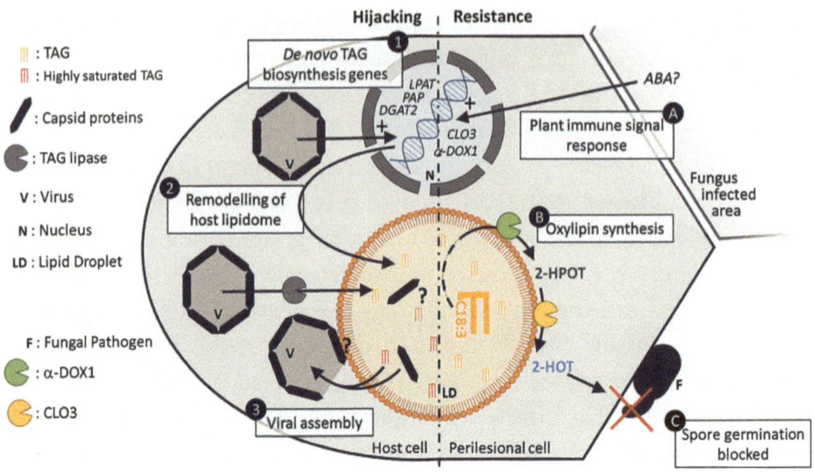

Fig. 4 Functions of LDs in pathogen response: two examples. Left panel: Hijacking of the host LD for the benefit of the pathogen: infection of the coccolithophore *Emiliana huxleyi* by the lytic *Emiliana huxleyi* virus (EhV201). At 24 h post infection, the expression of genes involved in *de novo* TAG biosynthesis such as *LPAT*, *PAP* and *DGAT2* is upregulated (1). This results in the remodeling of the host lipidome and the formation of lipid droplets (2), which could be associated with the storage of viral capsid proteins. The TAGs become progressively more saturated. In the later stage of the infection, high expression of virus-encoded lipases may lead to the disruption of the LD, and the formation of a new viral particle containing a high amount of highly saturated TAGs (3). Right panel: LD as a production platform for phytoalexins. Response of *Arabidopsis thaliana* leaves to fungi infection. Upon infection by fungi such as *Colletotrichum higginsianum*, the expression of Caleosin3 (CLO3) and α-Dioxygenase 1 (αDOX1) is induced in perilesional cells, possibly through the action of abscisic acid (ABA), an essential hormone involved in the plant immune response (A). α-DOX1 and CLO3 locate to the leaf LD and cooperatively produce the oxylipin 2-hydroxyoctadecatrienoic acid (2-HOT) from α-linolenic acid (C18:3, ω-3) (B). 2-HOT blocks the fungi (F) spore germination and its propagation to healthy tissues (C). A similar defense system is induced by senescence. *Left panel: model based on Malitsky, S., Ziv, C., Rosenwasser, S., Zheng, S., Schatz, D., Porat, Z., et al. (2016). Viral infection of the marine alga Emiliania huxleyi triggers lipidome remodeling and induces the production of highly saturated triacylglycerol.* New Phytologist, 210*(1), 88–96. https://doi.org/https://doi.org/10.1111/nph.13852 and inspired by Miyanari, Y., Atsuzawa, K., Usuda, N., Watashi, K., Hishiki, T., Zayas, M., et al. (2007). The lipid droplet is an important organelle for hepatitis C virus production.* Nature Cell Biology, 9*(9), 1089–1097. https://doi.org/https://doi.org/10.1038/ncb1631 Right panel: based on Shimada, T. L., Takano, Y., Shimada, T., Fujiwara, M., Fukao, Y., Mori, M., et al. (2014). Leaf oil body functions as a subcellular factory for the production of a phytoalexin in arabidopsis.* Plant Physiology, 164*(1), 105–118. https://doi.org/10.1104/pp.113.230185.*

Fernandez-Santos et al., 2020; Shimada, Hayashi, & Hara-Nishimura, 2018; Shimada et al., 2014). Indeed, proteomics have shown that several enzymes involved in production of phytoalexins, compounds that participate in plant defense, are located in LDs. The most prominent example is that of Caleosins (CLO), which are structural components of plant LDs and exhibit a peroxygenase activity. CLO3, which is also known as Responsive to Dehydratation20 (RD20) can be induced by abiotic stresses, notably drought (Aubert et al., 2010) or salt, but also in response to the pathogenic fungus *Colletotrichum higginsianum* (Shimada et al., 2014). Upon infection, the authors have shown that CLO3 colocalizes with α-Dioxygenase 1 (αDOX1) on LDs of perilesional cells of the infected leaf, where they coordinately use α-linolenic acid (C18:3, ω-3) to produce the antifungal oxylipin 2-hydroxy-octadecatrienoic acid (2-HOT). A similar effect is triggered by senescence and the authors propose that phytoalexin-producing LDs that develop in dying cells of perilesional tissues or senescent leaves could serve to form a shield in order to protect the young healthy parts of the plant from infection (Shimada et al., 2014) (Fig. 4, right panel). A recent proteomics analysis comparing LDs from leaves infected by the bacterium *Pseudomonas syringae pv. Tomato (Pst)* with senescent leaves has shown that they both contain enzymes involved in the biosynthesis of the antimicrobial compound camalexin, as well as GPAT4 and GPAT8, which are implicated in cutin biosynthesis (Fernandez-Santos et al., 2020). The overlap between LD proteomes in these two conditions confirms that they likely share common functions in cell defense against pathogens. Finally, the ectopic expression of stilbene synthases from *Vitis pseudoreticulata* (VpSTS29) in *Arabidopsis* shows that these enzymes, which play a critical role in powdery mildew resistance, can relocate from the cytoplasm to cytosolic LDs upon dark-induced senescence (Ma, Wang, & Wang, 2018). Intriguingly, stilbene synthase-containing LDs can be found both in the cytoplasm and in the vacuole, where stilbenes seem to be stored, thus implying a trafficking role for LDs, an hypothesis supported by the association of many cytoskeleton proteins with LDs (Brocard et al., 2017; Fernandez-Santos et al., 2020). Finally, AtCPK1, which is a Ca2+ dependent protein kinase involved in the response to fungal pathogens through the salicylic acid-mediated defense pathway, shows a dual localization to peroxisomes and LDs (Coca & Segundo, 2010). However, the function of this localization remains unknown.

Altogether, these results suggest that LDs of higher plants can be seen as production platforms for anti-pathogenic compounds involved in various plant defense mechanisms. However, we could not find in the literature any evidence that LDs play such a role in microalgae. The only potential link identified thus far lies in the antibacterial function of Eicosapentaenoic acid (EPA), a very long chain polyunsaturated fatty acid (VLC-PUFA) that is very abundant in many microalgae (Desbois, Mearns-Spragg, & Smith, 2009). However, VLC-PUFA are generally not very abundant in TAGs, and potential anti-pathogenic activities of LDs remain to be investigated in these organisms. This difference of functions, if it exists, may be linked with the multicellular nature of higher plants where an infected part of the plant can signal the danger to the rest of the organism and trigger defenses. It would be therefore very interesting to investigate whether similar signals could exist at the population scale in microalgae. Conversely, the hijacking of LDs has not yet been evidenced in plants even though it exists in multicellular organisms such as mammals. The function of LDs in defense against pathogens in plants and algae thus retains many mysteries to unravel.

5. Conclusion

In this review, we have explored the many roles of LDs, far from the simple fat stock that they initially represented (Fig. 5). It is not clear whether the same LD carries all functions or if a cell contains different types of LDs with distinct functions. Several lines of evidence suggest the latter. Analysis of LDs subpopulation in *Phaeodactylum* during nitrogen starvation has shown that there are three waves of LD formation and that these exhibit different behaviors (Jaussaud et al., 2020). The coexistence of several LDs populations with distinct functions has also been suggested in *Arabidopsis* aging leaves (Brocard et al., 2017). Finally, a refined analysis of LDs at different stages of seed formation, germination and seedling development has shown many changes in their proteome, which may reflect different subsets of LDs (Kretzschmar et al., 2020). Our understanding of LD biogenesis as well as of their functions comes mostly from the study of non photosynthetic organisms, animals and yeast. Yet, while there are many common points, evidences show that this knowledge cannot be simply extrapolated to photosynthetic organisms. The primary characteristic of plants and algae is the presence of a photosynthetic plastid that is central in the overall metabolism and in particular in the metabolism of lipids. Moreover, the evolutionary origin of this plastid can considerably diverge when we start looking at secondary endosymbionts, and its role in LD biogenesis requires further

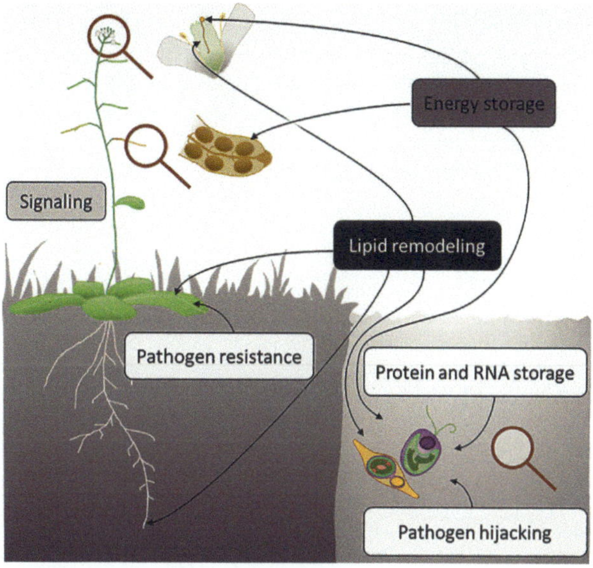

Fig. 5 The many functions of LDs. LDs have been studied in different organs from land plants, mainly seeds, pollen, leaves and to a lesser extent roots. They were also studied in different microalgae resulting from primary or secondary endosymbiosis, living in freshwater or marine environments. Functions that are common to most organs and organisms are indicated on a dark background. By contrast, functions that are only found in a group of organisms (plants or microalgae) or in a specific tissue are indicated on a light background.

investigations. The study of LD roles enlightens the specificities of the responses to abiotic stresses in one given organism, as well as the differences that can exist between a single cell organism or a pluricellular one, between a primary or secondary endosymbiont, or even between freshwater algae or marine ones (Fig. 5). Oleaginous microalgae are of great biotechnological interest, for the production of biofuels or high value-added molecules such as ω3-fatty acids or pigments, like EPA and astaxanthin, respectively. This has considerably increased the interest for these organisms. Unfortunately, many studies solely aim to increase the productivity of compounds of interest, without looking further into the biological or ecological relevance of the results. Therefore, knowledge remains sparse on many species and few studies compare the effects of different stresses. Yet, increasing our knowledge of LD functions and of their potential adaptive roles would be of great interest in the context of global change we are facing. The combination of genomics, proteomics, lipidomics and microscopy tools that are increasingly powerful provide exciting opportunities for future research.

Acknowledgements

The authors wish to thank Juliette Jouhet, Fabrice Rébeillé, Eric Maréchal and Alberto Amato for helpful discussions. This work was supported by GRAL, a programme from the Chemistry Biology Health (CBH) Graduate School of University Grenoble Alpes (ANR-17-EURE-0003), and programs from the CEA (FOCUS Fermeture du Cycle du Carbone (SL-DRF-20-0298)), and INRAE (BAP funding OLEuM).

References

Abida, H., Dolch, L. J., Mei, C., Villanova, V., Conte, M., Block, M. A., et al. (2015). Membrane glycerolipid remodeling triggered by nitrogen and phosphorus starvation in Phaeodactylum tricornutum. *Plant Physiology*, *167*(1), 118–136. https://doi.org/10.1104/pp.114.252395.

Altmann, R. (1894). *Die Elementarorganismen und ihre Beziehungen zu den Zellen*. Veit & comp.

Arisz, S. A., Heo, J.-Y., Koevoets, I. T., Zhao, T., van Egmond, P., Meyer, A. J., et al. (2018). Diacylglycerol acyltransferase1 contributes to freezing tolerance. *Plant Physiology*, *177*(4), 1410–1424. https://doi.org/10.1104/pp.18.00503.

Aubert, Y., Vile, D., Pervent, M., Aldon, D., Ranty, B., Simonneau, T., et al. (2010). RD20, a stress-inducible caleosin, participates in stomatal control, transpiration and drought tolerance in Arabidopsis thaliana. *Plant and Cell Physiology*, *51*(12), 1975–1987. https://doi.org/10.1093/pcp/pcq155.

Barbosa, A. D., Lim, K., Mari, M., Edgar, J. R., Gal, L., Sterk, P., et al. (2019). Compartmentalized synthesis of triacylglycerol at the inner nuclear membrane regulates nuclear organization. *Developmental Cell*, *50*(6), 755–766.e6. https://doi.org/10.1016/j.devcel.2019.07.009.

Baud, S. (2018). Seeds as oil factories. *Plant Reproduction*, *31*(3), 213–235. https://doi.org/10.1007/s00497-018-0325-6.

Baud, S., Dichow, N. R., Kelemen, Z., d'Andréa, S., To, A., Berger, N., et al. (2009). Regulation of HSD1 in seeds of Arabidopsis thaliana. *Plant and Cell Physiology*, *50*(8), 1463–1478. https://doi.org/10.1093/pcp/pcp092.

Beevers, H. (1979). Microbodies in higher plants. *Annual Review of Plant Physiology*, *30*(1), 159–193. https://doi.org/10.1146/annurev.pp.30.060179.001111.

Bersuker, K., & Olzmann, J. A. (2017). Establishing the lipid droplet proteome: Mechanisms of lipid droplet protein targeting and degradation. *Biochimica et Biophysica Acta (BBA) - Molecular and Cell Biology of Lipids*, *1862*(10, Pt. B), 1166–1177. https://doi.org/10.1016/j.bbalip.2017.06.006.

Blanc-Mathieu, R., Dahle, H., Hofgaard, A., Brandt, D., Ban, H., Kalinowski, J., et al. (2021). A persistent giant algal virus, with a unique morphology, encodes an unprecedented number of genes involved in energy metabolism. *Journal of Virology*, *95*(8), e02446–20. https://doi.org/10.1128/JVI.02446-20.

Bosch, M., Sánchez-Álvarez, M., Fajardo, A., Kapetanovic, R., Steiner, B., Dutra, F., et al. (2020). Mammalian lipid droplets are innate immune hubs integrating cell metabolism and host defense. *Science*, *370*(6514), eaay8085. https://doi.org/10.1126/science.aay8085.

Bose, D., Ngo, A. H., Nguyen, V. C., & Nakamura, Y. (2021). Non-specific phospholipases C2 and C6 redundantly function in pollen tube growth via triacylglycerol production in Arabidopsis. *The Plant Journal*, *106*(2), 409–418. https://doi.org/10.1111/tpj.15172.

Bougnères, L., Helft, J., Tiwari, S., Vargas, P., Chang, B. H.-J., Chan, L., et al. (2009). A role for lipid bodies in the cross-presentation of phagocytosed antigens by MHC Class I in dendritic cells. *Immunity*, *31*(2), 232–244. https://doi.org/10.1016/j.immuni.2009.06.022.

Brocard, L., Immel, F., Coulon, D., Esnay, N., Tuphile, K., Pascal, S., et al. (2017). Proteomic analysis of lipid droplets from Arabidopsis aging leaves brings new insight into their biogenesis and functions. *Frontiers in Plant Science*, *8*, 894. https://doi.org/10.3389/fpls.2017.00894.

Cai, Y., Goodman, J. M., Pyc, M., Mullen, R. T., Dyer, J. M., & Chapman, K. D. (2015). Arabidopsis SEIPIN proteins modulate triacylglycerol accumulation and influence lipid droplet proliferation. *Plant Cell*, *27*(9), 2616–2636. https://doi.org/10.1105/tpc.15.00588.

Castro, I. G., Eisenberg-Bord, M., Persiani, E., Rochford, J. J., Schuldiner, M., & Bohnert, M. (2019). Promethin is a conserved seipin partner protein. *Cell*, *8*(3). https://doi.org/10.3390/cells8030268.

Cermelli, S., Guo, Y., Gross, S. P., & Welte, M. A. (2006). The lipid-droplet proteome reveals that droplets are a protein-storage depot. *Current Biology*, *16*(18), 1783–1795. https://doi.org/10.1016/j.cub.2006.07.062.

Chapman, K. D., Aziz, M., Dyer, J. M., & Mullen, R. T. (2019). Mechanisms of lipid droplet biogenesis. *The Biochemical Journal*, *476*(13), 1929–1942. https://doi.org/10.1042/BCJ20180021.

Chapman, K. D., & Ohlrogge, J. B. (2012). Compartmentation of triacylglycerol accumulation in plants. *The Journal of Biological Chemistry*, *287*(4), 2288–2294. https://doi.org/10.1074/jbc.R111.290072.

Chen, E. C. F., Tai, S. S. K., Peng, C.-C., & Tzen, J. T. C. (1998). Identification of three novel unique proteins in seed oil bodies of sesame. *Plant and Cell Physiology*, *39*(9), 935–941. https://doi.org/10.1093/oxfordjournals.pcp.a029457.

Choudhary, V., Ojha, N., Golden, A., & Prinz, W. A. (2015). A conserved family of proteins facilitates nascent lipid droplet budding from the ER. *The Journal of Cell Biology*, *211*(2), 261–271. https://doi.org/10.1083/jcb.201505067.

Chung, J., Wu, X., Lambert, T. J., Lai, Z. W., Walther, T. C., & Farese, R. V. (2019). LDAF1 and seipin form a lipid droplet assembly complex. *Developmental Cell*, *51*(5), 551–563.e7. https://doi.org/10.1016/j.devcel.2019.10.006.

Coca, M., & Segundo, B. S. (2010). AtCPK1 calcium-dependent protein kinase mediates pathogen resistance in Arabidopsis. *The Plant Journal*, *63*(3), 526–540. https://doi.org/10.1111/j.1365-313X.2010.04255.x.

d'Andréa, S., Canonge, M., Beopoulos, A., Jolivet, P., Hartmann, M. A., Miquel, M., et al. (2007). At5g50600 encodes a member of the short-chain dehydrogenase reductase superfamily with 11β- and 17β-hydroxysteroid dehydrogenase activities associated with Arabidopsis thaliana seed oil bodies. *Biochimie*, *89*(2), 222–229. https://doi.org/10.1016/j.biochi.2006.09.013.

Davidi, L., Levin, Y., Ben-Dor, S., & Pick, U. (2015). Proteome analysis of cytoplasmatic and plastidic beta-carotene lipid droplets in Dunaliella bardawil. *Plant Physiology*, *167*(1), 60–79. https://doi.org/10.1104/pp.114.248450.

Degenkolbe, T., Giavalisco, P., Zuther, E., Seiwert, B., Hincha, D. K., & Willmitzer, L. (2012). Differential remodeling of the lipidome during cold acclimation in natural accessions of Arabidopsis thaliana. *The Plant Journal*, *72*(6), 972–982. https://doi.org/10.1111/tpj.12007.

Deruyffelaere, C., Bouchez, I., Morin, H., Guillot, A., Miquel, M., Froissard, M., et al. (2015). Ubiquitin-mediated proteasomal degradation of oleosins is involved in oil body mobilization during post-germinative seedling growth in Arabidopsis. *Plant and Cell Physiology*, *56*(7), 1374–1387. https://doi.org/10.1093/pcp/pcv056.

Deruyffelaere, C., Purkrtova, Z., Bouchez, I., Collet, B., Cacas, J.-L., Chardot, T., et al. (2018). PUX10 is a CDC48A adaptor protein that regulates the extraction of ubiquitinated oleosins from seed lipid droplets in arabidopsis. *The Plant Cell*, *30*(9), 2116–2136. https://doi.org/10.1105/tpc.18.00275.

Desbois, A. P., Mearns-Spragg, A., & Smith, V. J. (2009). A fatty acid from the diatom Phaeodactylum tricornutum is antibacterial against diverse bacteria including multi-resistant Staphylococcus aureus (MRSA). *Marine Biotechnology, 11*(1), 45–52. https://doi.org/10.1007/s10126-008-9118-5.

Dhiman, R., Caesar, S., Thiam, A. R., & Schrul, B. (2020). Mechanisms of protein targeting to lipid droplets: a unified cell biological and biophysical perspective. *Seminars in Cell & Developmental Biology, 108*, 4–13. https://doi.org/10.1016/j.semcdb.2020.03.004.

Dias, S. S. G., Soares, V. C., Ferreira, A. C., Sacramento, C. Q., Fintelman-Rodrigues, N., Temerozo, J. R., et al. (2020). Lipid droplets fuel SARS-CoV-2 replication and production of inflammatory mediators. *PLoS Pathogens, 16*(12), e1009127. https://doi.org/10.1371/journal.ppat.1009127.

Divi, U. K., & Krishna, P. (2009). Brassinosteroid: A biotechnological target for enhancing crop yield and stress tolerance. *New Biotechnology, 26*(3), 131–136. https://doi.org/10.1016/j.nbt.2009.07.006.

Eastmond, P. J. (2006). *SUGAR-DEPENDENT1* encodes a patatin domain triacylglycerol lipase that initiates storage oil breakdown in germinating *Arabidopsis* Seeds. *The Plant Cell, 18*(3), 665–675. https://doi.org/10.1105/tpc.105.040543.

Fan, J., Yan, C., Roston, R., Shanklin, J., & Xu, C. (2014). *Arabidopsis* lipins, PDAT1 acyltransferase, and SDP1 triacylglycerol lipase synergistically direct fatty acids toward β-oxidation, thereby maintaining membrane lipid homeostasis. *The Plant Cell, 26*(10), 4119–4134. https://doi.org/10.1105/tpc.114.130377.

Fan, J., Yu, L., & Xu, C. (2017). A central role for triacylglycerol in membrane lipid breakdown, fatty acid β-oxidation, and plant survival under extended darkness. *Plant Physiology, 174*(3), 1517–1530. https://doi.org/10.1104/pp.17.00653.

Fan, J., Yu, L., & Xu, C. (2019). Dual role for autophagy in lipid metabolism in Arabidopsis. *Plant Cell, 31*(7), 1598–1613. https://doi.org/10.1105/tpc.19.00170.

Fernandez-Santos, R., Izquierdo, Y., Lopez, A., Muniz, L., Martinez, M., Cascon, T., et al. (2020). Protein profiles of lipid droplets during the hypersensitive defense response of arabidopsis against pseudomonas infection. *Plant & Cell Physiology, 61*(6), 1144–1157. https://doi.org/10.1093/pcp/pcaa041.

Ferrer, A., Altabella, T., Arro, M., & Boronat, A. (2017). Emerging roles for conjugated sterols in plants. *Progress in Lipid Research, 67*, 27–37. https://doi.org/10.1016/j.plipres.2017.06.002.

Flori, S., Jouneau, P. H., Finazzi, G., Marechal, E., & Falconet, D. (2016). Ultrastructure of the periplastidial compartment of the diatom Phaeodactylum tricornutum. *Protist, 167*(3), 254–267. https://doi.org/10.1016/j.protis.2016.04.001.

Footitt, S., Dietrich, D., Fait, A., Fernie, A. R., Holdsworth, M. J., Baker, A., et al. (2007). The COMATOSE ATP-binding cassette transporter is required for full fertility in Arabidopsis. *Plant Physiology, 144*(3), 1467–1480. https://doi.org/10.1104/pp.107.099903.

Gaspar, M. L., Hofbauer, H. F., Kohlwein, S. D., & Henry, S. A. (2011). Coordination of storage lipid synthesis and membrane biogenesis: evidence for cross-talk between triacylglycerol metabolism and phosphatidylinositol synthesis. *Journal of Biological Chemistry, 286*(3), 1696–1708. https://doi.org/10.1074/jbc.M110.172296.

Gasulla, F., Vom Dorp, K., Dombrink, I., Zähringer, U., Gisch, N., Dörmann, P., et al. (2013). The role of lipid metabolism in the acquisition of desiccation tolerance in Craterostigma plantagineum: A comparative approach. *The Plant Journal, 75*(5), 726–741. https://doi.org/10.1111/tpj.12241.

Gidda, S. K., Park, S., Pyc, M., Yurchenko, O., Cai, Y., Wu, P., et al. (2016). Lipid droplet-associated proteins (LDAPs) are required for the dynamic regulation of neutral lipid compartmentation in plant cells. *Plant Physiology, 170*(4), 2052–2071. https://doi.org/10.1104/pp.15.01977.

Goepfert, S., & Poirier, Y. (2007). β-Oxidation in fatty acid degradation and beyond. *Current Opinion in Plant Biology*, *10*(3), 245–251. https://doi.org/10.1016/j.pbi.2007.04.007.

Goodson, C., Roth, R., Wang, Z. T., & Goodenough, U. (2011). Structural correlates of cytoplasmic and chloroplast lipid body synthesis in Chlamydomonas reinhardtii and stimulation of lipid body production with acetate boost. *Eukaryotic Cell*, *10*(12), 1592–1606. https://doi.org/10.1128/EC.05242-11.

Goold, H. D., Cuine, S., Legeret, B., Liang, Y., Brugiere, S., Auroy, P., et al. (2016). Saturating light induces sustained accumulation of oil in plastidal lipid droplets in Chlamydomonas reinhardtii. *Plant Physiology*, *171*(4), 2406–2417. https://doi.org/10.1104/pp.16.00718.

Graef, M. (2018). Lipid droplet-mediated lipid and protein homeostasis in budding yeast. *FEBS Letters*, *592*(8), 1291–1303. https://doi.org/10.1002/1873-3468.12996.

Graham, I. A. (2008). Seed storage oil mobilization. *Annual Review of Plant Biology*, *59*(1), 115–142. https://doi.org/10.1146/annurev.arplant.59.032607.092938.

Greer, M. S., Cai, Y., Gidda, S. K., Esnay, N., Kretzschmar, F. K., Seay, D., et al. (2020). SEIPIN isoforms interact with the membrane-tethering protein VAP27-1 for lipid droplet formation. *The Plant Cell*, *32*(9), 2932–2950. https://doi.org/10.1105/tpc.19.00771.

Guéguen, N., Le Moigne, D., Amato, A., Salvaing, J., & Maréchal, E. (2021). Lipid droplets in unicellular photosynthetic stramenopiles. *Frontiers in Plant Science*, *12*, 639276. https://doi.org/10.3389/fpls.2021.639276.

Haldar, A. K., Saka, H. A., Piro, A. S., Dunn, J. D., Henry, S. C., Taylor, G. A., et al. (2013). IRG and GBP host resistance factors target aberrant, "Non-self" vacuoles characterized by the missing of "self" IRGM proteins. *PLoS Pathogens*, *9*(6), e1003414. https://doi.org/10.1371/journal.ppat.1003414.

Han, B., Xu, H., Feng, Y., Xu, W., Cui, Q., & Liu, A. (2020). Genomic characterization and expressional profiles of autophagy-related genes (ATGs) in oilseed crop castor bean (Ricinus communis L.). *International Journal of Molecular Sciences*, *21*(2), 562. https://doi.org/10.3390/ijms21020562.

Hayashi, M., & Nishimura, M. (2003). Entering a new era of research on plant peroxisomes. *Current Opinion in Plant Biology*, *6*(6), 577–582. https://doi.org/10.1016/j.pbi.2003.09.012.

Henne, M., Goodman, J. M., & Hariri, H. (2020). Spatial compartmentalization of lipid droplet biogenesis. *Biochimica et Biophysica Acta (BBA) - Molecular and Cell Biology of Lipids*, *1865*(1), 158499. https://doi.org/10.1016/j.bbalip.2019.07.008.

Herker, E., Harris, C., Hernandez, C., Carpentier, A., Kaehlcke, K., Rosenberg, A. R., et al. (2010). Efficient hepatitis C virus particle formation requires diacylglycerol acyltransferase-1. *Nature Medicine*, *16*(11), 1295–1298. https://doi.org/10.1038/nm.2238.

Higashi, Y., Okazaki, Y., Myouga, F., Shinozaki, K., & Saito, K. (2015). Landscape of the lipidome and transcriptome under heat stress in Arabidopsis thaliana. *Scientific Reports*, *5*(1), 1–11. https://doi.org/10.1038/srep10533.

Higashi, Y., Okazaki, Y., Takano, K., Myouga, F., Shinozaki, K., Knoch, E., et al. (2018). HEAT INDUCIBLE LIPASE1 remodels chloroplastic monogalactosyldiacylglycerol by liberating α-linolenic acid in Arabidopsis leaves under heat stress. *The Plant Cell*, *30*(8), 1887–1905. https://doi.org/10.1105/tpc.18.00347.

Higashi, Y., & Saito, K. (2019). Lipidomic studies of membrane glycerolipids in plant leaves under heat stress. *Progress in Lipid Research*, *75*, 100990. https://doi.org/10.1016/j.plipres.2019.100990.

Hinson, E. R., & Cresswell, P. (2009). The antiviral protein, viperin, localizes to lipid droplets via its N-terminal amphipathic α-helix. *Proceedings of the National Academy of Sciences of the United States of America*, *106*(48), 20452–20457. https://doi.org/10.1073/pnas.0911679106.

Horn, P. J., James, C. N., Gidda, S. K., Kilaru, A., Dyer, J. M., Mullen, R. T., et al. (2013). Identification of a new class of lipid droplet-associated proteins in plants. *Plant Physiology*, *162*(4), 1926–1936. https://doi.org/10.1104/pp.113.222455.

Hsu, T.-H., Chen, R.-H., Cheng, Y.-H., & Wang, C.-W. (2017). Lipid droplets are central organelles for meiosis II progression during yeast sporulation. *Molecular Biology of the Cell*, *28*(3), 440–451. https://doi.org/10.1091/mbc.e16-06-0375.

Hu, X., Binns, D., & Reese, M. L. (2017). The coccidian parasites toxoplasma and neospora dysregulate mammalian lipid droplet biogenesis. *Journal of Biological Chemistry*, *292*(26), 11009–11020. https://doi.org/10.1074/jbc.M116.768176.

Huang, A. H. C. (1992). Oil bodies and oleosins in seeds. *Annual Review of Plant Physiology and Plant Molecular Biology*, *43*(1), 177–200. https://doi.org/10.1146/annurev.pp.43.060192.001141.

Huang, A. H. C. (2018). Plant lipid droplets and their associated proteins: Potential for rapid advances. *Plant Physiology*, *176*(3), 1894–1918. https://doi.org/10.1104/pp.17.01677.

Huang, B., Marchand, J., Thiriet-Rupert, S., Carrier, G., Saint-Jean, B., Lukomska, E., et al. (2019). Betaine lipid and neutral lipid production under nitrogen or phosphorus limitation in the marine microalga Tisochrysis lutea (Haptophyta). *Algal Research*, *40*, 101506. https://doi.org/10.1016/j.algal.2019.101506.

Hulatt, C. J., Smolina, I., Dowle, A., Kopp, M., Vasanth, G. K., Hoarau, G. G., et al. (2020). Proteomic and transcriptomic patterns during lipid remodeling in Nannochloropsis gaditana. *International Journal of Molecular Sciences*, *21*(18), 6946. https://doi.org/10.3390/ijms21186946.

Ischebeck, T. (2016). Lipids in pollen—They are different. *Biochimica et Biophysica Acta*, *1861*(9 Pt. B), 1315–1328. https://doi.org/10.1016/j.bbalip.2016.03.023.

Ischebeck, T., Krawczyk, H. E., Mullen, R. T., Dyer, J. M., & Chapman, K. D. (2020). Lipid droplets in plants and algae: Distribution, formation, turnover and function. *Seminars in Cell & Developmental Biology*, *108*, 82–93. https://doi.org/10.1016/j.semcdb.2020.02.014.

James, C. N., Horn, P. J., Case, C. R., Gidda, S. K., Zhang, D., Mullen, R. T., et al. (2010). Disruption of the Arabidopsis CGI-58 homologue produces Chanarin-Dorfman-like lipid droplet accumulation in plants. *Proceedings of the National Academy of Sciences of the United States of America*, *107*(41), 17833–17838. https://doi.org/10.1073/pnas.0911359107.

Jaussaud, A., Lupette, J., Salvaing, J., Jouhet, J., Bastien, O., Gromova, M., et al. (2020). Stepwise biogenesis of subpopulations of lipid droplets in nitrogen starved Phaeodactylum tricornutum cells. *Frontiers in Plant Science*, *11*(48). https://doi.org/10.3389/fpls.2020.00048.

Joshi, A. S., Nebenfuehr, B., Choudhary, V., Satpute-Krishnan, P., Levine, T. P., Golden, A., et al. (2018). Lipid droplet and peroxisome biogenesis occur at the same ER subdomains. *Nature Communications*, *9*(1), 2940. https://doi.org/10.1038/s41467-018-05277-3.

Joshi, A. S., Ragusa, J. V., Prinz, W. A., & Cohen, S. (2021). Multiple C2 domain–containing transmembrane proteins promote lipid droplet biogenesis and growth at specialized endoplasmic reticulum subdomains. *Molecular Biology of the Cell*, *32*(12), 1147–1157. https://doi.org/10.1091/mbc.E20-09-0590.

Kaup, M. T., Froese, C. D., & Thompson, J. E. (2002). A role for diacylglycerol acyltransferase during leaf senescence. *Plant Physiology*, *129*(4), 1616–1626. https://doi.org/10.1104/pp.003087.

Kelly, A. A., Quettier, A.-L., Shaw, E., & Eastmond, P. J. (2011). Seed storage oil mobilization is important but not essential for germination or seedling establishment in Arabidopsis. *Plant Physiology*, *157*(2), 866–875. https://doi.org/10.1104/pp.111.181784.

Kretzschmar, F. K., Doner, N. M., Krawczyk, H. E., Scholz, P., Schmitt, K., Valerius, O., et al. (2020). Identification of low-abundance lipid droplet proteins in seeds and seedlings. *Plant Physiology*, *182*(3), 1326–1345. https://doi.org/10.1104/pp.19.01255.

Kretzschmar, F. K., Mengel, L. A., Müller, A. O., Schmitt, K., Blersch, K. F., Valerius, O., et al. (2018). PUX10 is a lipid droplet-localized scaffold protein that interacts with CELL DIVISION CYCLE48 and is involved in the degradation of lipid droplet proteins. *The Plant Cell*, *30*(9), 2137–2160. https://doi.org/10.1105/tpc.18.00276.

Kumar, Y., Cocchiaro, J., & Valdivia, R. H. (2006). The obligate intracellular pathogen Chlamydia trachomatis targets host lipid droplets. *Current Biology*, *16*(16), 1646–1651. https://doi.org/10.1016/j.cub.2006.06.060.

Légeret, B., Schulz-Raffelt, M., Nguyen, H. M., Auroy, P., Beisson, F., Peltier, G., et al. (2016). Lipidomic and transcriptomic analyses of Chlamydomonas reinhardtii under heat stress unveil a direct route for the conversion of membrane lipids into storage lipids. *Plant, Cell & Environment*, *39*(4), 834–847. https://doi.org/10.1111/pce.12656.

Leyland, B., Boussiba, S., & Khozin-Goldberg, I. (2020). A review of diatom lipid droplets. *Biology (Basel)*, *9*, 38. https://doi.org/10.3390/biology9020038.

Leyland, B., Zarka, A., Didi-Cohen, S., Boussiba, S., & Khozin-Goldberg, I. (2020). High resolution proteome of lipid droplets isolated from the pennate diatom *Phaeodactylum tricornutum* (*Bacillariophyceae*) strain pt4 provides mechanistic insights into complex intracellular coordination during nitrogen deprivation. *Journal of Phycology*, *56*(6), 1642–1663. https://doi.org/10.1111/jpy.13063.

Li, F., Asami, T., Wu, X., Tsang, E. W. T., & Cutler, A. J. (2007). A putative hydroxysteroid dehydrogenase involved in regulating plant growth and development. *Plant Physiology*, *145*(1), 87–97. https://doi.org/10.1104/pp.107.100560.

Li, X., Moellering, E. R., Liu, B., Johnny, C., Fedewa, M., Sears, B. B., et al. (2012). A galactoglycerolipid lipase is required for triacylglycerol accumulation and survival following nitrogen deprivation in Chlamydomonas reinhardtii. *Plant Cell*, *24*(11), 4670–4686. https://doi.org/10.1105/tpc.112.105106.

Li, Z., Thiel, K., Thul, P. J., Beller, M., Kühnlein, R. P., & Welte, M. A. (2012). Lipid droplets control the maternal histone supply of Drosophila embryos. *Current Biology*, *22*(22), 2104–2113. https://doi.org/10.1016/j.cub.2012.09.018.

Li, D., Yang, S.-G., He, C.-W., Zhang, Z.-T., Liang, Y., Li, H., et al. (2020). Excess diacylglycerol at the endoplasmic reticulum disrupts endomembrane homeostasis and autophagy. *BMC Biology*, *18*(1), 107. https://doi.org/10.1186/s12915-020-00837-w.

Lin, L.-J., Tai, S. S. K., Peng, C.-C., & Tzen, J. T. C. (2002). Steroleosin, a sterol-binding dehydrogenase in seed oil bodies. *Plant Physiology*, *128*(4), 1200–1211. https://doi.org/10.1104/pp.010982.

Lu, Y., Wang, X., Balamurugan, S., Yang, W.-D., Liu, J.-S., Dong, H.-P., et al. (2017). Identification of a putative seipin ortholog involved in lipid accumulation in marine microalga Phaeodactylum tricornutum. *Journal of Applied Psychology*, *29*(6), 2821–2829. https://doi.org/10.1007/s10811-017-1173-8.

Lundquist, P. K., Shivaiah, K.-K., & Espinoza-Corral, R. (2020). Lipid droplets throughout the evolutionary tree. *Progress in Lipid Research*, *78*, 101029. https://doi.org/10.1016/j.plipres.2020.101029.

Lupette, J., Jaussaud, A., Seddiki, K., Morabito, C., Brugière, S., Schaller, H., et al. (2019). The architecture of lipid droplets in the diatom Phaeodactylum tricornutum. *Algal Research*, *38*, 101415. https://doi.org/10.1016/j.algal.2019.101415.

Lupette, J., & Maréchal, E. (2020). The puzzling conservation and diversification of lipid droplets from bacteria to eukaryotes. In M. Kloc (Ed.), *Vol. 69. Symbiosis: Cellular, molecular, medical and evolutionary aspects* (pp. 281–334). Springer International Publishing. https://doi.org/10.1007/978-3-030-51849-3_11.

Ma, F., Wang, L., & Wang, Y. (2018). Ectopic expression of VpSTS29, a stilbene synthase gene from Vitis pseudoreticulata, indicates STS presence in cytosolic oil bodies. *Planta, 248*(1), 89–103. https://doi.org/10.1007/s00425-018-2883-0.

Malitsky, S., Ziv, C., Rosenwasser, S., Zheng, S., Schatz, D., Porat, Z., et al. (2016). Viral infection of the marine alga Emiliania huxleyi triggers lipidome remodeling and induces the production of highly saturated triacylglycerol. *New Phytologist, 210*(1), 88–96. https://doi.org/10.1111/nph.13852.

Martin, B. A., & Wilson, R. F. (1984). Subcellular localization of triacylglycerol synthesis in spinach leaves. *Lipids, 19*(2), 117–121. https://doi.org/10.1007/bf02534501.

Miray, R., Kazaz, S., To, A., & Baud, S. (2021). Molecular control of oil metabolism in the endosperm of seeds. *International Journal of Molecular Sciences, 22*(4), 1621. https://doi.org/10.3390/ijms22041621.

Miyanari, Y., Atsuzawa, K., Usuda, N., Watashi, K., Hishiki, T., Zayas, M., et al. (2007). The lipid droplet is an important organelle for hepatitis C virus production. *Nature Cell Biology, 9*(9), 1089–1097. https://doi.org/10.1038/ncb1631.

Moellering, E. R., & Benning, C. (2010). RNA interference silencing of a major lipid droplet protein affects lipid droplet size in Chlamydomonas reinhardtii. *Eukaryotic Cell, 9*(1), 97–106. https://doi.org/10.1128/EC.00203-09.

Moellering, E. R., Muthan, B., & Benning, C. (2010). Freezing tolerance in plants requires lipid remodeling at the outer chloroplast membrane. *Science, 330*(6001), 226–228. https://doi.org/10.1126/science.1191803.

Mueller, S. P., Krause, D. M., Mueller, M. J., & Fekete, A. (2015). Accumulation of extra-chloroplastic triacylglycerols in Arabidopsis seedlings during heat acclimation. *Journal of Experimental Botany, 66*(15), 4517–4526. https://doi.org/10.1093/jxb/erv226.

Mueller, S. P., Unger, M., Guender, L., Fekete, A., & Mueller, M. J. (2017). Phospholipid: diacylglycerol acyltransferase-mediated triacylglyerol synthesis augments basal thermotolerance. *Plant Physiology, 175*(1), 486–497. https://doi.org/10.1104/pp.17.00861.

Mulders, K. J. M., Lamers, P. P., Wijffels, R. H., & Martens, D. E. (2015). Dynamics of biomass composition and growth during recovery of nitrogen-starved Chromochloris zofingiensis. *Applied Microbiology and Biotechnology, 99*(4), 1873–1884. https://doi.org/10.1007/s00253-014-6181-x.

Muller, A. O., & Ischebeck, T. (2018). Characterization of the enzymatic activity and physiological function of the lipid droplet-associated triacylglycerol lipase AtOBL1. *The New Phytologist, 217*(3), 1062–1076. https://doi.org/10.1111/nph.14902.

Murphy, D. J. (2012). The dynamic roles of intracellular lipid droplets: From archaea to mammals. *Protoplasma, 249*(3), 541–585. https://doi.org/10.1007/s00709-011-0329-7.

Næsted, H., Frandsen, G. I., Jauh, G.-Y., Hernandez-Pinzon, I., Nielsen, H. B., Murphy, D. J., et al. (2000). Caleosins: Ca2+-binding proteins associated with lipid bodies. *Plant Molecular Biology, 44*(4), 463–476. https://doi.org/10.1023/A:1026564411918.

Nguyen, H. M., Baudet, M., Cuine, S., Adriano, J. M., Barthe, D., Billon, E., et al. (2011). Proteomic profiling of oil bodies isolated from the unicellular green microalga Chlamydomonas reinhardtii: With focus on proteins involved in lipid metabolism. *Proteomics, 11*(21), 4266–4273. https://doi.org/10.1002/pmic.201100114.

Nojima, D., Yoshino, T., Maeda, Y., Tanaka, M., Nemoto, M., & Tanaka, T. (2013). Proteomics analysis of oil body-associated proteins in the oleaginous diatom. *Journal of Proteome Research, 12*(11), 5293–5301. https://doi.org/10.1021/pr4004085.

Nonoyama, T., Nojima, D., Maeda, Y., Noda, M., Yoshino, T., Matsumoto, M., et al. (2019). Proteomics analysis of lipid droplets indicates involvement of membrane trafficking proteins in lipid droplet breakdown in the oleaginous diatom Fistulifera solaris. *Algal Research, 44*, 101660. https://doi.org/10.1016/j.algal.2019.101660.

Olzmann, J. A., & Carvalho, P. (2019). Dynamics and functions of lipid droplets. *Nature Reviews. Molecular Cell Biology, 20*(3), 137–155. https://doi.org/10.1038/s41580-018-0085-z.

Parthibane, V., Rajakumari, S., Venkateshwari, V., Iyappan, R., & Rajasekharan, R. (2012). Oleosin is bifunctional enzyme that has both monoacylglycerol acyltransferase and phospholipase activities. *Journal of Biological Chemistry*, *287*(3), 1946–1954. https://doi.org/10.1074/jbc.M111.309955.

Pereira-Dutra, F. S., Teixeira, L., de Costa, M. F. S., & Bozza, P. T. (2019). Fat, fight, and beyond: The multiple roles of lipid droplets in infections and inflammation. *Journal of Leukocyte Biology*, *106*(3), 563–580. https://doi.org/10.1002/JLB.4MR0119-035R.

Petroutsos, D., Amiar, S., Abida, H., Dolch, L. J., Bastien, O., Rebeille, F., et al. (2014). Evolution of galactoglycerolipid biosynthetic pathways—From cyanobacteria to primary plastids and from primary to secondary plastids. *Progress in Lipid Research*, *54*, 68–85. https://doi.org/10.1016/j.plipres.2014.02.001.

Petschnigg, J., Wolinski, H., Kolb, D., Zellnig, G., Kurat, C. F., Natter, K., et al. (2009). Good fat, essential cellular requirements for triacylglycerol synthesis to maintain membrane homeostasis in yeast. *Journal of Biological Chemistry*, *284*(45), 30981–30993. https://doi.org/10.1074/jbc.M109.024752.

Pick, U., Zarka, A., Boussiba, S., & Davidi, L. (2019). A hypothesis about the origin of carotenoid lipid droplets in the green algae Dunaliella and Haematococcus. *Planta*, *249*(1), 31–47. https://doi.org/10.1007/s00425-018-3050-3.

Prasanna, X., Salo, V. T., Li, S., Ven, K., Vihinen, H., Jokitalo, E., et al. (2021). Seipin traps triacylglycerols to facilitate their nanoscale clustering in the endoplasmic reticulum membrane. *PLoS Biology*, *19*(1), e3000998. https://doi.org/10.1371/journal.pbio.3000998.

Pyc, M., Cai, Y., Gidda, S. K., Yurchenko, O., Park, S., Kretzschmar, F. K., et al. (2017). Arabidopsis lipid droplet-associated protein (LDAP)-Interacting protein (LDIP) influences lipid droplet size and neutral lipid homeostasis in both leaves and seeds. *The Plant Journal*, *92*(6), 1182–1201. https://doi.org/10.1111/tpj.13754.

Pyc, M., Gidda, S. K., Seay, D., Esnay, N., Kretzschmar, F. K., Cai, Y., et al. (2021). LDIP cooperates with SEIPIN and LDAP to facilitate lipid droplet biogenesis in Arabidopsis. *The Plant Cell*, koab179. https://doi.org/10.1093/plcell/koab179.

Quettier, A.-L., Shaw, E., & Eastmond, P. J. (2008). *SUGAR-DEPENDENT6* encodes a mitochondrial flavin adenine dinucleotide-dependent glycerol-3-P dehydrogenase, which is required for glycerol catabolism and postgerminative seedling growth in Arabidopsis. *Plant Physiology*, *148*(1), 519–528. https://doi.org/10.1104/pp.108.123703.

Rosenwasser, S., Mausz, M. A., Schatz, D., Sheyn, U., Malitsky, S., Aharoni, A., et al. (2014). Rewiring host lipid metabolism by large viruses determines the fate of Emiliania huxleyi, a bloom-forming alga in the ocean. *The Plant Cell*, *26*(6), 2689–2707. https://doi.org/10.1105/tpc.114.125641.

Santinho, A., Salo, V. T., Chorlay, A., Li, S., Zhou, X., Omrane, M., et al. (2020). Membrane curvature catalyzes lipid droplet assembly. *Current Biology*, *30*, 1–14. https://doi.org/10.1016/j.cub.2020.04.066.

Shimada, T. L., Hayashi, M., & Hara-Nishimura, I. (2018). Membrane dynamics and multiple functions of oil bodies in seeds and leaves. *Plant Physiology*, *176*(1), 199–207. https://doi.org/10.1104/pp.17.01522.

Shimada, T. L., Shimada, T., Takahashi, H., Fukao, Y., & Hara-Nishimura, I. (2008). A novel role for oleosins in freezing tolerance of oilseeds in *Arabidopsis thaliana*. *The Plant Journal*, *55*(5), 798–809. https://doi.org/10.1111/j.1365-313X.2008.03553.x.

Shimada, T. L., Takano, Y., Shimada, T., Fujiwara, M., Fukao, Y., Mori, M., et al. (2014). Leaf oil body functions as a subcellular factory for the production of a phytoalexin in Arabidopsis. *Plant Physiology*, *164*(1), 105–118. https://doi.org/10.1104/pp.113.230185.

Siloto, R. M. P., Findlay, K., Lopez-Villalobos, A., Yeung, E. C., Nykiforuk, C. L., & Moloney, M. M. (2006). The accumulation of oleosins determines the size of seed oilbodies in *Arabidopsis*. *The Plant Cell*, *18*(8), 1961–1974. https://doi.org/10.1105/tpc.106.041269.

Simionato, D., Block, M. A., La Rocca, N., Jouhet, J., Maréchal, E., Finazzi, G., et al. (2013). The response of Nannochloropsis gaditana to nitrogen starvation includes de novo biosynthesis of triacylglycerols, a decrease of chloroplast galactolipids, and reorganization of the photosynthetic apparatus. *Eukaryotic Cell*, *12*(5), 665–676.

Singh, A. P., & Savaldi-Goldstein, S. (2015). Growth control: Brassinosteroid activity gets context. *Journal of Experimental Botany*, *66*(4), 1123–1132. https://doi.org/10.1093/jxb/erv026.

Sorger, D., Athenstaedt, K., Hrastnik, C., & Daum, G. (2004). A yeast strain lacking lipid particles bears a defect in ergosterol formation. *Journal of Biological Chemistry*, *279*(30), 31190–31196. https://doi.org/10.1074/jbc.M403251200.

Tan, W.-J., Yang, Y.-C., Zhou, Y., Huang, L.-P., Xu, L., Chen, Q.-F., et al. (2018). DIACYLGLYCEROL ACYLTRANSFERASE and DIACYLGLYCEROL KINASE modulate triacylglycerol and phosphatidic acid production in the plant response to freezing stress. *Plant Physiology*, *177*(3), 1303–1318. https://doi.org/10.1104/pp.18.00402.

Tarazona, P., Feussner, K., & Feussner, I. (2015). An enhanced plant lipidomics method based on multiplexed liquid chromatography–mass spectrometry reveals additional insights into cold- and drought-induced membrane remodeling. *The Plant Journal*, *84*(3), 621–633. https://doi.org/10.1111/tpj.13013.

Tauchi-Sato, K., Ozeki, S., Houjou, T., Taguchi, R., & Fujimoto, T. (2002). The surface of lipid droplets is a phospholipid monolayer with a unique fatty acid composition. *Journal of Biological Chemistry*, *277*(46), 44507–44512. https://doi.org/10.1074/jbc.M207712200.

Taurino, M., Costantini, S., De Domenico, S., Stefanelli, F., Ruano, G., Delgadillo, M. O., et al. (2018). SEIPIN proteins mediate lipid droplet biogenesis to promote pollen transmission and reduce seed dormancy. *Plant Physiology*, *176*(2), 1531–1546. https://doi.org/10.1104/pp.17.01430.

Thiam, A. R., & Forêt, L. (2016). The physics of lipid droplet nucleation, growth and budding. *Biochimica et Biophysica Acta (BBA) - Molecular and Cell Biology of Lipids*, *1861*(8, Pt. A), 715–722. https://doi.org/10.1016/j.bbalip.2016.04.018.

Tsai, C.-H., Uygun, S., Roston, R., Shiu, S.-H., & Benning, C. (2018). Recovery from N deprivation is a transcriptionally and functionally distinct state in *Chlamydomonas*. *Plant Physiology*, *176*(3), 2007–2023. https://doi.org/10.1104/pp.17.01546.

Tsai, C. H., Zienkiewicz, K., Amstutz, C. L., Brink, B. G., Warakanont, J., Roston, R., et al. (2015). Dynamics of protein and polar lipid recruitment during lipid droplet assembly in Chlamydomonas reinhardtii. *The Plant Journal*, *83*(4), 650–660. https://doi.org/10.1111/tpj.12917.

Tzen, J.-T.-C., Cao, Y.-Z., Laurent, P., Ratnayake, C., & Huang, A.-H.-C. (1993). Lipids, proteins, and structure of seed oil bodies from diverse species. *Plant Physiology*, *101*(1), 267–276. https://doi.org/10.1104/pp.101.1.267.

Vardi, A., Mooy, B. A. S. V., Fredricks, H. F., Popendorf, K. J., Ossolinski, J. E., Haramaty, L., et al. (2009). Viral glycosphingolipids induce lytic infection and cell death in marine phytoplankton. *Science*, *326*(5954), 861–865. https://doi.org/10.1126/science.1177322.

Velázquez, A. P., Tatsuta, T., Ghillebert, R., Drescher, I., & Graef, M. (2016). Lipid droplet-mediated ER homeostasis regulates autophagy and cell survival during starvation. *Journal of Cell Biology*, *212*(6), 621–631. https://doi.org/10.1083/jcb.201508102.

Vevea, J. D., Garcia, E. J., Chan, R. B., Zhou, B., Schultz, M., Di Paolo, G., et al. (2015). Role for lipid droplet biogenesis and microlipophagy in adaptation to lipid imbalance in yeast. *Developmental Cell*, *35*(5), 584–599. https://doi.org/10.1016/j.devcel.2015.11.010.

Vieler, A., Brubaker, S. B., Vick, B., & Benning, C. (2012). A lipid droplet protein of nannochloropsis with functions partially analogous to plant oleosins. *Plant Physiology*, *158*(4), 1562–1569. https://doi.org/10.1104/pp.111.193029.

Walther, T. C., Chung, J., & Farese, R. V. (2017). Lipid droplet biogenesis. *Annual Review of Cell and Developmental Biology*, *33*, 491–510. https://doi.org/10.1146/annurev-cellbio-100616060608.

Wang, X., Hao, T.-B., Balamurugan, S., Yang, W.-D., Liu, J.-S., Dong, H.-P., et al. (2017). A lipid droplet-associated protein involved in lipid droplet biogenesis and triacylglycerol accumulation in the oleaginous microalga Phaeodactylum tricornutum. *Algal Research*, *26*, 215–224. https://doi.org/10.1016/j.algal.2017.07.028.

Wang, K., Hersh, H. L., & Benning, C. (2016). SENSITIVE TO FREEZING2 aids in resilience to salt and drought in freezing-sensitive tomato. *Plant Physiology*, *172*(3), 1432–1442. https://doi.org/10.1104/pp.16.01183.

Wang, S., Idrissi, F.-Z., Hermansson, M., Grippa, A., Ejsing, C. S., & Carvalho, P. (2018). Seipin and the membrane-shaping protein Pex30 cooperate in organelle budding from the endoplasmic reticulum. *Nature Communications*, *9*(1), 2939. https://doi.org/10.1038/s41467-018-05278-2.

Wang, X., Wei, H., Mao, X., & Liu, J. (2019). Proteomics analysis of lipid droplets from the oleaginous alga Chromochloris zofingiensis reveals novel proteins for lipid metabolism. *Genomics, Proteomics & Bioinformatics*, *17*(3), 260–272. https://doi.org/10.1016/j.gpb.2019.01.003.

Yoneda, K., Yoshida, M., Suzuki, I., & Watanabe, M. M. (2016). Identification of a major lipid droplet protein in a marine diatom Phaeodactylum tricornutum. *Plant & Cell Physiology*, *57*(2), 397–406. https://doi.org/10.1093/pcp/pcv204.

Yoon, K., Han, D., Li, Y., Sommerfeld, M., & Hu, Q. (2012). Phospholipid:diacylglycerol acyltransferase is a multifunctional enzyme involved in membrane lipid turnover and degradation while synthesizing triacylglycerol in the unicellular green microalga *Chlamydomonas reinhardtii*. *The Plant Cell*, *24*(9), 3708–3724. https://doi.org/10.1105/tpc.112.100701.

You, Z., Zhang, Q., Peng, Z., & Miao, X. (2019). Lipid droplets mediate salt stress tolerance in Parachlorella kessleri. *Plant Physiology*, *181*(2), 510–526. https://doi.org/10.1104/pp.19.00666.

Yu, L., Zhou, C., Fan, J., Shanklin, J., & Xu, C. (2021). Mechanisms and functions of membrane lipid remodeling in plants. *The Plant Journal*, *107*, 37–53. https://doi.org/10.1111/tpj.15273.

Zhang, M., Fan, J., Taylor, D. C., & Ohlrogge, J. B. (2010). *DGAT1* and *PDAT1* acyltransferases have overlapping functions in *Arabidopsis* triacylglycerol biosynthesis and are essential for normal pollen and seed development. *The Plant Cell*, *21*(12), 3885–3901. https://doi.org/10.1105/tpc.109.071795.

Zienkiewicz, K., Castro, A. J., de Alché, J. D., Zienkiewicz, A., Suárez, C., & Rodríguez-García, M. I. (2010). Identification and localization of a caleosin in olive (Olea europaea L.) pollen during in vitro germination. *Journal of Experimental Botany*, *61*(5), 1537–1546. https://doi.org/10.1093/jxb/erq022.

Zienkiewicz, A., Zienkiewicz, K., Rejón, J. D., Rodríguez-García, M. I., & Castro, A. J. (2013). New insights into the early steps of oil body mobilization during pollen germination. *Journal of Experimental Botany*, *64*(1), 293–302. https://doi.org/10.1093/jxb/ers332.

Zoni, V., Khaddaj, R., Lukmantara, I., Shinoda, W., Yang, H., Schneiter, R., et al. (2021). Seipin accumulates and traps diacylglycerols and triglycerides in its ring-like structure. *Proceedings of the National Academy of Sciences of the United States of America*, *118*(10), e2017205118. https://doi.org/10.1073/pnas.2017205118.

CHAPTER SIX

The plant lipid metabolic network for assembly of diverse triacylglycerol molecular species

Philip D. Bates*
Institute of Biological Chemistry, Washington State University, Pullman, WA, United States
*Corresponding author: e-mail address: phil_bates@wsu.edu

Contents

Abstract

Plant oils are a renewable natural resource providing food, feed, bio-fuels, and chemicals. The various applications of plant oils stem the highly diverse fatty acid compositions produced by plants. However, most crop plants have a very limited fatty acid composition, and with few acceptations there has been inadequate success in controlling plant oil fatty acid composition through bioengineering. Some of the bioengineering difficulties derive from an incomplete understanding of how plants control the flux of fatty acids through the glycolipid metabolic network for fatty acid modification and accumulation of diverse triacylglycerol molecular species. This chapter summarizes current knowledge and uncertainties for the metabolic pathways that compose the glycerolipid metabolic network and the gene products involved. Current hypotheses for the cellular organization of lipid metabolism into distinct metabolons that may contribute to the differential flux through the glycerolipid biosynthetic network by different plants are also discussed.

Advances in Botanical Research, Volume 101
ISSN 0065-2296
https://doi.org/10.1016/bs.abr.2021.07.003

1. Introduction

Triacylglycerols (TAGs) produced predominantly in plant seeds are used for food, feed, bio-fuels, and chemicals. The fatty acid building blocks of TAGs provide essential nutrients (*e.g.*, omega-6 and omega-3 fatty acids), energy dense carbon storage, and a wide range of chemical structures than can be used in production of a variety of lubricants, polymers, resins, cosmetics, and pharmaceuticals (Carlsson, Yilmaz, Green, Stymne, & Hofvander, 2011; Dyer, Stymne, Green, & Carlsson, 2008; Gunstone, Harwood, & Dijkstra, 2007; Harwood, Woodfield, Chen, & Weselake, 2017; Napier, Haslam, Tsalavouta, & Sayanova, 2019; Romsdahl et al., 2019; Ruiz-Lopez, Usher, Sayanova, Napier, & Haslam, 2015; Silva, Grilo, Gandini, & Lacerda, 2021; Vanhercke, Wood, Stymne, Singh, & Green, 2013). However, not all plant oils are alike and greater than 450 fatty acid structures are found in the plant kingdom that can produce a wide diversity of TAG molecular species in plant oils (Ohlrogge et al., 2018). In addition, many valuable fatty acids only accumulate to low amounts, or in plants with poor agronomic performance (Gunstone et al., 2007; Harwood et al., 2017). Yet, modern genetic engineering approaches hold the promise to optimize crop plant oil production for various uses (Ortiz et al., 2020; Pouvreau, Vanhercke, & Singh, 2018). This chapter focuses on summarizing the current knowledge of the plant oil glycerolipid assembly network (Fig. 1) leading to accumulation of TAG with various fatty acid compositions and highlighting areas of uncertainty, that together can be the basis for future research and developing bioengineering strategies to optimize plant oil crops.

2. Fatty acid and acyl-CoA production

Fatty acids are assembled from acetyl-CoA two carbons at a time up to chain lengths of 18 carbons while attached to acyl carrier proteins (ACP)s in the plastid stroma. The mechanisms of fatty acid synthesis, regulation, and associated gene products have been extensively reviewed (Bates, Stymne, & Ohlrogge, 2013; Chapman & Ohlrogge, 2012; Holzl & Dormann, 2019; Li-Beisson et al., 2013; Ohlrogge & Browse, 1995; Ohlrogge & Jaworski, 1997; Salie & Thelen, 2016). Total fatty acid synthesis and seed oil accumulation is dependent on the activity of acetyl-CoA carboxylase which is the first committed step and key regulatory point for fatty acid biosynthesis (Salie & Thelen, 2016; Sasaki & Nagano, 2004). Understanding the genetic,

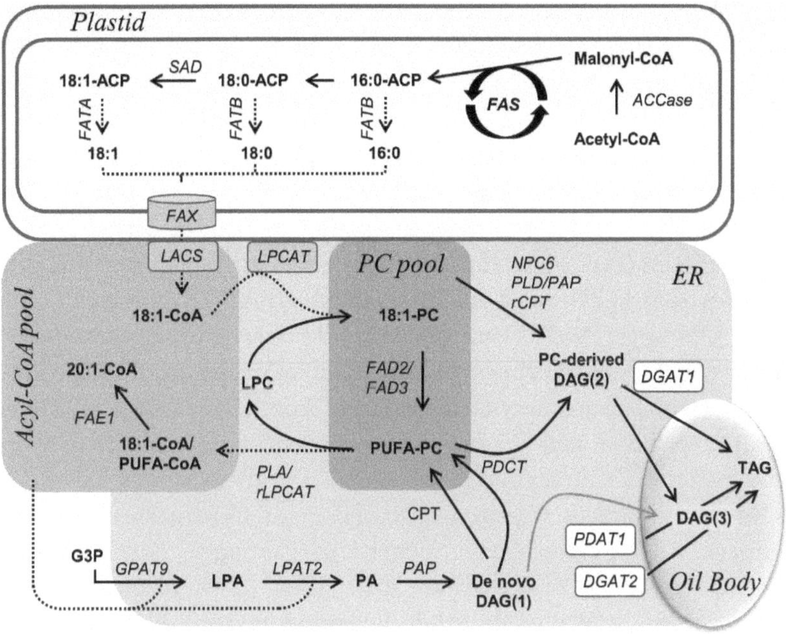

Fig. 1 Network of lipid flux toward oil biosynthesis in *Arabidopsis thaliana*. This diagram depicts the predominant PC-derived DAG pathway to TAG biosynthesis in Arabidopsis. Enzymatic reactions are italicized, specific genes have numbers (*e.g.*, GPAT9) otherwise the enzyme class is indicated. Solid arrows indicate transfer of fatty acids esterified to glycerol or other carrier; dashed arrows indicate acyl transfer reactions; gray arrows indicate uncertain amount of flux. Blue boxed enzymes highlight location to chloroplast envelope. Tag synthesizing acyltransferases are boxed in red with white background, utilize different DAG pools, and the acyl transfer is not shown, however DGATs utilize the acyl-CoA pool, and PDAT predominantly utilizes PC, producing LPC as a byproduct. Abbreviations: substrates are in bold; ACP, acyl carrier protein; DAG, diacylglycerol; G3P, glycerol-3-phosphate; LPA, lyso-phosphatidic acid; LPC, lysophosphatidylcholine; PA, phosphatidic acid; PC, phosphatidylcholine; PUFA, polyunsaturated fatty acids; TAG, triacylglycerol. Enzymes: ACCase, acetyl-CoA carboxylase; CPT, CDP-choline:DAG cholinephosphotransferase; DGAT, acyl-CoA:DAG acyltransferase; FAD, fatty acid desaturase; FAE1, fatty acid elongase 1; FAS, fatty acid synthase; FATA, acyl-ACP thioesterase A; FATB, acyl-ACP thioesterase B; FAX, fatty acid exporter; GPAT9, acyl-CoA:G3P acyltransferase 9; LACS, long chain acyl-CoA synthetase; LPAT2, acyl-CoA:LPA acyltransferase 2; LPCAT, acyl-CoA:LPC acyltransferase; NPC6, non-specific phospholipase C 6; PAP, PA phosphatase; PDCT, PC:DAG cholinephosphotransferase; PDAT, phospholipid:DAG acyltransferase; PLA, phospholipase A; PLD, phospholipase D; SAD, Stearoyl-ACP desaturase. Lowercase r before enzyme indicates reverse reaction.

molecular, and biochemical control of acetyl-CoA carboxylase regulation is an active field of current plant lipid research, and is essential to the breeding or engineering of total oil accumulation in plants (Bates et al., 2014; Caroca et al., 2021; Islam et al., 2019; Keereetaweep, Liu, Zhai, & Shanklin, 2018;

Kong, Yang, Guo, Yuan, & Ma, 2020; Parajuli et al., 2020; Salie, Zhang, Lancikova, Xu, & Thelen, 2016; Shivaiah, Ding, Upton, & Nikolau, 2020; Vanhercke et al., 2019; Ye, Fulcher, et al., 2020; Ye, Nikovics, et al., 2020; Zhou et al., 2020). Here we focus on the assembly of TAG in the endoplasmic reticulum (ER) of developing oil seed tissue, which starts with the export of free fatty acids (Koo, Ohlrogge, & Pollard, 2004; Pollard & Ohlrogge, 1999) from the plastid by way of fatty acid export (FAX) proteins (Li et al., 2015, 2020; Tian et al., 2019). The cleavage of free fatty acids from ACP by acyl-ACP thioesterases (FAT) (Kachroo et al., 2007; Salas & Ohlrogge, 2002; Zhang et al., 2014) marks the end point of *de novo* fatty acid synthesis and primes the acyl chain for export from the plastid. In most plants the major fatty acids exported from the plastid are oleic acid (18:1Δ9), palmitic acid (16:0), and stearic acid (18:0); however various plant species contain acyl-ACP thioesterases with specificity for short or medium chain length saturated fatty acids (*e.g.*, Doermann, Frentzen, & Ohlrogge, 1994; Dormann, Spener, & Ohlrogge, 1993; Kalinger, Pulsifer, Hepworth, & Rowland, 2020; Kim, Silva, Iskandarov, et al., 2015; Kim, Silva, Vu, et al., 2015; Pollard, Anderson, Fan, Hawkins, & Davies, 1991; Voelker et al., 1992), or contain acyl-ACP desaturases that produce monounsaturated fatty acids with alternative double bond locations (Cahoon, Shanklin, & Ohlrogge, 1992; Guy et al., 2011; Nguyen et al., 2010; Shanklin & Cahoon, 1998) leading to more diverse products of *de novo* fatty acid synthesis. Thus, the relative thioesterase and acyl-ACP desaturase specificities ultimately determine the composition of nascent fatty acids exported from the plastid. During the TAG accumulation phase in developing oilseeds most fatty acids exported from the plastid are newly synthesized fatty acids, but fatty acids released from plastid membrane lipids can also be incorporated into seed TAG (Aulakh & Durrett, 2019). Once exported from the plastid, free fatty acids are immediately activated to thioesters of coenzyme A (acyl-CoA) by long-chain acyl-CoA synthetases (LACS) (Koo et al., 2004; Schnurr, Shockey, de Boer, & Browse, 2002; Shockey & Browse, 2011; Shockey, Fulda, & Browse, 2002, 2003; Zhao, Kosma, & Lü, 2021). The *Arabidopsis thaliana* AtLACS9 localizes to the chloroplast envelop and may be the initial enzyme activating free fatty acids exported from the plastid, however mutational studies indicate a high degree of functional redundancy of AtLACS9 among various Arabidopsis LACS localized to the ER (Chapman & Ohlrogge, 2012; Jessen, Roth, Wiermer, & Fulda, 2015; Schnurr et al., 2002; Shockey et al., 2002; Zhao et al., 2021; Zhao, Haslam, Sonntag, Molina, & Kunst, 2019; Zhao, Katavic, Li, Haughn, & Kunst, 2010). The newly formed acyl-CoA can bind to acyl-CoA binding proteins (ACBP) which may be involved in acyl-CoA trafficking to lipid metabolic enzymes (Lung & Chye, 2016; Yurchenko et al., 2009), and the import of acyl-CoA into the ER lumen (or free fatty acid import with reactivation to acyl-CoA on the luminal side) through an ABC transporter may also be involved in TAG accumulation (Kim et al., 2013; Li-Beisson, Neunzig, Lee, & Philippar, 2017).

3. Acyl editing and fatty acid modification to provide diverse acyl-CoA substrates for TAG synthesis

In vivo isotopic tracing has indicated most acyl-CoA produced from newly synthesized fatty acids (predominantly 18:1Δ9) are first incorporated into the membrane lipid phosphatidylcholine (PC) through an acyl-CoA:PC acyl exchange cycle known as acyl editing (Allen, Bates, & Tjellström, 2015; Bates, 2016; Bates et al., 2012; Bates, Durrett, Ohlrogge, & Pollard, 2009; Bates, Ohlrogge, & Pollard, 2007; Karki, Johnson, & Bates, 2019; Tjellström, Yang, Allen, & Ohlrogge, 2012; Williams, Imperial, Khan, & Hodson, 2000; Yang et al., 2017; Zhou et al., 2020). Lysophosphatidylcholine acyltransferases (LPCATs) associated with outside of the chloroplast (Bessoule, Testet, & Cassagne, 1995; Karki et al., 2019; Tjellström et al., 2012) catalyze the forward reaction of the acyl editing cycle producing PC from lysophosphatidylcholine (LPC) and acyl-CoA, while the PC deacylation phase of the cycle may be through a phospholipase A (PLA) or the reverse action of LPCAT (rLPCAT) (Lager et al., 2013; Pan et al., 2015). The lipase based method is also referred to as the Lands Cycle (Lands, 1965). Stereochemical analysis of nascent fatty acid incorporation into PC from *in vivo* isotopic labeling demonstrates a preference for the *sn*-2 position, however the relative proportion *sn*-1 *vs.* *sn*-2 acyl editing appears to depend on species and tissue (Bates et al., 2007, 2009, 2012; Karki et al., 2019; Wang et al., 2012; Yang et al., 2017; Zhou et al., 2020). The *sn*-2 preference of acyl editing is also demonstrated by *in vitro* LPCAT assays where the *sn*-1 acyl transfer activity is ~15–70% that of the *sn*-2 acyl transfer activity (Lager et al., 2013). Arabidopsis contains two LPCAT enzymes AtLPCAT1 and AtLPCAT2 (Stahl, Stalberg, Stymne, & Ronne, 2008). The Arabidopsis *lpcat1*/*lpcat2* double mutant alters the route of newly synthesized fatty acid flux such that nascent fatty acids are not first incorporated into PC, but instead enter *de novo* glycerolipid assembly (Bates et al., 2012; Karki et al., 2019; Wang et al., 2012). PC is the main extra-plastidic site for fatty acid desaturation of oleic acid by the FAD2 (Okuley et al., 1994) and subsequently FAD3 (Arondel et al., 1992) enzymes producing the polyunsaturated fatty acids, linoleic acid (18:2Δ9,12) and linolenic acid (18:3Δ9,12,15), respectively. Many plants also accumulate various unusual fatty acids in seed TAG (Ohlrogge et al., 2018) that are similarly produced on PC by FAD2 homologs, for example, ricinoleic acid is produced by the castor delta-12 hydroxylase (RcFAH12) (Bafor, Smith, Jonsson, Stobart, & Stymne, 1991; Moreau & Stumpf, 1981; Vandeloo, Broun, Turner, & Somerville, 1995). In addition, fatty acids released from PC (mostly oleic acid) are the main substrate for fatty acid elongation to ≥20 carbons (Bao, Pollard, & Ohlrogge, 1998; Bates et al., 2012; Moon, Smith, & Kunst, 2001). Therefore, the LPCAT based acyl-CoA:PC acyl editing cycle produces a diverse acyl-CoA pool that can be used for glycerolipid assembly, and the coupling of acyl editing and TAG biosynthesis enhances the polyunsaturated fatty acid (or other PC-modified fatty acid) content of plant oils (Bates et al., 2012; Pan et al., 2015; Wang et al., 2012). Interestingly, sunflower contains a third LPCAT expressed specifically in seeds suggesting the high linolenic acid content of sunflower oils may be in part due to an enhanced acyl editing cycle for fatty acid desaturation on PC prior to incorporation into TAG (Mapelli-Brahm et al., 2020).

A major unknown within the PC acyl editing cycle is the relative proportion of PC-deacylation initiated by a PLA (as in the Lands Cycle), or by the reverse LPCAT reaction. Different acyl selectivity's of LPCATs in the forward and reverse directions suggest that LPCAT alone could be involved in modulation of the acyl-CoA pool fatty acid composition (Jasieniecka-Gazarkiewicz, Demski, Lager, Stymne, & Banas, 2016; Lager et al., 2013; Mapelli-Brahm et al., 2020). In the Arabidopsis *lpcat1/lpcat2* mutant, the polyunsaturated fatty content of TAG is reduced (Bates et al., 2012; Wang et al., 2012), however it is not clear if this is due to less 18:1 incorporation into PC for desaturation, and/or due to reduced removal of 18:2 and 18:3 from PC through acyl editing. The expression of multiple phospholipases is upregulated in *lpcat1/lpcat2*, and isotopic tracing indicates *de novo* PC synthesis and total PC turnover is also increased (Wang et al., 2012, 2014). However, it is not clear if the increased phospholipase activity or enhanced PC synthesis and turnover is an attempt by the plant to boost the endogenous pathway in the mutant background, or if it is a compensating mechanism to release polyunsaturated fatty acids from PC when the reverse reaction of LPCAT is not available. Arabidopsis contains more than 300 genes annotated as lipases, most of which have yet to be functionally characterized (McGlew et al., 2015; Wang, Durrett, & Benning, 2019). The Arabidopsis pPLAIIIδ can cleave fatty acids from both the *sn*-1 and *sn*-2 positions of PC, as well as acyl-CoA esters (Li et al., 2013). The AtpPLAIIIδ knockout mutant reduced seed oil content by ~5% while increasing the proportion of 18:1 in the oil. Additionally, seed specific over expression increased oil content by ~11% and increased the proportion of 20:1 in TAG (Li et al., 2013). These results suggest AtpPLPAIIIδ may be part of a Lands Cycle that removes 18:1 from PC for further elongation prior to incorporation into TAG (Bao et al., 1998; Bates et al., 2012). However, the small change in oil content of the AtpPLAIIIδ mutant suggests that other lipases and/or the reverse activity of LPCAT may also contribute to PC acyl editing. Many plant PLA type enzymes have been characterized to release fatty acids from PC or other membrane lipids as a response to biotic or abiotic stress, during which the PLA activity is involved in remodeling the lipid membrane and/or generating PUFA substrates for oxylipin signaling molecules (Wang, 2001; Wang et al., 2019; Yu, Zhou, Fan, Shanklin, & Xu, 2021). It is unclear if any of these stress induced lipases are also involved in the PC acyl editing cycle under non-stress conditions that provides modified-FA for seed TAG biosynthesis.

4. *De novo* glycerol lipid assembly for oil production

De novo glycerolipid assembly involves the esterification of fatty acids from acyl-CoA substrates to glycerol-3-phosphate producing phosphatidic acid (PA) and subsequent diacylglycerol (DAG) backbones utilized to produce ER membrane lipids and TAG (Fig. 1). Glycerolipid assembly in the ER utilizes a mixed acyl-CoA pool containing both nascent FA exported from the plastid, and those removed from PC by acyl-editing that may have been further modified while attached to PC or further elongated as acyl-CoAs (Bao et al., 1998; Bates et al., 2007, 2009, 2012; Karki et al., 2019; Williams et al., 2000). The first step of ER localized glycerolipid assembly is the acyl-CoA:glycerol-3-phosphate acyltransferase (GPAT) producing lysophosphatidic acid (LPA) encoded by *AtGPAT9* in Arabidopsis (Shockey et al., 2016; Singer et al., 2016) and homologs in other species (Lunn, Wallis, & Browse, 2019; Paya-Milans et al., 2016; Shockey et al., 2019). Second, acyl-CoA: lysophosphatidic acid acyltransferase (LPAT) adds a fatty acid to the *sn*-2 position of LPA producing phosphatidic acid (PA). There are five LPATs in Arabidopsis, and *AtLPAT2* appears to be the main LPAT involved in ER membrane lipid and TAG biosynthesis (Kim, Li, & Huang, 2005), similarly *RcLPAT2* and *VfLPAT2* are involved in TAG biosynthesis in castor bean and tung tree, respectively (Shockey et al., 2019).

Although the LPAT family is diverse (Arroyo-Caro et al., 2013) and the involvement of other LPAT enzymes for oil biosynthesis in different species is not clear. Most plant ER localized LPATs involved in membrane lipid production are selective for 18 carbon unsaturated fatty acids which leads to the characteristic 18C *sn*-2 composition of Eukaryotic pathway produced lipids (Frentzen, 1986, 1990; Ohlrogge & Browse, 1995), as opposed to the Prokaryotic pathway produced membrane lipids where plastid localized LPATs utilize acyl-ACPs as substrates and preferentially transfer 16 carbon fatty acids producing PA with *sn*-2 16:0 (Frentzen, Heinz, McKeon, & Stumpf, 1983). However, various oilseeds produce TAG in the ER containing high levels of unusual fatty acids (*e.g.*, *medium* chain length, hydroxylated, cyclopropane, or conjugated fatty acids) due to the selectivity of specialized plant LPAT2s for both LPA and acyl-CoA substrates containing the unusual fatty acids (Arroyo-Caro et al., 2013; Kim, Silva, Iskandarov, et al., 2015; Kim, Silva, Vu, et al., 2015; Shockey et al., 2019; Yu, Prakash, Sweet, & Shanklin, 2014). The subsequent dephosphorylation of PA by a phosphatidic acid phosphatase (PAP) produces *de novo* DAG, the penultimate metabolite for the *de novo* synthesis of TAG or ER membrane lipids PC or phosphatidylethanolamine (PE) by the Kennedy pathway (Kennedy, 1961; Weiss, Kennedy, & Kiyasu, 1960). *De novo* TAG biosynthesis is completed by the transfer of a third fatty acid to *de novo* DAG by an acyl-CoA:diacyglycerol acyltransferase (DGAT) or an acyl-CoA independent mechanism by phospholipid:diacylglycerol acyltransferase (PDAT). Plant PDAT can utilize multiple phospholipids as the acyl donor *in vitro* thus producing both TAG and a lysophospholipid (Dahlqvist et al., 2000; Stahl et al., 2004). Although *in vivo* PDAT activity is associated with enhanced accumulation of PC-modified fatty acids in TAG, and evidence for the coupling of PDAT activity with LPC re-acylation to PC by LPCATs together indicates that PC is the predominant *in vivo* substrate of PDAT during seed oil accumulation (Lager et al., 2020; Lunn, Le, Wallis, & Browse, 2020; Marmon et al., 2017; Pan, Siloto, Wickramarathna, Mietkiewska, & Weselake, 2013; van Erp, Bates, Burgal, Shockey, & Browse, 2011; Xu et al., 2012; Xu, Caldo, et al., 2018; Xu, Holic, et al., 2018).

Plants such as *Arabidopsis thaliana* contain three unrelated DGAT enzymes AtDGAT1, AtDGAT2, AtDGAT3 and two PDAT enzymes AtPDAT1 and AtPDAT2 (Li-Beisson et al., 2013), yet the endogenous role for each isozyme in plant TAG metabolism is still an area of active research. Plant DGAT1 and DGAT2 are structurally distinct integral membrane proteins that have demonstrated to produce TAG *in vitro* and *in planta* through mutation and/or overexpression analyses, DGAT1 and DGAT2 have non-overlapping localization within the ER membrane, different tissue expression patterns, different post-translational regulation, and utilize different substrate pools with different molecular species selectivity (Burgal et al., 2008; Caldo et al., 2017, 2018; Chen et al., 2016; Demski et al., 2019; Horn et al., 2016; Jeppson et al., 2019; Kroon, Wei, Simon, & Slabas, 2006; Lager et al., 2020; Li, Yu, & Hildebrand, 2010; Nguyen et al., 2013; Regmi, Shockey, Kotapati, & Bates, 2020; Shockey et al., 2006, 2019; Troncoso-Ponce et al., 2011; Winter et al., 2007; Xu, Caldo, et al., 2018; Xu, Holic, et al., 2018; Xu, Yang, Wang, & Liu, 2014; Zhou, Shrestha, Yin, Petrie, & Singh, 2013). The relative use of DGAT1 or DGAT2 to produce seed TAG also differs between plants. For example, AtDGAT1 is the major TAG synthesizing enzyme in wild-type Arabidopsis, and the *dgat1-1* mutant has a 20% reduction in oil content. The *AtDGAT2* mutant has no effect on seed oil accumulation in the wild-type or *dgat1-1* background (Zhang, Fan, Taylor, & Ohlrogge, 2009). Yet, over expression of *AtDGAT2* in leaves or seeds demonstrates AtDGAT2 is an active TAG producing enzyme (Regmi et al., 2020; Zhou et al., 2013). In contrast to Arabidopsis, RcDGAT2 is the major enzyme responsible for accumulating hydroxy-fatty acid containing TAG in castor endosperm (Bafor et al., 1991; Burgal et al., 2008; Kroon et al., 2006; Shockey et al., 2019). DGAT3 is a soluble enzyme first identified in peanut (Saha, Enugutti, Rajakumari, & Rajasekharan, 2006). AtDGAT3 is a [2Fe—2S] protein, possibly localized to the cytosol or chloroplast and has a suggested role in germinating seedling TAG metabolism (Aymé et al., 2018; Hernández et al., 2012). Overexpression of camelina CsDGAT3 can alter seed TAG profile in transgenic tobacco, however a substantial role in developing seed TAG production for endogenous plant DGAT3s has yet to be demonstrated (Gao et al., 2021).

The Arabidopsis *pdat1* mutant alone does not affect seed oil accumulation (Mhaske, Beldjilali, Ohlrogge, & Pollard, 2005). However, AtPDAT1 is responsible for the remaining 80% of TAG production in *dgat1-1* and the *dgat1/pdat1* double mutant is lethal, in contrast the *dgat1/pdat2* double mutant did not affect oil biosynthesis beyond that of *dgat1-1* (Zhang et al., 2009). Thus, Arabidopsis mutational analysis indicates that AtDGAT1 is predominantly responsible for oil biosynthesis, and AtPDAT1 compensates when AtDGAT1 is lacking, but it is unclear what is the role for AtDGAT2, AtDGAT3, and AtPDAT2 in developing Arabidopsis seed lipid metabolism. The relative amount of DGAT *vs.* PDAT activity responsible for seed TAG biosynthesis also varies between plant species, with species such as *Camelina sativa* (Aznar-Moreno & Durrett, 2017; Lager et al., 2020; Marmon et al., 2017) and flax (*Linum usitatissimum* L.) (Pan et al., 2013) demonstrating higher levels of PDAT activity under wild-type conditions than in Arabidopsis (Mhaske et al., 2005; Zhang et al., 2009). Therefore, seed oil composition in different plants is controlled in part by the relative tissue expression, enzymatic activities, and substrate selectivity's of DGAT1, DGAT2 and PDAT1 with the roles of DGAT3 and PDAT2 in seed oil production still unclear.

5. Overlap of TAG biosynthesis with ER membrane biosynthesis

The flux of fatty acids through both PC acyl editing and *de novo* TAG biosynthesis could in theory provide the acyl and DAG substrates required to produce the diverse TAG molecular species compositions found throughout the plant kingdom. However, the DAG substrate for TAG biosynthesis can also be provided by ER and plastid membrane lipid turnover during seed oil accumulation, or during vegetative tissue stress (Bates & Browse, 2012; Bates et al., 2013; Du & Benning, 2016; Li-Beisson et al., 2013; Roughan & Slack, 1982). During seed TAG accumulation PC is the predominant membrane lipid to provide a DAG scaffold for TAG assembly, and the relative amount of *de novo* DAG *vs.* PC-derived DAG utilized to produce TAG varies by plant species (Bates & Browse, 2012). Recent quantitative isotopic tracing of glycerol flux through *de novo* glycerolipid assembly into membrane lipids and/or TAG in soybean, Arabidopsis, *C. sativa*, and *Physaria fendleri* have estimated that $\geq$80% of seed TAG is synthesized from PC-derived DAG rather than *de novo* synthesized DAG (Bates & Browse, 2011; Bates et al., 2009; Bhandari & Bates, 2021; Yang et al., 2017). Therefore, in many species *de novo* glycerolipid assembly in developing oilseeds is predominantly utilized to first produce PC, and subsequently DAG is derived from PC for TAG synthesis. Because PC is the ER localized site for fatty acid modification and acyl editing, the molecular species of PC-derived DAG used to produce TAG can be substantially different than the molecular species of *de novo* DAG produced by the Kennedy pathway (Bates et al., 2009).

The membrane lipids PC and PE are produced through *de novo* glycerolipid assembly by the transfer of phosphocholine or phosphoethanolamine from CDP-choline or CDP-ethanolamine to the *sn*-3 hydroxyl of *de novo*

DAG by aminoalcoholphosphotransferases (AAPT) (also known as diacylglycerol:cholinephosphotransferase (CPT) and diacylglycerol:ethanolaminephosphotransferase (EPT)) (Li-Beisson et al., 2013). The biosynthesis of the CDP-choline and CDP-ethanolamine substrates has recently been reviewed (Nakamura, 2021). Arabidopsis contains two AAPT, and the *aapt1/aapt2* double mutant is embryo lethal (Liu, Wang, & Wang, 2015) indicating the essential nature of PC and PE to higher plant growth and reproduction. At least four different mechanisms may be involved in producing PC-derived DAG for TAG biosynthesis (Fig. 1). First, *in vitro* assays and *in vivo* labeling in developing safflower and linseed seeds indicate that CPT is reversible (Slack, Campbell, Browse, & Roughan, 1983; Slack, Roughan, Browse, & Gardiner, 1985), suggesting that the AAPT (CPT in Fig. 1) enzymes could be involved in both PC synthesis from *de novo* DAG and production of PC-derived DAG for TAG synthesis.

The second mechanism of PC-derived DAG production is through phosphatidylcholine:diacylglycerol cholinephosphotransferase (PDCT) which exchanges the phosphocholine headgroup between PC and DAG, producing new molecular species of PC and DAG (Lu, Xin, Ren, Miquel, & Browse, 2009). The PDCT reaction does not lead to net PC or DAG production, but allows the incorporation of fatty acids modified while attached to PC to enter the DAG pool by the equilibration of PC and DAG molecular species. In Arabidopsis, PDCT is encoded by the Reduced oleate desaturase 1 (*ROD1*) gene, and the polyunsaturated fatty acid content of seed TAG in the *rod1* mutant is reduced ~40%, suggesting that PC:DAG equilibration by ROD1 is a substantial mechanism of PC-derived DAG production in Arabidopsis seeds (Lu et al., 2009). Reductions in PC-modified fatty acid content of seed TAG was also found in homologous mutants of *rod1* in other species (Bai, Wallis, et al., 2020; Jarvis et al., 2021), but the proportional effect on TAG fatty acid composition appears to be species dependent suggesting possible differences in the relative flux of PC-derived DAG production through PDCT in different plants. Interestingly, when *rod1* is combined with a lack of PC acyl editing in the Arabidopsis *rod1/lpcat1/lpcat2* triple mutant the polyunsaturated fatty acid content of seed TAG is reduced ~66% and most of the remaining PUFA is at the *sn*-2 position of TAG (Bates et al., 2012). Considering that desaturation occurs predominantly at *sn*-2 PC, PC-derived DAG production likely continues to occur in the *rod1* background. PDCT selectivity for DAG or PC molecular species may also affect its ability to equilibrate the DAG and PC pools (Hu, Ren, & Lu, 2012; Lager et al., 2020; Wickramarathna et al., 2015). *C. sativa* PDCT did not effectively produce PC from DAG containing hydroxy fatty acids (HFA) *in vitro* (Lager et al., 2020). A similar result in Arabidopsis engineered to produce HFA found limited *in vivo* flux of HFA-containing *de novo* DAG into PC, indicating the pathway of TAG synthesis through PC-derived DAG is a bottleneck for accumulation of HFA-containing *de novo* DAG produced by the Kennedy pathway (Bates & Browse, 2011). In addition, when *RcFAH12* (*Ricinus communis* fatty acid hydroxylase 12) was expressed in the Arabidopsis *rod1* mutant, the accumulation of HFA in seed TAG was approximately half of that when *RcFAH12* was expressed in wild-type (Hu et al., 2012). Thus, it appears that Arabidopsis and Camelina PDCT may be selective against utilizing HFA-DAG for PC synthesis, but PDCT can utilize HFA-PC for production of HFA-containing PC-derived DAG destined for TAG synthesis. The expression of castor and flax PDCT in Arabidopsis also enhanced the accumulation of HFA and polyunsaturated fatty acids in TAG, respectively (Hu et al., 2012; Wickramarathna et al., 2015). Therefore, increasing PC:DAG equilibration can be a valuable strategy to engineer increased accumulation of PC-modified fatty acids in TAG within plant species that utilize PC-derived DAG for TAG biosynthesis.

The third and fourth possible mechanisms to produce PC-derived DAG are by direct cleavage of PC into DAG and phosphocholine by phospholipase C (PLC), and the combined action of phospholipase D (PLD) to produce PA and subsequent PAP activity to produce DAG. Unlike PDCT, both lipase-based mechanisms lead to the net turnover of PC and thus must be combined with *de novo* PC biosynthesis through AAPT to maintain membrane PC levels. Plants contain many PLC and PLD enzymes that are involved in the production of DAG and PA for signaling of developmental, environmental, stress, and symbiotic interaction responses (Cook, Lupette, & Benning, 2021; Liu et al., 2019; Meï et al., 2017; Nakamura et al., 2009; Wang, 2001; Wang et al., 2019; Zhang et al., 2021). However, multiple knockout, knockdown, or overexpression studies have demonstrated the involvement of PLC and PLD enzymes in seed TAG production. Knockouts of Arabidopsis Nonspecific phospholipase C6 (*AtNPC6*) displayed an ~7.5% decrease in seed oil content, and overexpression of *AtNPC5* increased oil ~6–8% in seeds of Arabidopsis and *C. sativa*. The Arabidopsis *npc6* knockouts had reduced seed 18:3 and 20:1, and increased 16:0, 18:0, and 18:1. While the *AtNPC6* over expression lines had the opposite phenotype in Arabidopsis (Cai et al., 2020). A *Brassica napus* gene association analysis also indicated *BnNPC6·C01* was associated with increased oil content (Cai et al., 2020). Interestingly, AtNPC6 appears to hydrolyze PC, monogalactosyldiacylglycerol (MGDG), and digalactosyldiacylglycerol (DGDG) to DAG (Cai et al., 2020) suggesting multiple membrane lipids may provide DAG for seed oil accumulation. Although the flux of fatty acids and glycerol from membranes into TAG is much higher from PC than from other membrane lipids in developing seeds of various Brassicaceae species (Bates & Browse, 2011; Bates et al., 2012; Bhandari & Bates, 2021; Cai et al., 2020; Yang et al., 2017). Single mutants of other NPCs in Arabidopsis did not affect seed oil content (Cai et al., 2020), but AtNPC6 and AtNPC2 function redundantly in pollen tube TAG production (Bose, Ngo, Nguyen, & Nakamura, 2021) indicating possible tissue specific effects of NPCs in cellular TAG metabolism. A castor bean phospholipase C-like protein (RcPLCL1) that can hydrolyze PC and phosphatidylinositol (PI) was expressed in transgenic HFA producing *C. sativa* and increased the HFA content of TAG while decreasing the HFA content of PC, suggesting PLC enzymes can selectively produce different DAG molecular species (Aryal & Lu, 2018). Together these results suggest that PLCs can produce PC-derived DAG molecular species that contribute to specific seed oil fatty acid compositions in various plants, and that other membrane lipids may also contribute some DAG to seed TAG biosynthesis.

Multiple pieces of evidence suggest PLDs are also involved in production of PC-derived DAG for seed TAG biosynthesis. RNAi suppression of PLDα in soybean reduced the level of polyunsaturated fatty acids in TAG, while increasing their content in PC (Lee, Welti, Schapaugh, & Trick, 2011). Overexpression of a soybean GmPLDγ in Arabidopsis that localizes to the mitochondria increased seed oil content containing 18:3 and very long chain fatty acids (20 and 22 carbons). In addition, the GmPLDγ OE lines had increased expression of *de novo* lipid biosynthetic enzymes in developing siliques (Bai, Jing, et al., 2020). The exact connection between a mitochondrial PLD and increased oil biosynthesis in the ER is not clear, it may be due to enhanced production of PC-derived DAG after PAP activity of the produced PA, or a possible signaling role from the PLD produced PA. The over expression of two Arabidopsis PLD (*AtPLDδ1* and *AtPLDδ2*) in *C. sativa* increased seed oil content by 2–3%, with enhances in 18:3 and 20:1 content. *In vivo* isotopic tracing of the AtPLDδ Camelina lines indicated enhanced conversion of PC to TAG supporting a role of AtPLDδ in production of PC-derived DAG for TAG biosynthesis (Yang et al., 2017). It is clear that multiple possible enzymatic mechanisms can produce PC-derived DAG for TAG biosynthesis (*e.g.*, PDCT, PLC, PLD/PAP) either as an endogenous mechanism or as a consequence of bioengineering. To better understand how to control seed oil accumulation and TAG fatty acid composition future research will need to understand the relative contribution of different enzymes to production of various DAG molecular species, and how each is regulated in relation to maintenance of membrane homeostasis and TAG production through the Kennedy pathway.

6. Complexity of the TAG biosynthetic network may stem from cellular organization of lipid biosynthesis

In vivo isotopic tracer studies have indicated developing oilseeds have a least three metabolically distinct pools of DAG involved in glycolipid biosynthesis (Figs. 1 and 2). *De novo* DAG(1) and an initially produced PC-derived DAG(2) represent two small and rapidly turned over pools. A third large DAG(3) pool (likely associated with the oil body) is more slowly turned over and is mostly composed of PC-derived DAG in Arabidopsis and soybean (Bates et al., 2009; Regmi et al., 2020). Multiple distinct metabolically active pools of PC and DAG have also been identified in leaf lipid metabolism (Karki et al., 2019). What controls TAG biosynthesis from different DAG pools is not clear, but a growing body of evidence suggests different parts of glycolipid assembly may be spatially separated in the ER, and possibly organized into distinct metabolons for substrate channeling or controlling metabolic flux between pathway branchpoints (Fig. 2) (Regmi et al., 2020; Shockey et al., 2016; Zhang & Fernie, 2021). While most enzymes involved in PC and TAG assembly are membrane proteins localized to the ER, not all enzymes are evenly distributed throughout the ER or interact with each other. Tung tree VfDGAT1 and VfDGAT2 do not

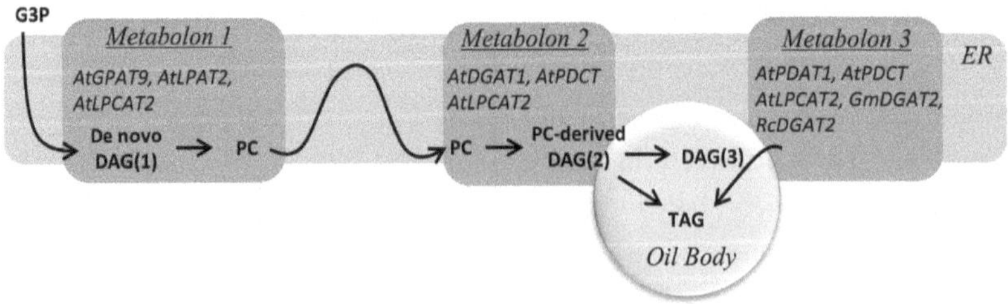

Fig. 2 Hypothesized model for organization of Arabidopsis lipid metabolism into discrete metabolons with PC as a DAG carrier through the membrane. The enzymes/genes listed in each metabolon are based on yeast two-hybrid results and *in vivo* flux results discussed in the text. Genes are italicized, metabolites are in bold. Abbreviations are as in Fig. 1.

colocalize in the ER, and VfGPAT9 colocalized with VfDGAT2 but not VfDGAT1 (Gidda et al., 2011; Shockey et al., 2006). Yeast two-hybrid assays with Arabidopsis acyltransferases indicated AtGPAT9 interacted with AtLPAT2 and AtLPCAT2, suggesting acyl editing can directly provide acyl groups to the first two steps of *de novo* glycerolipid assembly, but AtGPAT9 did not interact with AtDGAT1 (Shockey et al., 2016). AtDGAT1 does interact with AtPDCT and AtLPCAT2, but not with AtDGAT2 (Regmi et al., 2020). Both castor and soybean DGAT2s (RcDGAT2 and GmDGAT2, respectively) also interact with AtPDCT and AtLPCAT2 but not AtDGAT1 (Regmi et al., 2020). Similar protein-protein interactions were also found with flax acyltransferases (Xu et al., 2019). Isotopic tracing of the Arabidopsis *dgat1-1* mutant complemented with AtDGAT1, RcDGAT2, or GmDGAT2 further clarifies the protein-protein interaction results by demonstrating that AtDGAT1 utilizes the small initially produced PC-derived DAG(2) pool, whereas RcDGAT2, GmDGAT2 and AtPDAT1 utilize the larger and more slowly turned over PC-derived DAG(3) pool (Regmi et al., 2020). These results suggest that DGAT1s and DGAT2s may form separate TAG producing metabolons that can both utilize substrates from acyl editing and PDCT but DGAT1 and DGAT2 do not exist in the same metabolon, and the DGAT1 and DGAT2 containing metabolons utilize distinct pools of PC-derived DAG (Fig. 2). The results that AtDGAT1, AtPDAT1 and various DGAT2s do not utilize *de novo* DAG(1) in Arabidopsis (Regmi et al., 2020), and that AtGPAT9 and AtDGAT1 do not interact (Shockey et al., 2016) together suggest that the Kennedy pathway for *de novo* DAG(1) and PC biosynthesis represents a separate metabolon than the TAG producing metabolons in Arabidopsis. Spatially separated metabolons of *de novo* glycerolipid biosynthesis and TAG biosynthesis may explain the dominant flux of TAG biosynthesis from PC-derived DAG in many species by PC acting as a "DAG carrier" through the ER membrane between different lipid biosynthetic metabolons (Fig. 2) (Regmi et al., 2020; Shockey et al., 2016).

7. Current unknowns and future directions

The fatty acid compositions of plant oils are controlled by the availability of DAG and acyl substrates, acyltransferase enzymatic selectivity's, and the relative flux of glycerolipid assembly between different pathway branches, and/or substrate channeling through different lipid biosynthetic metabolons. Still not clear is what controls the relative flux of TAG biosynthesis from multiple distinct DAG pools in different species (Fig. 1), or what controls the enzymatic organization into possible distinct metabolons (Fig. 2). Understanding the metabolic control of glycerolipid assembly will be key to engineering biosynthesis of select TAG molecular species. However, recent research in *P. fendleri* suggests the fatty acid composition of initially produced TAG may be changed through TAG remodeling (Bhandari & Bates, 2021). It is not clear to what enzymatic mechanisms control TAG remodeling or to what extent TAG remodeling takes place in other oilseeds. Nevertheless, substantial TAG remodeling in plant oil crops would provide alternative mechanisms to produce diverse TAG compositions beyond that of the acyltransferases involved in initial TAG biosynthesis. Future research on understanding the organization of TAG biosynthesis needs to combine molecular biology, protein-protein interactions, and cell biology approaches with metabolic flux to understand both the cellular context and dynamics of lipid biosynthesis (Allen et al., 2015; Kotapati & Bates, 2021). In addition, complementary investigations between species that likely utilize different branches of the lipid metabolic network may provide valuable insights on the differential organization of lipid metabolism within the plant kingdom. Finally, a better understanding of the cellular organization of the lipid biosynthetic network in different oilseeds plants, and plant tissues will be essential for engineering plants to be lipid producing biorefineries for enhanced total oil and valuable fatty acid accumulation (Correa, Fernie, Nikoloski, & Brotman, 2020; Dyer et al., 2008; Vanhercke et al., 2019), or for the growth of bio-symbionts (Gargouri, Bates, & Declerck, 2021). However, plant metabolism is highly plastidic and can adapt in unexpected ways to engineered changes, examples include altered lipid metabolic flux, induced biochemical feedback regulation, and altered oil accumulation based on environmental conditions (Bates et al., 2014; Islam et al., 2019; Karki & Bates, 2018; Ye, Fulcher, et al., 2020; Ye, Nikovics, et al., 2020; Zhou et al., 2020). Therefore, understanding how engineered plants adapt metabolic flux and even the cellular organization of lipid metabolism to induced changes (regardless if by mutation, or over expression) will be an essential part of the design/build/test/learn cycle (Pouvreau et al., 2018) to develop the designer plant oil crops of the future.

Acknowledgments

This work is supported by the National Science Foundation grants PGRP IOS-1829365 and MCB-1716688; the United States Department of Agriculture National Institute of Food and Agriculture grants Nos. 2020-67013-30899 and 2021-67013-33778; the Hatch umbrella project #1015621; and the multi-state project #NC1203.

References

Allen, D. K., Bates, P. D., & Tjellström, H. (2015). Tracking the metabolic pulse of plant lipid production with isotopic labeling and flux analyses: Past, present and future. *Progress in Lipid Research*, *58*, 97–120. https://doi.org/10.1016/j.plipres.2015.02.002.

Arondel, V., Lemieux, B., Hwang, I., Gibson, S., Goodman, H. M., & Somerville, C. R. (1992). Map-based cloning of a gene controlling omega-3-fatty-acid desaturation in Arabidopsis. *Science*, *258*(5086), 1353–1355. https://doi.org/10.1126/science.1455229.

Arroyo-Caro, J. M., Chileh, T., Kazachkov, M., Zou, J., Alonso, D. L., & Garcia-Maroto, F. (2013). The multigene family of lysophosphatidate acyltransferase (LPAT)-related enzymes in *Ricinus communis*: Cloning and molecular characterization of two LPAT genes that are expressed in castor seeds. *Plant Science*, *199–200*, 29–40. S0168-9452(12)00223-3 [pii] https://doi.org/10.1016/j.plantsci.2012.09.015.

Aryal, N., & Lu, C. (2018). A phospholipase C-like protein from *Ricinus communis* increases hydroxy fatty acids accumulation in transgenic seeds of *Camelina sativa*. *Frontiers in Plant Science*, *9*, 29–40. https://doi.org/10.3389/fpls.2018.01576.

Aulakh, K., & Durrett, T. P. (2019). The plastid lipase PLIP1 is critical for seed viability in diacylglycerol acyltransferase 1 mutant seed. *Plant Physiology*, *180*(4), 1962–1974. https://doi.org/10.1104/pp.19.00600.

Aymé, L., Arragain, S., Canonge, M., Baud, S., Touati, N., Bimai, O., et al. (2018). *Arabidopsis thaliana* DGAT3 is a [2Fe-2S] protein involved in TAG biosynthesis. *Scientific Reports*, *8*(1), 17254. https://doi.org/10.1038/s41598-018-35545-7.

Aznar-Moreno, J. A., & Durrett, T. P. (2017). Simultaneous targeting of multiple gene homeologs to alter seed oil production in *Camelina sativa*. *Plant & Cell Physiology*, *58*(7), 1260–1267. https://doi.org/10.1093/pcp/pcx058.

Bafor, M., Smith, M. A., Jonsson, L., Stobart, K., & Stymne, S. (1991). Ricinoleic acid biosynthesis and triacylglycerol assembly in microsomal preparations from developing castor-bean (Ricinus communis) endosperm. *Biochemical Journal*, *280*, 507–514. Retrieved from <Go to ISI>://A1991GU90100035.

Bai, Y., Jing, G., Zhou, J., Li, S., Bi, R., Zhao, J., et al. (2020). Overexpression of soybean GmPLDgamma enhances seed oil content and modulates fatty acid composition in transgenic Arabidopsis. *Plant Science*, *290*, 110298. https://doi.org/10.1016/j.plantsci.2019.110298.

Bai, S., Wallis, J. G., Denolf, P., Engelen, S., Bengtsson, J. D., Van Thournout, M., et al. (2020). The biochemistry of headgroup exchange during triacylglycerol synthesis in canola. *The Plant Journal*, *103*(1), 83–94. https://doi.org/10.1111/tpj.14709.

Bao, X. M., Pollard, M., & Ohlrogge, J. (1998). The biosynthesis of erucic acid in developing embryos of *Brassica rapa*. *Plant Physiology*, *118*(1), 183–190. Retrieved from <Go to ISI>://000075955100018.

Bates, P. D. (2016). Understanding the control of acyl flux through the lipid metabolic network of plant oil biosynthesis. *Biochimica et Biophysica Acta (BBA)—Molecular and Cell Biology of Lipids*, *1861*(9 Part B), 1214–1225. https://doi.org/10.1016/j.bbalip.2016.03.021.

Bates, P. D., & Browse, J. (2011). The pathway of triacylglycerol synthesis through phosphatidylcholine in Arabidopsis produces a bottleneck for the accumulation of unusual fatty acids in transgenic seeds. *Plant Journal*, *68*(3), 387–399. https://doi.org/10.1111/j.1365-313X.2011.04693.x.

Bates, P. D., & Browse, J. (2012). The significance of different diacylgycerol synthesis pathways on plant oil composition and bioengineering. *Frontiers in Plant Science*, *3*, 147. https://doi.org/10.3389/fpls.2012.00147.

Bates, P. D., Durrett, T. P., Ohlrogge, J. B., & Pollard, M. (2009). Analysis of acyl fluxes through multiple pathways of triacylglycerol synthesis in developing soybean embryos. *Plant Physiology*, *150*(1), 55–72. https://doi.org/10.1104/pp.109.137737.

Bates, P. D., Fatihi, A., Snapp, A. R., Carlsson, A. S., Browse, J., & Lu, C. (2012). Acyl editing and headgroup exchange are the major mechanisms that direct polyunsaturated fatty acid flux into triacylglycerols. *Plant Physiology*, *160*(3), 1530–1539. https://doi.org/10.1104/pp.112.204438.

Bates, P. D., Johnson, S. R., Cao, X., Li, J., Nam, J.-W., Jaworski, J. G., et al. (2014). Fatty acid synthesis is inhibited by inefficient utilization of unusual fatty acids for glycerolipid assembly. *Proceedings of the National Academy of Sciences of the United States of America*, *111*(3), 1204–1209. https://doi.org/10.1073/pnas.1318511111.

Bates, P. D., Ohlrogge, J. B., & Pollard, M. (2007). Incorporation of newly synthesized fatty acids into cytosolic glycerolipids in pea leaves occurs via acyl editing. *Journal of Biological Chemistry*, *282*(43), 31206–31216. https://doi.org/10.1074/jbc.M705447200.

Bates, P. D., Stymne, S., & Ohlrogge, J. (2013). Biochemical pathways in seed oil synthesis. *Current Opinion in Plant Biology*, *16*(3), 358–364. https://doi.org/10.1016/j.pbi.2013.02.015.

Bessoule, J. J., Testet, E., & Cassagne, C. (1995). Synthesis of phosphatidylcholine in the chloroplast envelope after import of lysophosphatidylcholine from endoplasmic-reticulum membranes. *European Journal of Biochemistry*, *228*(2), 490–497. Retrieved from https://febs.onlinelibrary.wiley.com/doi/pdf/10.1111/j.1432-1033.1995.0490n.x.

Bhandari, S., & Bates, P. D. (2021). Triacylglycerol remodeling in *Physaria fendleri* indicates oil accumulation is dynamic and not a metabolic endpoint. *Plant Physiology*. https://doi.org/10.1093/plphys/kiab294.

Bose, D., Ngo, A. H., Nguyen, V. C., & Nakamura, Y. (2021). Non-specific phospholipases C2 and C6 redundantly function in pollen tube growth via triacylglycerol production in Arabidopsis. *The Plant Journal*, *106*(2), 409–418. https://doi.org/10.1111/tpj.15172.

Burgal, J., Shockey, J., Lu, C., Dyer, J., Larson, T., Graham, I., et al. (2008). Metabolic engineering of hydroxy fatty acid production in plants: RcDGAT2 drives dramatic increases in ricinoleate levels in seed oil. *Plant Biotechnology Journal*, *6*(8), 819–831. https://doi.org/10.1111/j.1467-7652.2008.00361.x.

Cahoon, E. B., Shanklin, J., & Ohlrogge, J. B. (1992). Expression of a coriander desaturase results in petroselinic acid production in transgenic tobacco. *Proceedings of the National Academy of Sciences of the United States of America*, *89*(23), 11184–11188.

Cai, G., Fan, C., Liu, S., Yang, Q., Liu, D., Wu, J., et al. (2020). Nonspecific phospholipase C6 increases seed oil production in oilseed Brassicaceae plants. *The New Phytologist*, *226*(4), 1055–1073. https://doi.org/10.1111/nph.16473.

Caldo, K. M. P., Acedo, J. Z., Panigrahi, R., Vederas, J. C., Weselake, R. J., & Lemieux, M. J. (2017). Diacylglycerol acyltransferase 1 is regulated by its N-terminal domain in response to allosteric effectors. *Plant Physiology*, *175*(2), 667–680. https://doi.org/10.1104/pp.17.00934.

Caldo, K. M. P., Shen, W., Xu, Y., Hanley-Bowdoin, L., Chen, G., Weselake, R. J., et al. (2018). Diacylglycerol acyltransferase 1 is activated by phosphatidate and inhibited by SnRK1-catalyzed phosphorylation. *The Plant Journal*, *96*(2), 287–299. https://doi.org/10.1111/tpj.14029.

Carlsson, A. S., Yilmaz, J. L., Green, A. G., Stymne, S., & Hofvander, P. (2011). Replacing fossil oil with fresh oil—With what and for what? *European Journal of Lipid Science and Technology*, *113*(7), 812–831. https://doi.org/10.1002/ejlt.201100032.

Caroca, R., Howell, K. A., Malinova, I., Burgos, A., Tiller, N., Pellizzer, T., et al. (2021). Knockdown of the plastid-encoded acetyl-CoA carboxylase gene uncovers functions in metabolism and development. *Plant Physiology*, *185*(3), 1091–1110. https://doi.org/10.1093/plphys/kiaa106.

Chapman, K. D., & Ohlrogge, J. B. (2012). Compartmentation of triacylglycerol accumulation in plants. *The Journal of Biological Chemistry*, *287*(4), 2288–2294. https://doi.org/10.1074/jbc.R111.290072.

Chen, B., Wang, J., Zhang, G., Liu, J., Manan, S., Hu, H., et al. (2016). Two types of soybean diacylglycerol acyltransferases are differentially involved in triacylglycerol biosynthesis and response to environmental stresses and hormones. *Scientific Reports, 6*, 28541. https://doi.org/10.1038/srep28541.

Cook, R., Lupette, J., & Benning, C. (2021). The role of chloroplast membrane lipid metabolism in plant environmental responses. *Cell, 10*(3), 706. Retrieved from https://www.mdpi.com/2073-4409/10/3/706.

Correa, S. M., Fernie, A. R., Nikoloski, Z., & Brotman, Y. (2020). Towards model-driven characterization and manipulation of plant lipid metabolism. *Progress in Lipid Research, 80*, 101051. https://doi.org/10.1016/j.plipres.2020.101051.

Dahlqvist, A., Stahl, U., Lenman, M., Banas, A., Lee, M., Sandager, L., et al. (2000). Phospholipid:diacylglycerol acyltransferase: An enzyme that catalyzes the acyl-CoA-independent formation of triacylglycerol in yeast and plants. *Proceedings of the National Academy of Sciences of the United States of America, 97*(12), 6487–6492.

Demski, K., Jeppson, S., Lager, I., Misztak, A., Jasieniecka-Gazarkiewicz, K., Walcron, M., et al. (2019). Isoforms of acyl-CoA:diacylglycerol acyltransferase-2 differ substantially in their specificities towards erucic acid. *Plant Physiology*. https://doi.org/10.1104/pp.19.01129. pp. 01129.02019.

Doermann, P., Frentzen, M., & Ohlrogge, J. B. (1994). Specificities of the acyl-acyl carrier protein (ACP) thioesterase and glycerol-3-phosphate acyltransferase for octadecenoyl-ACP isomers: Identification of a petroselinoyl-ACP thioesterase in Umbelliferae. *Plant Physiology (Rockville), 104*(3), 839–844. Retrieved from <Go to ISI>://BIOSIS:PREV199497244974.

Dormann, P., Spener, F., & Ohlrogge, J. B. (1993). Characterization of 2 acyl-acyl carrier protein thioesterases from developing cuphea seeds specific for medium-chain-acyl and oleoyl-acyl carrier protein. *Planta, 189*(3), 425–432. Retrieved from <Go to ISI>://WOS:A1993KP26100017.

Du, Z.-Y., & Benning, C. (2016). Triacylglycerol accumulation in photosynthetic cells in plants and algae. In Y. Nakamura, & Y. Li-Beisson (Eds.), *Lipids in plant and algae development* (pp. 179–205). Cham: Springer International Publishing.

Dyer, J. M., Stymne, S., Green, A. G., & Carlsson, A. S. (2008). High-value oils from plants. *The Plant Journal, 54*(4), 640–655. https://doi.org/10.1111/j.1365-313X.2008.03430.x.

Frentzen, M. (1986). Biosynthesis and desaturation of the different diacylglycerol moieties in higher plants. *Journal of Plant Physiology, 124*(3–4), 193–209.

Frentzen, M. (1990). Comparison of certain properties of membrane bound and solubilized acyltransferase activities of plant microsomes. *Plant Science, 69*(1), 39–48.

Frentzen, M., Heinz, E., McKeon, T. A., & Stumpf, P. K. (1983). Specificities and selectivities of glycerol-3-phosphate acyltransferase and monoacylglycerol-3-phosphate acyltransferase from pea and spinach chloroplasts. *European Journal of Biochemistry, 129*(3), 629–636. https://doi.org/10.1111/j.1432-1033.1983.tb07096.x.

Gao, H., Gao, Y., Zhang, F., Liu, B., Ji, C., Xue, J., et al. (2021). Functional characterization of an novel acyl-CoA:diacylglycerol acyltransferase 3-3 (CsDGAT3-3) gene from *Camelina sativa*. *Plant Science, 303*, 110752. https://doi.org/10.1016/j.plantsci.2020.110752.

Gargouri, M., Bates, P. D., & Declerck, S. (2021). Combinatorial reprogramming of lipid metabolism in plants: A way towards mass-production of bio-fortified arbuscular mycorrhizal fungi inoculants. *Microbial Biotechnology, 14*(1), 31–34. https://doi.org/10.1111/1751-7915.13684.

Gidda, S. K., Shockey, J. M., Falcone, M., Kim, P. K., Rothstein, S. J., Andrews, D. W., et al. (2011). Hydrophobic-domain-dependent protein–protein interactions mediate the localization of GPAT enzymes to ER subdomains. *Traffic, 12*(4), 452–472. https://doi.org/10.1111/j.1600-0854.2011.01160.x.

Gunstone, F. D., Harwood, J. L., & Dijkstra, A. J. (2007). *The lipid handbook with CD-ROM*. CRC Press.

Guy, J. E., Whittle, E., Moche, M., Lengqvist, J., Lindqvist, Y., & Shanklin, J. (2011). Remote control of regioselectivity in acyl-acyl carrier protein-desaturases. *Proceedings of the National Academy of Sciences of the United States of America*, *108*(40), 16594–16599. https://doi.org/10.1073/pnas.1110221108.

Harwood, J. L., Woodfield, H. K., Chen, G., & Weselake, R. J. (2017). Modification of oil crops to produce fatty acids for industrial applications. In *Fatty acids* (pp. 187–236). Elsevier.

Hernández, M. L., Whitehead, L., He, Z., Gazda, V., Gilday, A., Kozhevnikova, E., et al. (2012). A cytosolic acyltransferase contributes to triacylglycerol synthesis in sucrose-rescued Arabidopsis seed oil catabolism mutants. *Plant Physiology*, *160*(1), 215–225. https://doi.org/10.1104/pp.112.201541.

Holzl, G., & Dormann, P. (2019). Chloroplast lipids and their biosynthesis. *Annual Review of Plant Biology*, *70*, 51–81. https://doi.org/10.1146/annurev-arplant-050718-100202.

Horn, P. J., Liu, J., Cocuron, J. C., McGlew, K., Thrower, N. A., Larson, M., et al. (2016). Identification of multiple lipid genes with modifications in expression and sequence associated with the evolution of hydroxy fatty acid accumulation in *Physaria fendleri*. *The Plant Journal*, *86*(4), 322–348. https://doi.org/10.1111/tpj.13163.

Hu, Z., Ren, Z., & Lu, C. (2012). The phosphatidylcholine diacylglycerol cholinephosphotransferase is required for efficient hydroxy fatty acid accumulation in transgenic Arabidopsis. *Plant Physiology*, *158*(4), 1944–1954. https://doi.org/10.1104/pp.111.192153.

Islam, N., Bates, P. D., Maria John, K. M., Krishnan, H. B., Zhang, J. Z., Luthria, D. L., et al. (2019). Quantitative proteomic analysis of low linolenic acid transgenic soybean reveals perturbations of fatty acid metabolic pathways. *Proteomics*, *19*(7), e1800379. https://doi.org/10.1002/pmic.201800379.

Jarvis, B. A., Romsdahl, T. B., McGinn, M. G., Nazarenus, T. J., Cahoon, E. B., Chapman, K. D., et al. (2021). CRISPR/Cas9-induced fad2 and rod1 mutations stacked with fae1 confer high oleic acid seed oil in pennycress (*Thlaspi arvense* L.). *Frontiers in Plant Science*, *12*(652). https://doi.org/10.3389/fpls.2021.652319.

Jasieniecka-Gazarkiewicz, K., Demski, K., Lager, I., Stymne, S., & Banas, A. (2016). Possible role of different yeast and plant lysophospholipid:acyl-CoA acyltransferases (LPLATs) in acyl remodelling of phospholipids. *Lipids*, *51*(1), 15–23. https://doi.org/10.1007/s11745-015-4102-0.

Jeppson, S., Demski, K., Carlsson, A. S., Zhu, L.-H., Banaś, A., Stymne, S., et al. (2019). *Crambe hispanica subsp. abyssinica* diacylglycerol acyltransferase specificities towards diacylglycerols and acyl-CoA reveal combinatorial effects that greatly affect enzymatic activity and specificity. *Frontiers in Plant Science*, *10*(1442), 1442. https://doi.org/10.3389/fpls.2019.01442.

Jessen, D., Roth, C., Wiermer, M., & Fulda, M. (2015). Two activities of long-chain acyl-coenzyme A synthetase are involved in lipid trafficking between the endoplasmic reticulum and the plastid in Arabidopsis. *Plant Physiology*, *167*(2), 351–366. https://doi.org/10.1104/pp.114.250365.

Kachroo, A., Shanklin, J., Whittle, E., Lapchyk, L., Hildebrand, D., & Kachroo, P. (2007). The Arabidopsis stearoyl-acyl carrier protein-desaturase family and the contribution of leaf isoforms to oleic acid synthesis. *Plant Molecular Biology*, *63*(2), 257–271. https://doi.org/10.1007/s11103-006-9086-y.

Kalinger, R. S., Pulsifer, I. P., Hepworth, S. R., & Rowland, O. (2020). Fatty acyl synthetases and thioesterases in plant lipid metabolism: Diverse functions and biotechnological applications. *Lipids*, *55*(5), 435–455. https://doi.org/10.1002/lipd.12226.

Karki, N., & Bates, P. D. (2018). The effect of light conditions on interpreting oil composition engineering in Arabidopsis seeds. *Plant Direct*, *2*(6), e00067. https://doi.org/10.1002/pld3.67.

Karki, N., Johnson, B. S., & Bates, P. D. (2019). Metabolically distinct pools of phosphatidylcholine are involved in trafficking of fatty acids out of and into the chloroplast for membrane production. *Plant Cell*, *31*(11), 2768–2788. https://doi.org/10.1105/tpc.19.00121.

Keereetaweep, J., Liu, H., Zhai, Z., & Shanklin, J. (2018). Biotin attachment domain-containing proteins irreversibly inhibit acetyl CoA carboxylase. *Plant Physiology*, *177*(1), 208–215. https://doi.org/10.1104/pp.18.00216.

Kennedy, E. P. (1961). Biosynthesis of complex lipids. *Federation Proceedings*, *20*, 934–940. Retrieved from http://www.ncbi.nlm.nih.gov/pubmed/14455159.

Kim, H. U., Li, Y. B., & Huang, A. H. C. (2005). Ubiquitous and endoplasmic reticulum-located lysophosphatidyl acyltransferase, LPAT2, is essential for female but not male gametophyte development in Arabidopsis. *Plant Cell*, *17*(4), 1073–1089. https://doi.org/10.1105/104.030403.

Kim, H. J., Silva, J. E., Iskandarov, U., Andersson, M., Cahoon, R. E., Mockaitis, K., et al. (2015). Structurally divergent lysophosphatidic acid acyltransferases with high selectivity for saturated medium chain fatty acids from Cuphea seeds. *The Plant Journal*, *84*(5), 1021–1033. https://doi.org/10.1111/tpj.13063.

Kim, H. J., Silva, J. E., Vu, H. S., Mockaitis, K., Nam, J.-W., & Cahoon, E. B. (2015). Toward production of jet fuel functionality in oilseeds: Identification of FatB acyl-acyl carrier protein thioesterases and evaluation of combinatorial expression strategies in Camelina seeds. *Journal of Experimental Botany*, *66*(14), 4251–4265. https://doi.org/10.1093/jxb/erv225.

Kim, S., Yamaoka, Y., Ono, H., Kim, H., Shim, D., Maeshima, M., et al. (2013). AtABCA9 transporter supplies fatty acids for lipid synthesis to the endoplasmic reticulum. *Proceedings of the National Academy of Sciences of the United States of America*, *110*(2), 773–778. https://doi.org/10.1073/pnas.1214159110.

Kong, Q., Yang, Y., Guo, L., Yuan, L., & Ma, W. (2020). Molecular basis of plant oil biosynthesis: Insights gained from studying the WRINKLED1 transcription factor. *Frontiers in Plant Science*, *11*(24), 24. https://doi.org/10.3389/fpls.2020.00024.

Koo, A. J. K., Ohlrogge, J. B., & Pollard, M. (2004). On the export of fatty acids from the chloroplast. *Journal of Biological Chemistry*, *279*(16), 16101–16110.

Kotapati, H. K., & Bates, P. D. (2021). (14)C-tracing of lipid metabolism. *Methods in Molecular Biology*, *2295*, 59–80. https://doi.org/10.1007/978-1-0716-1362-7_5.

Kroon, J. T. M., Wei, W., Simon, W. J., & Slabas, A. R. (2006). Identification and functional expression of a type 2 acyl-CoA:diacylglycerol acyltransferase (DGAT2) in developing castor bean seeds which has high homology to the major triglyceride biosynthetic enzyme of fungi and animals. *Phytochemistry*, *67*(23), 2541–2549. Retrieved from http://www.sciencedirect.com/science/article/B6TH7-4M936P1-1/1/402875a144ab34c43a7dfa798574a21c.

Lager, I., Jeppson, S., Gippert, A.-L., Feussner, I., Stymne, S., & Marmon, S. (2020). Acyltransferases regulate oil quality in *Camelina sativa* through both acyl donor and acyl acceptor specificities. *Frontiers in Plant Science*, *11*(1144). https://doi.org/10.3389/fpls.2020.01144.

Lager, I., Yilmaz, J. L., Zhou, X.-R., Jasieniecka, K., Kazachkov, M., Wang, P., et al. (2013). Plant acyl-CoA:lysophosphatidylcholine acyltransferases (LPCATs) have different specificities in their forward and reverse reactions. *Journal of Biological Chemistry*, *288*(52), 36902–36914. https://doi.org/10.1074/jbc.M113.521815.

Lands, W. E. M. (1965). Lipid metabolism. *Annual Review of Biochemistry*, *34*, 313. Retrieved from http://www.annualreviews.org/doi/abs/10.1146/annurev.bi.34.070165.001525.

Lee, J., Welti, R., Schapaugh, W. T., & Trick, H. N. (2011). Phospholipid and triacylglycerol profiles modified by PLD suppression in soybean seed. *Plant Biotechnology Journal, 9*(3), 359–372. https://doi.org/10.1111/j.1467-7652.2010.00562.x.

Li, M., Bahn, S. C., Fan, C., Li, J., Phan, T., Ortiz, M., et al. (2013). Patatin-related phospholipase pPLAIIIδ increases seed oil content with long-chain fatty acids in Arabidopsis. *Plant Physiology, 162*(1), 39–51. https://doi.org/10.1104/pp.113.216994.

Li, N., Gügel, I. L., Giavalisco, P., Zeisler, V., Schreiber, L., Soll, J., et al. (2015). FAX1, a novel membrane protein mediating plastid fatty acid export. *PLoS Biology, 13*(2), e1002053. https://doi.org/10.1371/journal.pbio.1002053.

Li, N., Meng, H., Li, S., Zhang, Z., Zhao, X., Wang, S., et al. (2020). Two plastid fatty acid exporters contribute to seed oil accumulation in Arabidopsis. *Plant Physiology, 182*(4), 1910–1919. https://doi.org/10.1104/pp.19.01344.

Li, R. Z., Yu, K. S., & Hildebrand, D. F. (2010). DGAT1, DGAT2 and PDAT expression in seeds and other tissues of epoxy and hydroxy fatty acid accumulating plants. *Lipids, 45*(2), 145–157. https://doi.org/10.1007/s11745-010-3385-4.

Li-Beisson, Y., Neunzig, J., Lee, Y., & Philippar, K. (2017). Plant membrane-protein mediated intracellular traffic of fatty acids and acyl lipids. *Current Opinion in Plant Biology, 40*, 138–146. https://doi.org/10.1016/j.pbi.2017.09.006.

Li-Beisson, Y., Shorrosh, B., Beisson, F., Andersson, M. X., Arondel, V., Bates, P. D., et al. (2013). Acyl-lipid metabolism. *Arabidopsis Book, 11*, e0161. https://doi.org/10.1199/tab. 0161.

Liu, X., Ma, D., Zhang, Z., Wang, S., Du, S., Deng, X., et al. (2019). Plant lipid remodeling in response to abiotic stresses. *Environmental and Experimental Botany, 165*, 174–184. https://doi.org/10.1016/j.envexpbot.2019.06.005.

Liu, Y., Wang, G., & Wang, X. (2015). Role of aminoalcoholphosphotransferases 1 and 2 in phospholipid homeostasis in Arabidopsis. *The Plant Cell, 27*(5), 1512–1528. https://doi.org/10.1105/tpc.15.00180.

Lu, C., Xin, Z., Ren, Z., Miquel, M., & Browse, J. (2009). An enzyme regulating triacylglycerol composition is encoded by the ROD1 gene of Arabidopsis. *Proceedings of the National Academy of Sciences of the United States of America, 106*(44), 18837–18842. https://doi.org/10.1073/pnas.0908848106.

Lung, S.-C., & Chye, M.-L. (2016). The binding versatility of plant acyl-CoA-binding proteins and their significance in lipid metabolism. *Biochimica et Biophysica Acta (BBA)—Molecular and Cell Biology of Lipids, 1861*(9), 1409–1421.

Lunn, D., Le, A., Wallis, J. G., & Browse, J. (2020). Castor LPCAT and PDAT1A act in concert to promote transacylation of hydroxy-fatty acid onto triacylglycerol. *Plant Physiology, 184*(2), 709–719. https://doi.org/10.1104/pp.20.00691.

Lunn, D., Wallis, J. G., & Browse, J. (2019). Tri-hydroxy-triacylglycerol is efficiently produced by position-specific castor acyltransferases. *Plant Physiology, 179*(3), 1050–1063. https://doi.org/10.1104/pp.18.01409.

Mapelli-Brahm, A., Sanchez, R., Pan, X., Moreno-Perez, A. J., Garces, R., Martinez-Force, E., et al. (2020). Functional characterization of lysophosphatidylcholine: Acyl-CoA acyltransferase genes from sunflower (*Helianthus annuus* L.). *Frontiers in Plant Science, 11*(403), 403. https://doi.org/10.3389/fpls.2020.00403.

Marmon, S., Sturtevant, D., Herrfurth, C., Chapman, K., Stymne, S., & Feussner, I. (2017). Two acyltransferases contribute differently to linolenic acid levels in seed oil. *Plant Physiology, 173*(4), 2081–2095. https://doi.org/10.1104/pp.16.01865.

McGlew, K., Shaw, V., Zhang, M., Kim, R., Yang, W., Shorrosh, B., et al. (2015). An annotated database of Arabidopsis mutants of acyl lipid metabolism. *Plant Cell Reports, 34*(4), 519–532. https://doi.org/10.1007/s00299-014-1710-8.

Meï, C. E., Cussac, M., Haslam, R. P., Beaudoin, F., Wong, Y.-S., Maréchal, E., et al. (2017). C1 metabolism inhibition and nitrogen deprivation trigger triacylglycerol accumulation in *Arabidopsis thaliana* cell cultures and highlight a role of NPC in phosphatidylcholine-to-triacylglycerol pathway. *Frontiers in Plant Science*, 7(2014). https://doi.org/10.3389/fpls.2016.02014.

Mhaske, V., Beldjilali, K., Ohlrogge, J., & Pollard, M. (2005). Isolation and characterization of an *Arabidopsis thaliana* knockout line for phospholipid:diacylglycerol transacylase gene (At5g13640). *Plant Physiology and Biochemistry*, *43*(4), 413–417. https://doi.org/10.1016/j.plaphy.2005.01.013.

Moon, H., Smith, M. A., & Kunst, L. (2001). A condensing enzyme from the seeds of *Lesquerella fendleri* that specifically elongates hydroxy fatty acids. *Plant Physiology*, *127*(4), 1635–1643. Retrieved from <Go to ISI>://WOS:000172824500045. https://www.ncbi.nlm.nih.gov/pmc/articles/PMC133568/pdf/pp1201001635.pdf.

Moreau, R. A., & Stumpf, P. K. (1981). Recent studies of the enzymic-synthesis of ricinoleic acid by developing castor beans. *Plant Physiology*, *67*(4), 672–676. Retrieved from <Go to ISI>://A1981LM91100014.

Nakamura, Y. (2021). Headgroup biosynthesis of phosphatidylcholine and phosphatidylethanolamine in seed plants. *Progress in Lipid Research*, *82*, 101091. https://doi.org/10.1016/j.plipres.2021.101091.

Nakamura, Y., Koizumi, R., Shui, G., Shimojima, M., Wenk, M. R., Ito, T., et al. (2009). Arabidopsis lipins mediate eukaryotic pathway of lipid metabolism and cope critically with phosphate starvation. *Proceedings of the National Academy of Sciences of the United States of America*, *106*(49), 20978–20983. https://doi.org/10.1073/pnas.0907173106.

Napier, J. A., Haslam, R. P., Tsalavouta, M., & Sayanova, O. (2019). The challenges of delivering genetically modified crops with nutritional enhancement traits. *Nature Plants*, *5*(6), 563–567. https://doi.org/10.1038/s41477-019-0430-z.

Nguyen, H. T., Mishra, G., Whittle, E., Bevan, S. A., Merlo, A. O., Walsh, T. A., et al. (2010). Metabolic engineering of seeds can achieve levels of {omega}-7 fatty acids comparable with the highest levels found in natural plant sources. *Plant Physiology*, *154*(4), 1897–1904. https://doi.org/10.1104/pp.110.165340.

Nguyen, H. T., Silva, J. E., Podicheti, R., Macrander, J., Yang, W., Nazarenus, T. J., et al. (2013). Camelina seed transcriptome: A tool for meal and oil improvement and translational research. *Plant Biotechnology Journal*, *11*, 759–769. https://doi.org/10.1111/pbi.12068.

Ohlrogge, J., & Browse, J. (1995). Lipid biosynthesis. *Plant Cell*, 7(7), 957–970.

Ohlrogge, J. B., & Jaworski, J. G. (1997). Regulation of fatty acid synthesis. *Annual Review of Plant Physiology and Plant Molecular Biology*, *48*, 109–136.

Ohlrogge, J., Thrower, N., Mhaske, V., Stymne, S., Baxter, M., Yang, W., et al. (2018). PlantFAdb: A resource for exploring hundreds of plant fatty acid structures synthesized by thousands of plants and their phylogenetic relationships. *The Plant Journal*, *96*(6), 1299–1308. https://doi.org/10.1111/tpj.14102.

Okuley, J., Lightner, J., Feldmann, K., Yadav, N., Lark, E., & Browse, J. (1994). Arabidopsis FAD2 gene encodes the enzyme that is essential for polyunsaturated lipid-synthesis. *Plant Cell*, *6*(1), 147–158. Retrieved from <Go to ISI>://A1994MZ10100014. http://www.plantcell.org/content/plantcell/6/1/147.full.pdf.

Ortiz, R., Geleta, M., Gustafsson, C., Lager, I., Hofvander, P., Lofstedt, C., et al. (2020). Oil crops for the future. *Current Opinion in Plant Biology*, *56*, 181–189. https://doi.org/10.1016/j.pbi.2019.12.003.

Pan, X., Chen, G., Kazachkov, M., Greer, M. S., Caldo, K. M., Zou, J., et al. (2015). In vivo and in vitro evidence for biochemical coupling of reactions catalyzed by lysophosphatidylcholine acyltransferase and diacylglycerol acyltransferase. *The Journal of Biological Chemistry*, *290*(29), 18068–18078. https://doi.org/10.1074/jbc.M115.654798.

Pan, X., Siloto, R. M. P., Wickramarathna, A. D., Mietkiewska, E., & Weselake, R. J. (2013). Identification of a pair of phospholipid:diacylglycerol acyltransferases from developing flax (*Linum usitatissimum* L.) seed catalyzing the selective production of Trilinolenin. *Journal of Biological Chemistry*, *288*(33), 24173–24188. https://doi.org/10.1074/jbc.M113.475699.

Parajuli, S., Kannan, B., Karan, R., Sanahuja, G., Liu, H., Garcia-Ruiz, E., et al. (2020). Towards oilcane: Engineering hyperaccumulation of triacylglycerol into sugarcane stems. *GCB Bioenergy*, *12*(7), 476–490. https://doi.org/10.1111/gcbb.12684.

Paya-Milans, M., Aznar-Moreno, J. A., Balbuena, T. S., Haslam, R. P., Gidda, S. K., Perez-Hormaeche, J., et al. (2016). Sunflower HaGPAT9-1 is the predominant GPAT during seed development. *Plant Science*, *252*, 42–52. https://doi.org/10.1016/j.plantsci.2016.07.002.

Pollard, M. R., Anderson, L., Fan, C., Hawkins, D. J., & Davies, H. M. (1991). A specific acyl-ACP thioesterase implicated in medium-chain fatty-acid production in immature cotyledons of Umbellularia-californica. *Archives of Biochemistry and Biophysics*, *284*(2), 306–312. Retrieved from http://www.sciencedirect.com/science/article/pii/0003986191903008.

Pollard, M., & Ohlrogge, J. (1999). Testing models of fatty acid transfer and lipid synthesis in spinach leaf using in vivo oxygen-18 labeling. *Plant Physiology*, *121*(4), 1217–1226. Retrieved from http://www.ncbi.nlm.nih.gov/pmc/articles/PMC59488/pdf/pp001217.pdf.

Pouvreau, B., Vanhercke, T., & Singh, S. (2018). From plant metabolic engineering to plant synthetic biology: The evolution of the design/build/test/learn cycle. *Plant Science*, *273*, 3–12. https://doi.org/10.1016/j.plantsci.2018.03.035.

Regmi, A., Shockey, J., Kotapati, H. K., & Bates, P. D. (2020). Oil-producing metabolons containing DGAT1 use separate substrate pools from those containing DGAT2 or PDAT. *Plant Physiology*, *184*(2), 720–737. https://doi.org/10.1104/pp.20.00461.

Romsdahl, T., Shirani, A., Minto, R. E., Zhang, C., Cahoon, E. B., Chapman, K. D., et al. (2019). Nature-guided synthesis of advanced bio-lubricants. *Scientific Reports*, *9*(1), 11711. https://doi.org/10.1038/s41598-019-48165-6.

Roughan, P. G., & Slack, C. R. (1982). Cellular-organization of glycerolipid metabolism. *Annual Review of Plant Physiology and Plant Molecular Biology*, *33*, 97–132.

Ruiz-Lopez, N., Usher, S., Sayanova, O. V., Napier, J. A., & Haslam, R. P. (2015). Modifying the lipid content and composition of plant seeds: Engineering the production of LC-PUFA. *Applied Microbiology and Biotechnology*, *99*(1), 143–154. https://doi.org/10.1007/s00253-014-6217-2.

Saha, S., Enugutti, B., Rajakumari, S., & Rajasekharan, R. (2006). Cytosolic triacylglycerol biosynthetic pathway in oilseeds. Molecular cloning and expression of peanut cytosolic diacylglycerol acyltransferase. *Plant Physiology*, *141*(4), 1533–1543. https://doi.org/10.1104/pp.106.082198.

Salas, J. J., & Ohlrogge, J. B. (2002). Characterization of substrate specificity of plant FatA and FatB acyl-ACP thioesterases. *Archives of Biochemistry and Biophysics*, *403*(1), 25–34. Retrieved from <go to ISI>://000176822100003.

Salie, M. J., & Thelen, J. J. (2016). Regulation and structure of the heteromeric acetyl-CoA carboxylase. *Biochimica et Biophysica Acta (BBA)—Molecular and Cell Biology of Lipids*, *1861*(9 Part B), 1207–1213. https://doi.org/10.1016/j.bbalip.2016.04.004.

Salie, M. J., Zhang, N., Lancikova, V., Xu, D., & Thelen, J. J. (2016). A family of negative regulators targets the committed step of de novo fatty acid biosynthesis. *Plant Cell*, *28*(9), 2312–2325. https://doi.org/10.1105/tpc.16.00317.

Sasaki, Y., & Nagano, Y. (2004). Plant acetyl-CoA carboxylase: Structure, biosynthesis, regulation, and gene manipulation for plant breeding. *Bioscience Biotechnology and Biochemistry*, *68*(6), 1175–1184. Retrieved from https://www.jstage.jst.go.jp/article/bbb/68/6/68_6_1175/_pdf.

Schnurr, J. A., Shockey, J. M., de Boer, G. J., & Browse, J. A. (2002). Fatty acid export from the chloroplast. Molecular characterization of a major plastidial acyl-coenzyme a synthetase from Arabidopsis. *Plant Physiology*, *129*(4), 1700–1709. https://doi.org/10.1104/pp.003251.

Shanklin, J., & Cahoon, E. B. (1998). Desaturation and related modifications of fatty acids. *Annual Review of Plant Physiology and Plant Molecular Biology*, *49*, 611–641. Retrieved from <Go to ISI>://000074266700025. https://www.annualreviews.org/doi/pdf/10.1146/annurev.arplant.49.1.611.

Shivaiah, K. K., Ding, G., Upton, B., & Nikolau, B. J. (2020). Non-catalytic subunits facilitate quaternary organization of plastidic acetyl-CoA carboxylase. *Plant Physiology*, *182*(2), 756–775. https://doi.org/10.1104/pp.19.01246.

Shockey, J., & Browse, J. (2011). Genome-level and biochemical diversity of the acyl-activating enzyme superfamily in plants. *The Plant Journal*, *66*(1), 143–160. https://doi.org/10.1111/j.1365-313X.2011.04512.x.

Shockey, J. M., Fulda, M. S., & Browse, J. A. (2002). Arabidopsis contains nine long-chain acyl-coenzyme a synthetase genes that participate in fatty acid and glycerolipid metabolism. *Plant Physiology*, *129*(4), 1710–1722.

Shockey, J. M., Fulda, M. S., & Browse, J. (2003). Arabidopsis contains a large superfamily of acyl-activating enzymes. Phylogenetic and biochemical analysis reveals a new class of acyl-coenzyme A synthetases. *Plant Physiology*, *132*(2), 1065–1076. Retrieved from https://www.ncbi.nlm.nih.gov/pmc/articles/PMC167044/pdf/1321065.pdf.

Shockey, J. M., Gidda, S. K., Chapital, D. C., Kuan, J.-C., Dhanoa, P. K., Bland, J. M., et al. (2006). Tung tree DGAT1 and DGAT2 have nonredundant functions in triacylglycerol biosynthesis and are localized to different subdomains of the endoplasmic reticulum. *The Plant Cell*, *18*(9), 2294–2313. https://doi.org/10.1105/tpc.106.043695.

Shockey, J., Lager, I., Stymne, S., Kotapati, H. K., Sheffield, J., Mason, C., et al. (2019). Specialized lysophosphatidic acid acyltransferases contribute to unusual fatty acid accumulation in exotic Euphorbiaceae seed oils. *Planta*, *249*(5), 1285–1299. https://doi.org/10.1007/s00425-018-03086-y.

Shockey, J., Regmi, A., Cotton, K., Adhikari, N., Browse, J., & Bates, P. D. (2016). Identification of Arabidopsis GPAT9 (At5g60620) as an essential gene involved in triacylglycerol biosynthesis. *Plant Physiology*, *170*(1), 163–179. https://doi.org/10.1104/pp.15.01563.

Silva, J. A. C., Grilo, L. M., Gandini, A., & Lacerda, T. M. (2021). The prospering of macromolecular materials based on plant oils within the blooming field of polymers from renewable resources. *Polymers*, *13*(11), 1722. https://doi.org/10.3390/polym13111722.

Singer, S. D., Chen, G., Mietkiewska, E., Tomasi, P., Jayawardhane, K., Dyer, J. M., et al. (2016). Arabidopsis GPAT9 contributes to synthesis of intracellular glycerolipids but not surface lipids. *Journal of Experimental Botany*, *67*(15), 4627–4638. https://doi.org/10.1093/jxb/erw242.

Slack, C. R., Campbell, L. C., Browse, J. A., & Roughan, P. G. (1983). Some evidence for the reversibility of the cholinephosphotransferase-catalysed reaction in developing linseed cotyledons in vivo. *Biochimica et Biophysica Acta*, *754*(1), 10–20. Retrieved from <Go to ISI>://A1983RR63700002.

Slack, C. R., Roughan, P. G., Browse, J. A., & Gardiner, S. E. (1985). Some properties of cholinephosphotransferase from developing safflower cotyledons. *Biochimica et Biophysica Acta*, *833*(3), 438–448. Retrieved from <Go to ISI>://A1985AFA6000011.

Stahl, U., Carlsson, A. S., Lenman, M., Dahlqvist, A., Huang, B. Q., Banas, W., et al. (2004). Cloning and functional characterization of a phospholipid:diacylglycerol acyltransferase from Arabidopsis. *Plant Physiology*, *135*(3), 1324–1335. Retrieved from <Go to ISI>://000222692700017.

Stahl, U., Stalberg, K., Stymne, S., & Ronne, H. (2008). A family of eukaryotic lysophospholipid acyltransferases with broad specificity. *FEBS Letters, 582*(2), 305–309. Retrieved from <Go to ISI>://MEDLINE:18154737.

Tian, Y., Lv, X., Xie, G., Wang, L., Dai, T., Qin, X., et al. (2019). FAX2 mediates fatty acid export from plastids in developing Arabidopsis seeds. *Plant & Cell Physiology, 60*(10), 2231–2242. https://doi.org/10.1093/pcp/pcz117.

Tjellström, H., Yang, Z., Allen, D. K., & Ohlrogge, J. B. (2012). Rapid kinetic labeling of Arabidopsis cell suspension cultures: Implications for models of lipid export from plastids. *Plant Physiology, 158*(2), 601–611. https://doi.org/10.1104/pp.111.186122.

Troncoso-Ponce, M. A., Kilaru, A., Cao, X., Durrett, T. P., Fan, J., Jensen, J. K., et al. (2011). Comparative deep transcriptional profiling of four developing oilseeds. *The Plant Journal, 68*(6), 1014–1027. https://doi.org/10.1111/j.1365-313X.2011.04751.x.

van Erp, H., Bates, P. D., Burgal, J., Shockey, J., & Browse, J. (2011). Castor phospholipid: diacylglycerol acyltransferase facilitates efficient metabolism of hydroxy fatty acids in transgenic Arabidopsis. *Plant Physiology, 155*(2), 683–693. https://doi.org/10.1104/pp.110.167239.

Vandeloo, F. J., Broun, P., Turner, S., & Somerville, C. (1995). An oleate 12-hydroxylase from *Ricinus communis* L. is a fatty acyl desaturase homolog. *Proceedings of the National Academy of Sciences of the United States of America, 92*(15), 6743–6747. https://doi.org/10.1073/pnas.92.15.6743.

Vanhercke, T., Dyer, J. M., Mullen, R. T., Kilaru, A., Rahman, M. M., Petrie, J. R., et al. (2019). Metabolic engineering for enhanced oil in biomass. *Progress in Lipid Research, 74*, 103–129. https://doi.org/10.1016/j.plipres.2019.02.002.

Vanhercke, T., Wood, C. C., Stymne, S., Singh, S. P., & Green, A. G. (2013). Metabolic engineering of plant oils and waxes for use as industrial feedstocks. *Plant Biotechnology Journal, 11*(2), 197–210. https://doi.org/10.1111/pbi.12023.

Voelker, T., Worrell, A., Anderson, L., Bleibaum, J., Fan, C., Hawkins, D., et al. (1992). Fatty acid biosynthesis redirected to medium chains in transgenic oilseed plants. *Science, 257*(5066), 72–74. https://doi.org/10.1126/science.1621095.

Wang, X. M. (2001). Plant phospholipases. *Annual Review of Plant Physiology and Plant Molecular Biology, 52*, 211.

Wang, K., Durrett, T. P., & Benning, C. (2019). Functional diversity of glycerolipid acylhydrolases in plant metabolism and physiology. *Progress in Lipid Research, 75*, 100987. https://doi.org/10.1016/j.plipres.2019.100987.

Wang, L., Kazachkov, M., Shen, W., Bai, M., Wu, H., & Zou, J. (2014). Deciphering the roles of Arabidopsis LPCAT and PAH in phosphatidylcholine homeostasis and pathway coordination for chloroplast lipid synthesis. *The Plant Journal, 80*(6), 965–976. https://doi.org/10.1111/tpj.12683.

Wang, L., Shen, W., Kazachkov, M., Chen, G., Chen, Q., Carlsson, A. S., et al. (2012). Metabolic interactions between the lands cycle and the Kennedy pathway of glycerolipid synthesis in Arabidopsis developing seeds. *The Plant Cell, 24*(11), 4652–4669. https://doi.org/10.1105/tpc.112.104604.

Weiss, S. B., Kennedy, E. P., & Kiyasu, J. Y. (1960). Enzymatic synthesis of triglycerides. *Journal of Biological Chemistry, 235*(1), 40–44.

Wickramarathna, A. D., Siloto, R. M., Mietkiewska, E., Singer, S. D., Pan, X., & Weselake, R. J. (2015). Heterologous expression of flax phospholipid:diacylglycerol cholinephosphotransferase (PDCT) increases polyunsaturated fatty acid content in yeast and Arabidopsis seeds. *BMC Biotechnology, 15*(1), 63. https://doi.org/10.1186/s12896-015-0156-6.

Williams, J. P., Imperial, V., Khan, M. U., & Hodson, J. N. (2000). The role of phosphatidylcholine in fatty acid exchange and desaturation in *Brassica napus* L. leaves. *Biochemical Journal, 349*, 127–133.

Winter, D., Vinegar, B., Nahal, H., Ammar, R., Wilson, G. V., & Provart, N. J. (2007). An "electronic fluorescent pictograph" browser for exploring and analyzing large-scale biological data sets. *PLoS One*, *2*(8), e718. https://doi.org/10.1371/journal.pone.0000718.

Xu, Y., Caldo, K. M. P., Jayawardhane, K., Ozga, J. A., Weselake, R. J., & Chen, G. (2019). A transferase interactome that may facilitate channeling of polyunsaturated fatty acid moieties from phosphatidylcholine to triacylglycerol. *Journal of Biological Chemistry*, *294*, 14838–14844. jbc. AC119. 010601.

Xu, Y., Caldo, K. M. P., Pal-Nath, D., Ozga, J., Lemieux, M. J., Weselake, R. J., et al. (2018). Properties and biotechnological applications of acyl-CoA:diacylglycerol acyltransferase and phospholipid:diacylglycerol acyltransferase from terrestrial plants and microalgae. *Lipids*, *53*(7), 663–688. https://doi.org/10.1002/lipd.12081.

Xu, J., Carlsson, A., Francis, T., Zhang, M., Hoffmann, T., Giblin, M., et al. (2012). Triacylglycerol synthesis by PDAT1 in the absence of DGAT1 activity is dependent on re-acylation of LPC by LPCAT2. *BMC Plant Biology*, *12*(1), 4. Retrieved from http://www.biomedcentral.com/1471-2229/12/4.

Xu, Y., Holic, R., Li, D., Pan, X., Mietkiewska, E., Chen, G., et al. (2018). Substrate preferences of long-chain acyl-CoA synthetase and diacylglycerol acyltransferase contribute to enrichment of flax seed oil with α-linolenic acid. *Biochemical Journal*, *475*(8), 1473–1489. https://doi.org/10.1042/bcj20170910.

Xu, R., Yang, T., Wang, R., & Liu, A. (2014). Characterisation of DGAT1 and DGAT2 from *Jatropha curcas* and their functions in storage lipid biosynthesis. *Functional Plant Biology*, *41*(3), 321–329. https://doi.org/10.1071/FP12388.

Yang, W., Wang, G., Li, J., Bates, P. D., Wang, X., & Allen, D. K. (2017). Phospholipase Dzeta enhances diacylglycerol flux into triacylglycerol. *Plant Physiology*, *174*(1), 110–123. https://doi.org/10.1104/pp.17.00026.

Ye, Y., Fulcher, Y. G., Sliman, D. J., Day, M. T., Schroeder, M. J., Koppisetti, R. K., et al. (2020). The BADC and BCCP subunits of chloroplast acetyl-CoA carboxylase sense the pH changes of the light–dark cycle. *Journal of Biological Chemistry*, *295*(29), 9901–9916. https://doi.org/10.1074/jbc.RA120.012877.

Ye, Y., Nikovics, K., To, A., Lepiniec, L., Fedosejevs, E. T., Van Doren, S. R., et al. (2020). Docking of acetyl-CoA carboxylase to the plastid envelope membrane attenuates fatty acid production in plants. *Nature Communications*, *11*(1), 6191. https://doi.org/10.1038/s41467-020-20014-5.

Yu, X.-H., Prakash, R. R., Sweet, M., & Shanklin, J. (2014). Coexpressing *Escherichia coli* cyclopropane synthase with *Sterculia foetida* lysophosphatidic acid acyltransferase enhances cyclopropane fatty acid accumulation. *Plant Physiology*, *164*(1), 455–465. https://doi.org/10.1104/pp.113.230953.

Yu, L., Zhou, C., Fan, J., Shanklin, J., & Xu, C. (2021). Mechanisms and functions of membrane lipid remodeling in plants. *The Plant Journal*. https://doi.org/10.1111/tpj.15273.

Yurchenko, O. P., Nykiforuk, C. L., Moloney, M. M., Stahl, U., Banas, A., Stymne, S., et al. (2009). A 10-kDa acyl-CoA-binding protein (ACBP) from *Brassica napus* enhances acyl exchange between acyl-CoA and phosphatidylcholine. *Plant Biotechnology Journal*, 7(7), 602–610. https://doi.org/10.1111/j.1467-7652.2009.00427.x.

Zhang, M., Fan, J., Taylor, D. C., & Ohlrogge, J. B. (2009). DGAT1 and PDAT1 acyltransferases have overlapping functions in Arabidopsis triacylglycerol biosynthesis and are essential for normal pollen and seed development. *Plant Cell*, *21*(12), 3885–3901. https://doi.org/10.1105/tpc.109.071795.

Zhang, Y., & Fernie, A. R. (2021). Metabolons, enzyme–enzyme assemblies that mediate substrate channeling, and their roles in plant metabolism. *Plant Communications*, *2*(1), 100081. https://doi.org/10.1016/j.xplc.2020.100081.

Zhang, J., Li, J., Garcia-Ruiz, H., Bates, P. D., Mirkov, T. E., & Wang, X. (2014). A stearoyl-acyl carrier protein desaturase, NbSACPD-C, is critical for ovule development in Nicotiana benthamiana. *The Plant Journal*, *80*(3), 489–502. https://doi.org/10.1111/tpj.12649.

Zhang, G., Yang, J., Chen, X., Zhao, D., Zhou, X., Zhang, Y., et al. (2021). Phospholipase D- and phosphatidic acid-mediated phospholipid metabolism and signaling modulate symbiotic interaction and nodulation in soybean (*Glycine max*). *The Plant Journal*, *106*(1), 142–158. https://doi.org/10.1111/tpj.15152.

Zhao, H., Kosma, D. K., & Lü, S. (2021). Functional role of long-chain acyl-CoA synthetases in plant development and stress responses. *Frontiers in Plant Science*, *12*(276), 640996. https://doi.org/10.3389/fpls.2021.640996.

Zhao, L., Haslam, T. M., Sonntag, A., Molina, I., & Kunst, L. (2019). Functional overlap of long-chain acyl-CoA synthetases in Arabidopsis. *Plant and Cell Physiology*, *60*(5), 1041–1054.

Zhao, L., Katavic, V., Li, F., Haughn, G. W., & Kunst, L. (2010). Insertional mutant analysis reveals that long-chain acyl-CoA synthetase 1 (LACS1), but not LACS8, functionally overlaps with LACS9 in Arabidopsis seed oil biosynthesis. *The Plant Journal*, *64*(6), 1048–1058. https://doi.org/10.1111/j.1365-313X.2010.04396.x.

Zhou, X. R., Bhandari, S., Johnson, B. S., Kotapati, H. K., Allen, D. K., Vanhercke, T., et al. (2020). Reorganization of acyl flux through the lipid metabolic network in oil-accumulating tobacco leaves. *Plant Physiology*, *182*(2), 739–755. https://doi.org/10.1104/pp.19.00667.

Zhou, X. R., Shrestha, P., Yin, F., Petrie, J. R., & Singh, S. P. (2013). AtDGAT2 is a functional acyl-CoA:diacylglycerol acyltransferase and displays different acyl-CoA substrate preferences than AtDGAT1. *FEBS Letters*, *587*(15), 2371–2376. https://doi.org/10.1016/j.febslet.2013.06.003.

CHAPTER SEVEN

Plant polyunsaturated fatty acids: Biological roles, regulation and biotechnological applications

Thibault Geneste[a] **and Jean-Denis Faure**[a,b,*]
[a]Institut Jean-Pierre Bourgin, INRAE, AgroParisTech, Université Paris-Saclay, Versailles, France
[b]AgroParisTech, Paris, France
*Corresponding author: e-mail address: jean-denis.faure@inrae.fr

Contents

Abstract

Polyunsaturated fatty acids (PUFAs) are particularily enriched in land plants and are the products of the editing of fatty acids synthetized in the plastids. The desaturases FAD2, FAD3, FAD6, FAD7 and FAD8 enables the synthesis of linoleic and linolenic acids (C18:2 and C18:3, respectively) that are the most abundant PUFAs in land plants. PUFA content is regulated by environmental conditions mainly temperature and stresses associated with oxidative conditions. Genetic regulation of the FAD desaturases is occurring at the transcriptional and post-transcriptional levels allowing fine tuning of PUFAs levels in changing environment but also during seed development to modulate

Advances in Botanical Research, Volume 101
ISSN 0065-2296
https://doi.org/10.1016/bs.abr.2021.09.003

seed oil content. This review aims in describing PUFA synthesis, their biological roles and the different factors involved in their regulation with a specific focus on omega-3 linolenic acid. It will also detail in several crop species the different genetic leverage that were used to modify omega-3 content for plant breeding.

1. Introduction

The polyunsaturated fatty acids (PUFAs) are fatty acids with multiple double bonds (>1). Plants are mostly harboring the dienoic omega-6 linoleic acid (C18:2) and omega-3 alpha-linolenic acid or ALA (C18:3) PUFAs that are essentially found in membranes (Harwood, 1980). Common aspects of the function of PUFAs in animals and plants include the maintenance of membrane fluidity (Prades, Funari, Escribá, & Barceló, 2003) in particular for low temperature adaptation but also for the synthesis of functional and/or signaling derivatives, such as oxylipins.

In plants, PUFAs represent a major part of the fatty acids bound to triglycerides in seed oil and are thus readily available for nutrition or industrial applications (Iii, Fitzpatrick, & Kamal-Eldin, 2009). PUFAs are also referred to essential fatty acids, since most animals with the notable exception of the worm *Caenorhabditis elegans* lack the enzymes enabling the introduction of additional desaturation (desaturases) in omega 6 and 3 positions in fatty acids (Spychalla, Kinney, & Browse, 1997). PUFA consumption are often associated with health benefits such as lower risk for heart disease, for example (Clifton & Keogh, 2017). In animals, ALA can serve as a precursor for very long chain PUFAs that are important fatty acids associated with inflammatory responses and critical physiological roles in neural development (Dyall, 2015). The efficiency of conversion of ALA in long chain omega-3 fatty acids is significantly demonstrated for food intake in fish species, making plant sourced food of use in this application (Glencross, 2009).

The review will mostly focus on the specificity of omega-3 linolenic acid synthesis and function relative to omega-6 linoleic acid. We will also discuss how different stresses and physiological processes regulate PUFA synthesis in plants, and finally how the understanding of desaturation metabolism can improve crop selection and breeding.

2. Synthesis and biochemistry of polyunsaturated fatty acids in plants

2.1 *De novo* fatty acid synthesis occurs in plastids

Unlike other eukaryotes, plants synthesize fatty acids exclusively within plastids using acetyl-CoA as carbon donor. The simplest pathway to generate acetyl-CoA is from pyruvate produced after photosynthesis and glycolysis, through the action of the plastidial pyruvate dehydrogenase complex (PDHC). The acetyl-CoA is first converted into malonyl-CoA in plastids through the action of the heteromeric Acyl-CoA-Carboxylase, an enzyme composed of 3 nuclear-encoded subunits (BCCP, BC, alpha-CT) and the plastid-encoded beta-CT (Baud et al., 2003). The malonyl moiety is then transferred to the Acyl-Carrier protein (ACP) by the Malonyl-Coa:ACP malonyltransferase (MCMT) to form malonyl-ACP that is finally used to elongate acyl-ACP chains by three Ketoacyl-ACP-Synthases (KAS I, II, III) (Li-Beisson et al., 2013).

The malonyl-ACP (3 carbons chain) is first extended by KAS III using acetyl-CoA (2 carbons) to generate ACP-4:0 releasing 1 molecule of CO2 in the process. Then, KASI produces ACP-16:0 from ACP-4:0 through six sequential condensation reactions using malonyl-ACP as substrate and also releasing free ACP in the process. Finally, KASII synthesizes ACP-18:0 through the last condensation reaction. ACP-18:0 and ACP-C16:0 can then be desaturated by AADs, plastid localized enzymes, to yield ACP-C18:1 or ACP-C16:1. Upon completion of the elongation cycles and desaturation reactions, the fatty acids can be either used for plastid lipid synthesis or exported to the cytosol and ER.

In plastids, the "prokaryotic" synthesis pathway uses acyl-ACP to produce glycerolipids by transferring fatty acid chains to glycerol-3-phosphate (G3P) or to acyl-G3P. In the ER, the "eukaryotic" synthesis pathway uses acyl chains released from ACP by the thioesterases FATA or FATB enzymes and coupled to Coenzyme-A by thioesterification. The acyl-CoA lipid pool localized in the cytoplasm will be used subsequently by most of the ER based lipid reactions.

2.2 PUFA are synthesized from monounsaturated fatty acids in the endoplasmic reticulum and plastids

Desaturases add double bonds in a sequential manner on omega-9 oleic acid (C18:1). First, omega-6 desaturases yield dienoic omega-6 fatty acids C18:2 (Falcone, Gibson, Lemieux, & Somerville, 1994; Miquel & Browse, 1992). The omega-6 desaturases are coded by two genes in Arabidopsis that have a distinct localization and function. FAD2 is present in the endoplasmic reticulum outer membrane leaflet where it is inserted co-translationally. FAD2 desaturates C18:1 esterified to phosphatidylcholine (PC) (McCartney et al., 2004) to yield C18:2-PC. The omega-6 desaturase FAD6 is localized within the chloroplast membrane (Ferro et al., 2003) where it regulates plastid lipid composition. FAD6 has a different substrate specificity compared to FAD2 since it desaturates galactolipids and phosphatidylglycerol (PG) C16:1 and C18:1 (Browse, Kunst, Anderson, Hugly, & Somerville, 1989).

The omega-6 class fatty acids can then be desaturated by omega-3 desaturases to yield trienoic fatty acids (TFAs). The omega-3 desaturases exist as three isoforms in *Arabidopsis*, FAD3, localized in the ER, and FAD7 and FAD8 which are two plastidial desaturases. Like FAD2, FAD3 acts on PC-C18:2 to yield PC-C18:3 and FAD7 and FAD8 like FAD6, acts on galactolipids and PG C16:2 and C18:2 (Román et al., 2015). The sequential desaturation reactions are illustrated in Fig. 1.

The chloroplastic desaturases FAD6, FAD7 and FAD8 are transmembrane enzymes with four integral transmembrane helices and possess three histidine box motifs (Fig. 2A and B) (Shanklin, Whittle, & Fox, 1994) which are conserved and necessary elements among all desaturases identified so far. The histidine boxes stabilizes a di-iron center essential for the desaturation reaction and which has been well characterized in the closely related mammalian SCD1 (Bai et al., 2015). Plastidial desaturase substrate specificity is restricted by the acyl chain length (Broadwater, Whittle, &

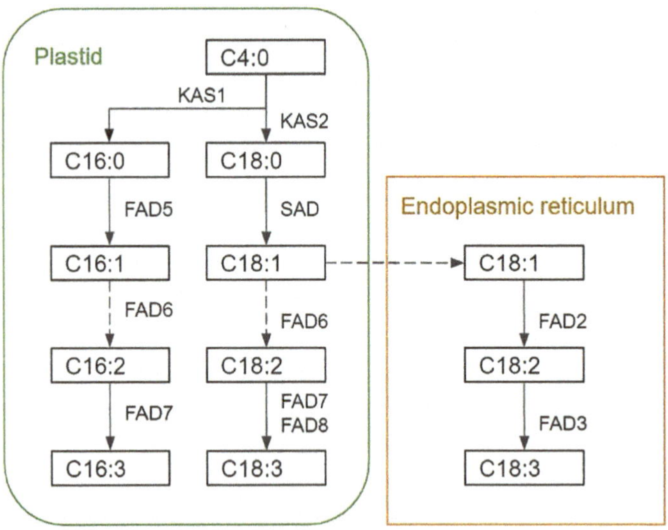

Fig. 1 Simplified biosynthesis pathway of polyunsaturated fatty acids in the plastids and the endoplasmic reticulum. C4:0 elongation is carried out by the FAS complex and fatty acids are extended by KAS consuming malonyl-ACP. The desaturation reactions of the desaturases are carried out on specific substrates in organelles and on lipid classes described in Table 1.

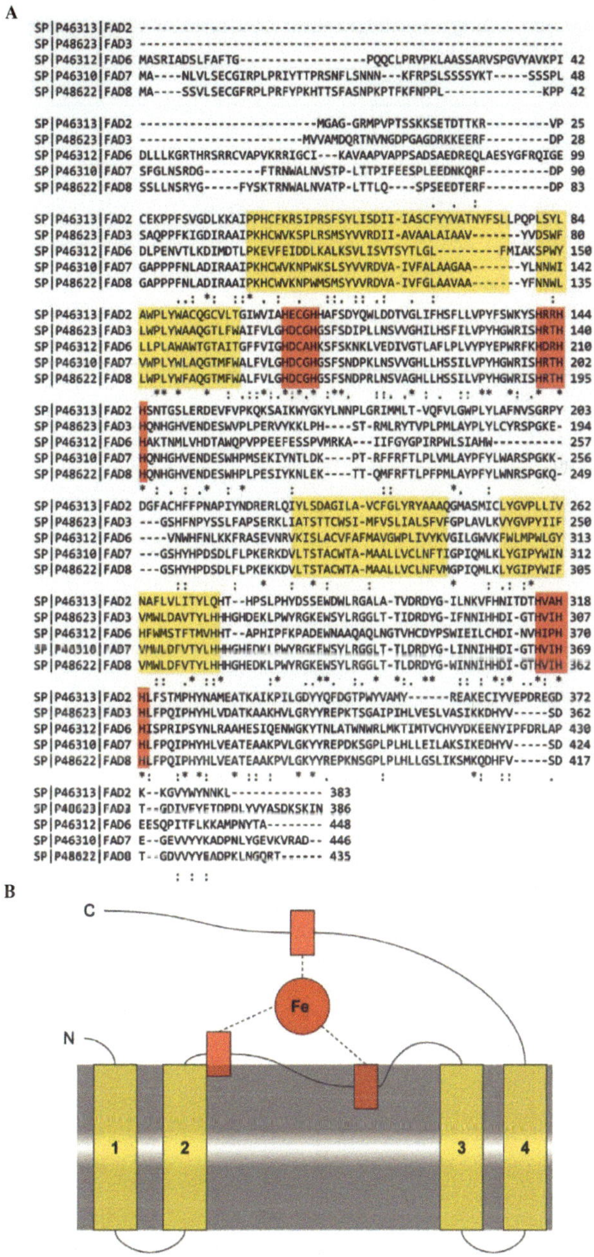

Fig. 2 Protein sequence and structure analysis of Arabidopsis FAD2, FAD3, FAD6, FAD7 and FAD8. (A) Alignment of *Arabidopsis* FAD2, FAD3, FAD6, FAD7 and FAD8 protein sequences using clustal omega and accession numbers described in Table 1. Integral transmembrane regions are highlighted in yellow and histidine boxes highlighted in red. (B) Schematic representation of FAD desaturase structure. Gray box represents membrane lipid bilayers, Fe indicates the stabilized di-iron center, yellow bars represent transmembrane domain. (For interpretation of the references to color in this figure legend, the reader is referred to the web version of this article.)

Table 1 *Arabidopsis* Omega-6 and Omega-3 desaturase characteristics.

	Substrate	Product	Cofactor	Localisation	Uniprot accession
FAD2	PC-C18:1	PC-C18:2	Cyt-B5	ER, external leaflet.	P46313
FAD3	PC-C18:2	PC-C18:3	Cyt-B5	ER, external leaflet.	P48623
FAD6	MGDG-C18:1 DGDG-C18:1 SQDG-C18:1 PG-C18:1 MGDG-C16:1 DGDG-C16:1 SQDG-C16:1 PG-C16:1	MGDG-C18:2 DGDG-C18:2 SQDG-C18:2 PG-C18:2 MGDG-C16:2 DGDG-C16:2 SQDG-C16:2 PG-C16:2	Ferredoxin	Plastid	P46312
FAD7	MGDG-C18:2 DGDG-C18:2 SQDG-C18:2 PG-C18:2 MGDG-C16:2 DGDG-C16:2 SQDG-C16:2 PG-C16:2	MGDG-C18:3 DGDG-C18:3 SQDG-C18:3 PG-C18:3 MGDG-C16:3 DGDG-C16:3 SQDG-C16:3 PG-C16:3	Ferredoxin	Plastid	P46310
FAD8	MGDG-C18:2 DGDG-C18:2 SQDG-C18:2 PG-C18:2 MGDG-C16:2 DGDG-C16:2 SQDG-C16:2 PG-C16:2	MGDG-C18:3 DGDG-C18:3 SQDG-C18:3 PG-C18:3 MGDG-C16:3 DGDG-C16:3 SQDG-C16:3 PG-C16:3	Ferredoxin	Plastid	P48622

Shanklin, 2002) and their activity require ferredoxin as electron donor (Schmidt & Heinz, 1990). Table 1 summarizes desaturases accession numbers, localization and respective cofactors. The ER localized FAD2 and FAD3 are also transmembrane proteins and possess three histidine boxes but require cytochrome B5 as electron donors (Kumar, Tran, Neelakandan, & Nguyen, 2012). The omega-6 and 3 desaturases can act as homodimers or heterodimers associating FAD2/FAD3, FAD6/FAD7 and FAD6/FAD8 complexes while FAD7 and FAD8 were not reported to associate (Lou, Schwender, & Shanklin, 2014).

2.3 PUFAs are substrates for oxylipin synthesis

The PUFAs are molecules sensitive to oxidation either through the action of biological enzymes or by chemical reaction to produce a large variety of lipid derivatives named oxylipins.

Oxylipins are involved in many stress responses ranging from abiotic (freezing) and biotic stresses (pathogens) to mechanical stress (wounding). Among oxylipins, the signaling molecule jasmonic acid (JA) is the most studied (Routaboul, Skidmore, Wallis, & Browse, 2012). Jasmonate synthesis requires first the release of linolenate by lipases. This step is of critical importance and varies between tissues. For example, DAD1 is a phospholipase A1 required for JA synthesis specifically in stamen filaments of *Arabidopsis* (Ishiguro, Kawai-Oda, Ueda, Nishida, & Okada, 2001).

The jasmonate pathway involves first different lipoxygenases (LOX) in plastids that metabolize free trienoic fatty acids. Following their oxidation by LOX, a series of sequential reactions involving the allene oxide synthase (AOS) and the allene oxide cyclase (AOC) in plastids, followed by the 12-OPDA reductase (OPR) in peroxisomes and 3 rounds of beta-oxidation leads eventually to JA (Schaller & Stintzi, 2009). However, trienoic fatty acids are also implied in pathogen stress tolerance independently of JA since they are essential to plant survival even in *aos* mutants defective in JA (Mène-Saffrané et al., 2009). This result would be worth investigating in other mutants like *lox* to check if C18:3 lipoxygenation products are also involved in biotic stress tolerance.

The non-enzymatic and enzymatic oxylipin generation and subsequent acyl chain breakdown also contributes to the synthesis of malondialdehyde (MDA), a major PUFA by-product that is an oxidation marker involved in stress response as well as carbonyl species that have been well summarized recently (Tsikas, 2017; Yalcinkaya, Uzilday, Ozgur, Turkan, & Mano, 2019). MDA seems to be involved in large variety of processes. It was involved in post-translational modification of proteins that are regulated by PUFA (Yamauchi et al., 2008). MDA could also be a major sink for reactive oxygen species (ROS) since they are produced by the oxidation of PUFAs (mostly C18:3) and could be re-incorporated into lipids preferentially in trienoic or dienoic galactolipids by a still unknown mechanism (Schmid-Siegert, Stepushenko, Glauser, & Farmer, 2016). PUFA oxidative catabolism is complex but recent works suggest that many intermediates of these pathways could be involved in several biological processes related to PUFA regulation and functions.

2.4 Endoplasmic reticulum and plastid desaturase crosstalks

Genetic analyses of omega-6 and -3 desaturases have demonstrated the existence of a rather complex metabolic network leading to C18:3 accumulation that depends on the tissue and subcellular localization but also the specificity of the enzymes (Table 2).

In seeds, the loss of FAD2 function (*fad2–1* mutant) led as expected to a 10- and 4-fold reduction of C18:2 and C18:3 contents, respectively, mostly in triacylglycerol. However in leaves, *fad2–1* mutation is characterized by a 70% reduction of C18:2 content (from 14 to 4 mol%) but only a 20% decrease in C18:3 content in total lipids (Lemieux, Miquel, Somerville, & Browse, 1990). The loss of function of the plastidial omega-6 desaturase FAD6 has, however, a more important impact on leaf lipid composition since it resulted in 50% decreased in C18:3 content associated with an important increase in C16:1 and C18:1, but with little effect on C18:2 composition. So, the accumulation of C18:3 is more dependent on plastidial activity of FAD6 than FAD2 in photosynthetic tissue. Interestingly, leaf C18:2 content is not much affected in *fad6* mutant suggesting that ER FAD2 activity could still compensate for the loss of FAD6 activity and produce C18:2 by using plastidial C18:1 as substrate. But FAD2 does not fully compensate for the function of FAD6 as C18:3 content never reached wild type levels in *fad6* mutant (Falcone et al., 1994). The *fad6* mutation did not affect seed and roots C18:3 content in contrast to *fad2–1*, confirming the importance of FAD6 mainly for photosynthetic tissues and the major role of the microsomal desaturation in non-photosynthetic tissues (Browse et al., 1989).

As expected, *fad2 fad6* double mutants have negligible levels of PUFAs in general and C18:3 in particular and are not capable of autotrophic growth, highlighting the importance of PUFAs for photosynthesis (McConn & Browse, 1998). Interestingly, the light harvesting complexes are quantitatively not much altered and the presence of a single functional FAD2 allele is sufficient to recover autotrophy and normal development (McConn & Browse, 1998).

Omega 3 desaturation in seeds and roots depends mostly on FAD3 activity since *fad3* mutation drastically reduces C18:3 levels in those tissues with a two-thirds reduction (29–10 mol%) in roots and a 10-fold reduction in dry seeds (20–2 mol%) compared to leaf tissues (only 3 mol% reduction) (Lemieux et al., 1990). Omega 3 desaturation in plastids is carried out by the two desaturases FAD7 and FAD8. The double mutant *fad7 fad8* could

Table 2 Fatty acid content in desaturases mutants compared to wild type expressed as log2-fold change from published work (reference).

Tissue	Mutant	C16:0	C16:1	C16:3	C18:0	C18:1	C18:2	C18:3	Reference
	fad2-1	0,03	-0,26	0,21		3,18	-1,86	-0,34	Lemieux 1990
	fad6	0,23	1,46	nd	0,15	1,55	0,22	-0,10	Falcone 1994
leaves	*fad3-1*	-0,40	0,37	0,19		-1,06	0,43	-0,11	Lemieux 1990
	fad7	0,14	1,27	-2,42	0,11	1,31	1,21	-1,39	Browse 1986
	fad8-1								McConn 1994
	fad2-1	-0,78	-0,48		-0,91	3,04	-2,23	-1,16	Lemieux 1990
	fad6	-0,01	0,58		0,00	0,17	-0,14	-0,01	Browse 1989
roots	*fad3-1*	-0,52	-0,36		-0,48	0,82	0,88	-1,47	Lemieux 1990
	fad7	-0,04	-0,39		0,08	-0,05	0,02	0,05	Browse 1986
	fad8								McConn 1994
	fad2-1	-0,22			0,78	1,82	-3,32	-1,86	Lemieux 1990
	fad6	0,15			0,00	-0,15	-0,02	-0,58	Browse 1989
dry seeds	*fad3-1*	-0,38			0,06	0,58	0,49	-3,14	Lemieux 1990
	fad7	0,00	-0,32		-0,09	0,02	0,01	-0,02	Browse 1986
	fad8								McConn 1994

therefore only synthesize C18:3 bound to PC thanks to ER FAD3. *FAD3* activity is sufficient to synthesize about one-third (15 mol%) of C18:3 in total leaf lipids (Román et al., 2015). Interestingly, FAD7 and FAD8 are not completely redundant. FAD7 appears to be more active at high temperature since C18:3 content decreased by 65% in leaves of *fad7* mutants grown between 26 °C and 32 °C and decreased only by 25% in mutants grown between 10 °C and 23 °C (Browse, McCourt, & Somerville, 1986). On the contrary, FAD8 showed increased expression upon cold application (Gibson, Arondel, Iba, & Somerville, 1994; McConn, Hugly, Browse, & Somerville, 1994).

Finally, genetic studies highlighted the importance of fatty acid exchange between plastids and ER compartments (McConn & Browse, 1998). This is well illustrated by the absence of phenotype of the single mutants *fad6* or *fad2–1* while the double mutant showed chloroplast defects and chlorosis. Interestingly, the double mutant between *fad6* and *tgd1*, a plastid lipid carrier is also extremely chlorotic like *fad2fad6*. While *tgd1–1* mutation caused a strong decrease in C18:3, the double mutations *tgd1–1fad2* did not further reduce C18:3 levels within plastids confirming that an important part of plastidial C18:3 are from ER origin and that TGD1 is one of the main entry gate for trienoic trafficking (Xu, Moellering, Muthan, Fan, & Benning, 2010).

3. PUFA physiological roles

3.1 PUFA are required for temperature tuning of photosynthesis

The desaturase mutants showed the importance of PUFAs for plant photosynthesis as detailed above but also the importance of dienoic and trienoic C18 fatty acids for temperature adaptation. The critical requirement for PUFA synthesis for sustaining cold temperature is obvious in *Arabidopsis fad2–2* mutants as they fail to germinate if the siliques and seeds develop at 6 °C for more than 10 days. Even seeds of *fad2–2* mutants developed at permissive temperature showed delayed germination at 6 °C (Miquel & Browse, 1994). At low temperatures after 3 weeks of culture at 5 °C, the *fad6* mutant exhibited a two-fold reduction of the chloroplast area associated with slow growth phenotype and chlorotic appearance (Hugly & Somerville, 1992). Similarly, *fad2–2* mutant showed chlorotic phenotype and an impaired growth at 12 °C that could not recover even when re-exposed to normal temperatures. The *fad2–2* plants would even die after a long exposure (50 days) to cold (6 °C). But a residual FAD2 activity is enough for cold adaptation since the *fad2–1* allele which was shown to be a leaky mutation exhibited a wild type phenotype at cold temperatures (Miquel, James, Dooner, & Browse, 1993). These results indicated that for omega-6 desaturation, the plastidial and the ER activities are redundant at normal temperatures but are both necessary for the plant to cope with low temperatures.

For trienoic FAs, FAD7 activity is also critical for plastid function since *fad7* mutation affects chloroplast size and also its structure but with minimal physiological impact because this alteration is compensated by a 45% increase in chloroplast number (Mccourt, Kunst, Browse, & Somerville, 1987). Both effects are directly related to C18:3 synthesis since these phenotypes are complemented by FAD7 ectopic expression (Iba et al., 1993). To our knowledge, no specific role on chloroplast structure and function had been identified for FAD8 but the modified expression of FAD7 and FAD3 genes in *fad8* mutant as well as the specific FA profile of ER and plastid lipids suggested an alternative function. Interestingly, the overexpression of FAD8 in wild type Arabidopsis did not modify plastid lipid composition but caused a significant increase of C18:3 and a mirrored decrease of C18:2 in ER synthesized glycolipids (Román et al., 2015). These results suggest that ER-plastid PUFA trafficking is important for plastidial homeostasis in particular at low temperature. The different desaturases are required to work in a coordinated manner to tune and adapt photosynthesis to environmental changes. In depth analysis of plastid functions in *fad8* and *fad3* mutants and combination of *fad3 fad7* or *fad3 fad8* mutants are still missing to better understand the interplay between these genes and the precise role of the omega-6 and -3 FA in plastid function but also in non photosynthetic tissues like roots. ER-plastid PUFA trafficking is also important for cold adaptation during seed development and germination. Desaturase mutants suggest that seed germination, particularly at cold temperatures, requires minimal PUFA synthesis from the activity of plastidial desaturases even if their activities have most probably a marginal role during late seed development since chloroplasts are undergoing dedifferentiation (Allorent, Courtois, Chevalier, & Lerbs-Mache, 2013). For instance, during germination of *fad2–2* mutants, PUFAs can not be provided by storage lipids that are strongly impacted but their synthesis relied mainly on FAD6, FAD7 and FAD8 activities. Plastidial desaturases use the reducing power of ferredoxins which have high cost on photosynthesis and possibly impact seedling viability. Another explanation could be a direct impairment of photosynthesis in the plastids caused by a low-PUFA membrane content (Allorent et al., 2015).

Mirrored effect could be seen with PUFA accumulation and its effect on temperature adaptation. PUFA accumulation is clearly associated with low temperature in seeds of various oilseed crops (Canvin, 2011; Menard et al., 2017, p. 017). Omega-3 desaturases mutants and overexpressors exhibit only subtle phenotypes and are quite difficult to study. The over-expression of FAD8 and to a lesser extent FAD3, reduced growth of tobacco BY-2 cells at high temperature (Zhang et al., 2005), indicating that omega-3 desaturation levels should be tuned down at elevated temperatures to avoid deleterious effects. Nonetheless, trienoic fatty acids are also required for long-term thermotolerance, since the inactivation of the three desaturases FAD3, FAD7 and FAD8 led to plant death after 15 days at 33 °C. This lethality is not caused by a deficiency in jasmonate or JA response since *aos*, *opr3* and *coi1* mutants are thermotolerant (Routaboul et al., 2012).

3.2 PUFA involvement in abiotic and biotic stress responses

3.2.1 Salt, drought and oxidative stresses

High salt medium is known to induce oxidative stress delaying *fad2* germination. Interestingly, *fad2* mutants grown in presence of high salt showed 30% increase of MDA (Zhang et al., 2012). Similarly, *fad6* mutant is also sensitive to salt stress and exhibited increased MDA levels (Zhang et al., 2009). The increase in MDA production is relatively counterintuitive since lower levels of C18:2 and C18:3 in *fad2* and *fad6* mutants should have been associated with lower lipid peroxidation (Ayala, Muñoz, & Argüelles, 2014). As expected, the suppression of all trienoic fatty acids in the *fad3 fad7 fad8* led to reduced steady-state levels of MDA (Mène-Saffrané, Davoine, Stolz, Majcherczyk, & Farmer, 2007). The link between trienoic fatty acids and oxidative stress was clearly observed in tomato over-expressing LeFAD3 with an increase of ROS detoxification with the activity of superoxide dismutase. Conversely, the antisense expression of LeFAD3 reduced ROS detoxification and resulted in a net decrease in photosynthetic efficiency (Wang et al., 2014). Similarly, the expression in tobacco of anti-sense *AtFAD7* that most probably impaired the expression of both FAD7 and FAD8, enhanced salt and drought sensitivity (Im, Han, Chung, & Cho, 2002).

3.2.2 Metals and heavy metals

Oxidative stress could also be induced by heavy metals leading to the production of lipid hydroxyperoxides and MDA from PUFAs (Skórzyńska-Polit, Drążkiewicz, & Krupa, 2009). Nickel accumulation was shown to cause a drastic reduction of PUFAs in favor of saturated fatty acids in roots and shoots of wheat seedlings (Gajewska, Bernat, Długoński, & Skłodowska, 2012), and in young leaves of tomato seedlings (Verdoni, Mench, Cassagne, & Bessoule, 2001). In *Arabidopsis,* copper application led to a similar decrease of PUFAs particularly in roots (Yuan, He, Huang, Hua, & Sheng, 2016). In *Camelina,* the exposure to Cadmium and Copper particularly reduces the accumulation of C18:3 in dry seeds and the overexpression of the Heavy-Metal ATPase CsHMA3 could partially alleviate this defect reducing the associated oxidative stress (Park, Feng, Kim, Suh, & Ahn, 2015).

Contrary to heavy metals, limitation of iron bioavailability was shown to induce preferential accumulation of PUFAs in whole plants. Even more importantly, it was shown that the addition of C18:2 or C18:3 can increase the activity of the Ca^{2+} ATPases in isolated wild-type plasma membrane vesicles of *Arabidopsis* roots. The rate of this activity was only significantly reduced in *fad2–2* but not in *fad3–2* mutants, suggesting that C18:2 is sufficient to maintain membrane dynamics for Ca^{2+}-dependent ATPases activity (Oh et al., 2016). Both C18:2 and C18:3 seems to be important for optimal ATPases activities but C18:3 content was more specifically modified by cadmium- or copper-induced stress in camelina.

Interestingly, Boron uptake was shown to be reduced in *fad7–2, fad7 fad8* mutants compared to wild type when the plants were exposed to boric acid. However, the chilling sensitive mutant *chs1–1* that exhibits an absence of FAD8 expression (see below) showed an increased Boron uptake (Dordas & Brown, 2000). The link with PUFA content remains to be demonstrated.

3.2.3 Biotic stresses

Linolenic acid is directly involved in biotic stress responses by the fact that it is a direct substrate for jasmonate synthesis. However, several jasmonate independent effects could be associated with C18:3 during pathogen response. C18:3 was found to be important in the tolerance toward *Botrytis cinerea* and the oomycete *Hyaloperonospora* in Arabidopsis. Indeed, the triple *fad3 fad7 fad8* mutants showed increased sensitivity toward pathogens compared with *opr3* mutant that is impaired only in JA synthesis suggesting a JA-independent effect of trienoic acid probably based on the alleviation of oxidative stress (Mène-Saffrané et al., 2009).

The subcellular synthesis of C18:3 is also important for pathogen response The *fad3–2* mutant showed for instance a 45% decrease in crown gall tumor induced by *Agrobacterium tumefaciens* compared to wild type or *fad7 fad8* mutant (Klinkenberg et al., 2014). However, plastidial synthesis of C18:3 is important in *Arabidopsis* to induce Systemic Acquired Resistance (SAR). FAD7 in combination with SUPPRESSOR OF FATTY ACID DEFICIENCY 1 (SFD1 or GLY1) and SFD2 were necessary for the production of a SAR signal in petiole exudate of leaves challenged with an avirulent strain of *Pseudomonas syringae* that promoted subsequent resistance to Arabidopsis, tomato and wheat. The SAR signal is independent from JA and is most probably derived from plastid glycerolipid synthesis since the three proteins FAD7, SFD1 and SFD2 are involved in this metabolic pathway and the SAR effect is dependent on the lipid transfer protein DIR1 (Chaturvedi et al., 2008).

FAD7 was also involved in aphid resistance since antisense inactivation of tomato LeFAD7 reduced aphid infestations in a JA-independent way but with increased SA accumulation. Aphid infestations were also reduced in Arabidopsis *fad7–2* and *fad7–1 fad8* mutants (Avila et al., 2012). It should be noted that the works involving *fad7–1* alleles and pathogen responses have to be analyzed cautiously since it was conducted in the *glabra1* background known to be involved in JA and Gibberellic signaling pathways (Xia et al., 2010).

3.2.4 Concluding remark

The management of ROS could be a common link between trienoic acid and biotic or abiotic stresses since the trienoic fatty acids were proposed to act as a sink for ROS. The production of MDA could act as a non-enzymatic reaction to protect cells from oxidative stress (Mène-Saffrané et al., 2009). While free MDA is a marker of oxidative damage, its accumulation after a long time has deleterious effects for the cells, but a transient accumulation could be the sign of oxidative stress adaptation and detoxification (Morales & Munné-Bosch, 2019). A hypothesis could be the incorporation of MDA back into membrane lipids. Application of labeled MDA to plants showed specific carbon flux into C18:3-MGDG as a terminal product, and in a lesser extent in C18:2-MGDG through an unknown metabolic pathway most probably to detoxify this product (Schmid-Siegert et al., 2016).

4. Regulation of omega-6 and omega-3 desaturases

4.1 Regulation of microsomal desaturases

In Arabidopsis, FAD2 and FAD3 desaturases are responsible for the accumulation of C18:2 and C18:3 in seeds. These fatty acids are constantly accumulated from pollination to full seed maturation without any plateau but C18:3 accumulates at a slower rate and with a delay of 3 days compared to C18:2 (Baud, Boutin, Miquel, Lepiniec, & Rochat, 2002; Rossak, Smith, & Kunst, 2001). C18:2 and C18:3 accumulation is correlated with enhanced expression of FAD2 and FAD3 that, respectively, peaked 5 to 11 and 8 and 11 days after flowering and decreased sharply afterward (Ruuska, Girke, Benning, & Ohlrogge, 2002; Schmid et al., 2005).

Seed C18:3 content is tightly correlated with FAD3 expression levels. The duplication of the FAD3 locus in two Arabidopsis mutants resulted in an increase of FAD3 expression and C18:3 content in seed oil (O'Neill, Baker, Bennett, Clarke, & Bancroft, 2011). The overexpression of Arabidopsis FAD3 with seed specific promoters caused at least a twofold increase in C18:3 levels compared to wild type plants (Puttick, Dauk, Lozinsky, & Smith, 2009; Yadav et al., 1993). The overexpression of FAD3 using the constitutive 35S promoter increased only by 50% the levels of C18:3 in seeds, stressing the importance of transcriptional regulation of FAD3 expression (Hu et al., 2015). Conversely, the *fad3* SALK-042036 mutant which has a T-DNA insertion between its 5′UTR and promoter which impaired its transcriptional activity but not the enzyme activity showed strong reduction of C18:3 content in seed oil (Claessens, 2012.).

4.1.1 Regulation of FAD3 expression during seed development

Unbiased genetic analysis of omega 3 desaturation regulation carried out by Genome wide association study (GWAS) demonstrated that C18:3 content is controlled by several factors but most predominantly by the FAD3 locus (Menard et al., 2017). FAD3 promoter defined at least as 1138 bp upstream of the FAD3 transcription starting site (TSS) was shown to be sufficient to induce a luciferase activity in shoot apical meristems, flowers and developing siliques in Arabidopsis, as well as providing response to auxin, cytokinins and abscisic acid especially in roots (Matsuda, Watanabe, & Iba, 2001). Several known transcription factor binding sites were found in this region and in particular two G-boxes and one CCAAT-box (respectively 174, 400 and 191 bp upstream of FAD3 TSS). This promoter region was also able to be transactivated by several transcription factor complexes including LEC1, LEC-1-Like (L1L), NFYC2 and one of the following bZIP family transcription factors: bZIP67 or AREB3 or EEL or ABI5. The transcription factor bZIP67 is able to bind the G-box1 (−174) *in vitro* and is required with the G-boxes for the transactivation by the complex L1L/NFYC2. However, only a limited effect has been reported on seed lipid composition in the *bzip67* mutant suggesting that other transcription factors are important for FAD3 expression in the seed (Mendes et al., 2013). EEL, AREB3 and bZIP67 are homologous transcription factors coexpressed with FAD3 in embryos that could be also direct regulators (Bensmihen, Giraudat, & Parcy, 2005). Furthermore, it has been shown that bZIP12 could directly bind FAD3 promoters in Camelina (Na, Mu, Grabowski, Schmutz, & Lu, 2019). Data also suggest that the histone acetyltransferase GCN5 could promote FAD3 expression in a direct manner within the 0–200 bp region upstream of the TSS, suggesting a potential transcriptional regulatory hub near FAD3 TATA Box (Hu et al., 2015, p. 5). The fact that *gcn5* mutation has a larger effect on seed lipid profile and in particular on C18:3/C18:2 ratio compared to individual *bzip* mutants like *bzip67*, indicates the existence of redundant or additive transcriptional effects. The histone methyltransferase CLF28 that is involved in the PRC2 complex appears also to be important for FAD3 expression. The *clf28* mutants showed a 2.5-fold increased FAD3 expression in 5-week-old plant siliques followed by a significant decrease in mature green isolated embryos. In consequence the *clf28* mutation resulted in a reduced C18:2 and C18:3 content but an increase in C20:1. The effect of the *clf28* mutation could be linked to small RNA that would then deregulate other master regulators of seed development like FUS3, ABI3, LEC1 and LEC2 (Liu et al., 2016).

The transcription factors LEC1 and L1L, NFYB factors alone cannot *trans*-activate FAD3 expression but are nonetheless necessary to fully activate FAD3. Transcriptomic data during Arabidopsis silique development showed that FAD3 expression is reduced in *lec-1* mutants compared to WT (Yamamoto et al., 2014). The inducible overexpression of LEC1 in Arabidopsis increased FAD3 expression and the C18:3 content. This effect is conditional to the presence of functional ABI3 and FUS3 activity confirming the importance of these three master regulators during seed development (Mu et al., 2008). BnLEC1 and BnL1L overexpression in *Brassica napus* led also to significant increase in FAD3 expression and C18:3 content during seed development (Tan et al., 2011). These data suggest that LEC1 and potentially LEC1-Like are sufficient for inducing FAD3 expression even in ectopic tissues. Therefore, these two transcription factors could be considered as a critical integrator of developmental cues required for the correct expression of FAD3 during seed development.

Other transcription factors from the MYB and/or BhLh family have also been shown to have an effect on FAD3 expression. TT2 and TT8, for instance, are potentially integrating inhibitory signals since *tt2* and *tt8* mutants showed an increased expression of FAD3. These mutations resulted in modified seed fatty acid composition in specific arabidopsis accessions but with decreased C18:3 accumulation, which is counter-intuitive. The effect on FAD3 expression appeared to be indirect in the case of TT8 since the expression of LEC-1 ABI1 FUS3 LEC2 genes (LAFL) was also modified in *tt8*. However, TT2 did not appear to directly impair master regulator expressions and could therefore be a downstream regulator in the FAD3 transcriptional regulatory network (Chen et al., 2012, 2014). MYB76 is also a candidate for PUFA regulation since FAD2 and FAD3 expressions were decreased as well as PUFA levels in *myb76* mutant (Duan et al., 2017).

Interestingly, the overexpression of the transcription factor AmDREB2C from *Ammopiptanthus mongolicus* in *Arabidopsis* increased FAD3 expression and C18:3 levels during seed development. However, the overexpression of other DREB transcription factors resulted in downregulation of FAD3 and FAD8 indicating complex regulatory pattern of the DREB family of transcription factors (Yin et al., 2018).

4.1.2 Regulation of FAD3 expression in response to stresses

Cold stress appears to have a positive effect on FAD3, since FAD3 expression can be transiently induced during cold stress in birch (Martz, Kiviniemi, Palva, & Sutinen, 2006), oilseed rape (Tasseva, de Virville, Cantrel, Moreau, & Zachowski, 2004), Arabidopsis (Kilian et al., 2007), tomato roots (Yu et al., 2009) and in wheat (Golizadeh & Kumleh, 2019). Those reports mostly point toward a rapid and transient upregulation of FAD3 in vegetative tissues. Conversely, heat stress downregulates FAD3 expression in FAD3 expression roots during a 3h exposure to 38°C and subsequent recovery (Kilian et al., 2007). LeFAD3 expression in tomato roots is also markedly reduced upon exposure to high temperature 40°C (Yu et al., 2009).

FAD3 can be transiently induced in cowpea by drought (Torres-Franklin et al., 2009) and transcriptomic analysis showed that hypoxia induces FAD3 as well in *Arabidopsis* (Klinkenberg et al., 2014). Other transcriptomic reports showed that FAD3 can be downregulated during drought in *Arabidopsis* (de Simone et al., 2017; Perera, Hung, Moore, Stevenson-Paulik, & Boss, 2008).

The application of DMTU, a ROS scavenger molecule, in guard cell suspension cultures caused a twofold increase in *FAD3* expression, an effect that could be blocked by the addition of ABA with DMTU. These results, if confirmed, could be a first step toward a mechanistic link between C18:3, FAD3 expression and ROS-mediated stresses discussed above (Böhmer & Schroeder, 2011).

Very few reports described an effect of soil metals on FAD3 expression which is surprising considering their effect on PUFA levels (see above). FAD3 expression was nonetheless reported to be induced with FAD2 and CBR1 in seedlings after application of the iron chelatant ferrozine (Oh et al., 2016).

Similarly, to metals, FAD3 expression was modified by biotic stresses in crown gall infection where its expression correlated with tumor development (Klinkenberg et al., 2014).

4.1.3 FAD3 post-transcriptional regulation

While transcriptional regulation of FAD3 was clearly associated with tissue specific expression and with several abiotic and biotic stresses, post-transcriptional regulation of FAD3 was mostly associated with temperature control. The *B. napus* FAD3 protein could be degraded faster at 30 °C than at 20 °C by the ER associated degradation pathway while tung tree FAD3 desaturase was more stable (O'Quin et al., 2010). It has later been confirmed that the N-termini of both FAD3 proteins could induce proteasome mediated degradation, potentially coupling transcription with protein degradation (Khuu, Gidda, Shockey, Dyer, & Mullen, 2011). Surprisingly, BnFAD3 expressed in yeast showed lower enzymatic activity at cold temperatures suggesting the existence of plant specific factors stabilizing FAD3 at low temperature (Dyer, Chapital, Cary, & Pepperman, 2001). In wheat roots, accumulation of C18:3 in the cold was linked to higher levels of FAD3 protein compared to RNA (Horiguchi, Kodama, & Iba, 2001). Wheat 5′ untranslated region FAD3 was found to slightly increase the translation efficiency of *Arabidopsis* FAD3 during cold stress (Wang & Xu, 2010) but that effect was not sufficient to explain the accumulation of FAD3 protein in cold-treated wheat.

Another mechanism enhancing desaturase activity is metabolite channeling. FAD2 and FAD3 are able to interact and to form stable heterodimer that enable sequential desaturation reactions from C18:1 to C18:3 with limited release of C18:2 (Lou et al., 2014). The positive effect of this regulation on C18:3 production could be further modulated by ALA10, a phospholipid flippase that also physically interacts with both microsomal desaturases enhancing the availability of the C18:1 or C18:2 substrates from the plastid lipids (Botella et al., 2016).

Finally, FAD3 activity is also involved in a post-translational regulatory loop associating membrane fluidity and stress response since *fad3* mutation triggers the proteolytic cleavage of NTL6 (Seo et al., 2010), a transcription factor involved in drought response. Processed NTL6 moves in the nucleus where it regulates ABA and ER stress related genes (Yang et al., 2014). The NTL6 proteolytic cleavage could be directly inhibited by C18:3 application or activated by cold exposure. Cold activation was enhanced in *fad3* background but not in the triple *fad3 fad7 fad8* mutation suggesting that only ER synthesis of C18:3 are involved in this process (Seo et al., 2010). Similarly, the addition of C18:3 can also induce the nuclear localization of the transcription RAP2.12 linked to hypoxia tolerance (Zhou et al., 2019).

4.2 Regulation of plastidial PUFA desaturases

The *Arabidopsis FAD7* promoter expressed in tobacco showed that the −362/−166 region containing two putative boxes II could induce GUS expression after light exposition in leaves (Nishiuchi et al., 1995). Similar induction was also found in *Arabidopsis* cotyledon (Soria-García et al., 2019). Furthermore, the FAD7 promoter is also responsive to wounding in leaves, stems and to a lesser extent in roots of tobacco transgenic plants, and that effect could be reduced by the application of JA synthesis inhibitors or salicylic acid (Nishiuchi, Hamada, Kodama, & Iba, 1997). The wounding response was shown to be dependent on JA-biosynthesis genes and SCF^{COI1}-mediated signaling and was also enhanced by exogenous Me-JA applications. FAD7 promoter is able to integrate both JA but also JA unrelated signaling since wounding response can also be induced in *aos* mutants (Matsuda, Sakamoto, Nakao, Oda, & Iba, 2009). FAD7 promoter wounding response was confirmed in Arabidopsis transgenic plants and compared with FAD8 promoter (Soria-García et al., 2019). FAD8 promoter induced steady-state GUS reporter expression at lower levels than FAD7 in cotyledons but did not respond to wounding. FAD8 promoter activity in root tips was also undetectable. Interestingly, GFP translational fusion showed that FAD8 proteins are more accumulated than FAD7 after JA application suggesting that JA regulation of FAD8 promoter is not identical as for FAD7 promoter (Soria-García et al., 2019). Arabidopsis FAD7 is expressed constitutively in leaves while FAD8 is essentially cold inducible. The correlations between the relative amount of FAD7 and FAD8 protein and transcript levels suggests that these genes are essentially regulated at the transcriptional level (Soria-García et al., 2019). Table 3 summarizes the different studies describing transcriptional regulation of the plastidial desaturases.

Altogether these studies confirmed the involvement of transcriptional regulation of plastid desaturase expression by abiotic stresses to fine tune PUFA accumulation in vegetative tissues. However, the transcriptional factors involved in plastidial desaturases regulation are still not known, with the exception of a potential role of AmDREB2C which upregulated FAD7 expression in Arabidopsis. However, the existence of an Arabidopsis ortholog is not known (Yin et al., 2018). It can be noted that several chilling-sensitive *chs* mutants did show altered expression of omega-3 desaturases upon cold application. In particular, the cold inducible desaturase FAD8, was not induced in *chs1, chs2* and *chs3* mutants potentially contributing to the lethality phenotype (Provart et al., 2003).

Post-translational regulation was also observed in plastidial PUFA desaturases. Similar to FAD2/FAD3, FAD6 was found to interact *in vivo*

Table 3 Reports of differential regulation of FAD7 and FAD8 in response to stress related signals.

Organism	Gene	Stress	Regulation	References
Olive tree	FAD7	Cold	Up	Hashempour, Ghasemnezhad, Sohani, Ghazvini, and Abedi (2019)
Soybean	FAD7–1	Cold	Up	Román et al. (2012)
Arabidopsis	FAD7	Cold	Down	Román et al. (2015)
Birch	FAD7	Cold	Down	Martz et al. (2006)
Maize	FAD7	Cold	Up	Berberich et al. (1998)
Arabidopsis	FAD8	Cold	Up	Román et al. (2015)
Birch	FAD8	Cold	Up	Martz et al. (2006)
Rice	FAD8	Cold	Up	Wang et al. (2006)
Cowpea	FAD8	Drought	Down	Torres-Franklin et al. (2009)

with FAD7 or FAD8 resulting in trienoic fatty acids channeling (Lou et al., 2014; Soria-García et al., 2019). In addition, temperature-dependent control of plastidial desaturase activity occurs also at the post-transcriptional level since the C-terminal of Arabidopsis *FAD8* was shown to reduce protein stability at 27 °C which in combination with reduced FAD8 transcriptional activity led to the decrease of trienoic FAs at high temperatures (Matsuda, Sakamoto, Hashimoto, & Iba, 2005).

5. Issues and strategies for breeding high omega-3 crops

5.1 High omega 3 oilseed crops

High omega-3 oilseed crops are bred mainly for the high nutritional value of linolenate for food and feed uses but also for its physical properties (fluidity and oxidation) that is of interest for the lubricant, solvent and deperlant industry (Iii et al., 2009).

The most popular omega 3 crop is linseed or Flax (*Linum usitatissimum)* for its high linolenate content and generally high yields. For instance, M5791 is a recent flax cultivar containing the highest fraction of linolenate in seeds, reaching almost 70 mol% in field conditions. Higher expression of FAD2A, FAD3A and FAD3B microsomal desaturases in this cultivar was proposed to explain the high omega-3 content in seed oil (Thambugala & Cloutier, 2014). Further breeding enabled Agriculture and Agri Food Canada (AAFC) to generate plants with ALA seed oil contents greater than 70 mol% of total FAs (Kenaschuk, 2010). Conversely, the very low linolenic acid flax cultivar (1.3 mol%) Solin, showed stop mutations in both FAD3A and FAD3B homeologs confirming that the microsomal FAD3 are the main source of seed linolenate in this crop (Vrinten, Hu, Munchinsky, Rowland, & Qiu, 2005).

Chia (*Salvia hispanica)* is an oilseed crop from South America that belongs to the Lamiacea family. This crop is characterized by a high linolenate content in seed oil (up to 69%) most probably because its two FAD3 homeologs are particularly highly expressed during early seed development (Xue et al., 2018). Chia shows seed yields that can reach up to 2500 kg/ha in Chile (Baginsky et al., 2016) but with a high variability. Chia was introduced in Europe with seed yield peaking around 1200 kg/Ha (Grimes, Phillips, Hahn, Capezzone, & Graeff-Hönninger, 2018). Another interesting feature of Chia is that its major triglyceride is trilinolenin where all three FA positions are filled by C18:3 (Ixtaina et al., 2011). Chia is therefore able to transfer efficiently C18:3 acyl moieties to sn1 or 2 positions of TAG with specific acyltransferases that remain to be characterized.

Perilla (*Perilla frutescens*) is another omega-3 rich oilseed crop also belonging to the lamiaceae family and originating from South East Asia. Perilla seed has a high oil content (more than 40%) in which ALA can represent more than 60% of total fatty acids (Lee et al., 2016). Recent varieties can reach grain yield over 2200 kg/ha in China (Qi, Sen, Ping, & JiXian, 2019). Perilla FAD3 expression was found to be 192-fold higher in developing seeds than in other tissues, and correlated with the highest 18:3 contents of perilla seed fatty acids confirming the importance of the expression of this desaturase during seed development (Lee et al., 2016).

Finally, camelina (*Camelina sativa)* albeit with lower linolenate seed content (30%) turnt out to be a valuable omega-3 crop. As a Brassicacea, this ancient European crop is close to the model organism *Arabidopsis thaliana* and the widely cultivated *B. napus*, which has a long history of breeding. *C. sativa* is also very easy to transform by flower dip and its genome is sequenced in many accessions and cultivars (Kagale et al., 2014; Li et al., 2021; Luo et al., 2019). Camelina is a polyploid species with three closely related sub-genomes coding three FAD3 homeologs (Kang, Snapp, & Lu, 2011).

5.2 Crop improvement through gene engineering of omega-3 desaturases

5.2.1 Seed oil tailoring

Oilseed crop breeding focused primarily in modifying seed oil profile and yield. The discovery of desaturase genes immediately provided tools to change monounsaturated *vs* polyunsaturated profiles in seed oils. Mutations in FAD2 genes resulted in plants producing seed oil essentially devoided of polyunsaturated fatty acids as described above The FAD2 gene was found in many plant species but its loss of function seemed to have varied effects. While Arabidopsis *fad2* mutants showed mainly reduced tolerance to several stresses (Zhang et al., 2012), camelina *fad2* mutants showed stunted growth and altered development (Lee et al., 2021; Morineau et al., 2017). The different engineering strategies involving FAD2 has been covered recently in detail (Dar, Choudhury, Kancharla, & Arumugam, 2017).

FAD3 engineering was more specifically used to modify omega-3 content in seed oil. For instance, overexpression of FAD3 using a constitutive promoter enables the accumulation of omega-3 fatty acids in rice seeds that normally accumulate almost no ALA but mainly C18:1 and C18:2. ALA reached about 30% of total fatty acids (Anai et al., 2003). The overexpression of FAD3 with seed specific promoter caused a dramatic increase in C18:3 content in whole seeds of plants accumulating C18:1 and C18:2 causing severe depletion of these fatty acids, suggesting a significant pull effect. Furthermore, the association of microalgae elongases with desaturases, including FAD2 and FAD3 in order to cause a metabolic push effect in *C. sativa* and *B. napus* allowed the production of very long chain polyunsaturated fatty acids: EPA and DHA (Petrie et al., 2020; Usher et al., 2017). Similarly, soybean could be engineered to produce stearidonic acid that can efficiently increase EPA accumulation in blood cells (Harris et al., 2008). These fatty acids have significant health benefits and can be immediately metabolized by mammals (Schuchardt & Hahn, 2013). Specific desaturases were also used to improve ALA content in crops. The identification of a bifunctional desaturase from the three fungi *Fusarium moniliforme, Fusarium graminearum*, and *Magnaporthe grisea* that can act on both C18:1 to produce C18:2 and on C18:2 to produce C18:3 led to the possibility of generating soybean producing high linolenic plants that could produce 71 mol% of total fatty acid of ALA in dry seeds, while decreasing oleic acid (Damude et al., 2006). Two clusters of amino acids were shown to be responsible for the combined delta 12 and 15 activities, but the latter activity was clearly dependent on the former in *F. moniliforme* (Cai, Yu, Liu, Liu, & Shanklin, 2018). Furthermore, a recent field trial of T1 soybean plants overexpressing an isoform of FAD3 under the control of a seed specific promoter improved ALA content in seed along with plant growth and pod number as well as grain yield (Yeom et al., 2020). Even if more field trials in different environmental conditions are still needed, these results suggest that seed specific modification of PUFA desaturase activity seem to affect plant fitness.

5.2.2 Stress tolerance

It is now admitted that the overexpression of omega-3 desaturases genes improves cold resistance and tolerance, while loss-of-function mutations increases heat tolerance. For example, the overexpression of LeFAD3 improved chilling tolerance in tomato (Yu et al., 2009) but also improve salt stress tolerance in seedlings (Wang et al., 2014). LeFAD7 overexpression led to an increased photosynthetic activity in tomatoes exposed to 4 °C for 6 h providing a possible explanation for cold tolerance (Liu, Teng, Li, & Meng, 2013). The combined expression of FAD3 and FAD7 further enhanced plant cold tolerance and doubles the content in omega-3 FAs in ripe tomatoes compared to their single overexpression and as a consequence led to an important acumulation of cis-3-hexenal at the expense of hexanal, produced by the lipoxygenase activity and endoperoxydation of C18:3 (Domínguez et al., 2010). Conversely, the depletion of FAD3 expression enhanced heat tolerance (Wang et al., 2010) while FAD7 loss of function in tomato improved photosynthetic activity (Wickramanayake, Goss, Zou, & Goggin, 2020).

In tobacco, FAD3 and FAD8 overexpression improved drought tolerance and recovery (Zhang et al., 2005) while antisense expression had the opposite effect (Im et al., 2002). A more recent work showed that the overexpression of the cryophite *Chorispora bungeana CbFAD3* in tobacco improved resistance to drought, salt and cold stresses (Shi, Yue, & An, 2018). The expression of *CbFAD3* reduced electrolyte leakage and MDA formation while increasing photosynthetic yield in these different stress conditions (Shi et al., 2018). More recently, it has been shown in cotton that the overexpression of a *B. napus* FAD3 also accelerated seed germination in suboptimal condition while slightly improving photosynthesis and fiber resistance.(Gao et al., 2020).

In conclusion, omega-3 desaturases appear to represent an interesting leverage to modify agronomic traits including seed oil quality as well as resistance to various stresses. However, the pleiotropic nature of PUFAs in general and omega-3 in particular limits their biotechnological applications. Editing tissue-specific isoforms or the use of tissue-specific promoters could circumvent these limitations by restricting PUFA modification to seed development for instance.

References

Allorent, G., Courtois, F., Chevalier, F., & Lerbs-Mache, S. (2013). Plastid gene expression during chloroplast differentiation and dedifferentiation into non-photosynthetic plastids during seed formation. *Plant Molecular Biology*, *82*(1–2), 59–70. https://doi.org/10.1007/s11103-013-0037-0.

Allorent, G., Osorio, S., Vu, J. L., Falconet, D., Jouhet, J., Kuntz, M., et al. (2015). Adjustments of embryonic photosynthetic activity modulate seed fitness in Arabidopsis thaliana. *New Phytologist*, *205*(2), 707–719. https://doi.org/10.1111/nph.13044.

Anai, T., Koga, M., Tanaka, H., Kinoshita, T., Rahman, S. M., & Takagi, Y. (2003). Improvement of rice (Oryza sativa L.) seed oil quality through introduction of a soybean microsomal omega-3 fatty acid desaturase gene. *Plant Cell Reports*, *21*(10), 988–992. https://doi.org/10.1007/s00299-003-0609-6.

Avila, C. A., Arévalo-Soliz, L. M., Jia, L., Navarre, D. A., Chen, Z., Howe, G. A., et al. (2012). Loss of function of FATTY ACID DESATURASE7 in tomato enhances basal aphid resistance in a salicylate-dependent manner. *Plant Physiology*, *158*(4), 2028–2041. https://doi.org/10.1104/pp.111.191262.

Ayala, A., Muñoz, M. F., & Argüelles, S. (2014). Lipid peroxidation: Production, metabolism, and signaling mechanisms of malondialdehyde and 4-Hydroxy-2-Nonenal. *Oxidative Medicine and Cellular Longevity, 2014*, 31. https://doi.org/10.1155/2014/360438. Article ID 360438.

Baginsky, C., Arenas, J., Escobar, H., Garrido, M., Valero, N., Tello, D., et al. (2016). Growth and yield of chia (Salvia hispanica L.) in the Mediterranean and desert climates of Chile. *Chilean Journal of Agricultural Research, 76*(3), 255–264. https://doi.org/10.4067/S0718-58392016000300001.

Bai, Y., McCoy, J. G., Levin, E. J., Sobrado, P., Rajashankar, K. R., Fox, B. G., et al. (2015). X-ray structure of a mammalian stearoyl-CoA desaturase. *Nature, 524*(7564), 252–256. https://doi.org/10.1038/nature14549.

Baud, S., Boutin, J.-P., Miquel, M., Lepiniec, L., & Rochat, C. (2002). An integrated overview of seed development in Arabidopsis thaliana ecotype WS. *Plant Physiology and Biochemistry, 40*(2), 151–160. https://doi.org/10.1016/S0981-9428(01)01350-X.

Baud, S., Guyon, V., Kronenberger, J., Wuillème, S., Miquel, M., Caboche, M., et al. (2003). Multifunctional acetyl-CoA carboxylase 1 is essential for very long chain fatty acid elongation and embryo development in Arabidopsis. *The Plant Journal, 33*(1), 75–86. https://doi.org/10.1046/j.1365-313X.2003.016010.x.

Bensmihen, S., Giraudat, J., & Parcy, F. (2005). Characterization of three homologous basic leucine zipper transcription factors (bZIP) of the ABI5 family during Arabidopsis thaliana embryo maturation. *Journal of Experimental Botany, 56*(412), 597–603. https://doi.org/10.1093/jxb/eri050.

Berberich, T., Harada, M., Sugawara, K., Kodama, H., Iba, K., & Kusano, T. (1998). Two maize genes encoding omega-3 fatty acid desaturase and their differential expression to temperature. *Plant Molecular Biology, 36*(2), 297–306.

Böhmer, M., & Schroeder, J. I. (2011). Quantitative transcriptomic analysis of abscisic acid-induced and reactive oxygen species-dependent expression changes and proteomic profiling in Arabidopsis suspension cells. *The Plant Journal, 67*(1), 105–118. https://doi.org/10.1111/j.1365-313X.2011.04579.x.

Botella, C., Sautron, E., Boudiere, L., Michaud, M., Dubots, E., Yamaryo-Botté, Y., et al. (2016). ALA10, a phospholipid Flippase, controls FAD2/FAD3 desaturation of phosphatidylcholine in the ER and affects chloroplast lipid composition in Arabidopsis thaliana. *Plant Physiology, 170*(3), 1300–1314. https://doi.org/10.1104/pp.15.01557.

Broadwater, J. A., Whittle, E., & Shanklin, J. (2002). Desaturation and hydroxylation: Residues 148 and 324 of arabidopsis fad2, in addition to substrate chain length, exert a major influence in partitioning of catalytic specificity*. *Journal of Biological Chemistry, 277*(18), 15613–15620. https://doi.org/10.1074/jbc.M200231200.

Browse, J., Kunst, L., Anderson, S., Hugly, S., & Somerville, C. (1989). A mutant of Arabidopsis deficient in the chloroplast 16:1/18:1 desaturase 1. *Plant Physiology, 90*(2), 522–529.

Browse, J., McCourt, P., & Somerville, C. (1986). A mutant of Arabidopsis deficient in C18:3 and C16:3 leaf lipids 1. *Plant Physiology, 81*(3), 859–864.

Cai, Y., Yu, X.-H., Liu, Q., Liu, C.-J., & Shanklin, J. (2018). Two clusters of residues contribute to the activity and substrate specificity of Fm1, a bifunctional oleate and linoleate desaturase of fungal origin. *The Journal of Biological Chemistry, 293*(51), 19844–19853. https://doi.org/10.1074/jbc.RA118.005972.

Canvin, D. T. (2011). The effect of temperature on the oil content and fatty acid composition of the oils from several oil seed crops. *Canadian Journal of Botany, 43*(1), 63–69. https://doi.org/10.1139/b65-008.

Chaturvedi, R., Krothapalli, K., Makandar, R., Nandi, A., Sparks, A. A., Roth, M. R., et al. (2008). Plastid ω3-fatty acid desaturase-dependent accumulation of a systemic acquired resistance inducing activity in petiole exudates of Arabidopsis thaliana is independent of jasmonic acid. *The Plant Journal*, *54*(1), 106–117. https://doi.org/10.1111/j.1365-313X.2007.03400.x.

Chen, M., Wang, Z., Zhu, Y., Li, Z., Hussain, N., Xuan, L., et al. (2012). The effect of TRANSPARENT TESTA2 on seed fatty acid biosynthesis and tolerance to environmental stresses during young seedling establishment in Arabidopsis. *Plant Physiology*, *160*(2), 1023–1036. https://doi.org/10.1104/pp.112.202945.

Chen, M., Xuan, L., Wang, Z., Zhou, L., Li, Z., Du, X., et al. (2014). TRANSPARENT TESTA8 inhibits seed fatty acid accumulation by targeting several seed development regulators in Arabidopsis. *Plant Physiology*, *165*(2), 905–916. https://doi.org/10.1104/pp.114.235507.

Claessens, S. M. C. (2012). *Dormancy cycling in seeds: Mechanisms and regulation*. Wageningen University and Research.

Clifton, P. M., & Keogh, J. B. (2017). A systematic review of the effect of dietary saturated and polyunsaturated fat on heart disease. *Nutrition, Metabolism and Cardiovascular Diseases*, *27*(12), 1060–1080. https://doi.org/10.1016/j.numecd.2017.10.010.

Damude, H. G., Zhang, H., Farrall, L., Ripp, K. G., Tomb, J.-F., Hollerbach, D., et al. (2006). Identification of bifunctional Δ12/ω3 fatty acid desaturases for improving the ratio of ω3 to ω6 fatty acids in microbes and plants. *Proceedings of the National Academy of Sciences*, *103*(25), 9446–9451. https://doi.org/10.1073/pnas.0511079103.

Dar, A. A., Choudhury, A. R., Kancharla, P. K., & Arumugam, N. (2017). The FAD2 gene in plants: Occurrence, regulation, and role. *Frontiers in Plant Science*, *8*, 1789. https://doi.org/10.3389/fpls.2017.01789.

de Simone, A., Hubbard, R., de la Torre, N. V., Velappan, Y., Wilson, M., Considine, M. J., et al. (2017). Redox changes during the cell cycle in the embryonic root meristem of *Arabidopsis thaliana*. *Antioxidants & Redox Signaling*, *27*(18), 1505–1519. https://doi.org/10.1089/ars.2016.6959.

Domínguez, T., Hernández, M. L., Pennycooke, J. C., Jiménez, P., Martínez-Rivas, J. M., Sanz, C., et al. (2010). Increasing ω-3 desaturase expression in tomato results in altered aroma profile and enhanced resistance to cold stress. *Plant Physiology*, *153*(2), 655–665. https://doi.org/10.1104/pp.110.154815.

Dordas, C., & Brown, P. H. (2000). Permeability of boric acid across lipid bilayers and factors affecting it. *The Journal of Membrane Biology*, *175*(2), 95–105. https://doi.org/10.1007/s002320001058.

Duan, S., Jin, C., Li, D., Gao, C., Qi, S., Liu, K., et al. (2017). MYB76 inhibits seed fatty acid accumulation in Arabidopsis. *Frontiers in Plant Science*, *8*. https://doi.org/10.3389/fpls.2017.00226.

Dyall, S. C. (2015). Long-chain omega-3 fatty acids and the brain: A review of the independent and shared effects of EPA, DPA and DHA. *Frontiers in Aging Neuroscience*, 7, 52. https://doi.org/10.3389/fnagi.2015.00052.

Dyer, J. M., Chapital, D. C., Cary, J. W., & Pepperman, A. B. (2001). Chilling-sensitive, post-transcriptional regulation of a plant fatty acid desaturase expressed in yeast. *Biochemical and Biophysical Research Communications*, *282*(4), 1019–1025. https://doi.org/10.1006/bbrc.2001.4667.

Falcone, D. L., Gibson, S., Lemieux, B., & Somerville, C. (1994). Identification of a gene that complements an Arabidopsis mutant deficient in chloroplast [omega]6 desaturase activity. *Plant Physiology*, *106*(4), 1453–1459. https://doi.org/10.1104/pp.106.4.1453.

Ferro, M., Salvi, D., Brugière, S., Miras, S., Kowalski, S., Louwagie, M., et al. (2003). Proteomics of the chloroplast envelope membranes from Arabidopsis thaliana. *Molecular & Cellular Proteomics: MCP*, *2*(5), 325–345. https://doi.org/10.1074/mcp.M300030-MCP200.

Gajewska, E., Bernat, P., Długoński, J., & Skłodowska, M. (2012). Effect of nickel on membrane integrity, lipid peroxidation and fatty acid composition in wheat seedlings: Ni-induced alteration of fatty acid composition. *Journal of Agronomy and Crop Science, 198*(4), 286–294. https://doi.org/10.1111/j.1439-037X.2012.00514.x.

Gao, L., Chen, W., Xu, X., Zhang, J., Singh, T. K., Liu, S., et al. (2020). Engineering Trienoic fatty acids into cottonseed oil improves low-temperature seed germination, plant photosynthesis and cotton fiber quality. *Plant and Cell Physiology, 61*(7), 1335–1347. https://doi.org/10.1093/pcp/pcaa062.

Gibson, S., Arondel, V., Iba, K., & Somerville, C. (1994). Cloning of a temperature-regulated gene encoding a chloroplast omega-3 desaturase from Arabidopsis thaliana. *Plant Physiology, 106*(4), 1615–1621.

Glencross, B. D. (2009). Exploring the nutritional demand for essential fatty acids by aquaculture species. *Reviews in Aquaculture, 1*(2), 71–124. https://doi.org/10.1111/j.1753-5131.2009.01006.x.

Golizadeh, F., & Kumleh, H. H. (2019). Physiological responses and expression changes of fatty acid metabolism–related genes in wheat (Triticum aestivum) under cold stress. *Plant Molecular Biology Reporter, 37*(3), 224–236. https://doi.org/10.1007/s11105-019-01150-9.

Grimes, S. J., Phillips, T. D., Hahn, V., Capezzone, F., & Graeff-Hönninger, S. (2018). Growth, yield performance and quality parameters of three early flowering chia (Salvia hispanica L.) genotypes cultivated in southwestern Germany. *Agriculture, 8*(10), 154. https://doi.org/10.3390/agriculture8100154.

Harris, W. S., Lemke, S. L., Hansen, S. N., Goldstein, D. A., DiRienzo, M. A., Su, H., et al. (2008). Stearidonic acid-enriched soybean oil increased the omega-3 index, an emerging cardiovascular risk marker. *Lipids, 43*(9), 805–811. https://doi.org/10.1007/s11745-008-3215-0.

Harwood, J. L. (1980). 1 - Plant acyl lipids: Structure, distribution, and analysis. In P. K. Stumpf (Éd.), Lipids: Structure and function (pp. 1–55). Academic Press. https://doi.org/10.1016/B978-0-12-675404-9.50007-2.

Hashempour, A., Ghasemnezhad, M., Sohani, M. M., Ghazvini, R. F., & Abedi, A. (2019). Effects of freezing stress on the expression of fatty acid desaturase (FAD2, FAD6 and FAD7) and Beta-glucosidase (BGLC) genes in tolerant and sensitive olive cultivars. *Russian Journal of Plant Physiology, 66*(2), 214–222. https://doi.org/10.1134/S1021443719020079.

Horiguchi, G., Kodama, H., & Iba, K. (2001). Characterization of Arabidopsis mutants that are associated with altered C18 unsaturated fatty acid metabolism. *Plant Science, 161*(6), 1117–1123. https://doi.org/10.1016/S0168-9452(01)00523-4.

Hu, Z., Song, N., Zheng, M., Liu, X., Liu, Z., Xing, J., et al. (2015). Histone acetyltransferase GCN5 is essential for heat stress-responsive gene activation and thermotolerance in Arabidopsis. *The Plant Journal: For Cell and Molecular Biology, 84*(6), 1178–1191. https://doi.org/10.1111/tpj.13076.

Hugly, S., & Somerville, C. (1992). A role for membrane lipid Polyunsaturation in chloroplast biogenesis at low temperature. *Plant Physiology, 99*(1), 197–202.

Iba, K., Gibson, S., Nishiuchi, T., Fuse, T., Nishimura, M., Arondel, V., et al. (1993). A gene encoding a chloroplast omega-3 fatty acid desaturase complements alterations in fatty acid desaturation and chloroplast copy number of the fad7 mutant of Arabidopsis thaliana. *Journal of Biological Chemistry, 268*(32), 24099–24105. https://doi.org/10.1016/S0021-9258(20)80498-0.

Iii, C. H., Fitzpatrick, K. C., & Kamal-Eldin, A. (2009). Flax, Perilla, and Camelina seed oils: α-linolenic acid-rich oils. In *Gourmet and health-promoting specialty oils* (pp. 151–183). Elsevier. https://doi.org/10.1016/B978-1-893997-97-4.50010-3.

Im, Y. J., Han, O., Chung, G. C., & Cho, B. H. (2002). Antisense expression of an Arabidopsis omega-3 fatty acid desaturase gene reduces salt/drought tolerance in transgenic tobacco plants. *Molecules and Cells*, *13*(2), 264–271.

Ishiguro, S., Kawai-Oda, A., Ueda, J., Nishida, I., & Okada, K. (2001). The defective in anther dehiscience gene encodes a novel phospholipase A1 catalyzing the initial step of jasmonic acid biosynthesis, which synchronizes pollen maturation, anther dehiscence, and flower opening in Arabidopsis. *The Plant Cell*, *13*(10), 2191–2209.

Ixtaina, V. Y., Martínez, M. L., Spotorno, V., Mateo, C. M., Maestri, D. M., Diehl, B. W. K., et al. (2011). Characterization of chia seed oils obtained by pressing and solvent extraction. *Journal of Food Composition and Analysis*, *24*(2), 166–174. https://doi.org/10.1016/j.jfca.2010.08.006.

Kagale, S., Koh, C., Nixon, J., Bollina, V., Clarke, W. E., Tuteja, R., et al. (2014). The emerging biofuel crop Camelina sativa retains a highly undifferentiated hexaploid genome structure. *Nature Communications*, *5*, 3706. https://doi.org/10.1038/ncomms4706.

Kang, J., Snapp, A. R., & Lu, C. (2011). Identification of three genes encoding microsomal oleate desaturases (FAD2) from the oilseed crop Camelina sativa. *Plant Physiology and Biochemistry*, *49*(2), 223–229. https://doi.org/10.1016/j.plaphy.2010.12.004.

Kenaschuk, E. O. (2010). *High linolenic acid flax*. United States Patent No US7763303B2 https://patents.google.com/patent/US7763303B2/en.

Khuu, N., Gidda, S., Shockey, J. M., Dyer, J. M., & Mullen, R. T. (2011). The N termini of brassica and tung omega-3 fatty acid desaturases mediate proteasome-dependent protein degradation in plant cells. *Plant Signaling & Behavior*, *6*(3), 422–425.

Kilian, J., Whitehead, D., Horak, J., Wanke, D., Weinl, S., Batistic, O., et al. (2007). The AtGenExpress global stress expression data set : Protocols, evaluation and model data analysis of UV-B light, drought and cold stress responses. *The Plant Journal: For Cell and Molecular Biology*, *50*(2), 347–363. https://doi.org/10.1111/j.1365-313X.2007.03052.x.

Klinkenberg, J., Faist, H., Saupe, S., Lambertz, S., Krischke, M., Stingl, N., et al. (2014). Two fatty acid desaturases, stearoyl-acyl carrier protein Δ9-desaturase6 and fatty acid desaturase3, are involved in drought and hypoxia stress signaling in Arabidopsis crown galls. *Plant Physiology*, *164*(2), 570–583. https://doi.org/10.1104/pp.113.230326.

Kumar, R., Tran, L.-S. P., Neelakandan, A. K., & Nguyen, H. T. (2012). Higher plant cytochrome b5 polypeptides modulate fatty acid desaturation. *PLoS One*, *7*(2), e31370. https://doi.org/10.1371/journal.pone.0031370.

Lee, K.-R., Jeon, I., Yu, H., Kim, S.-G., Kim, H.-S., Ahn, S.-J., et al. (2021). Increasing monounsaturated fatty acid contents in Hexaploid Camelina sativa seed oil by FAD2 gene knockout using CRISPR-Cas9. *Frontiers in Plant Science*, *12*, 1279. https://doi.org/10.3389/fpls.2021.702930.

Lee, K.-R., Lee, Y., Kim, E.-H., Lee, S.-B., Roh, K. H., Kim, J.-B., et al. (2016). Functional identification of oleate 12-desaturase and ω-3 fatty acid desaturase genes from Perilla frutescens var Frutescens. *Plant Cell Reports*, *35*(12), 2523–2537. https://doi.org/10.1007/s00299-016-2053-4.

Lemieux, B., Miquel, M., Somerville, C., & Browse, J. (1990). Mutants of Arabidopsis with alterations in seed lipid fatty acid composition. *Theoretical and Applied Genetics*, *80*(2), 234–240. https://doi.org/10.1007/BF00224392.

Li, H., Hu, X., Lovell, J. T., Grabowski, P. P., Mamidi, S., Chen, C., et al. (2021). Genetic dissection of natural variation in oilseed traits of camelina by whole-genome resequencing and QTL mapping. *The Plant Genome*, *14*(2), e20110. https://doi.org/10.1002/tpg2.20110.

Li-Beisson, Y., Shorrosh, B., Beisson, F., Andersson, M. X., Arondel, V., Bates, P. D., et al. (2013). Acyl-lipid metabolism. *The Arabidopsis Book*, *11*, e0161. https://doi.org/10.1199/tab.0161.

Liu, J., Deng, S., Wang, H., Ye, J., Wu, H.-W., Sun, H.-X., et al. (2016). CURLY LEAF regulates gene sets coordinating seed size and lipid Biosynthesis1[OPEN]. *Plant Physiology*, *171*(1), 424–436. https://doi.org/10.1104/pp.15.01335.

Liu, X. Y., Teng, Y. B., Li, B., & Meng, Q. W. (2013). Enhancement of low-temperature tolerance in transgenic tomato plants overexpressing Lefad7 through regulation of trienoic fatty acids. *Photosynthetica*, *51*(2), 238–244. https://doi.org/10.1007/s11099-013-0014-5.

Lou, Y., Schwender, J., & Shanklin, J. (2014). FAD2 and FAD3 desaturases form heterodimers that facilitate metabolic channeling in vivo. *The Journal of Biological Chemistry*, *289*(26), 17996–18007. https://doi.org/10.1074/jbc.M114.572883.

Luo, Z., Brock, J., Dyer, J. M., Kutchan, T., Schachtman, D., Augustin, M., et al. (2019). Genetic diversity and population structure of a Camelina sativa spring panel. *Frontiers in Plant Science*, *10*, 184. https://doi.org/10.3389/fpls.2019.00184.

Martz, F., Kiviniemi, S., Palva, T. E., & Sutinen, M.-L. (2006). Contribution of omega-3 fatty acid desaturase and 3-ketoacyl-ACP synthase II (KASII) genes in the modulation of glycerolipid fatty acid composition during cold acclimation in birch leaves. *Journal of Experimental Botany*, *57*(4), 897–909. https://doi.org/10.1093/jxb/erj075.

Matsuda, O., Sakamoto, H., Hashimoto, T., & Iba, K. (2005). A temperature-sensitive mechanism that regulates post-translational stability of a plastidial omega-3 fatty acid desaturase (FAD8) in Arabidopsis leaf tissues. *The Journal of Biological Chemistry*, *280*(5), 3597–3604. https://doi.org/10.1074/jbc.M407226200.

Matsuda, O., Sakamoto, H., Nakao, Y., Oda, K., & Iba, K. (2009). CTD phosphatases in the attenuation of wound-induced transcription of jasmonic acid biosynthetic genes in Arabidopsis. *The Plant Journal*, *57*(1), 96–108. https://doi.org/10.1111/j.1365-313X.2008.03663.x.

Matsuda, O., Watanabe, C., & Iba, K. (2001). Hormonal regulation of tissue-specific ectopic expression of an Arabidopsis endoplasmic reticulum-type omega-3 fatty acid desaturase (FAD3) gene. *Planta*, *213*(6), 833–840.

McCartney, A. W., Dyer, J. M., Dhanoa, P. K., Kim, P. K., Andrews, D. W., McNew, J. A., et al. (2004). Membrane-bound fatty acid desaturases are inserted co-translationally into the ER and contain different ER retrieval motifs at their carboxy termini. *The Plant Journal: For Cell and Molecular Biology*, *37*(2), 156–173.

McConn, M., & Browse, J. (1998). Polyunsaturated membranes are required for photosynthetic competence in a mutant of Arabidopsis. *The Plant Journal*, *15*(4), 521–530. https://doi.org/10.1046/j.1365-313X.1998.00229.x.

McConn, M., Hugly, S., Browse, J., & Somerville, C. (1994). A mutation at the fad8 locus of Arabidopsis identifies a second chloroplast [omega]-3 desaturase. *Plant Physiology*, *106*(4), 1609–1614.

Mccourt, P., Kunst, L., Browse, J., & Somerville, C. (1987). The effects of reduced amounts of lipid unsaturation on chloroplast ultrastructure and photosynthesis in a mutant of Arabidopsis. *Plant Physiology*, *84*, 353–360. https://doi.org/10.1104/pp.84.2.353.

Menard, G. N., Moreno, J. M., Bryant, F. M., Munoz-Azcarate, O., Kelly, A. A., Hassani-Pak, K., et al. (2017). Genome wide analysis of fatty acid desaturation and its response to temperature. *Plant Physiology*, *173*(3), 1594–1605. https://doi.org/10.1104/pp.16.01907.

Mendes, A., Kelly, A. A., van Erp, H., Shaw, E., Powers, S. J., Kurup, S., et al. (2013). BZIP67 regulates the omega-3 fatty acid content of Arabidopsis seed oil by activating fatty acid desaturase3. *The Plant Cell*, *25*(8), 3104–3116. https://doi.org/10.1105/tpc.113.116343.

Mène-Saffrané, L., Davoine, C., Stolz, S., Majcherczyk, P., & Farmer, E. E. (2007). Genetic removal of tri-unsaturated fatty acids suppresses developmental and molecular phenotypes of an Arabidopsis tocopherol-deficient mutant. Whole-body mapping of malondialdehyde pools in a complex eukaryote. *The Journal of Biological Chemistry*, *282*(49), 35749–35756. https://doi.org/10.1074/jbc.M706838200.

Mène-Saffrané, L., Dubugnon, L., Chételat, A., Stolz, S., Gouhier-Darimont, C., & Farmer, E. E. (2009). Nonenzymatic oxidation of trienoic fatty acids contributes to reactive oxygen species management in Arabidopsis. *The Journal of Biological Chemistry*, *284*(3), 1702–1708. https://doi.org/10.1074/jbc.M807114200.

Miquel, M., & Browse, J. (1992). Arabidopsis mutants deficient in polyunsaturated fatty acid synthesis. Biochemical and genetic characterization of a plant oleoyl-phosphatidylcholine desaturase. *Journal of Biological Chemistry*, *267*(3), 1502–1509. https://doi.org/10.1016/S0021-9258(18)45974-1.

Miquel, M. F., & Browse, J. A. (1994). High-Oleate oilseeds fail to develop at low temperature. *Plant Physiology*, *106*(2), 421–427.

Miquel, M., James, D., Dooner, H., & Browse, J. (1993). Arabidopsis requires polyunsaturated lipids for low-temperature survival. *Proceedings of the National Academy of Sciences of the United States of America*, *90*(13), 6208–6212.

Morales, M., & Munné-Bosch, S. (2019). Malondialdehyde : Facts and artifacts. *Plant Physiology*, *180*(3), 1246–1250. https://doi.org/10.1104/pp.19.00405.

Morineau, C., Bellec, Y., Tellier, F., Gissot, L., Kelemen, Z., Nogué, F., et al. (2017). Selective gene dosage by CRISPR-Cas9 genome editing in hexaploid Camelina sativa. *Plant Biotechnology Journal*, *15*(6), 729–739. https://doi.org/10.1111/pbi.12671.

Mu, J., Tan, H., Zheng, Q., Fu, F., Liang, Y., Zhang, J., et al. (2008). Leafy cotyledon1 is a key regulator of fatty acid biosynthesis in Arabidopsis. *Plant Physiology*, *148*(2), 1042–1054. https://doi.org/10.1104/pp.108.126342.

Na, G., Mu, X., Grabowski, P., Schmutz, J., & Lu, C. (2019). Enhancing microRNA167A expression in seed decreases the α-linolenic acid content and increases seed size in Camelina sativa. *The Plant Journal*, *98*(2), 346–358. https://doi.org/10.1111/tpj.14223.

Nishiuchi, T., Hamada, T., Kodama, H., & Iba, K. (1997). Wounding changes the spatial expression pattern of the arabidopsis plastid omega-3 fatty acid desaturase gene (FAD7) through different signal transduction pathways. *The Plant Cell*, *9*(10), 1701–1712. https://doi.org/10.1105/tpc.9.10.1701.

Nishiuchi, T., Nakamura, T., Abe, T., Kodama, H., Nishimura, M., & Iba, K. (1995). Tissue-specific and light-responsive regulation of the promoter region of the Arabidopsis thaliana chloroplast omega-3 fatty acid desaturase gene (FAD7). *Plant Molecular Biology*, *29*(3), 599–609. https://doi.org/10.1007/BF00020987.

Oh, Y. J., Kim, H., Seo, S. H., Hwang, B. G., Chang, Y. S., Lee, J., et al. (2016). Cytochrome b5 reductase 1 triggers serial reactions that Lead to iron uptake in plants. *Molecular Plant*, *9*(4), 501–513. https://doi.org/10.1016/j.molp.2015.12.010.

O'Neill, C. M., Baker, D., Bennett, G., Clarke, J., & Bancroft, I. (2011). Two high linolenic mutants of Arabidopsis thaliana contain megabase-scale genome duplications encompassing the FAD3 locus. *The Plant Journal: For Cell and Molecular Biology*, *68*(5), 912–918. https://doi.org/10.1111/j.1365-313X.2011.04742.x.

O'Quin, J. B., Bourassa, L., Zhang, D., Shockey, J. M., Gidda, S. K., Fosnot, S., et al. (2010). Temperature-sensitive post-translational regulation of plant Omega-3 fatty-acid desaturases is mediated by the endoplasmic reticulum-associated degradation pathway. *The Journal of Biological Chemistry*, *285*(28), 21781–21796. https://doi.org/10.1074/jbc.M110.135236.

Park, W., Feng, Y., Kim, H., Suh, M. C., & Ahn, S.-J. (2015). Changes in fatty acid content and composition between wild type and CsHMA3 overexpressing Camelina sativa under heavy-metal stress. *Plant Cell Reports*, *34*(9), 1489–1498. https://doi.org/10.1007/s00299-015-1801-1.

Perera, I. Y., Hung, C.-Y., Moore, C. D., Stevenson-Paulik, J., & Boss, W. F. (2008). Transgenic Arabidopsis plants expressing the type 1 inositol 5-phosphatase exhibit

increased drought tolerance and altered abscisic acid signaling. *The Plant Cell, 20*(10), 2876–2893. https://doi.org/10.1105/tpc.108.061374.
Petrie, J. R., Zhou, X.-R., Leonforte, A., McAllister, J., Shrestha, P., Kennedy, Y., et al. (2020). Development of a Brassica napus (canola) crop containing fish oil-like levels of DHA in the seed oil. *Frontiers in Plant Science, 11*, 727. https://doi.org/10.3389/fpls.2020.00727.
Prades, J., Funari, S. S., Escribá, P. V., & Barceló, F. (2003). Effects of unsaturated fatty acids and triacylglycerols on phosphatidylethanolamine membrane structure. *Journal of Lipid Research, 44*(9), 1720–1727. https://doi.org/10.1194/jlr.M300092-JLR200.
Provart, N. J., Gil, P., Chen, W., Han, B., Chang, H.-S., Wang, X., et al. (2003). Gene expression phenotypes of Arabidopsis associated with sensitivity to low temperatures. *Plant Physiology, 132*(2), 893–906. https://doi.org/10.1104/pp.103.021261.
Puttick, D., Dauk, M., Lozinsky, S., & Smith, M. A. (2009). Overexpression of a FAD3 desaturase increases synthesis of a Polymethylene-interrupted Dienoic fatty acid in seeds of Arabidopsis thaliana L. *Lipids, 44*(8), 753–757. https://doi.org/10.1007/s11745-009-3315-5.
Qi, S., Sen, Y., Ping, G., & JiXian, Z. (2019). Responses of seed yield of Perilla frutescens to fertilization rates of nitrogen, phosphorus and potassium. *Southwest China Journal of Agricultural Sciences, 32*(12), 2885–2889.
Román, Á., Andreu, V., Hernández, M. L., Lagunas, B., Picorel, R., Martínez-Rivas, J. M., et al. (2012). Contribution of the different omega-3 fatty acid desaturase genes to the cold response in soybean. *Journal of Experimental Botany, 63*(13), 4973–4982. https://doi.org/10.1093/jxb/ers174.
Román, Á., Hernández, M. L., Soria-García, Á., López-Gomollón, S., Lagunas, B., Picorel, R., et al. (2015). Non-redundant contribution of the Plastidial FAD8 ω-3 desaturase to Glycerolipid unsaturation at different temperatures in Arabidopsis. *Molecular Plant, 8*(11), 1599–1611. https://doi.org/10.1016/j.molp.2015.06.004.
Rossak, M., Smith, M., & Kunst, L. (2001). Expression of the FAE1 gene and FAE1 promoter activity in developing seeds of Arabidopsis thaliana. *Plant Molecular Biology, 46*(6), 717–725. https://doi.org/10.1023/a:1011603923889.
Routaboul, J.-M., Skidmore, C., Wallis, J. G., & Browse, J. (2012). Arabidopsis mutants reveal that short- and long-term thermotolerance have different requirements for trienoic fatty acids. *Journal of Experimental Botany, 63*(3), 1435–1443. https://doi.org/10.1093/jxb/err381.
Ruuska, S. A., Girke, T., Benning, C., & Ohlrogge, J. B. (2002). Contrapuntal networks of gene expression during Arabidopsis seed filling. *The Plant Cell, 14*(6), 1191–1206. https://doi.org/10.1105/tpc.000877.
Schaller, A., & Stintzi, A. (2009). Enzymes in jasmonate biosynthesis – Structure, function, regulation. *Phytochemistry, 70*(13–14), 1532–1538. https://doi.org/10.1016/j.phytochem.2009.07.032.
Schmid, M., Davison, T. S., Henz, S. R., Pape, U. J., Demar, M., Vingron, M., et al. (2005). A gene expression map of Arabidopsis thaliana development. *Nature Genetics, 37*(5), 501–506. https://doi.org/10.1038/ng1543.
Schmid-Siegert, E., Stepushenko, O., Glauser, G., & Farmer, E. E. (2016). Membranes as structural antioxidants: Recycling of malondialdehyde to its source in oxidation-sensitive chloroplast fatty acids. *The Journal of Biological Chemistry, 291*(25), 13005–13013. https://doi.org/10.1074/jbc.M116.729921.
Schmidt, H., & Heinz, E. (1990). Involvement of ferredoxin in desaturation of lipid-bound Oleate in chloroplasts 1. *Plant Physiology, 94*(1), 214–220.
Schuchardt, J. P., & Hahn, A. (2013). Bioavailability of long-chain omega-3 fatty acids. *Prostaglandins, Leukotrienes and Essential Fatty Acids, 89*(1), 1–8. https://doi.org/10.1016/j.plefa.2013.03.010.

Seo, P. J., Kim, M. J., Song, J.-S., Kim, Y.-S., Kim, H.-J., & Park, C.-M. (2010). Proteolytic processing of an Arabidopsis membrane-bound NAC transcription factor is triggered by cold-induced changes in membrane fluidity. *The Biochemical Journal*, *427*(3), 359–367. https://doi.org/10.1042/BJ20091762.

Shanklin, J., Whittle, E., & Fox, B. G. (1994). Eight histidine residues are catalytically essential in a membrane-associated iron enzyme, stearoyl-CoA desaturase, and are conserved in alkane hydroxylase and xylene monooxygenase. *Biochemistry*, *33*(43), 12787–12794.

Shi, Y., Yue, X., & An, L. (2018). Integrated regulation triggered by a cryophyte ω-3 desaturase gene confers multiple-stress tolerance in tobacco. *Journal of Experimental Botany*, *69*(8), 2131–2148. https://doi.org/10.1093/jxb/ery050.

Skórzyńska-Polit, E., Drążkiewicz, M., & Krupa, Z. (2009). Lipid peroxidation and antioxidative response in Arabidopsis thaliana exposed to cadmium and copper. *Acta Physiologiae Plantarum*, *32*(1), 169. https://doi.org/10.1007/s11738-009-0393-1.

Soria-García, Á., Rubio, M. C., Lagunas, B., López-Gomollón, S., & Luján, M. de los Á., Díaz-Guerra, R., Picorel, R., & Alfonso, M. (2019). Tissue distribution and specific contribution of Arabidopsis FAD7 and FAD8 plastid desaturases to the JA- and ABA-mediated cold stress or defense responses. *Plant and Cell Physiology*, *60*(5), 1025–1040. https://doi.org/10.1093/pcp/pcz017.

Spychalla, J. P., Kinney, A. J., & Browse, J. (1997). Identification of an animal ω-3 fatty acid desaturase by heterologous expression in Arabidopsis. *Proceedings of the National Academy of Sciences*, *94*(4), 1142–1147.

Tan, H., Yang, X., Zhang, F., Zheng, X. L., Qu, C., Mu, J., et al. (2011). Enhanced seed oil production in canola by conditional expression of Brassica napus LEAFY COTYLEDON1 and LEC1-LIKE in developing seeds. *Plant Physiology*, *156*(3), 1577–1588. https://doi.org/10.1007/BF00021523.

Tasseva, G., de Virville, J. D., Cantrel, C., Moreau, F., & Zachowski, A. (2004). Changes in the endoplasmic reticulum lipid properties in response to low temperature in Brassica napus. *Plant Physiology and Biochemistry: PPB*, *42*(10), 811–822. https://doi.org/10.1016/j.plaphy.2004.10.001.

Thambugala, D., & Cloutier, S. (2014). Fatty acid composition and desaturase gene expression in flax (Linum usitatissimum L.). *Journal of Applied Genetics*, *55*(4), 423–432. https://doi.org/10.1007/s13353-014-0222-0.

Torres-Franklin, M.-L., Repellin, A., Huynh, V.-B., & d'Arcy-Lameta, A., Zuily-Fodil, Y., & Pham-Thi, A.-T. (2009). Omega-3 fatty acid desaturase (FAD3, FAD7, FAD8) gene expression and linolenic acid content in cowpea leaves submitted to drought and after rehydration. *Environmental and Experimental Botany*, *65*(2), 162–169. https://doi.org/10.1016/j.envexpbot.2008.12.010.

Tsikas, D. (2017). Assessment of lipid peroxidation by measuring malondialdehyde (MDA) and relatives in biological samples : Analytical and biological challenges. *Analytical Biochemistry*, *524*, 13–30. https://doi.org/10.1016/j.ab.2016.10.021.

Usher, S., Han, L., Haslam, R. P., Michaelson, L. V., Sturtevant, D., Aziz, M., et al. (2017). Tailoring seed oil composition in the real world : Optimising omega-3 long chain polyunsaturated fatty acid accumulation in transgenic Camelina sativa. *Scientific Reports*, 7(1), 6570. https://doi.org/10.1038/s41598-017-06838-0.

Verdoni, N., Mench, M., Cassagne, C., & Bessoule, J.-J. (2001). Fatty acid composition of tomato leaves as biomarkers of metal-contaminated soils. *Environmental Toxicology and Chemistry*, *20*(2), 382–388. https://doi.org/10.1002/etc.5620200220.

Vrinten, P., Hu, Z., Munchinsky, M.-A., Rowland, G., & Qiu, X. (2005). Two FAD3 desaturase genes control the level of linolenic acid in flax seed. *Plant Physiology*, *139*(1), 79–87. https://doi.org/10.1104/pp.105.064451.

Wang, J., Ming, F., Pittman, J., Han, Y., Hu, J., Guo, B., et al. (2006). Characterization of a rice (Oryza sativa L.) gene encoding a temperature-dependent chloroplast ω-3 fatty acid desaturase. *Biochemical and Biophysical Research Communications*, *340*(4), 1209–1216. https://doi.org/10.1016/j.bbrc.2005.12.126.

Wang, C.-T., & Xu, Y.-N. (2010). The 5′ untranslated region of the FAD3 mRNA is required for its translational enhancement at low temperature in Arabidopsis roots. *Plant Science*, *179*(3), 234–240. https://doi.org/10.1016/j.plantsci.2010.05.008.

Wang, H.-S., Yu, C., Tang, X.-F., Wang, L.-Y., Dong, X.-C., & Meng, Q.-W. (2010). Antisense-mediated depletion of tomato endoplasmic reticulum omega-3 fatty acid desaturase enhances thermal tolerance. *Journal of Integrative Plant Biology*, *52*(6), 568–577. https://doi.org/10.1111/j.1744-7909.2010.00957.x.

Wang, H.-S., Yu, C., Tang, X.-F., Zhu, Z.-J., Ma, N.-N., & Meng, Q.-W. (2014). A tomato endoplasmic reticulum (ER)-type omega-3 fatty acid desaturase (LeFAD3) functions in early seedling tolerance to salinity stress. *Plant Cell Reports*, *33*(1), 131–142. https://doi.org/10.1007/s00299-013-1517-z.

Wickramanayake, J. S., Goss, J. A., Zou, M., & Goggin, F. L. (2020). Loss of function of fatty acid desaturase 7 in tomato enhances photosynthetic carbon fixation efficiency. *Frontiers in Plant Science*, *11*, 932. https://doi.org/10.3389/fpls.2020.00932.

Xia, Y., Yu, K., Navarre, D., Seebold, K., Kachroo, A., & Kachroo, P. (2010). The glabra1 mutation affects cuticle formation and plant responses to Microbes1[C][W][OA]. *Plant Physiology*, *154*(2), 833–846. https://doi.org/10.1104/pp.110.161646.

Xu, C., Moellering, E. R., Muthan, B., Fan, J., & Benning, C. (2010). Lipid transport mediated by Arabidopsis TGD proteins is unidirectional from the endoplasmic reticulum to the plastid. *Plant and Cell Physiology*, *51*(6), 1019–1028. https://doi.org/10.1093/pcp/pcq053.

Xue, Y., Chen, B., Win, A. N., Fu, C., Lian, J., Liu, X., et al. (2018). Omega-3 fatty acid desaturase gene family from two ω-3 sources, Salvia hispanica and Perilla frutescens: Cloning, characterization and expression. *PLoS One*, *13*(1), e0191432. https://doi.org/10.1371/journal.pone.0191432.

Yadav, N. S., Wierzbicki, A., Aegerter, M., Caster, C. S., Pérez-Grau, L., Kinney, A. J., et al. (1993). Cloning of higher plant omega-3 fatty acid desaturases. *Plant Physiology*, *103*(2), 467–476.

Yalcinkaya, T., Uzilday, B., Ozgur, R., Turkan, I., & Mano, J. (2019). Lipid peroxidation-derived reactive carbonyl species (RCS): Their interaction with ROS and cellular redox during environmental stresses. *Environmental and Experimental Botany*, *165*, 139–149. https://doi.org/10.1016/j.envexpbot.2019.06.004.

Yamamoto, A., Yoshii, M., Murase, S., Fujita, M., Kurata, N., Hobo, T., et al. (2014). Cell-by-cell developmental transition from embryo to post-germination phase revealed by Heterochronic gene expression and ER-body formation in Arabidopsis leafy cotyledon mutants. *Plant and Cell Physiology*, *55*(12), 2112–2125. https://doi.org/10.1093/pcp/pcu139.

Yamauchi, Y., Furutera, A., Seki, K., Toyoda, Y., Tanaka, K., & Sugimoto, Y. (2008). Malondialdehyde generated from peroxidized linolenic acid causes protein modification in heat-stressed plants. *Plant Physiology and Biochemistry*, *46*(8), 786–793. https://doi.org/10.1016/j.plaphy.2008.04.018.

Yang, Z.-T., Lu, S.-J., Wang, M.-J., Bi, D.-L., Sun, L., Zhou, S.-F., et al. (2014). A plasma membrane-tethered transcription factor, NAC062/ANAC062/NTL6, mediates the unfolded protein response in Arabidopsis. *The Plant Journal: For Cell and Molecular Biology*, *79*(6), 1033–1043. https://doi.org/10.1111/tpj.12604.

Yeom, W. W., Kim, H. J., Lee, K.-R., Cho, H. S., Kim, J.-Y., Jung, H. W., et al. (2020). Increased production of α-linolenic acid in soybean seeds by overexpression of

Lesquerella FAD3-1. *Frontiers in Plant Science, 10*. https://doi.org/10.3389/fpls.2019.01812, 1812.

Yin, Y., Jiang, X., Ren, M., Xue, M., Nan, D., Wang, Z., et al. (2018). AmDREB2C, from Ammopiptanthus mongolicus, enhances abiotic stress tolerance and regulates fatty acid composition in transgenic Arabidopsis. *Plant Physiology and Biochemistry, 130*, 517–528. https://doi.org/10.1016/j.plaphy.2018.08.002.

Yu, C., Wang, H.-S., Yang, S., Tang, X.-F., Duan, M., & Meng, Q.-W. (2009). Overexpression of endoplasmic reticulum omega-3 fatty acid desaturase gene improves chilling tolerance in tomato. *Plant Physiology and Biochemistry: PPB, 47*(11 – 12), 1102–1112. https://doi.org/10.1016/j.plaphy.2009.07.008.

Yuan, J., He, A., Huang, S., Hua, J., & Sheng, G. D. (2016). Internalization and phytotoxic effects of CuO nanoparticles in *Arabidopsis thaliana* as revealed by fatty acid profiles. *Environmental Science & Technology, 50*(19), 10437–10447. https://doi.org/10.1021/acs.est.6b02613.

Zhang, M., Barg, R., Yin, M., Gueta-Dahan, Y., Leikin-Frenkel, A., Salts, Y., et al. (2005). Modulated fatty acid desaturation via overexpression of two distinct omega-3 desaturases differentially alters tolerance to various abiotic stresses in transgenic tobacco cells and plants. *The Plant Journal: For Cell and Molecular Biology, 44*(3), 361–371. https://doi.org/10.1111/j.1365-313X.2005.02536.x.

Zhang, J., Liu, H., Sun, J., Li, B., Zhu, Q., Chen, S., et al. (2012). Arabidopsis fatty acid desaturase FAD2 is required for salt tolerance during seed germination and early seedling growth. *PLoS One*, 7(1). https://doi.org/10.1371/journal.pone.0030355.

Zhang, J.-T., Zhu, J.-Q., Zhu, Q., Liu, H., Gao, X.-S., & Zhang, H. (2009). Fatty acid desaturase-6 (Fad6) is required for salt tolerance in Arabidopsis thaliana. *Biochemical and Biophysical Research Communications, 390*, 469–474. https://doi.org/10.1016/j.bbrc.2009.09.095.

Zhou, Y., Tan, W.-J., Xie, L.-J., Qi, H., Yang, Y.-C., Huang, L.-P., et al. (2019). Polyunsaturated linolenoyl-CoA modulates ERF-VII-mediated hypoxia signaling in Arabidopsis. *Journal of Integrative Plant Biology, 62*(3), 330–348. https://doi.org/10.1111/jipb.12875.

CHAPTER EIGHT

Phytosterol metabolism in plants and algae, time to dive into the unknown

Aurélien Blary, Sylvain Darnet, Anne Berna, and Hubert Schaller*
Institut de biologie moléculaire des plantes, Plant Isoprenoid Biology team, CNRS, Université de Strasbourg, Strasbourg, France
*Corresponding author: e-mail address: hubert.schaller@ibmp-cnrs.unistra.fr

Contents

Abstract

The chapter will provide a current understanding of the biosynthesis pathways of sterols in photosynthetic eukaryotes and most importantly of their metabolism into various sterol conjugates. The diversification of enzymes and sterol structures in simple and

Advances in Botanical Research, Volume 101
ISSN 0065-2296
https://doi.org/10.1016/bs.abr.2021.09.006

multicellular eukaryotes from various phylogenetic groups establishes main pathways in algae and plants. A still missing subset of enzymes appealing further research efforts gives this hot topic an unbroken interest. It is crucial indeed to describe how sterol homeostasis is achieved in plants and algae in the context of growth and metabolism coordination, which is of tremendous importance for the success of biotech and agro-industrial programs.

1. Introduction

Algae and plants are sterol autotrophs that synthesize tetracyclic isoprenoid lipids according to the isoprenoid biogenetic rule (Bloch, 1965; Eschenmoser, Ruzicka, Jeger, & Arigoni, 1955). Relevant historical and educational considerations on sterol biological chemistry since the initial discovery and naming of cholesterol by Chevreul in the nineteen century have been summarized in an authoritative reference review on the biosynthesis of cholesterol (Nes, 2011). Sterols are mandatory for cellular membrane biogenesis, organization and dynamics (Dufourc, 2008a; Grosjean, Mongrand, Beney, Simon-Plas, & Gerbeau-Pissot, 2015; Pinkwart et al., 2019; Takahashi et al., 2021) and also for biological and signaling functions, notably as precursors of steroid hormones in most eukaryotic groups (reviewed in Payne & Hales, 2004; Schaller, 2010).

2. Phytosterols and oxysterols in algae and plants

2.1 Structure and distribution of phytosterols

2.1.1 Diversity of the sterol side chain

Contrasting with fungal and animal sterol autotrophs, which both produce a single major sterol, ergosterol and cholesterol, respectively, *Viridiplantae* (green algae and plants of all water or land ecosystems) and also algae from other taxa like chromalveolates (dinoflagellates, diatoms, brown algae) produce a very rich and diverse sterolome that includes also ergosterol and cholesterol (Darnet, Blary, Chevalier, & Schaller, 2021). Sterols from algae and plants are collectively defined as "phytosterols" herein (Benveniste, 2004; Darnet et al., 2021; Nes & McKean, 1977) (Fig. 1). Major sterols found in algae are often cholesterol, brassicasterol, ergosterol, fucosterol, clionasterol, and poriferasterol while cholesterol, campesterol, isofucosterol, sitosterol, and stigmasterol are the major sterols in plants (Figs. 1 and 2) (Nes & McKean, 1977 and all references therein). Phytosterols are made of a 1,2-cyclopentanoperhydrophenanthrene ring forming the sterol nucleus

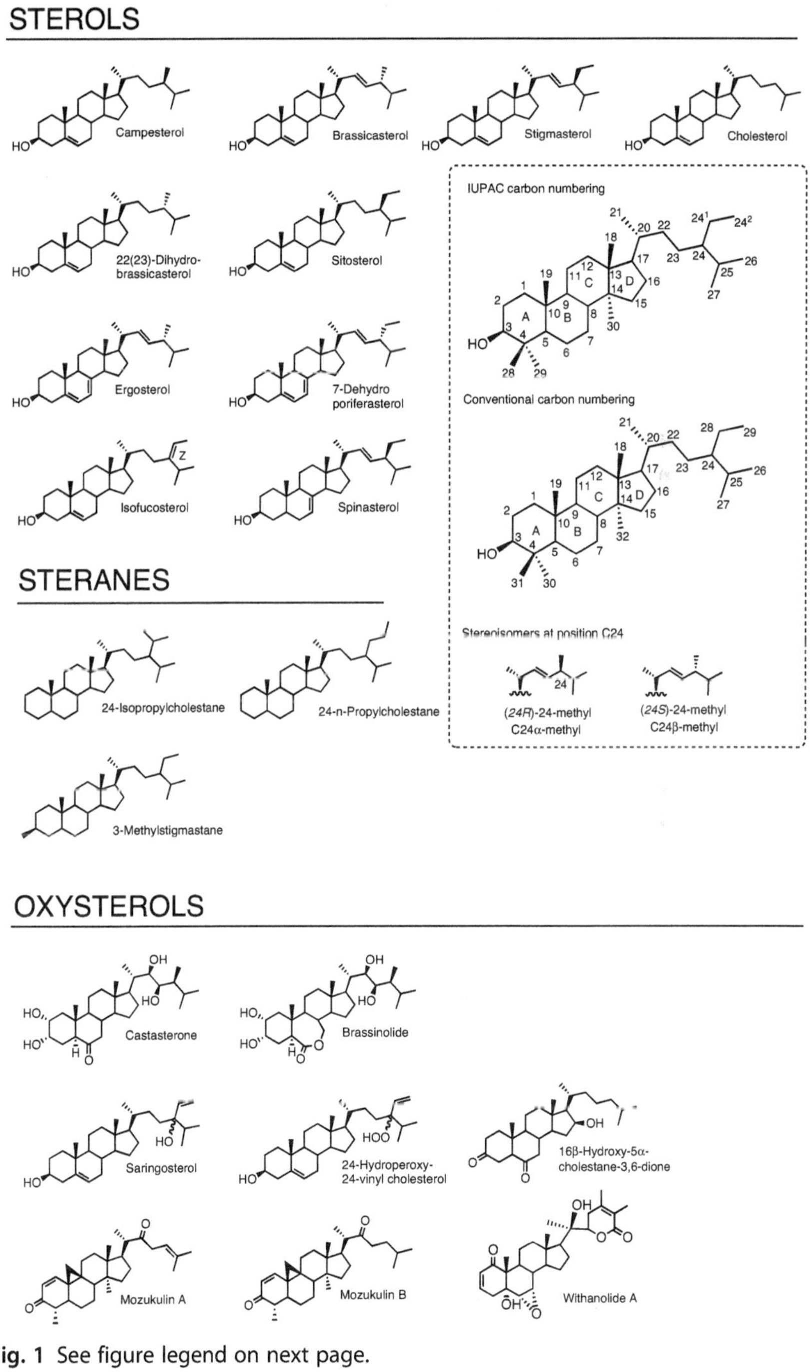

Fig. 1 See figure legend on next page.

and a C17-side chain of 8–10 carbons (Fig. 1). Triterpenic $C_{30}H_{50}O$ sterol precursors possess methyl substituents at C4, C10, C13, and C14 and a hydroxy group at C3 on the sterol nucleus. The astonishing diversity of the sterolome of photosynthetic organisms is achieved by structural variations of the C17 sterol side chain and the number of unsaturations of the cycle B of the sterol nucleus. Phytosterols differ by the number of carbons at C24 and fall into three major categories: 24-desmethylsterols (C_8 side chain, like cholesterol), 24-methylsterols (C_9 side chain, like campesterol, brassicasterol, 22(23)-dihydrobrassicasterol, and ergosterol), or 24-ethylsterols (C_{10} side chain, like stigmasterol, sitosterol, 7-dehydroporiferasterol, isofucosterol, and spinasterol). An important feature of sterol structural diversity is the epimeric carbon substituents at position C24 that enables to distinguish series of C24-methylsterols present in plants as racemic mixtures like for instance (24*R*,*S*)-C24 methylcholesterol (formed by the 24*R* epimer (or 24α, depending on the nomenclature) campesterol, and the 24*S* epimer (24β) 22(23)-dihydrobrassicasterol). Unlike C24-methylsterols, C24-ethylsterols are usually found as unique epimers in algae (7-dehydroporiferasterol) and plants (sitosterol, stigmasterol) (Fig. 1; Fig. 2) (Goad & Akihisa, 1997; Nes & McKean, 1977). Phytosterols exhibiting a side chain with a cyclopropanic ring (gorgosterol, 23-demethylgorgosterol) are known in symbiotic dinoflagellates (Withers, Kokke, Fenical, & Djerassi, 1982). Sterols with a C_{11} side chain like 24-isopropylcholesterol and 24-propylidenecholesterol have been found in picoplanktonic algae (*Aureococcus anophagefferens*, *Aureoumbra lagunensis*, from the *Chrysophyceae* family) otherwise responsible for the formation of blooms called "brown tides" (Giner & Boyer, 1998; Giner, Li, & Boyer, 2001). Sterols bearing such branched side chains are usually described in

Fig. 1 Prominent phytosterols, steranes, and oxysterols. *Phytosterols* bear an aliphatic side chain of generally 8–10 carbon atoms. Ten widely distributed sterols in photosynthetic organisms are shown. Carbon numbering and the particular options of nomenclature are given in the inset. Comprehensive nomenclature of sterols and steroids are found in Moss (1989) and Goad and Akihisa (1997). *Steranes* are saturated hydrocarbons derived from sterols via diagenetic degradation. These compounds are considered molecular fossils or biogeochemical markers. *Oxysterols* form a vast group of compounds generated by biological oxygenation of sterols. Brassinosteroids like castasterone and brassinolide are well characterized plant growth regulators. Oxysterols like saringosterol or 24-hydroperoxy-24-vinylcholesterol occur via chemical oxidization. A wealth of oxidized sterols like the mozukulins in algae or withanolides in certain Solanaceae are currently classified as specialized metabolites.

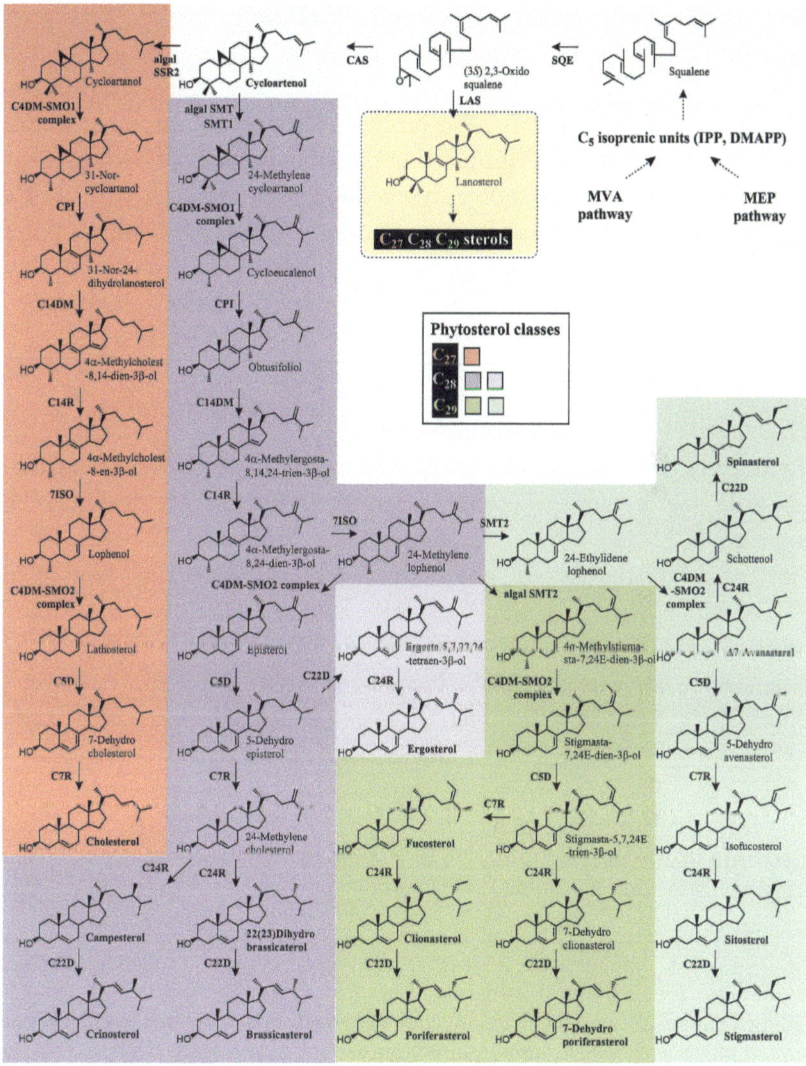

Fig. 2 Phytosterol pathways. Post-squalene pathways are shown in colorized frames. Cycloartenol is the product of the enzymatic cyclization of (3*S*) 2,3-oxidosqualene in almost all photosynthetic and also in some non-photosynthetic unicellular organisms (Godzina et al., 2000). (3*S*)-2,3-oxidosqualene is cyclized into lanosterol in dinoflagellates (Giner, Wünsche, Andersen, & Djerassi, 1991; Lu et al., 2020) and in some plant species (Forestier et al., 2019; Ohyama, Suzuki, Kikuchi, Saito, & Muranaka, 2009). *Yellow frame*, lanosterol pathways to cholesterol (C_{27} carbon) in metazoans, to ergosterol (C_{28} carbon) in fungi, and 24-ethylsterols (C_{29} carbon) and in some plant species. *Pink frame*, the side chain $\Delta^{24(25)}$ reduction of cycloartenol defines the cholesterol pathway in photosynthetic organisms (Sawai et al., 2014). *Purple frame*, the C24-methylation of cycloartenol *(Continued)*

non-photosynthetic marine organisms (Rohmer, Kokke, Fenical, & Djerassi, 1980) but their widespread presence in some groups of algae reinforces the view of the corresponding C_{11} side chain-steranes (24-isopropylcholestane and 24-n-propylcholestane besides other steranes like 3-methylstigmastane, Fig. 1) being probable markers of photosynthetic fossilized eukaryotes and not only of animal biomarkers. In fact, C24 alkylated sterols (of the ergostane and stigmastane types) are all potential markers of the earliest eukaryotic acquisition of a sterol side chain alkylation capacity (Bobrovskiy et al., 2021; Kodner, Pearson, Summons, & Knoll, 2008).

2.1.2 Sterol side chain unsaturations

Phytosterols differ by the degree of unsaturation of the side chain at C22 or C24 like for instance sitosterol (stigmast-5-en-3β-ol) and stigmasterol (stigmasta-5,22-dien-3β-ol) or isofucosterol (stigmasta-5,24-dien-3β-ol). Sitosterol and stigmasterol are ubiquitously distributed in seed plants (Nes & McKean, 1977). Genetic studies have revealed the crucial role of sitosterol in plant development and polarized cell division (De Storme et al., 2013; Hase et al., 2005; Nakamoto, Schmit, Heintz, Schaller, & Ohta, 2015; Schaeffer, Bronner, Benveniste, & Schaller, 2001). Other studies using loss-of-function mutants have revealed that a defect in the capacity to produce stigmasterol from sitosterol has a very limited effect on plant development (Arnqvist, Persson, Jonsson, Dutta, & Sitbon, 2008; Morikawa et al., 2006), raising then the question of the biological function of stigmasterol. Stigmasterol was however found as a marker of

Fig. 2—Cont'd defines the 24-methylcholesterol pathways in algae and plants, as well as the ergosterol pathway (*blue frame*) in fungi. *Green frames*, the C28-methylation of 24-methylenelophenol defines distinct 24-ethylcholesterol pathways in algae (*light green frame*) yielding fucosterol, clionasterol, poriferasterol, and 7-dehydroporiferasterol, and plants (*bright green frame*) yielding sitosterol and stigmasterol. MVA, mevalonate; MEP, methylerythritol 4-phosphate ; SQE, squalene monooxygenase (epoxidase); CAS, cycloartenol synthase (or 2,3-oxidosqualene cycloartenol cyclase); LAS, lanosterol synthase; SMT1, cycloartenol-C24-methyltransferase; SSR, sterol side chain reductase; C4DM, C4-demethylase complex; SMO, sterol methyloxidase; CPI, cyclopropylisomerase; C14DM, sterol-14α-demethylase; C14R, sterol-14-reductase; 7ISO, sterol-8-isomerase; SMT2, 24-methylenelophenol-C28-methylstransferase; C5D, sterol-5-desaturase; C7R, sterol-7-reductase; C24R, sterol-24-reductase; C22D, sterol-22-desaturase. Enzymes forming the C4DM complex are shown in Fig. 3. Gene and enzyme accessions are given in Table 1 for selected species *Arabidopsis thaliana (L.) Heynh*, *Chlamydomonas reinhardtii (P.A. Dang)*, and *Chondrus crispus (L.) Stackhouse.*

plant disease susceptibility due to the stimulation of the sitosterol-to-stigmasterol conversion in leaves from the model plant *Arabidopsis thaliana* challenged by microbial pathogenic bacteria (Aboobucker & Suza, 2019; Griebel & Zeier, 2010). Isofucosterol is found in small amounts in almost all plant species (Nes & McKean, 1977) but a few displays rather large amounts of isofucosterol like strawberries or potato tubers (Arnqvist, Dutta, Jonsson, & Sitbon, 2003; Couture, Arul, Gosselin, & Willemot, 1989). In fact, the complexity of a phytosterol profile comprising several Δ^5-sterols is partly understood from a biophysical perspective like an adaptation to heat-stress (Dufourc, 2008b).

2.1.3 Diversity of the sterol tetracyclic skeleton

Sterols from plants have generally one C5(6) double bond (or Δ^5–sterols) whereas the dienic system C5(6)/C7(8) (or $\Delta^{5,7}$-sterols) is frequently found in sterols from algae and predominantly found in fungi (Miller, Haubrich, Wang, Snell, & Nes, 2012). Sterols bearing a C7(8) bond (Δ^7-sterols, like spinasterol, Fig. 1) are the major products found in a small number of plant families such as the Cucurbitaceae, Caryophilaceae, Cactaceae and Chenopodiaceae (Akihisa, Thakur, Rosenstein, & Matsumoto, 1986). Compounds with a fully saturated tetracyclic nucleus and a 3β-hydroxy group are phytostanols referred to as Δ^0, by far less abundant than sterols in plants and algae (Moreau et al., 2018; Moreau, Whitaker, & Hicks, 2002). Seed plants with $\Delta^{5,7}$-sterols as physiological sterols (like ergosterol in fungi and green algae) are not described. *A. thaliana* loss-of-function mutant of the enzyme sterol-7-reductase (Fig. 2) (which is absent from yeast and consequently allows the biosynthesis of ergosterol) leads to the production of 7-dehydrostigmasterol as a major product, which could be photo-converted into the corresponding secosterol or vitamin D_5 (Silvestro et al., 2018).

2.2 Oxidized phytosterol derivatives

2.2.1 Brassinosteroids

The most popular phytosterol-derived compounds in plants are brassinosteroids like castasterone and brassinolide (Fig. 1), which have been initially described in chestnut (Yokota, Arima, & Takahashi, 1982) and rapeseed pollen (Grove et al., 1979; Mitchell, Mandava, Worley, Plimmer, & Smith, 1970), respectively. The vital role of campesterol-derived brassinosteroids in seed plants was discovered in the late nineties thanks to the power of

Arabidopsis genetics (Clouse, 1996; Li, Nagpal, Vitart, McMorris, & Chory, 1996; Szekeres et al., 1996). The biosynthetic pathway from campesterol to brassinolide via C22 oxidation is firmly established (Fujita et al., 2006) as well as the mechanisms by which the brassinosteroid receptor BRI1 (brassinosteroid-insensitive 1) and the signaling cascade downstream to BRI1 control gene expression implied in cellular elongation, photosynthesis, and skotomorphogenesis mainly (Vert, Nemhauser, Geldner, Hong, & Chory, 2005). The detection of a same class of compounds in a handful of non-seed plant (Yokota et al., 2017) or algae species (Bajguz & Hayat, 2009; Stirk et al., 2013) has led to the hypothesis that these molecules act on growth or development in the entire *Viridiplantae*.

In algae, castasterone and brassinolide were quantified in 24 microalgae strains (Stirk et al., 2013). In the green microalgae *Hydrodictyon reticulatum*, C24β epimers of these brassinosteroids were also detected (Yokota et al., 1987). *H. reticulatum* was found to contain both C24 epimers of a C24-ethylsterol: stigmasterol and poriferasterol (Akihisa, Kojima, Yokota, & Tamura, 1991). This was the first occurrence of epimeric 24-ethylsterols in the Chlorophyta. The presence of C24β epimers of brassinosteroids is consistent with the C24β oriented alkyl groups of sterols like 22(23)-dihydrobrassicasterol, clionasterol and chondrillasterol in Chlorella species and other green microalgae (Patterson & Krauss, 1965). Brassinosteroids in algae were proposed to act in response to abiotic stresses (Stirk et al., 2018). Further investigations will certainly determine the features of a putative steroid signaling pathway in algae, which must be different from that of conifers, gnetales and angiosperms—the taxa in which a true BRI1 receptor is described (Wang & Mao, 2014).

2.2.2 Oxysterols

Oxysterols in mammals including dietary oxyphytosterols (Olkkonen, Gylling, & Ikonen, 2017; Plat, Brzezinka, Lütjohann, & von Bergmann, 2001) have attracted considerable research in human physiopathology and in analytical chemistry (Griffiths & Wang, 2019). These mammalian oxysterols (7-hydroxycholesterol, 27-hydroxycholesterol, 24-hydroxycholesterol) are formed by the action of specific cholesterol hydroxylases (CYP7A1, CYP27A1, CYP46A1) or by autooxidation, and have not been identified in plants and algae. Oxysterol-binding proteins have been apparently overlooked in plants. One of the twelve oxysterol-binding proteins of *A. thaliana* (called ORP3a) was shown to bind sitosterol (but not stigmasterol), which suggested a function in sterol trafficking between the ER and the Golgi endomembrane systems (Saravanan et al., 2009).

There are only a very few reports about oxysterols in algae. A cytotoxic oxysterol 16β-hydroxy-5α-cholestane-3,6-dione was described in red algae, in the context of a pharmacological screening (Ktari, Blond, & Guyot, 2000). Cycloartane-type oxidized products called mozukulins were found in the brown algae *Cladosiphon okamuranus* (Cheng et al., 2016) and in another brown algae *Ectocarpus siliculosus* (Girard et al., 2021). These cycloartanes exhibit anti-inflammatory activity when considered as possible pharmaceuticals but their physiological or ecological properties have not been unveiled yet. Algae are also a rich source of fucosterol derivatives: 24-hydroperoxy-24-vinyl cholesterol (Fig. 1) was identified in a cytotoxicity screening with extracts from *Padina pavonica* (Ktari & Guyot, 1999). The photoconversion of fucosterol into saringosterol (Fig. 1) was recently confirmed as a non-enzymatic transformation (Vanbrabant et al., 2021). Saringosterol was initially identified in the seaweed *Sargassum fusiforme*, which is consumed for antiatherosclerotic properties described in traditional medicine. Remarkably, the C24β epimer of saringosterol was by far a more potent agonist of liver X receptors (LXRα and LXRβ) than the C24α epimer in a reporter assay (Chen et al., 2014). Furthermore, as a very strong lead compound in searching for medications against Alzheimer disease (AD), 24(*S*)-saringosterol taken as a dietary supplementation with the seaweed *Sargassum fusiforme* was shown to prevent cognitive decline in an AD mouse model in exhibiting marked anti-inflammatory effects (Martens et al., 2021).

2.3 Saponins, withanolides, cardenolides

Phytosterol oxidized derivatives are found as highly specialized metabolites in particular plant taxa. The steroidal glycoalkaloids solanine and tomatine found in Solanaceae (potato, tomato, eggplant) are antinutrients, which have attracted interest more than two-hundred years ago (Desfosses, 1820) until the elucidation of the biosynthesis pathway leading from cholesterol to the aglycone solanidine and tomatidine in potato and tomato (Itkin et al., 2013). Steroidal saponins like diosgenin in *Dioscorea* are found in species from a few families (*Disocoreaceae, Liliaceae, Solanaceae*). These steroidal saponins comprise a vast array of pharmacologically active compounds, some being of particular interest as steroid precursors in hemi-synthetic processes. Withanolides such as withanolide A (Fig. 1) are found in the solanaceous species from the *Withania* and *Physalis* genera. These steroids derive from 24-methylenecholesterol by the action of a specialized paralogous DWARF1 enzyme (sterol-Δ^{24}-reductase) producing 24-methyldesmosterol, which in turn is converted through succeeding enzymatic steps into the highly oxidized withaferin A (Knoch et al., 2018). Cardiac steroidal glucosides are found in plants (cardenolides) and in amphibians (bufadienolides). One of the most popular cardiac steroidal glucosides is digitoxin (or digitalin) found in *Digitalis* species. Biosynthetic, genetic, and cell technology studies in *Digitalis lanata* provided the identification of key enzymes of the cardenolide pathway (3β-hydroxysteroid dehydrogenase, progesterone 5β-reductase) (Thorn et al., 2008), the modulation of the cardenolide content by an RNAi approach in *D. lanata* (Klein et al., 2021) and combinatorial expression in yeast to generate cardenolides from pregnenolone (Rieck et al., 2019).

2.4 Sourcing of phytosterols for industrial applications

Algae and plants (as primary producers) provide primary consumers including humans with an abundant source of phytosterols that exert biological effects as intact phytosterols or as oxyphytosterols generated by enzymatic or non-enzymatic transformations (Hovenkamp et al., 2008; Olkkonen et al., 2017). Bioactive sterols have been shown to prevent cardiovascular disease, and most importantly to exert antioncogenic properties (Cioccoloni et al., 2020). The latter property of phytosterols was demonstrated at the subcellular level by the impairment of an oncoprotein recruitment and activation at the plasma membrane, where phytosterols perturb cholesterol-rich domains and alter phosphatidylinositolphosphate-mediated recruitment of that oncoprotein (Cioccoloni et al., 2020; Myeong, Park, Suh, & Hille, 2021; Soteriou, Kalli, Connell, Tyler, & Thorne, 2021). Phytosterols are known as efficient blood cholesterol-lowering agents for patients with metabolic syndromes or familial hypercholesterolemia (Moreau et al., 2002; Moreau et al., 2018; Plat et al., 2019; Ras, Geleijnse, & Trautwein, 2014). In fact, an estimate of cardiovascular disease deaths is about 17.3 million per year in the world (especially in low and middle-income countries), among which 80% could be preventable (World-Heart-Federation, 2019). Current health policies recommend the consumption of phytosterols for lowering low-density lipoprotein associated cholesterol, as a replacement of the massive use of drugs like the statins that inhibit cholesterol biosynthesis (Ras et al., 2014). Food or beverage products with phytosterols, phytosterol esters, or phytostanol esters have been approved as nutraceuticals of functional foods by authorities as well as tablets or capsules based on these health claims (EFSA, 2010; FDA, 2000). For these reasons, a ~1 billion USD global market forecast (years 2019-2024) for phytosterols as ingredients represent a production of 40 kilotons per year (Mordor Intelligence, 2019-2024).

Agro-industrial products like vegetable oils or residues from the wood industries are the major supplies of crude materials from which phytosterols are extracted as by-products in very large quantities for food, pharmaceutical, and cosmetic uses (Fernandes & Cabral, 2007). Phytosterols account for 5%–10 % of soybean oil distillate. Pure soybean oil phytosterols comprise campesterol, sitosterol, stigmasterol as main compounds and brassicasterol and sitostanol in lesser amounts (Yang et al., 2010). The largest industrial source of phytosterols is the tall oil, which is a by-product from the wood pulp mills. The tall oil pitch (or distillate) contains about 10%–15% of phytosterols, mostly campesterol, campestanol, sitosterol, and sitostanol. Scots pine (*Pinus sylvestris*) and spruce (*Picea abies*) contain also some amount of 4,4-dimethylsterols (citrostadienol, 24-methylenecycloartanol) but lack stigmasterol (Vikström, Holmbom, & Hamunen, 2005). Microalgae are currently considered as a potential source of phytosterols (Randhir, Laird, Maker, Trengove, & Moheimani, 2020).

The industrial production of steroids (androsteroids, progesteroids) and vitamins by the pharmaceutical sector is based on hemi-synthetic processes using phytosterols or phytosterol derivatives like steroidal saponins as starting molecules. Diosgenin, a steroidal saponine from yam (*Dioscorea* spp.), is a common reagent for hemi-syntheses; stigmasterol, which is one of the most abundant phytosterols from agro-industrial sources describes above is also used as starting material for the synthesis of estrogens (progesterone), androgens and other steroids (Lindow, 2016).

3. Pathways for phytosterol biosynthesis

3.1 From phytochemistry to genomics

The current understanding of phytosterol biogenetic pathways in the post-genomic era is based on structural elucidation of purified sterol molecules, biosynthetic and biochemical studies using radiolabeled sterol precursors fed to cells or developing seedlings, molecular enzymology, forward and reverse genetics and comparative functional genomics. Methods for extraction, purification, analysis and nomenclature of phytosterols, and their occurrence, can be found in comprehensive series of reviews and book chapters (Goad & Akihisa, 1997; Nes & McKean, 1977; Rahier & Benveniste, 1989). Biosynthetic studies implying labelled precursors, to show for instance the conversion of isofucosterol (a 24-ethylidenesterol) into sitosterol (a C24α-ethylsterol) in barley seedlings (Largeau, Goad, & Goodwin, 1977) are methods of interest for determining biosynthetic capacities in cells and organs: incorporation of labelled sterol precursors into

4-methylsterols was in fact demonstrated in tobacco pollen tubes in an original study unveiling a truncation of the phytosterol pathway in the male gametophyte of flowering plants (Villette, Berna, Compagnon, & Schaller, 2015). Heterologous expression of steroidogenic enzymes in yeast or their overexpression or gene silencing in *Arabidopsis thaliana* or *Nicotiana benthamiana* (Burger, Rondet, Benveniste, & Schaller, 2003; Darnet & Rahier, 2004; Silvestro, Andersen, Schaller, & Jensen, 2013; Sonawane et al., 2016) are mandatory for the identification of accurate substrate-to-product conversions in an intricate metabolic grid (Benveniste, 2004; Darnet & Schaller, 2019a). Likewise, the use of inhibitors of post-squalene enzymes in plants (Benveniste, 1986; Burden, Cooke, & Carter, 1989; Schaller, Maillot-Vernier, Gondet, Belliard, & Benveniste, 1993) or algae (Chan, Patterson, Dutky, & Cohen, 1974) provided major contributions in deciphering succeeding steps in sterol pathways. In very recent years such approaches developed with microalgae (Fabris et al., 2014) have proven instrumental to unveil the existence of new enzymes especially a non-canonical squalene epoxidase (alternative squalene epoxidase, or AltSQE) of the fatty acid hydroxylase type in *Phaeodactylum tricornatum* and even more in several groups of overlooked eukaryotes. The initial observation was the absence of sensitivity of *P. dactylum* cells to the well-known squalene epoxidase (of the canonical flavoprotein type) inhibitor terbinafine (Pollier et al., 2019). The rational design and synthesis of chemical inhibitors of enzymes is of major interest to link specific sterol or groups of sterols to a biological process: the inhibition of sterol-C24-methyltransferases of the SMT2 type (and not SMT1) by synthetic 23-azasterols (Renard, Perruchon, Giera, Müller, & Bracher, 2009) was shown to be associated with root growth inhibition (Darnet et al., 2020). Combinations of approaches briefly depicted above are readily integrated into functional genomics aimed at deciphering novel sterol pathways in photosynthetic organisms including the bloom-forming diatoms microalgae *Skeletonema marinoi* and *Cyclotella cryptica* (Gallo et al., 2020), the brown algae *Ectocarpus silicolosus* (Girard et al., 2021), the green microalgae *Chorella sorokiniana, Chlorella variabilis, Coccomyxa subellipsoidea,* and *Chlamydomonas rheinhardtii* (Voshall et al., 2021). These studies showed that almost all microalgae species produce phytosterols from cycloartenol, like it is the case of land plants.

3.2 Phytosterol isoprenic precursors

A sterol biosynthesis pathway for algae and plants is proposed in this chapter with a very brief description of the origin of the C_5 isoprenic building blocks necessary to produce squalene and (3*S*)-2,3-oxidosqualene, the committed precursor of sterols. A more comprehensive analysis of post-squalene phytosterol pathways is then described (Fig. 2). Photosynthetic organisms exhibit two distinct pathways for the production of the C_5 building blocks isopentenyl diphosphate (IPP) and its isomer dimethylallyl diphosphate necessary to generate all isoprenoids including sterols: the so-called mevalonate (MVA) and methylerythritol phosphate (MEP) pathways (Rohmer, 1999). Algae from the green lineage (chlorophyta) are devoid of a MVA pathway and therefore make sterols from the MEP pathway, also called the plastidial pathway (Lichtenthaler, 1999). Mining the genomes of diatoms like *P. tricornutum* led to the identification of a MVA pathway (Fabris et al., 2014); an identical situation was found in *Skeletonema marinoi* and *Cyclotella cryptica* (Gallo et al., 2020). Brown algae like *E. siliculosus* exhibit also both the MVA and MEP functional pathways (Cock et al., 2010). Higher plants have been shown in numerous studies to use the MVA pathway for the biogenesis of sterols (Bouvier, Rahier, & Camara, 2005; Rohmer, 1999). In plants, the conversion of MEP pathway intermediates (deoxyxylulose 5-phosphate, methylerythritol 4-phosphate) into phytosterols was proven possible by tracing labelled ^{14}C or ^{13}C deoxyxylulose 5-phosphate in a tobacco BY-2 cell culture system (Hemmerlin et al., 2003). The contribution of the plastidial pathway to the constitutive production of phytosterols in the model *A. thaliana* is at best very limited under standard laboratory conditions since loss-of-function 3-hydroxy-3-methylglutaryl-CoA-reductase (HMGR, the key limiting enzyme of the MVA pathway) or loss-of-function alleles of MEP pathway genes exhibit very strong to lethal developmental schemes (Estévez et al., 2000; Suzuki et al., 2009). In algae, an inventory of the distribution of MVA or MEP pathways was proposed (Sasso, Pohnert, Lohr, Mittag, & Hertweck, 2012). This survey confirmed the existence of both pathways in several algal groups in addition to plants, except for the chlorophyta that carry exclusively the plastidial pathway. In fact, streptophyta, rhodophyta, heterokonts (stramenopiles, diatoms), haptophyta exhibit all the dual biosynthetic capacity for the production of isoprenic building blocks, except *Cyanidioschyzon merolae* (Sasso et al., 2012).

The prenyltransferase-mediated elongation of the allylic diphosphate C_5 precursor dimethylallyl diphosphate (DMAPP) by serial condensations with isopentenyl diphosphate (IPP) forms a pool of farnesyl diphosphate (C_{15}) that undergo head-to-head enzymatic condensation (Kellogg & Poulter, 1997) to produce the C_{30} linear sterol precursors squalene and its oxidized derivative (3*S*)-2,3-oxidosqualene (Fig. 2). The biotransformation of a unique steroidal precursor (3*S*)-2,3-oxidosqualene (Corey, Russey, & de Montellano, 1966) into phytosterols strongly suggest an evolutionary convergence of structural sterol requirements in organisms that may use alternative enzymatic transformations, like for instance distinct squalene epoxidases (canonical and alternative, *vide supra*) (Pollier et al., 2019).

3.3 Organization of the phytosterol pathways

The general organization of the phytosterol pathways in algae and plants includes three biosynthetic segments from cycloartenol to cholesterol (8 carbon atoms side chain, or C_{27} sterols), to 24-methylsterols like campesterol or ergosterol (9 carbon atoms side chain, or C_{28} sterols), and to 24-ethylsterols like sitosterol, stigmasterol, clionasterol and poriferasterol (10 carbon atoms side chain, C_{29} sterols) (Fig. 2). A comparative analysis of phytosterol profiles, genes and enzymes of a large set of eukaryotes allowed the classification of those species in distinct sterolotypes, which are described in details in Darnet et al. (2021). Remarkably, this classification gathers angiosperms (e.g., *Solanum lycopersicum*) and red algae (e.g., *Chondrus crispus*) in the same Solanum-type of high cholesterol producers because both species possess a sterol side chain reductase called SSR that reduces cycloartenol into cycloartanol and favors the cholesterol pathway (Calegario et al., 2016; Kodner et al., 2008; Sawai et al., 2014; Sonawane et al., 2016). Phytosterol pathways of three species of choice are illustrated in Fig. 2 (gene and enzyme accessions are given in Table 1).

Chondrus crispus, which contains over 90% of cholesterol (Tasende, 2000) was shown to possess such an early SSR pathway shown in Fig. 2 (left, pink frame) (Belcour et al., 2020). *Chlamydomomas reinhardtii*, a microalgae from the green lineage, exhibit a 24-methylsterol pathway shown in Fig. 2 (middle left, blue frame) and an algal 24-ethylsterol pathway (middle right, light green frame) (Miller et al., 2012), which defines a Chlorella-chlamydo-sterolotype (Darnet et al., 2021). *Arabidopsis thaliana* possess a well characterized sterol metabolism that includes a 24-methylsterol pathway shown in Fig. 2 (middle left, blue frame) and a higher plant 24-ethylsterol pathway (right, bright green frame) identified as a canonical Arabido-sterolotype (Darnet et al., 2021). All pathways require the removal of C4 methyl substituents of cycloartenol by distinct and non-consecutive C4-demethylation complexes of enzymes detailed in Fig. 3.

Table 1 Gene accessions for sterol biosynthetic enzymes in selected species.

Ab.	Description	*A. thaliana*	*C. reinhardtii*	*C. crispus*
SQS	Squalene synthase	AT4G34640 (SQS1) AT4G34650 (SQS2)	Cre03. g175250.t1.2	CHC_T0000158100
SQE	Squalene monooxygenase/squalene epoxidase	AT1G58440 (SQE1, XF1) AT2G22830 (SQE2) AT4G37760 (SQE3) AT5G24140 (SQE4, SQP2) AT5G24150 (SQE5, SQP1) AT5G24160 (SQE6)	g17770.t1	scaffolds 90, 20, 57
CAS	2,3-oxidosqualene-cycloartenol cyclase	AT2G07050 (CAS1)	Cre01.g011100	CHC_T00008265001
LAS	Lanosterol synthase	AT3G45130 (LAS1)	–	–
SSR2	Sterol side chain reductase	–	–	CHC_T00002789001
SMT1	C24 sterol methyltransferase	AT5G13710 (SMT1)	Cre12.g500500	CHC_T00009101001
SMT2	C28 sterol methyltransferase	AT1G20330 (SMT2-1, SMT2) AT1G76090 (SMT2-2, SMT3)	–	CHC_T00000837001
SMO	C-4 methyl sterol oxidase	–	Cre06. g261200.t1.2	CHC_T00010320001
SMO1	24-methylenecycloartanol 4α-methyl oxidase	AT4G12110 (SMO1-1) AT4g22753 (SMO1-2) AT4g22756 (SMO1-3)	–	–
SMO2	4α-methyl-Δ7-sterol-4α-methyl oxidase	AT1G07420 (SMO2-1) AT2G29390 (SMO2-2)	–	CHC_T00010320001
βHSD	3β-hydroxysteroid dehydrogenase/ C4-decarboxylase	AT2G33630 (AT2G33630.1) AT2G43420 AT1G47290 (AT3BETAHSD/D1, CSD) AT2G26260 (AT3BETAHSD/D2)	Cre10. g442200.t1.2	–

Continued

Table 1 Gene accessions for sterol biosynthetic enzymes in selected species.—cont'd

Ab.	Description	*A. thaliana*	*C. reinhardtii*	*C. crispus*
SR	Sterone ketoreductase	At5g18210 (SKR)	–	–
CPI	Cyclopropyl sterol isomerase	AT5G50375 (CPI1)	Cre16.g657300.t1.2	CHC_T00002985001
C14DM	Obtusifoliol-14α-demethylase	AT1G11680 (OBT14DM1, CYP51G1) AT2G17330 (OBT14DM1, CYP51G2)	Cre02.g092350.t1.2	CHC_T00009303001
7ISO	Δ8-Δ7-sterol isomerase	AT1G20050 (HYD1)	Cre12.g557900.t1.2	CHC_T00001257001
C14R	Δ8,14-sterol-Δ14-reductase	AT3G52940 (FK)	Cre02.g076800.t1.2	CHC_T00003466001
C5D	Δ7-sterol-C5-desaturase	AT3G02580 (C5DES1, STE1, DWF7, BUL1) AT3G02590 (C5DES2, STE2)	Cre16.g663950.t1.2	CHC_T00006481001
C7R	Δ5,7-sterol-Δ7-reductase	AT1G50430 (DWF5)	–	CHC_T00006492-3001
C24R	Δ5-sterol-Δ24-reductase (isomerase)	AT3G19820 (DWF1)	Cre02.g076800	
C22D	C-22 sterol desaturase	AT2G28860 (AT2G28860, CYP710A4) AT2G28850 (AT2G28850, CYP710A3) AT2G34500 (CYP710A1) AT2G34490 (CYP710A2)	g11457.t1	CYP805A1-C1 CYP808A1-H1

The locus numbers were retrieved from TAIR10 version of *Arabidopsis thaliana* genome (https://arabidopsis.org), R64.2.1 version of *Saccharomyces cerevisiae* genome database (SGD, Saccharomyces Genome Database—SGD https://yeastgenome.org), 5.6 version of *C. reinhardtii* genome (https://phytozome-next.jgi.doe.gov/) and ASM35022v2 version of *Chondrus crispus* (https://plants.ensembl.org/Chondrus_crispus/).

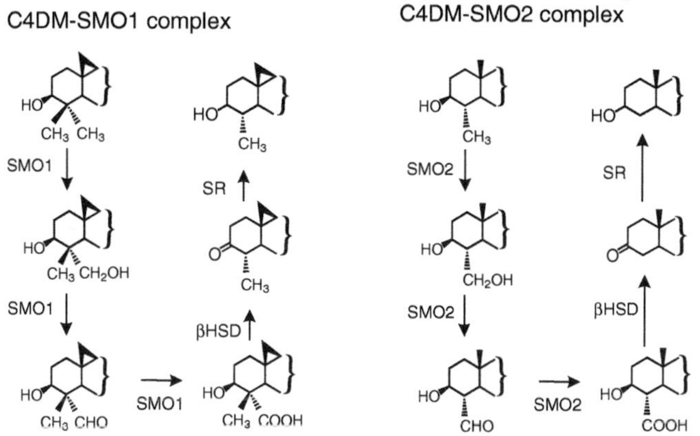

Fig. 3 Sterol-C4-demethylation (DM) reaction in plants. Two distinct C4 demethylation complexes use 4,4-dimethyl-9β,19-cyclopropylsterol intermediates (left, C4 DM complex (SMO1)) or 4-methylsterols (right, C4 DM complex (SMO2)) as substrates (Darnet & Rahier, 2004). SMO, sterol methyl oxidase; βHSD, 3β-hydroxysteroid dehydrogenase/decarboxylase; SR, sterone reductase.

4. Genetic and biotechnological approaches to enhance phytosterol production

4.1 Genetic analysis of phytosterol composition

Increasing knowledge gained on genes, enzymes and key regulatory steps involved in sterol biosynthesis has opened prospects for increasing sterol amounts or to modify its composition in plant (crops) and algae. Manipulation of sterol levels can be achieved in multiple ways. In plants, statistical methods have been used to identify regions in the genome associated with natural variation for sterol content and composition. Introgression of these genomic regions into elite cultivars through "traditional" plant breeding approaches could result in the development of new varieties with increased sterol content. Alternatively, forwards or reverse genetic strategies have yielded mutant and transgenic lines displaying sterol accumulation.

Genetically controlled variation for sterol content and composition is a pre-requisite for the development of high sterol cultivars. In plants, intraspecific variation for total sterol content and sterol composition has been mainly accessed in seed oil crops (a main source of sterol for sterol enriched food), nut tree oil and cereal grains (a main dietary source of sterol). In oleaginous plants, total and individual phytosterol content is highly variable.

Up to 2.5-fold increase in total sterol content is observed in seeds (or seed oils) of rapeseed (Amar, Becker, & Möllers, 2009), sunflower (Roche et al., 2010) and soybean (Vlahakis & Hazebroek, 2000). A similar intraspecific variation for total sterol content is found in almond (2.5-fold) (Font i Forcada, Velasco, Socias i Company, & Fernández i Martí, 2015) and walnuts (1.5-fold) (Abdallah et al., 2015). In cereals, the reported range of variation for sterol levels is less. In wheat, intraspecific variation for total sterol content is up to 0.3-fold in winter, spring and other wheat types (Nurmi, Nyström, Edelmann, Lampi, & Piironen, 2008). This range of variation is consistent with intraspecific variation for sterol content reported in tetraploid wheat species (Merah & Mouloungui, 2019) but also in other cereals like barley (Andersson et al., 2008), oat (Shewry et al., 2008) or rye (Nyström et al., 2008). Only a handful of studies (summarized in Table 2) have been performed to identify Quantitative Trait Loci (QTL) or genetic variants associated with alteration of sterol content. QTLs analysis were notably successful in the two oilseed crops barley and sunflower. In those crops, high heritability for total sterol content was observed ($h^2 > 0.7$) although environment was a significant source of variation (Merah et al., 2012; Teh & Möllers, 2016). In barley, approximately the same number of QTLs were detected in two distinct double haploid segregation populations. A minor QTL ($R^2 < 25\%$) for total sterol content was notably detected in both studies on chromosome C08 and a major QTL for brassicasterol level on chromosome A04 (Amar et al., 2008; Teh & Möllers, 2016). In sunflower, a minor but stable QTLs for total sterol content was reported on LG1 in two independent studies (Haddadi et al., 2012; Merah et al., 2012). Analysis of the genetic variability of sterol content is rare outside of oilseed crops (Table 2). Association mapping in almonds yielded significant associations for individual or total sterol content despite a low-density genotyping (Font i Forcada et al., 2015). In all studies, positional candidate genes involved in sterol biosynthesis were investigated (Table 2), however the exact nature and identity of the causal polymorphisms remain unknown.

4.2 Trangenesis approaches to enhance phytosterols

Tickling with sterol biosynthesis using forward and reverse genetic approach first started in the nineties with the characterization of high sterol tobacco plants isolated after combined mutagenesis and selection against inhibitors of sterol biosynthesis (Maillot-Vernier, Gondet, Schaller, Benveniste, & Belliard, 1991; Schaller, Maillot-Vernier, Belliard, & Benveniste, 1992).

Table 2 Genetic variability studies for phytosterol content in plants.

Method	Species	Population	QTL for total sterol content/composition	Notes	References
QTL mapping	*Brassica napus*	148 doubled haploid lines derived from the cross "Mansholt's Hamburger Raps" x "Samourai"	– Two major QTLs on A08 and C03 and a minor QTL on C08 for total phytosterol content (not found in Teh & Möllers, 2016) – Majors QTL for individual phytosterol content (Sitosterol, Campesterol, Brassicasterol, Avenasterol)	The major QTLs on A08 and C03 likely reflect pleiotropic effect of the erucic acid genes on phytosterol content	Amar, Ecke, Becker, and Möllers (2008)
QTL mapping	*Brassica napus*	226 doubled haploid lines derived from the cross "Sansibar" x "Oase" (SODH population)	– Two minor QTLs located on A07 and C08 for total sterol content. – Major QTLs for individual phytosterol content (brassicasterol on A04, campesterol to sitosterol ratio and 24-methyl to 24-ethyl sterol ratio on A06)	Orthologs of *CYP710A2* and *SMT2* were found within the genomic region of the major QTLs on A04 and A06, respectively	Teh and Möllers (2016)
QTL mapping	*Helianthus annuus*	200 recombinant inbred lines (RIL) derived from the cross "XRQ" x "PSC8"	– Minor QTLs for total and individual phytosterol content on LG1 (sitosterol, desmethylsterols and total sterol), LG4 (campesterol and squalene), LG7 (campesterol and avenasterol)		Alignan et al. (2008), Merah et al. (2012)
QTL mapping	*Helianthus annuus*	123 RILs derived from the cross ("RHA266" x "PAC2")	– Minor QTLs for total phytosterol content on LG1, LG2, LG16 and LG17		Haddadi et al. (2012)

Continued

Table 2 Genetic variability studies for phytosterol content in plants.—cont'd

Method	Species	Population	QTL for total sterol content/composition	Notes	References
QTL mapping	*Oryza sativa*	5 segregating populations derived from crosses between *japonica* and *indica*-type	– Minor QTLs and candidate regions for total γ-oryzanol content	Detection of QTLs and candidate regions dependent on the cross combination	Kato, Matsukawa, and Horibata (2017)
Association mapping	*Prunus amygdalus Batsch*	71 almond cultivars	13 significant associations between seven markers and eight phytosterol traits: – 2 Markers associated with both total phytosterol content and individual phytosterol content – 2 Markers associated with individual phytosterol content only	High percentage of explained phenotypic variation	Font i Forcada et al. (2015)

In those mutants, increased HMGR activity resulted in dramatic accumulation of cycloartenol, the first committed intermediate in plant sterol biosynthesis pathway and substrate of the SMT1 enzyme (Fig. 2, Table 3). Interestingly, all extra produced sterols were found as conjugates of fatty acid (steryl esters). The underlying mechanism leading to sterol over-accumulation in those "historical" mutants remains up to this date unknown. Although only a few altered sterol phenotypes have been identified over the years through genetic screens (Table 3), forward genetic approaches remain relevant to identify new players in regulation of sterol biosynthesis as exemplified by the recent characterization of HISE1, a proposed negative regulator of HMGR (Shimada et al., 2019).

Almost concomitantly with the mutagenic approach, transgenic tobacco lines overproducing sterol were obtained through overexpression (OE) of heterologous HMGR in two independent studies (Chappell, Wolf, Proulx, Cuellar, & Saunders, 1995; Schaller et al., 1995), highlighting the rate limiting activity of HMGR in sterol biosynthesis. Following the seminal work of Chappell and Schaller, numerous research groups generated transgenic lines overexpressing genes encoding limiting enzymes of the sterol biosynthesis pathway. The outcome of these studies is summarized in Table 3. A first wave of studies, exemplified by a series of papers from Safford's lab, focused on overexpressing *HMGR* and/or *SMT1* from different sources in tobacco and Arabidopsis. OE of native or recombinant *HMGR* recapitulates the initial findings of Chappell and Schaller: an overall increase in total sterol content driven by a dramatic accumulation of cycloartenol (Harker et al., 2003). OE of SMT1 led to a reduction in cycloartenol content concomitant with an increase in downstream intermediates content (Holmberg et al., 2002). Finally, OE of both *HMGR* and *SMT1* resulted in a 2.5-fold increase in total sterol content whereas 2.2- or 1.2-fold increase were achieved with individual OE of HMGR or SMT1, respectively (Holmberg et al., 2003). High residual levels of sterol intermediates in those OE lines suggests the existence of additional bottlenecks in the sterol biosynthesis pathway. For example, SMO1 and DWF1 in addition to SMT1 enzymatic activities were shown to be rate limiting in transgenic lines overexpressing *HMGR* (Lange, Poirier, Herron, & Lange, 2015). Moreover, OE of SMT2 led to dramatic increase of sitosterol (Schaeffer, Bouvier-Navé, Benveniste, & Schaller, 2000; Sitbon & Jonsson, 2001) and OE of CYP710A8 to an increase of stigmasterol at the expense of sitosterol (Tang et al., 2011). Finally, development of molecular cloning system allowing efficient multigene engineering has opened the way to the

Table 3 Main outcomes of forward (top) and reverse (bottom) genetic approaches to find high sterol plants and algae.

● 1 to 2, ● 2 to 5, ● 5 to 10

Mutant/clone	Species / Mutagenesis and selection method	Analyzed tissue(s) for sterol content	Change in total sterol content	Highest changes in individual sterol content (compared to wt, in absolute value)	Reference
3 clones R3-7, 7-4, R20-3	*Chlamydomonas reinhardtii* random plasmid integration and selection of resistant cells against a growth inhibitor (polyene antibiotic nystatin)	Cell culture	na	Accumulation of ergosterol and 7-dehydroporiferasterol	Salimova et al., 1999
lab1-4	*Nicotiana tabacum L. var Xanthi* UV-mutagenesis of dihaploid protoplast cultures and selection for resistance to lethal dose of triazole drug	Callus	●	Cycloartenol : increase (up to 8-fold) 24-methylene cycloartanol : increase (not detected in wild-type) 24-ethylidene-lophenol : increase (not detected in wild-type)	Maillot-Vernier et al., 1991
lab1-4	*Nicotiana tabacum L. var Xanthi* UV-mutagenesis of dihaploid protoplast cultures and selection for resistance to lethal dose of triazole drug	Leaves	●	Cycloartenol : increase (up to 10-fold) 24-methylene cycloartanol : increase (up to 5-fold) 24-ethylidene-lophenol : increase (not detected in wild-type)	Maillot-Vernier et al., 1991
26B	*Nicotiana tabacum L. var Xanthi* UV-mutagenesis of dihaploid protoplast cultures and selection for resistance to lethal dose of triazole drug	Callus	●	Cycloartenol : increase (up to 59-fold) 24-methylene cycloartanol : increase (up to 43-fold) 24-ethylidene-lophenol : increase (not detected in wild-type)	Schaller et al., 1992
hise1	*Arabidopsis thaliana* mutagenesis of transgenic lines expressing a GFP fusion of the oil-body membrane protein CALEOSIN3 (CLO3) and screen for CLO3–GFP fluorescence (presence of oil bodies)	Leaves	na	Accumulation of sterol esters / proliferation of oil bodies	Shimada et al., 2019

OE gene(s)	Species	Analyzed tissue(s) for sterol content	Change in total sterol content	Highest changes in individual sterol content (compared to wt. in absolute value)	Reference
Chlamydomonas reinhardtii SQS	*Chlamydomonas reinhardtii*	Cell culture	no change	squalene not detected	Kajikawa et al., 2015
Phaeodactylum tricornutum HMGR or *tHMGR*	*Phaeodactylum tricornutum*	Cell culture	no change	Squalene : increase (up to 10-fold) Cycloartenol : increase (up to 3-fold) Obtusifoliol : increase (up to 2.5-fold) Campesterol : decrease (up to 2-fold)	Jaramillo-Madrid et al., 2020
Brassica juncea wild-type or mutant (replacement of conserved amino acid residues) *HMGS1*	*Arabidopsis thaliana*	Leaves and Seedlings	●	Stigmasterol : increase (up to 2.5-fold)	Wang et al., 2012
Brassica juncea wild-type or mutant (replacement of conserved amino acid residues) *HMGS1*	*Solanum lycopersicum*	Leaves and Seedlings	●	Squalene : increase (up to 2.1-fold) Cycloartenol : increase (up to 2-fold)	Liao et al., 2018
N-terminal truncated *Hamster HMGR (t-HMGR)*	*Nicotiana tabacum*	Callus	●	Cycloartenol : increase (up to 80-fold) developmental stage dependent increase	Chappell et al., 1995
		Leaf	●	Cycloartenol : increase (up to 68-fold) developmental stage dependent increase	
		Root	●	Cycloartenol : increase (up to 17-fold) developmental stage dependent increase	

Continued

Table 3 Main outcomes of forward (top) and reverse (bottom) genetic approaches to find high sterol plants and algae.—cont'd

Hevea brasiliensis HMGR1	*Nicotiana tabacum*	Leaves	●	Cycloartenol : increase (up to 142-fold) Increase of the level of sterol during development	Schaller et al., 1995
		Callus	●	same pattern as in leaf tissues (data not shown)	
Hevea brasiliensis HMGR1	*Nicotiana tabacum*	Seed	●	na	Harker et al., 2003
Hevea brasiliensis N-terminal truncated HMGR1 (t-HMGR1)	*Nicotiana tabacum*	Seed	●	Cycloartenol : increase (up to 5-fold)	
		Leaves	●	Cycloartenol : increase (up to 100-fold)	
Arabidopsis thaliana HMGR1	*Solanum lycopersicum*	Leaves	●	na	Enfissi et al., 2005
		Fruits	●	Campesterol : increase (up to 5.7 fold) Cycloartenol : increase (up to 4.2 fold) Sitosterol : increase (up to 3 fold) increase in total sterol content not inherited in the next generation	
Arabidopsis thaliana HMGR1 and two modified At*HMGR1* lacking the target site for phosphorylation by SnRK1	*Nicotiana tabacum*	Seed	●	Sitosterol : increase (up to 1.2 fold)	Hey et al., 2006
		Leaf	no change	na	
Arabidopsis thaliana HMGR1	*Lavandula latifolia*	Leaves	na	Sitosterol and stigmasterol : increase (up to 1.8 and 1.9-fold) only stigmasterol and sitosterol quantified	Muñoz-Bertomeu et al., 2007

Arabidopsis thaliana HMGR1	*Digitalis minor*	Leaves (in vitro grown)	na	Sitosterol : increase (up to 4.8-fold)	Sales et al., 2007
		Leaves (greenhouse condition)	na	Stigmasterol : increase (up to 7.2-fold) only stigmasterol and sitosterol quantified	
Panax ginseng HMGR1	*Arabidopsis thaliana*	Leaves	●	Sitosterol : increase (up to 2-fold)	Kim et al., 2014
		Inflorescences	●	Cycloartenol : increase (up to 2.6 fold)	
Arabidopsis thaliana HMGR1	*Taraxacum brevicorniculatum*	Roots	●	Squalene and 2–3-oxidosqualene : increase (more than 15-fold) Campesterol, stigmasterol and sitosterol : increase (up to 5-fold)	Pütter et al., 2017
Arabidopsis thaliana ACL (ATP citrate lyase)			●	Pentacyclic triterpenes (taraxasterol, α-amyrin, β-amyrin and lupeol) : increase (up to 1.6-fold)	
Arabidopsis thaliana AACT (acetoacetyl-CoA thiolase)			●	Pentacyclic triterpenes (taraxasterol, α-amyrin, β-amyrin and lupeol) : increase (up to 1.7-fold)	
AtHMGR1, AtACL, AtAACT (combined)			●	Squalene and 2–3-oxidosqualene : increase (more than 100-fold) Campesterol, stigmasterol and sitosterol : increase (up to 5-fold)	
Nicotiana tabacum SMT1	*Nicotiana tabacum*	Seed	●	Cycloartenol : decrease (up to 2.1-fold) Cholesterol : decrease (up to 1.5-fold)	Holmberg et al., 2002
		Leaf	no change	Cycloartenol : decrease (at least 25-fold, under the limit of detection) Cholesterol : decrease (up to 5.4-fold)	

Continued

Table 3 Main outcomes of forward (top) and reverse (bottom) genetic approaches to find high sterol plants and algae.—cont'd

Hevea brasiliensis t-HMGR1	*Nicotiana tabacum*	Seed	●	Cycloartenol : increase (up to 6.2-fold) 24-methylene cycloartenol : increase (up to 19-fold)	Holmberg et al., 2003
Nicotiana tabacum SMT1	*Nicotiana tabacum*	Seed	●	Cycloartenol : decrease (up to 1.7-fold)	
Hevea brasiliensis t-HMGR1 and *Nicotiana tabacum SMT1* (combined)	*Nicotiana tabacum*	Seed	●	Cycloartenol : increase (up to 3.4-fold) 24-methylene cycloartenol : increase (up to 54-fold)	
Arabidopsis thaliana SMT2	*Nicotiana tabacum*	Leaves	no change	Sitosterol : increase (around 3-fold) 24-methyl cholesterol : decrease (around 50-fold)	Schaller et al., 1998
Arabidopsis thaliana SMT2-1	*Nicotiana tabacum*	Shoot meristematic tissue	no change	24-methyl cholesterol : decrease (up to 3.5-fold) Cholesterol : decrease (up to 2.4-fold) Sitosterol : increase (up to 1.5-fold) Stigmasterol : increase (up to 1.8-fold)	Schaeffer et al., 2000
Nicotiana tabacum SMT1-1			no change	Cycloartenol : decrease (up to 8-fold) Sitosterol : increase (up to 1.5-fold)	
Glycine max SMT1	*Nicotiana tabacum*	Leaves	no change	Cholesterol : decrease (up to 2.1-fold)	Sitbon & Jonsson, 2001
Arabidopsis thaliana SMT2			no change	Sitosterol : increase (up to 1.6-fold) 24-methyl cholesterol : decrease (up to 1.68-fold)	
Arabidopsis thaliana t-HMGR1	*Glycine max*	Seed	●	Cycloartenol : increase (around 3-fold)	Neelakandan et al., 2012
At t-HMGR1 combined with AtSMT1			less than 1.1-fold increase	Cycloartenol : decrease (under the limit of detection)	
Glycine max SMT2-2			less than 1.1-fold increase	Cycloartenol : decrease (under the limit of detection)	

Arabidopsis thaliana HMG1	*Arabidopsis thaliana*	Leaves	●	Cycloartenol : increase (around 90-fold) Isofucosterol : increase (around 10-fold) Sitosterol : increase (around 1.5-fold)	Lange et al., 2015
Arabidopsis thaliana SMO1			●	na	
Arabidopsis thaliana HMGR and SMO1 (combined)			●	Cycloartenol : increase (around 25-fold)	
Arabidopsis thaliana DWF1			no change	Isofucosterol : increase (around 1.4-fold)	
Arabidopsis thaliana HMG1 and DWF1 (combined)			●	Isofucosterol : increase (around 6-fold) Cycloartenol : increase (around 35-fold)	
Arabidopsis thaliana CYP710A1	*Arabidopsis thaliana*	Seedlings	na (no change)	Stigmasterol : increase (up to 32-fold) Sitosterol : decrease (up to 3-fold)	Morikawa et al., 2006
Arabidopsis thaliana CYP710A2			na (no change)	Stigmasterol : increase (up to 24-fold) Brassicasterol / crinosterol : increase (up to 18-fold) 24-epi-campesterol/campesterol : decrease (up to 9.2-fold) Sitosterol : decrease (up to 2.3 fold)	
Solanum lycopersicum CYP710A11			na (no change)	Stigmasterol : increase (up to 28-fold) Sitosterol : decrease (up to 1.8-fold)	
Arabidopsis thaliana CYP710A1	*Arabidopsis thaliana*	Leaves	na (no change)	Stigmasterol : increase (up to 33-fold) Sitosterol : decrease (up to 1.7-fold)	Arnqvist et al., 2008
Arabidopsis thaliana CYP710A4			na (no change)	Stigmasterol : increase (up to 1.8-fold) Sitosterol : decrease (up to 1.3-fold)	
Physcomitrella patens CYP710A14	*Arabidopsis thaliana*	Cell culture	na (no change)	Stigmasterol : increase (up to 72-fold)	Morikawa et al., 2009
Triticum aestivum CYP710A8	*Arabidopsis thaliana*	Seedlings	no change	Stigmasterol : increase (up to 10 to 20-fold) Sitosterol : decrease (up to 4 to 26-fold)	Tang et al., 2011
Hordeum vulgare CYP710A8			no change		

Overexpression of wheat and barley *DWARF5 in arabidopsis* led to no detectable change in sterol content

References: Salimova, Boschetti, Eichenberger, and Lutova (1999), Maillot-Vernier et al. (1991), Schaller et al. (1992), Shimada et al. (2019), Kajikawa et al. (2015), Jaramillo-Madrid et al. (2020), Wang et al. (2012), Liao, Chen, Wang, Bach, and Chye (2018), Chappell, Wolf, Proulx, Cuellar, and Saunders (1995), Schaller et al. (1995), Harker et al. (2003), Enfissi et al. (2005), Hey et al. (2006), Muñoz-Bertomeu et al. (2007), Sales et al. (2007), Kim et al. (2014), Pütter et al. (2017), Holmberg et al. (2002), Holmberg et al. (2003), Schaller et al. (1998), Schaeffer et al. (2000), Sitbon & Jonsson (2001), Neelakandan et al. (2012), Lange et al. (2015), Morikawa et al. (2006), Arnqvist et al. (2008), Morikawa et al. (2009), and Tang et al. (2011).

production of "high value" sterols plants as exemplified with stable OE of 11 tomato cholesterol pathway genes to increase cholesterol production in *Arabidopsis thaliana* (Sonawane et al., 2016).

Whether orthologs of these genes constitute relevant targets for similar metabolic engineering approaches to increase sterol content in algae is questionable. Although scarce, the results obtained so far suggest the existence of different or additional regulatory mechanisms to limit the accumulation of end-product sterols. OE of *SQS* in the green microalgae *Chlamydomonas reinhardtii* led to no significant accumulation of squalene and left the amounts of sterols unchanged (Kajikawa et al., 2015) while OE of *HMGR* in the diatom *Phaeodactylum tricornutum* led to significant accumulations of downstream sterol pathway intermediates but no change in total sterol levels (Jaramillo-Madrid et al., 2020).

5. Conjugated sterols

The C3 hydroxy group of sterols is the target of enzymes catalyzing the formation of conjugated form of sterols. A non-negligible fraction of sterol lipids in plant and algae is actually composed of steryl esters (SEs), steryl glucosides (SGs) and acyl steryl glycoside (ASGs) where the sugar C6 hydroxy group of SGs is further esterified via a fatty acid. In addition to fatty acids and sugars, sterol can also be linked to other classes of moiety. In cereal grains, a significant amount of sterol is found as esters of phenolic acid or steryl ferulate (SFs) while sterols sulfates (StSFs) are occasionally found in algae and marine invertebrates (Fig. 4).

5.1 Occurrence, diversity and biological roles of phytosterol glucosides and esters of fatty acids

5.1.1 Distribution in plants and algae

Since they were first isolated, SEs (MacLachlan, 1936), SGs (Power & Salway, 1913) and ASGs (Lepage, 1964) have been detected in a variety of plant tissues (Dyas & Goad, 1993; Nyström, Schär, & Lampi, 2012) as well as in various group of algae (Duperon, Thiersault, & Duperon, 1983). In contrast to free sterols (FSs), conjugates of sugars and fatty acids are generally viewed as minor although common components of the plant and algae sterol lipid fractions. In the model plant species *A. thaliana*, 20% of sterols are either esterified (9%) or glycosylated (11%) (SGs + ASGs) (Wewer, Dombrink, & vom Dorp, & Dörmann, 2011). The proportion of SGs that is further esterified into ASGs is variable in plants. Whereas a majority of glycosylated

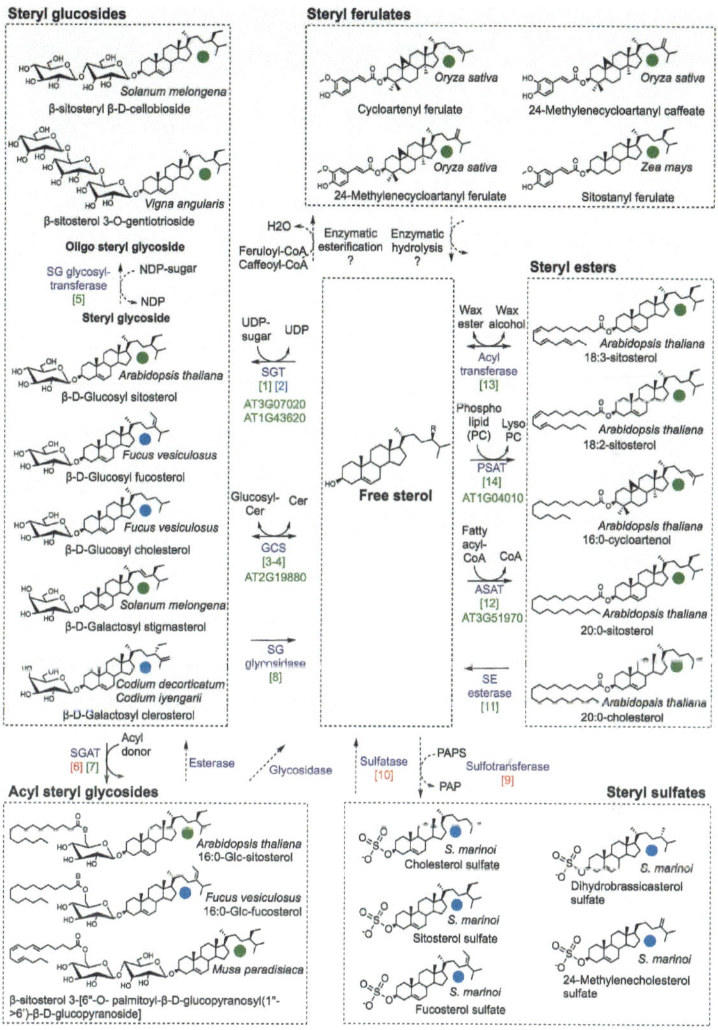

Fig. 4 Conjugated forms of sterols in plants and algae. The C3 hydroxy group of "free" sterols is the target of multiple enzymes catalyzing the formation of sterol conjugates. The nature of the compound linked to the sterol moiety (sugar, fatty acids, phenol or sulfates) define different classes of conjugated form of sterols. Steryl glucosides, acyl steryl glycosides and sterol esters are ubiquitously found in plants and algae while steryl ferulates are restricted to *Poaceae* in plants and steryl sulfates to diatoms in algae. Examples of compounds detected in the land plant *Arabidopsis thaliana* and the brown algae *Fucus vesiculosus* as well as in other plant or algae species (filled green or blue circle, respectively) are shown (more details in the text). Interconversions between free and conjugated forms of sterols are represented with a plain or dashed arrow depending on whether the enzymatic reaction is supported by experimental evidence in plants, algae or in other models (references in green, blue or red, respectively).

(Continued)

sterols is found as "free" SGs in *A. thaliana* (Wewer et al., 2011), ASGs are predominant over SGs in Solanum species (Furt, Simon-Plas, & Mongrand, 2011; Nyström et al., 2012). These general proportions are rather similar in algae, for example, 11% of sterol is on average either esterified (6%) or glycosylated (5%) in a subset of green, red and brown algae (Duperon et al., 1983). Nevertheless, large interspecific variation in the relative content of conjugated form of sterol is observed in plants and algae (Dyas & Goad, 1993; Nyström et al., 2012). SGs or SEs are even reported as the dominant sterol fraction in algae species grown under standard condition. For example, more than 60% of sterols are glycosylated in the marine diatoms *Chaetoceros calcitrans* and *Phaeodactylum tricornutum* while SEs are dominant in the eustigmatophyte *Nannochloropsis oculata* and the diatom *Thalassiosira pseudonana* (Véron, Billard, Dauguet, & Hartmann, 1996; Véron, Dauguet, & Billard, 1998). Detailed sterol profiling in plants have revealed enrichment of SGs or SEs in specific tissues or organs. For example, glycosylated sterols are highly dominant in tomato fruit, in which, depending on the ripening stage, 86% to 66% of sterol are glycosylated (Whitaker, 1988). SEs are notably enriched in seeds, flowers and senescing tissues (Debolt et al., 2009; Dyas & Goad, 1993). Intriguingly, SGs and SEs, are enriched in the phloem sap of *Phaseolus vulgaris* and *Nicotiana tabacum* (Behmer et al., 2013) and accumulate in the pollen coat (Choi et al., 2014; Hernández-Pinzón, Ross, Barnes, Damant, & Murphy, 1999; Wu, Moreau, Whitaker, & Huang, 1999).

5.1.2 Nature of the sterol moiety and the C3 substituents

Sterol moieties in fractions of conjugates usually reflects the mixture of free sterols found in plants or algae (Duperon et al., 1983; Kovganko & Kashkan, 1999). Sugar moiety in plants and algae SGs or ASGs is predominantly glucose while fatty acids moiety in plants SEs is commonly palmitic, stearic, oleic, linoleic or linolenic acid (Dyas & Goad, 1993). Little information is

Fig. 4—Cont'd When cloned in plants, the corresponding gene accession is given in *Arabidopsis thaliana*. SGT, sterol glucosyltransferase; GCS, glucosylceramide synthase; PSAT, phospholipid:sterol acyltransferases; ASAT, acyl-CoA:sterol acyltransferases; SGAT, steryl glycoside acyltransferase; PAPS, 3′-phosphoadenosine 5′-phospho-sulfate; PAP, 3′-phosphoadenosine 5-phosphate. *References: [1] Warnecke and Heinz (1994), [2] Hopp et al. (1978), [3] Lynch et al. (1997), [4] Hillig et al. (2003), [5] Peng et al. (2002), [6] Jan et al. (2020), [7] Heinz et al. (1975), [8] Kalinowska and Wojciechowski (1978), [9] Chapman et al. (2004), [10] Reed et al. (2005), [11] Kalinowska and Wojciechowski (1983), [12] Chen et al. (2007), [13] Kalinowska et al. (1989), [14] Banaś et al. (2005).*

available on the exact nature of the fatty acid moiety in algal SE. These general observations are exemplified in plants and brown algae with the sterol profiles established in A. *thaliana* leaves (Wewer et al., 2011) and in *Fucus vesiculosus* (Duperon et al., 1983) with example of highly abundant compounds given in Fig. 4. In *A. thaliana*, SG and ASG composition is dominated by Glc-sitosterol and 16:0-Glc-sitosterol, respectively while the most abundant SEs are 18:3-sitosterol and 18:2-sitosterol. In *Fucus vesiculosus*, consistent with sterol composition of the free form, the most common SEs, SG and ASG are conjugates of fucosterol (and to a lesser extent of cholesterol).

Nevertheless, when accessed over a wide variety of species and tissues, deviation in sterol profile is found between free and conjugated sterol fractions. For example, in plant foods, sterol moiety in SGs does not reflect free sterol composition in black eyed beans, aubergine, flaxseed, sunflower seeds and wheat flour (Nyström et al., 2012). This also holds true in algae where differences in profile were noted between sterol fractions, notably in microalgae (Mohammady, 2004; Véron et al., 1998). For example, the SG fraction of the marine unicellular algae *Tetraselmis chiu* is dominated by cholesterol and 22-dehydrocholesterol while these compounds are virtually absent from the FS fraction (Mohammady, 2004). Differences in sterol composition is also found between the FS and SE fraction, notably in pollen grains of angiosperms species. Whereas the FS profile is dominated by sterols with a double bond at C5 (like sitosterol for example), SEs profile is dominated by cycloartenol derivatives, an example of which, 16:0-cycloartenol is represented in Fig. 4 (Villette et al., 2015).

Unusual sugar and fatty acid moiety are also found in sterol conjugates. Sugar moiety other than glucose (e.g., galactose, mannose, xylose and others) have been detected in plants (Grille, Zaslawski, Thiele, Plat, & Warnecke, 2010; Kovganko & Kashkan, 1999) and in algae. For example, in eggplant (*Solanum melongena),* galactosides of sitosterol and stigmasterol were quantified and found 100-times less abundant than their glucosides counterparts (Heinz & Glomb, 2018) (Fig. 4). In algae, a β-D-galactoside cholesterol has been detected in the marine green algae *Codium decorticatum* and *Codium iyengarii* (Fig. 4) (Ahmad, Aliya, Perveen, & Shameel, 1992, 1993). In addition, occurrence of steryl and acyl steryl oligosaccharides have been sporadically reported in plants. For example, conjugated of sitosterol with multiple sugars residues have been isolated in adzuki bean (*Vigna angulari)* (Kojima, Ohnishi, Ito, & Fujino, 1989) and more recently in eggplant (Heinz & Glomb, 2018) while acylated forms have been reported in cooking banana *(Musa paradisiaca)* (reviewed in Kovganko & Kashkan, 1999) (Fig. 4). SE with unusual fatty acid moiety have also been reported in *A. thaliana* where a non-negligible amount of cholesterol and sitosterol were found acylated with very-long-chain (C20:0) saturated fatty acids (Fig. 4) (Wewer et al., 2011).

5.1.3 Biological roles

The biological roles of sterol conjugates have been extensively reviewed in Ferrer, Altabella, Arró, and Boronat (2017). While the significance of tissue specific enrichment in sterol conjugates is not always fully understood, genetic studies in plants have revealed central roles of SGs and SEs in membrane ordering and free sterol homeostasis related process, respectively. Like free sterols, SGs accumulate in the outer leaflet of the cellular membrane (Cacas et al., 2016) where they contribute to the regulation of its physico-chemical properties (Mamode Cassim et al., 2019). As such, SGs are viewed as modulators of membrane function and variation of SGs levels has been associated with differential plasma membrane (PM) mediated response, notably to abiotic and biotic stress. Enhanced tolerance or sensitivity to abiotic stress have been observed in various species depending on whether the levels in SGs have been experimentally increased or decreased, respectively (Mishra et al., 2013, 2015). In addition, SGs levels have been shown to vary in response to various stress. For example, cold stress in rye *(Secale cereale L.)* resulted in drastic change in membrane lipid composition and notably in a decrease in the proportion of SG and ASG (Lynch & Steponkus, 1987). More recently, lipidomic analysis of *Arabidopsis thaliana* leaves revealed an accumulation of ASG and SG in response to water deprivation (Tarazona, Feussner, & Feussner, 2015). The role of plant SGs in response to biotic stress has been less studied. Recently, deficiency in SGs has been linked through a yet to be discovered pathway to increased resistance against the pathogen *Botrytis cinerea* in *Arabidopsis thaliana* (Castillo et al., 2019).

In contrast to glycosylated form of sterols, SEs are excluded from the PM. Like in yeast and animals, plant SEs are preferentially found in cytoplasmic lipid bodies, commonly referred to as lipid droplet (LD). The role played by SEs in the regulation of free sterols homeostasis in plants has notably been revealed in overproducing sterol mutants (Table 3). In those mutants, excess of free sterols is converted into SEs and stored into LDs (Maillot-Vernier et al., 1991; Schaller et al., 1992; Shimada et al., 2019). Similarly, accumulation of SEs has been observed during senescence and seed maturation.

For example, rapid increase in sterol content during seed development is mainly driven by accumulation of SEs while the level of FSs remains fairly constant in tobacco and oilseed rape seeds (Harker et al., 2003). SEs are thus generally considered as a storage pool of sterols when produced in excess.

No biological role for steryl esters or steryl glucosides has been evidenced in algae yet. Light spectra and temperature have been showed to drastically impact the relative proportion of SGs, ASGs and in a minor extent SEs in the diatom *Phaeodactylum tricornutum* (Véron et al., 1996). Functional analysis studies are necessary to gain additional insights on the link between conjugated sterols and response to changes in environmental conditions.

5.2 Occurrence, diversity and biological roles of phytosterol ferulates and sulfates

In contrast to the ubiquitous above-mentioned SEs, SGs and ASGs, the distribution of SFs and StSFs is restricted to a subset of plant and algae species, respectively.

Occurrence of phenolic conjugates was first reported in rice bran (i.e., the kernel outer layer) (Kaneko & Tsuchiya, 1955) and SFs were since commonly reported in other cereal grains. SFs can represent a non-negligible fraction of total sterol in cereal grains. In corn, when accessed over 49 cultivars, around 10% of total sterols are phenolic conjugates (Moreau, Singh, & Hicks, 2001). Although SFs are suggested to have health promoting properties in humans (notably through their antioxidant and cholesterol lowering properties) (Nyström, Achrenius, Lampi, Moreau, & Piironen, 2007; Wilson, Nicolosi, Woolfrey, & Kritchevsky, 2007), their function in plant remains unknown. Interestingly, significant differences are found between the sterol profile of free and phenolic conjugates of sterol in cereals. For example, while the FS fraction of corn, wheat, rye and spelt grains is dominated by sitosterol, the SF fraction is dominated by sitostanols (Fig. 4) in corn and a mixture of campestanols and sitostanols in wheat, rye and spelt (Esche, Barnsteiner, Scholz, & Engel, 2012). This difference is even more pronounced in rice bran where the FS fraction is dominated by sitosterol, campesterol and stigmasterol while the corresponding SF fraction (also called oryzanol) is dominated by 24-methylenecycloartenol and cycloartenol esters of ferulate (Fig. 4) (Ha et al., 2006). Ferulic acid, a derivative of cinnamic acid (4-hydroxy-3-methoxycinnamic acid) is the most common phenol moiety in SFs. Nevertheless, coumaric, sinapic and caffeic acids are also found esterified to sterol in rice (Fang, Yu, & Badger, 2003), corn (Seitz, 1989) or barley (Ito et al., 2019) (Fig. 4).

Despite being present in other marine organisms like starfish and sponges, StSFs have only been sporadically detected in diatoms and in the haptophyte *Hymenomonas* sp. (Gallo, D'Ippolito, Nuzzo, Sardo, & Fontana, 2017 and references within). A wide range of variation is found for StSFs content in diatoms whereas they are not found in dinoflagellates, another major component of the marine phytoplankton (Nuzzo et al., 2018). In the diatom species *Skeletonema marinoi*, StSFs act as a signaling compound in the regulation of cell death (Gallo et al., 2017). An overall good correlation is reported between sterol composition of free and sulfated sterols (Gallo et al., 2017; Nuzzo et al., 2018). Accordingly, sterol composition in sulfated sterols reflected the mixture of free sterols reported in the diatom species *Skeletonema marinoi* with a StSFs fraction dominated by cholesterol, dihydrobrassicasterol, and sitosterol while sulfate derivatives of fucosterol and 24-methylenecholesterol are minor compounds (Fig. 4). More generally, 24-methylene cholesterol and 24-methylcholesterol are the most frequently detected sterol moiety in diatoms (Gallo et al., 2017; Nuzzo et al., 2018).

5.3 Genes and enzymes of sterol conjugation

5.3.1 Sterol acyltransferases

The genes encoding sterol acyl-transferases cloned and characterized in *A. thaliana* encode for a phospholipid:sterol acyltransferase (PSAT) and a acyl-CoA:sterol acyltransferase (ASAT) (Fig. 4). *AtPSAT (AT1G04010)* and *AtASAT (AT3G51970)*, were identified and biochemically characterized in the context of a candidate gene approach for homologs of the animal LCAT (lecithin cholesterol acyltransferase) (Banaś et al., 2005) and ACAT (acyl-CoA:cholesterol acyltransferase) type of enzyme (Chen, Steinhauer, Hammerlindl, Keller, & Zou, 2007). AtPSAT preferentially transfers fatty acid from the sn-2 position of phosphatidylethanolamine to the sterol acceptors campesterol and sitosterol while AtASAT preferentially transfers fatty acid from acyl-CoAs to cycloartenol. While AtPSAT is a major SE synthetizing enzyme in *A. thaliana* seeds and leaves, only a minor AtASAT contribution to SE formation in leaves was observed (Bouvier-Navé et al., 2010).

Orthologs of *AtPSAT* and *AtASAT* have been since cloned and characterized in other plant species and notably in tomato (*StPSAT* and *StASAT*) (Lara et al., 2018). Transient expression of the corresponding proteins in

Nicotiana benthamiana leaves gave additional insights on their subcellular localization: AtPSAT was associated to ER micro-domains (Shimada et al., 2019) while StPSAT localized to cytosolic lipid droplets (Lara et al., 2018). Interestingly, a plasma membrane and not an endoplasmic reticulum (ER) localization (like it is the case for AtASAT) was found for StASAT (Lara et al., 2018).

Residual level of SEs in *Arabidopsis thaliana psat/asat* double mutants, notably in flowers, suggests activity of additional sterol acyltransferases or an alternative esterification mechanism (Bouvier-Navé et al., 2010). On the basis of biochemical studies, a production of steryl ester through reversible transesterification between wax esters and sterols has been proposed (Kalinowska, Zimowski, Bucholc, & Wojciechowski, 1989) (Fig. 4).

In algae, no gene information for acyl-transferase is available. A genome mining approach using *Arabidopsis thaliana* sequence as a query yielded candidate genes for *ASAT* but not for *PSAT* in *Chlamydomonas reinhardtii* (Darnet et al., 2021).

5.3.2 Sterol glucosyltransferases

The first sterol glucosyltransferases (SGT) gene cloned in plants was identified from protein sequence information after purification of the corresponding enzyme in *Avena sativa* (Warnecke & Heinz, 1994). After cloning of the gene in *Avena sativa* (*UGT80A1)* and in *A. thaliana (UGT80A2-At3g07020)*, SGT activity was characterized in *E. coli* (Warnecke, Baltrusch, Buck, Wolter, & Heinz, 1997). A second SGT gene *UGT80B1 (At1g43620)* was identified in *A. thaliana* through homology search (Debolt et al., 2009) (Fig. 4). When accessed in vitro, both UGT80A2 and UGT80B1 preferentially use UDP-glucose as a glucose donor and can transfer sugar to 3-hydroxy groups of a wide range of Δ^5-sterols acceptors (Stucky, Arpin, & Schrick, 2015; Warnecke et al., 1997). Analysis of *ugt80A2, ugt80B1* and *ugt80A2/ugt80B1* double mutants suggests that while both UGT80A2 and UGT80B1 contribute in an additive way to sterol glycosylation (Debolt et al., 2009), there is some level of functional diversification between the two SGTs: in seeds, while *ugt80A2* is mainly depleted in the most common SGs (sitosteryl, stigmasteryl, and campesteryl glucoside), *ugt80B1* is depleted in the less abundant SG brassicasteryl glucoside and in ASGs (Stucky et al., 2015).

Based on homology search using UGT80A2 and UGT80B1 sequences, other SGT genes have been since identified and cloned in plants, where they are usually found as small gene families (Chaturvedi et al., 2012;

Li et al., 2014; Ramirez-Estrada et al., 2017; Sharma, Madina, Chaturvedi, Sangwan, & Tuli, 2007; Tiwari et al., 2014). For example, in tomato, four *SGT* genes were shown to encode functional sterol glucosyltransferases (Ramirez-Estrada et al., 2017). Subcellular localization of tomato and Arabidopsis recombinant SGT accessed using imaging and immunoblot based techniques suggested a peripheral membrane behavior of SGT (Ramirez-Estrada et al., 2017).

Whether the synthesis of (rare) steryl oligoglycosides is catalyzed by a specific steryl glucoside glycosyltransferase (SG glycosyltransferase) or by SGT is not known (Fig. 4). Evidence for an in vitro UDP-glucose-dependent SG glucosyltransferase activity has been detected in cotton fiber membranes (Peng, Kawagoe, Hogan, & Delmer, 2002).

Residual SGs content in the Arabidopsis double mutant *ugt80A2/ugt80B1* (6 to 10% of the wild-type SG levels) (Schrick et al., 2012) is indicative of additional SGTs or alternative ways to produce SGs. Based on homology with UGT80 enzyme, UGT713B has been suggested as a possible additional SGT but functional assays showed that UGT713B was unable to produce SG (Stucky et al., 2015). An alternative mechanism based on glucosylceramide synthase (GCS) activity has been suggested to be involved in SG production on the basis of biochemical studies (Fig. 4). Glucose transfer from SG to ceramide to form glucosylceramide has been evidenced in *Phaseolus vulgaris* (Lynch, Criss, Lehoczky, & Bui, 1997). Heterologous expression of *Gossypium arboretum* (cotton) GCS in *Pichia pastoris* mutants defective for both SGT and GCS activity resulted in some accumulation of SG in addition to glucosylceramide production (Hillig et al., 2003). In Arabidopsis, whether and the extent to which the only GCS identified so far (encoded by *At2g19880*) contribute to SGs synthesis in vivo is not known (Msanne et al., 2015).

In algae, gene information for SGT is scarce. Enzymatic glucosylation of sterols in the presence of UDP-glucose has been evidenced in the green algae *Chlorophyta Prototheca zopfii* and partial SGT purification has been reported (Hopp, Romero, Daleo, & Lezica, 1978). SGT functional candidates have been identified in diatoms (Stonik & Stonik, 2018) and chlorophyte based on genome mining (Darnet et al., 2021).

5.3.3 Sterol sulfotransferases

Sulfonation, the transfer of a sulfuryl group from a donor molecule to a variety of substrates, is catalyzed by members of the sulfotransferase (SULT) protein family (Fig. 4). Although putative sulfotransferases have been identified in a wide variety of organisms, most of the structural and mechanistic studies have been performed in mammalian models (Chapman, Best, Hanson, & Wong, 2004; Gamage et al., 2006). Sulfonation of cholesterol in mammals is catalyzed by the cytosolic sulfotransferases SULT2B1. Cholesterol sulfate in humans plays important role in skin integrity and adhesion but it is also involved in other biological processes (Strott & Higashi, 2003). In plants, AtSOT10 and AtSOT12 were shown to be active towards brassinosteroids in vitro and were hypothesized to be involved in brassinosteroid catabolism (reviewed in Hirschmann, Krause, & Papenbrock, 2014). Using data mining approaches, four putative proteins with a conserved SULT *Sulfotransfer-1* domain have been identified in diatoms (Gallo et al., 2017) but no functional characterization has been performed. StSFs are desulfated in mammals by a member of the sulfatase superfamily, a steroid sulfatase (Reed, Purohit, Woo, Newman, & Potter, 2005). Whether no counterpart for sulfatase has been reported in plant genomes, putative sulfatases are massively predicted in the genome of marine organisms. Their precise functions however, in the absence of experimental validation, remain unknown (Fig. 4).

5.3.4 A wealth of unknown conjugating enzymes

Although sterol acyl and glycosyltransferase activity have been since many years object of biochemical studies, the major breakthrough towards their functional characterization was the cloning of the corresponding genes. Similar success is yet to be achieved for the enzymes responsible for ASGs, SFs and StSFs biosynthesis as well for the hydrolysis of conjugated sterol.

In plants like in algae, no gene information is available for the enzymatic activity responsible for the acylation of SGs to form ASGs. Although steryl glycoside acyltransferase (SGAT) activity from various plant tissues has been characterized in vitro and partially purified (reviewed in Ferrer et al., 2017), our understanding of ASG biosynthesis remains limited (Fig. 4). The recent characterization of a cholesteryl α-D-glucopyranoside 6′-acyltransferase (CGAT) in *Helicobacter pylori* (Jan et al., 2020) might open new perspective for the identification and characterization of SGAT in plants and algae.

In contrast to the wealth of information available on chemical and enzymatic procedure for SFs synthesis and hydrolysis in vitro (Nyström, Moreau, Lampi, Hicks, & Piironen, 2008; Schär & Nyström, 2016), no gene information is available for the enzyme involved in SFs metabolism in cereal grains (Fig. 4).

Finally, little is known on the hydrolysis of conjugated sterols. Both esterase and glycosidase activities able to hydrolyze SEs and SGs (but not ASGs) have been detected in vitro in plant tissues (reviewed in Ferrer et al., 2017) (Fig. 4). Homologs of genes encoding hydrolase activities in the mammalians and yeast models have been searched for in plants. Orthologs of *Saccharomyces cerevisiae* SE hydrolases (Yeh1p, Yeh2p and Tgl1p), mammalian cholesteryl ester hydrolase but not of yeast ergosteryl-β-glucosidase Egh1 have notably been reported (Ferrer et al., 2017) but the biological relevance *in planta* of these functional candidates remain to be tested.

6. Molecular regulation of phytosterol metabolism

6.1 Genetic and biochemical aspects

Phytosterol production in algae and plants implies its regulation at the genetic and biochemical levels to maintain appropriate quantities of free sterols and sterol conjugates in cellular compartments (plasma membrane, endomembrane systems, lipid droplets) in the context of a general metabolic homeostasis. In plants, the enzyme 3-hydroxy-3-methylglutaryl CoA reductase (HMGR) is a limiting step that controls the production of sterol precursors (farnesyl diphosphate, squalene, 2,3-oxidosqualene, cycloartenol) and subsequently of phytosterols (Chappell et al., 1995; Schaller et al., 1995) (Fig. 5). Upon up-regulation of the activity of HMGR, cycloartenol and phytosterols accumulate mostly as esters of fatty acids generated by the enzyme PSAT and are stored in cellular lipid droplets that contain also triacylglycerols (Bouvier-Navé et al., 2010; Chapman, Aziz, Dyer, & Mullen, 2019). A gene called *High Sterol Ester 1* (*HiSE*) was identified from an Arabidopsis line isolated in a genetic screen in which the lipid droplet protein called caleosin 3 fused to the reporter GFP was dramatically increased upon the mutagenesis process (Shimada et al., 2019). The hise1 mutant was shown to contain high amounts of sterol esters caused by a strong increase of the HMGR activity. HISE1, which is similar to a mammalian membralin protein, is therefore considered as a negative regulator of the phytosterol rate-limiting enzyme HMGR, possibly implied in the control of the endoplasmic reticulum associated protein degradation of the plant enzyme HMGR (Darnet & Schaller, 2019b; Shimada et al., 2019) (Fig. 5). Interestingly, plant genomes do not contain homologs of mammalian regulatory components acting on cholesterol homeostasis such as the sterol regulatory element-binding protein (SREBP) and SREBP-cleavage

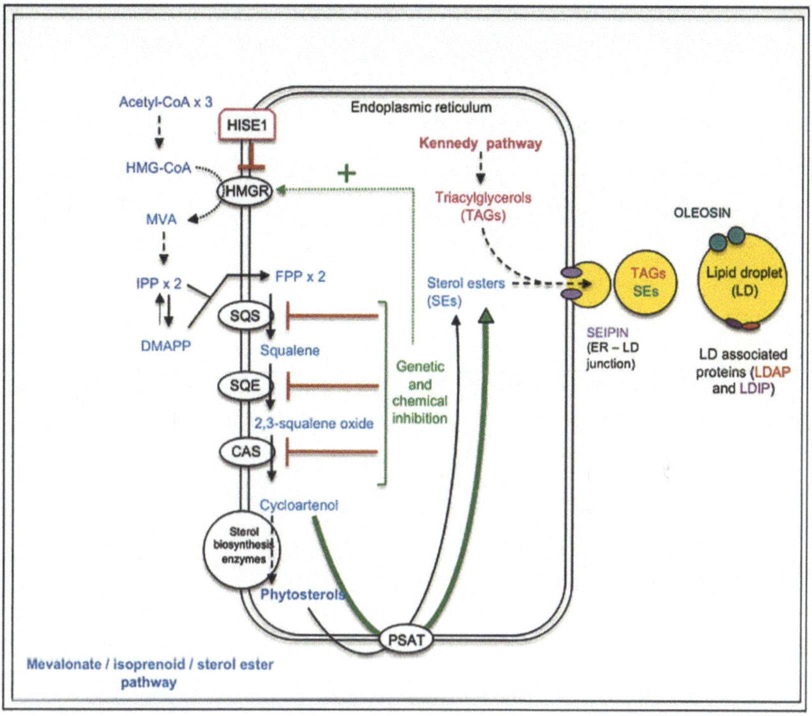

Fig. 5 Regulation of sterol accumulation in higher plant cells: key enzymes and regulatory loops are indicated. Increase in the metabolic flux towards phytosterol production leads to the conversion of sterol intermediates (notably cycloartenol, green thick arrow) and sterol end-products (although in a minor extent, thin black arrow) into sterol esters (SEs). Along with triacylglycerols (TAGs), SEs constitute the hydrophobic core of lipid droplet (LDs). The hydrophobic core and protein composition of LDs can vary between species in plant and algae. Here is represented a theoretical lipid droplet in *Arabidopsis thaliana* (Ischebeck, Krawczyk, Mullen, Dyer, & Chapman, 2020).

activating protein (SCAP) that control the mevalonate pathway (Brown & Goldstein, 1997) and must therefore use an alternative regulatory circuit that is just coming out of the dark. The chemical inhibition of squalene synthase and squalene epoxidase by squalestatin and terbinafine, respectively, in tobacco cells (Wentzinger, Bach, & Hartmann, 2002), the genetic inhibition of squalene epoxidase in an *Arabidopsis thaliana* mutant called *dry2* (Doblas et al., 2013), and the genetic inhibition of cycloartenol synthase (CAS1) in a weak *cas1* allele (Babiychuk et al., 2008) result in these biological materials in a marked increase of the activity of the enzyme HMGR, pointing out the existence of a yet unknown positive feedback mechanism between squalene metabolism and the upstream mevalonate pathway (Fig. 5).

This is reminiscent of the action of other chemicals found to enhance the production of triacylgycerols in diatoms (Conte et al., 2018).

6.2 Effect of the environment

Light strongly influences the biosynthesis of all isoprenoid biosynthetic segments including phytosterol pathways in plants (Rodríguez-Concepción et al., 2004). The same holds true in the case of algae of diverse taxa for which the production of sterols has been monitored upon varying light intensity and quality. Exhaustive sterol analysis carried out with the diatom *Phaeodactylum tricornutum* grown under different light spectral quality demonstrated that red light enhanced sterol production but most importantly an increase in temperature (from 13 to 23°C) considerably decreased sterol production (Véron et al., 1996). Light abiotic stress like UV-C treatment was shown to trigger sterol accumulation in the microalgae *Pavlova lutheri* with a dose-response production of all types of sterols that compose the profile of *P. lutheri* (Ahmed & Schenk, 2017). A pilot study on the effect of nutrient availability and light conditions in interaction as two major factors governing the production of phytosterol in the brown algae *Saccharina latissima* showed very clearly that combinations of light quality and nutrient availability had a tremendous action on phytosterol production (de Jong, Timmermans, de Winter, & Goverdina, 2021). In fact, an increase in sterol accumulation was concomitant to decreased photosynthetic performances of *S. latissima*, but high light and low nutrient availability was found the optimal combination to enhance sterol production. The control of these parameters in algae harvesting could be of economic relevance.

7. Concluding remarks

Sterol metabolism in algae and plants has not been approached with equivalent efforts so far with respect to gene and enzyme functions and regulation. Whereas some key regulatory components acting on the biosynthesis and accumulation of phytosterols are identified in plants, cognate elements acting similarly in algae have not been identified yet. Moreover, only a fragmentary vision of the genes and enzymes catalyzing the interconversions between free and conjugated sterols is available in algae and to a lesser extent in plants. Homology search-based and enzyme purification studies in plants led to cloning of genes encoding sterol glucosides and sterol esters, however, the genes and enzymes at play in the biosynthesis of sterol acylglucosides, sterol sulfates and sterol ferulates, as well as in the degradation of those conjugated sterols, remain unknown. Photosynthetic sterol autotrophic eukaryotes use distinct homeostasis determinants as those identified in other taxa, as master switches controlling sterol homeostasis are just coming out of the dark concerning seed plants.

Algae and plants are considered as a rich source of functional phytosterols. Metabolic engineering initiatives of selected species is "all about" manipulating the source (biosynthesis pathways, reception of environmental features acting on gene expression) and the sink capacities (like lipid droplets, Lupette et al., 2019) of cells and cell-types capable of converting photosynthetic carbon to compounds of interest to the bioeconomy (D'Adamo et al., 2019). Alternatively, screening natural isolates of microalgae for the production of valuable compounds and optimizing culture conditions may also lead to promising results, like it is the case for distantly related organisms (Zhang et al., 2019).

References

Abdallah, I. B., Tlili, N., Martinez-Force, E., Rubio, A. G. P., Perez-Camino, M. C., Albouchi, A., et al. (2015). Content of carotenoids, tocopherols, sterols, triterpenic and aliphatic alcohols, and volatile compounds in six walnuts (Juglans regia L.) varieties. *Food Chemistry, 173*, 972–978. https://doi.org/10.1016/j.foodchem.2014.10.095.

Aboobucker, S. I., & Suza, W. P. (2019). Why do plants convert sitosterol to stigmasterol? *Frontiers in Plant Science, 10*(March). https://doi.org/10.3389/fpls.2019.00354.

Ahmad, U. V., Aliya, R., Perveen, S., & Shameel, M. (1992). A sterol glycoside from marine green alga *Codium iyengarii*. *Phytochemistry, 31*(4), 1429–1431. https://doi.org/10.1016/0031-9422(92)80311-2.

Ahmad, V. U., Aliya, R., Perveen, S., & Shameel, M. (1993). Sterols from marine green alga *Codium decorticatum*. *Phytochemistry, 33*(5), 1189–1192. https://doi.org/10.1016/0031-9422(93)85047-U.

Ahmed, F., & Schenk, P. M. (2017). UV-C radiation increases sterol production in the microalga *Pavlova lutheri*. *Phytochemistry, 139*, 25–32. https://doi.org/10.1016/j.phytochem.2017.04.002.

Akihisa, T., Kojima, S., Yokota, T., & Tamura, T. (1991). 24-Methylene-25-methylcholesterol and both C-24 epimers of 24-ethyl-22-dehydrocholesterol in a freshwater green alga *Hydrodictyon reticulatum*. *Phytochemistry, 30*(11), 3621–3624. https://doi.org/10.1016/0031-9422(91)80079-G.

Akihisa, T., Thakur, S., Rosenstein, F. U., & Matsumoto, T. (1986). Sterols of cucurbitaceae: The configurations at C-24 of 24-Alkyl-Δ5-,Δ7- and Δ8-sterols. *Lipids, 21*(1), 39–47. https://doi.org/10.1007/BF02534301.

Alignan, M., Roche, J., Vear, F., Vincourt, P., Bouniols, A., Cerny, M., et al. (2008). Variability and genetic analysis of sterol content in sunflower seeds. In *Proceedings of the 17th International Sunflower Conference, 8–12/6/2008 Cordoba, Spain* (pp. 751–756).

Amar, S., Becker, H. C., & Möllers, C. (2009). Genetic variation in phytosterol content of winter rapeseed (Brassica napus L.) and development of NIRS calibration equations. *Plant Breeding, 128*(1), 78–83. https://doi.org/10.1111/j.1439-0523.2008.01531.x.

Amar, S., Ecke, W., Becker, H. C., & Möllers, C. (2008). QTL for phytosterol and sinapate ester content in Brassica napus L. collocate with the two erucic acid genes. *Theoretical and Applied Genetics, 116*(8), 1051–1061. https://doi.org/10.1007/s00122-008-0734-2.

Andersson, A. A. M., Lampi, A. M., Nyström, L., Piironen, V., Li, L., Ward, J. L., et al. (2008). Phytochemical and dietary fiber components in barley varieties in the HEALTHGRAIN diversity screen. *Journal of Agricultural and Food Chemistry, 56*(21), 9767–9776. https://doi.org/10.1021/jf802037f.

Arnqvist, L., Dutta, P. C., Jonsson, L., & Sitbon, F. (2003). Reduction of cholesterol and glycoalkaloid levels in transgenic potato plants by overexpression of a type 1 sterol methyltransferase cDNA. *Plant Physiology, 131*(4), 1792–1799. https://doi.org/10.1104/pp.102.018788.

Arnqvist, L., Persson, M., Jonsson, L., Dutta, P. C., & Sitbon, F. (2008). Overexpression of CYP710A1 and CYP710A4 in transgenic Arabidopsis plants increases the level of stigmasterol at the expense of sitosterol. *Planta, 227*(2), 309–317. https://doi.org/10.1007/s00425-007-0618-8.

Babiychuk, E., Bouvier-Nave, P., Compagnon, V., Suzuki, M., Muranaka, T., Van Montagu, M., et al. (2008). Allelic mutant series reveal distinct functions for Arabidopsis cycloartenol synthase 1 in cell viability and plastid biogenesis. *Proceedings of the National Academy of Sciences of the United States of America, 105*(8), 3163–3168. https://doi.org/10.1073/pnas.0712190105.

Bajguz, A., & Hayat, S. (2009). Effects of brassinosteroids on the plant responses to environmental stresses. *Plant Physiology and Biochemistry, 47*(1), 1–8. https://doi.org/10.1016/j.plaphy.2008.10.002.

Banaś, A., Carlsson, A. S., Huang, B., Lenman, M., Banaś, W., Lee, M., et al. (2005). Cellular sterol ester synthesis in plants is performed by an enzyme (phospholipid:Sterol acyltransferase) different from the yeast and mammalian acyl-CoA:Sterol acyltransferases. *Journal of Biological Chemistry, 280*(41), 34626–34634. https://doi.org/10.1074/jbc.M504459200.

Behmer, S. T., Olszewski, N., Sebastiani, J., Palka, S., Sparacino, G., Sciarrno, E., et al. (2013). Plant phloem sterol content: Forms, putative functions, and implications for phloem-feeding insects. *Frontiers in Plant Science, 4*(SEP), 1–7. https://doi.org/10.3389/fpls.2013.00370.

Belcour, A., Girard, J., Aite, M., Delage, L., Trottier, C., Marteau, C., et al. (2020). Inferring biochemical reactions and metabolite structures to understand metabolic pathway drift. *iScience, 23*(2), 100849. https://doi.org/10.1016/j.isci.2020.100849.

Benveniste, P. (1986). Sterol biosynthesis. *Annual Review of Plant Physiology, 37*(1), 275–308. https://doi.org/10.1146/annurev.pp.37.060186.001423.

Benveniste, P. (2004). Biosynthesis and accumulation of sterols. *Annual Review of Plant Biology, 55*, 429–457. https://doi.org/10.1146/annurev.arplant.55.031903.141616.

Bloch, K. (1965). The biological synthesis of cholesterol. *Science, 150*(3692), 19–28. https://doi.org/10.1126/science.150.3692.19.

Bobrovskiy, I., Hope, J. M., Nettersheim, B. J., Volkman, J. K., Hallmann, C., & Brocks, J. J. (2021). Algal origin of sponge sterane biomarkers negates the oldest evidence for animals in the rock record. *Nature Ecology & Evolution, 5*(2), 165–168. https://doi.org/10.1038/s41559-020-01334-7.

Bouvier, F., Rahier, A., & Camara, B. (2005). Biogenesis, molecular regulation and function of plant isoprenoids. *Progress in Lipid Research, 44*(6), 357–429. https://doi.org/10.1016/j.plipres.2005.09.003.

Bouvier-Navé, P., Berna, A., Noiriel, A., Compagnon, V., Carlsson, A. S., Banas, A., et al. (2010). Involvement of the phospholipid sterol acyltransferase1 in plant sterol homeostasis and leaf senescence. *Plant Physiology, 152*(1), 107–119. https://doi.org/10.1104/pp.109.145672.

Brown, M. S., & Goldstein, J. L. (1997). The SREBP pathway: regulation of cholesterol metabolism by proteolysis of a membrane-bound transcription factor. *Cell, 89*(3), 331–340. https://doi.org/10.1016/S0092-8674(00)80213-5.

Burden, R. S., Cooke, D. T., & Carter, G. A. (1989). Inhibitors of sterol biosynthesis and growth in plants and fungi. *Phytochemistry*, *28*(7), 1791–1804. https://doi.org/10.1016/S0031-9422(00)97862-2.

Burger, C., Rondet, S., Benveniste, P., & Schaller, H. (2003). Virus-induced silencing of sterol biosynthetic genes: Identification of a *Nicotiana tabacum* L. obtusifoliol-14-demethylase (CYP51) by genetic manipulation of the sterol biosynthetic pathway in *Nicotiana benthamiana* L. *Journal of Experimental Botany*, *54*(388), 1675–1683. https://doi.org/10.1093/jxb/erg184.

Cacas, J. L., Buré, C., Grosjean, K., Gerbeau-Pissot, P., Lherminier, J., Rombouts, Y., et al. (2016). Revisiting plant plasma membrane lipids in tobacco: A focus on sphingolipids. *Plant Physiology*, *170*(1), 367–384. https://doi.org/10.1104/pp.15.00564.

Calegario, G., Pollier, J., Arendt, P., de Oliveira, L. S., Thompson, C., Soares, A. R., et al. (2016). Cloning and functional characterization of cycloartenol synthase from the red seaweed Laurencia dendroidea. *PLoS One*, *11*(11), e0165954. https://doi.org/10.1371/journal.pone.0165954.

Castillo, N., Pastor, V., Chávez, Á., Arró, M., Boronat, A., Flors, V., et al. (2019). Inactivation of UDP-glucose sterol glucosyltransferases enhances arabidopsis resistance to Botrytis cinerea. *Frontiers in Plant Science*, *10*(September), 1–15. https://doi.org/10.3389/fpls.2019.01162.

Chan, J. T., Patterson, G. W., Dutky, S. R., & Cohen, C. F. (1974). Inhibition of sterol biosynthesis in chlorella sorokiniana by triparanol. *Plant Physiology*, *53*(2), 244–249. https://doi.org/10.1104/pp.53.2.244.

Chapman, K. D., Aziz, M., Dyer, J. M., & Mullen, R. T. (2019). Mechanisms of lipid droplet biogenesis. *Biochemical Journal*, *476*(13), 1929–1942. https://doi.org/10.1042/BCJ20180021.

Chapman, E., Best, M. D., Hanson, S. R., & Wong, C. H. (2004). Sulfotransferases: Structure, mechanism, biological activity, inhibition, and synthetic utility. *Angewandte Chemie—International Edition*, *43*(27), 3526–3548. https://doi.org/10.1002/anie.200300631.

Chappell, J., Wolf, F., Proulx, J., Cuellar, R., & Saunders, C. (1995). Is the reaction catalyzed by 3-hydroxy-3-methylglutaryl coenzyme A reductase a rate-limiting step for isoprenoid biosynthesis in plants? *Plant Physiology*, *109*(4), 1337–1343. https://doi.org/10.1104/pp.109.4.1337.

Chaturvedi, P., Mishra, M., Akhtar, N., Gupta, P., Mishra, P., & Tuli, R. (2012). Sterol glycosyltransferases-identification of members of gene family and their role in stress in Withania somnifera. *Molecular Biology Reports*, *39*(10), 9755–9764. https://doi.org/10.1007/s11033-012-1841-3.

Chen, Z., Liu, J., Fu, Z., Ye, C., Zhang, R., Song, Y., et al. (2014). 24(S)-Saringosterol from edible marine seaweed Sargassum fusiforme is a novel selective LXRβ agonist. *Journal of Agricultural and Food Chemistry*, *62*(26), 6130–6137. https://doi.org/10.1021/jf500083r.

Chen, Q., Steinhauer, L., Hammerlindl, J., Keller, W., & Zou, J. (2007). Biosynthesis of phytosterol esters: Identification of a sterol O-acyltransferase in arabidopsis. *Plant Physiology*, *145*(3), 974–984. https://doi.org/10.1104/pp.107.106278.

Cheng, K.-C., Kuo, P.-C., Hung, H.-Y., Yu, K.-H., Hwang, T.-L., Shieh, P.-C., et al. (2016). Four new compounds from edible algae Cladosiphon okamuranus and Chlorella sorokiniana and their bioactivities. *Phytochemistry Letters*, *18*, 113–116. https://doi.org/10.1016/j.phytol.2016.09.008.

Choi, H., Ohyama, K., Kim, Y. Y., Jin, J. Y., Lee, S. B., Yamaoka, Y., et al. (2014). The role of Arabidopsis ABCG9 and ABCG31 ATP binding cassette transporters in pollen fitness and the deposition of steryl glycosides on the pollen coat. *Plant Cell*, *26*(1), 310–324. https://doi.org/10.1105/tpc.113.118935.

Cioccoloni, G., Soteriou, C., Websdale, A., Wallis, L., Zulyniak, M. A., & Thorne, J. L. (2020). Phytosterols and phytostanols and the hallmarks of cancer in model organisms: A systematic review and meta-analysis. *Critical Reviews in Food Science and Nutrition*, 1–21. https://doi.org/10.1080/10408398.2020.1835820.

Clouse, S. D. (1996). Molecular genetic studies confirm the role of brassinosteroids in plant growth and development. *Plant Journal*, *10*(1), 1–8. https://doi.org/10.1046/j.1365-313X.1996.10010001.x.

Cock, J. M., Sterck, L., Rouzé, P., Scornet, D., Allen, A. E., Amoutzias, G., et al. (2010). The Ectocarpus genome and the independent evolution of multicellularity in brown algae. *Nature*, *465*(7298), 617–621. https://doi.org/10.1038/nature09016.

Conte, M., Lupette, J., Seddiki, K., Meï, C., Dolch, L. J., Gros, V., et al. (2018). Screening for biologically annotated drugs that trigger triacylglycerol accumulation in the diatom *Phaeodactylum*. *Plant Physiology*, *177*(2), 532–552. https://doi.org/10.1104/pp.17.01804.

Corey, E. J., Russey, W. E., & de Montellano, P. R. O. (1966). 2,3-Oxidosqualene, an intermediate in the biological synthesis of sterols from squalene. *Journal of the American Chemical Society*, *88*(20), 4750–4751. https://doi.org/10.1021/ja00972a056.

Couture, R., Arul, J., Gosselin, C., & Willemot, C. (1989). The sterols of strawberry fruit. *Phytochemistry*, *28*(4), 1276–1277. https://doi.org/10.1016/0031-9422(89)80233-X.

D'Adamo, S., Schiano di Visconte, G., Lowe, G., Szaub-Newton, J., Beacham, T., Landels, A., et al. (2019). Engineering the unicellular alga Phaeodactylum tricornutum for high-value plant triterpenoid production. *Plant biotechnology journal*, *17*(1), 75–87. https://doi.org/10.1111/pbi.12948.

Darnet, S., Blary, A., Chevalier, Q., & Schaller, H. (2021). Phytosterol profiles, genomes and enzymes—An overview. *Frontiers in Plant Science*, *12*(May), 1–18. https://doi.org/10.3389/fpls.2021.665206.

Darnet, S., Martin, L. B. B., Mercier, P., Bracher, F., Geoffroy, P., & Schaller, H. (2020). Inhibition of phytosterol biosynthesis by azasterols. *Molecules*, *25*(5), 1–14. https://doi.org/10.3390/molecules25051111.

Darnet, S., & Rahier, A. (2004). Plant sterol biosynthesis: Identification of two distinct families of sterol 4α-methyl oxidases. *Biochemical Journal*, *378*(3), 889–898. https://doi.org/10.1042/BJ20031572.

Darnet, S., & Schaller, H. (2019a). Metabolism and biological activities of 4-methyl-sterols. *Molecules*, *24*(3). https://doi.org/10.3390/molecules24030451.

Darnet, S., & Schaller, H. (2019b). Unleashed sterol production in thale cress. *Nature Plants*, *5*(11), 1112–1113. https://doi.org/10.1038/s41477-019-0541-6.

de Jong, D. L. C., Timmermans, K. R., de Winter, J. M., & Goverdina, G. C. H. (2021). Effects of nutrient availability and light intensity on the sterol content of *Saccharina latissima* (Laminariales, Phaeophyceae). *Journal of Applied Phycology*, *33*, 1101–1113. https://doi.org/10.1007/s10811-020-02359-y.

De Storme, N., De Schrijver, J., Van Criekinge, W., Wewer, V., Dörmann, P., & Geelena, D. (2013). GLUCAN SYNTHASE-LIKE8 and STEROL METHYLTRANSFERASE2 are required for ploidy consistency of the sexual reproduction system in Arabidopsis. *Plant Cell*, *25*(2), 387–403. https://doi.org/10.1105/tpc.112.106278.

Debolt, S., Scheible, W. R., Schrick, K., Auer, M., Beisson, F., Bischoff, V., et al. (2009). Mutations in UDP-glucose:Sterol glucosyltransferase in Arabidopsis cause transparent testa phenotype and suberization defect in seeds. *Plant Physiology*, *151*(1), 78–87. https://doi.org/10.1104/pp.109.140582.

Desfosses, M. (1820). Extrait d'une lettre à M. Robiquet. *Journal of Pharmacy*, *6*, 374–376.

Doblas, V. G., Amorim-Silva, V., Posé, D., Rosado, A., Esteban, A., Arró, M., et al. (2013). The SUD1 gene encodes a putative E3 ubiquitin ligase and is a positive regulator of 3-hydroxy-3-methylglutaryl coenzyme A reductase activity in arabidopsis. *The Plant Cell*, *25*(2), 728–743. https://doi.org/10.1105/tpc.112.108696.

Dufourc, E. J. (2008a). Sterols and membrane dynamics. *Journal of Chemical Biology, 1*(1–4), 63–77. https://doi.org/10.1007/s12154-008-0010-6.

Dufourc, E. J. (2008b). The role of phytosterols in plant adaptation to temperature. *Plant Signaling and Behavior, 3*(2), 133–134. https://doi.org/10.4161/psb.3.2.5051.

Duperon, R., Thiersault, M., & Duperon, P. (1983). Occurrence of steryl glycosides and acylated steryl glycosides in some marine algae. *Phytochemistry, 22*(2), 535–538. https://doi.org/10.1016/0031-9422(83)83040-4.

Dyas, L., & Goad, L. J. (1993). Steryl fatty acyl esters in plants. *Phytochemistry, 34*(1), 17–29. https://doi.org/10.1016/S0031-9422(00)90777-5.

EFSA. (2010). https://efsa.onlinelibrary.wiley.com/doi/epdf/10.2903/j.efsa.2010.1813.

Enfissi, E. M. A., Fraser, P. D., Lois, L. M., Boronat, A., Schuch, W., & Bramley, P. M. (2005). Metabolic engineering of the mevalonate and non-mevalonate isopentenyl diphosphate-forming pathways for the production of health-promoting isoprenoids in tomato. *Plant Biotechnology Journal, 3*(1), 17–27. https://doi.org/10.1111/j.1467-7652.2004.00091.x.

Esche, R., Barnsteiner, A., Scholz, B., & Engel, K. H. (2012). Simultaneous analysis of free phytosterols/phytostanols and intact phytosteryl/phytostanyl fatty acid and phenolic acid esters in cereals. *Journal of Agricultural and Food Chemistry, 60*(21), 5330–5339. https://doi.org/10.1021/jf300878h.

Eschenmoser, A., Ruzicka, L., Jeger, O., & Arigoni, D. (1955). Zur Kenntnis der Triterpene. 190. Mitteilung. Eine stereochemische Interpretation der biogenetischen Isoprenregel bei den Triterpenen. *Helvetica Chimica Acta, 38*(7), 1890–1904. https://doi.org/10.1002/hlca.19550380728.

Estévez, J. M., Cantero, A., Romero, C., Kawaide, H., Jiménez, L. F., Kuzuyama, T., et al. (2000). Analysis of the expression of CLA1, a gene that encodes the 1-deoxyxylulose 5-phosphate synthase of the 2-C-methyl-d-erythritol-4-phosphate pathway in arabidopsis. *Plant Physiology, 124*(1), 95–104. https://doi.org/10.1104/pp.124.1.95.

Fabris, M., Matthijs, M., Carbonelle, S., Moses, T., Pollier, J., Dasseville, R., et al. (2014). Tracking the sterol biosynthesis pathway of the diatom P haeodactylum tricornutum. *New Phytologist, 204*(3), 521–535. https://doi.org/10.1111/nph.12917.

Fang, N., Yu, S., & Badger, T. M. (2003). Characterization of triterpene alcohol and sterol ferulates in rice bran using LC-MS/MS. *Journal of Agricultural and Food Chemistry, 51*(11), 3260–3267. https://doi.org/10.1021/jf021162c.

FDA. (2000). Food labeling: Health claims; plant sterol/stanol esters and coronary heart disease. *Food and Drug Administration, HHS, 65*(175), 54686–54739.

Fernandes, P., & Cabral, J. M. S. (2007). Phytosterols: Applications and recovery methods. *Bioresource Technology, 98*(12), 2335–2350. https://doi.org/10.1016/j.biortech.2006.10.006.

Ferrer, A., Altabella, T., Arró, M., & Boronat, A. (2017). Emerging roles for conjugated sterols in plants. *Progress in Lipid Research, 67*, 27–37. https://doi.org/10.1016/j.plipres.2017.06.002.

Font i Forcada, C., Velasco, L., Socias i Company, R., & Fernández i Martí, Á. (2015). Association mapping for kernel phytosterol content in almond. *Frontiers in Plant Science, 6*(July), 1–12. https://doi.org/10.3389/fpls.2015.00530.

Forestier, E., Romero-Segura, C., Pateraki, I., Centeno, E., Compagnon, V., Preiss, M., et al. (2019). Distinct triterpene synthases in the laticifers of Euphorbia lathyris. *Scientific Reports, 9*(1), 4840. https://doi.org/10.1038/s41598-019-40905-y.

Fujita, S., Ohnishi, T., Watanabe, B., Yokota, T., Takatsuto, S., Fujioka, S., et al. (2006). Arabidopsis CYP90B1 catalyses the early C-22 hydroxylation of C 27, C 28 and C 29 sterols. *The Plant Journal, 45*(5), 765–774. https://doi.org/10.1111/j.1365-313X.2005.02639.x.

Furt, F., Simon-Plas, F., & Mongrand, S. (2011). *Lipids of the Plant Plasma Membrane* (pp. 3–30). https://doi.org/10.1007/978-3-642-13431-9_1.

Gallo, C., D'Ippolito, G., Nuzzo, G., Sardo, A., & Fontana, A. (2017). Autoinhibitory sterol sulfates mediate programmed cell death in a bloom-forming marine diatom. *Nature Communications*, *8*(1), 1–11. https://doi.org/10.1038/s41467-017-01300-1.

Gallo, C., Landi, S., D'Ippolito, G., Nuzzo, G., Manzo, E., Sardo, A., et al. (2020). Diatoms synthesize sterols by inclusion of animal and fungal genes in the plant pathway. *Scientific Reports*, *10*(1), 4204. https://doi.org/10.1038/s41598-020-60993-5.

Gamage, N., Barnett, A., Hempel, N., Duggleby, R. G., Windmill, K. F., Martin, J. L., et al. (2006). Human sulfotransferases and their role in chemical metabolism. *Toxicological Sciences*, *90*(1), 5–22. https://doi.org/10.1093/toxsci/kfj061.

Giner, J. L., & Boyer, G. L. (1998). Sterols of the brown tide alga Aureococcus anophagefferens. *Phytochemistry*, *48*(3), 475–477. https://doi.org/10.1016/S0031-9422(97)00860-1.

Giner, J.-L., Li, X., & Boyer, G. L. (2001). Sterol composition of Aureoumbra lagunensis, the Texas brown tide alga. *Phytochemistry*, *57*(5), 787–789. https://doi.org/10.1016/S0031-9422(01)00135-2.

Giner, J., Wünsche, L., Andersen, R. A., & Djerassi, C. (1991). Dinoflagellates cyclize squalene oxide to lanosterol. *Biochemical Systematics and Ecology*, *19*(2), 143–145. https://doi.org/10.1016/0305-1978(91)90037-Z.

Girard, J., Lanneau, G., Delage, L., Leroux, C., Belcour, A., Got, J., et al. (2021). Semi-quantitative targeted gas chromatography-mass spectrometry profiling supports a late side-chain reductase cycloartenol-to-cholesterol biosynthesis pathway in brown algae. *Frontiers in Plant Science*, *12*. https://doi.org/10.3389/fpls.2021.648426.

Goad, L. J., & Akihisa, T. (1997). *Analysis of sterols*. Springer Netherlands. https://doi.org/10.1007/978-94-009-1447-6.

Godzina, S. M., Lovato, M. A., Meyer, M. M., Foster, K. A., Wilson, W. K., Gu, W., et al. (2000). Cloning and characterization of the Dictyostelium discoideum cycloartenol synthase cDNA. *Lipids*, *35*(3), 249–255. https://doi.org/10.1007/s11745-000-0520-3.

Griebel, T., & Zeier, J. (2010). A role for β-sitosterol to stigmasterol conversion in plant-pathogen interactions. *Plant Journal*, *63*(2), 254–268. https://doi.org/10.1111/j.1365-313X.2010.04235.x.

Griffiths, W. J., & Wang, Y. (2019). Oxysterol research: A brief review. *Biochemical Society Transactions*, *47*(2), 517–526. https://doi.org/10.1042/BST20180135.

Grille, S., Zaslawski, A., Thiele, S., Plat, J., & Warnecke, D. (2010). The functions of steryl glycosides come to those who wait: Recent advances in plants, fungi, bacteria and animals. *Progress in Lipid Research*, *49*(3), 262–288. https://doi.org/10.1016/j.plipres.2010.02.001.

Grosjean, K., Mongrand, S., Beney, L., Simon-Plas, F., & Gerbeau-Pissot, P. (2015). Differential effect of plant lipids on membrane organization specificities of phytosphingolipids and phytosterols. *Journal of Biological Chemistry*, *290*(9), 5810–5825. https://doi.org/10.1074/jbc.M114.598805.

Grove, M. D., Spencer, G. F., Rohwedder, W. K., Mandava, N., Worley, J. F., Warthen, J. D., et al. (1979). Brassinolide, a plant growth-promoting steroid isolated from Brassica napus pollen. *Nature*, *281*(5728), 216–217. https://doi.org/10.1038/281216a0.

Ha, T.-Y., Ko, S.-N., Lee, S.-M., Kim, H.-R., Chung, S.-H., Kim, S.-R., et al. (2006). Changes in nutraceutical lipid components of rice at different degrees of milling. *The European Journal of Lipid Science and Technology*, *108*, 175–181. https://doi.org/10.1002/ejlt.200500250.

Haddadi, P., Ebrahimi, A., Langlade, N. B., Yazdi-samadi, B., Berger, M., Calmon, A., et al. (2012). Genetic dissection of tocopherol and phytosterol in recombinant inbred lines of sunflower through quantitative trait locus analysis and the candidate gene approach. *Molecular Breeding*, *29*(3), 717–729. https://doi.org/10.1007/s11032-011-9585-7.

Harker, M., Hellyer, A., Clayton, J. C., Duvoix, A., Lanot, A., & Safford, R. (2003). Co-ordinate regulation of sterol biosynthesis enzyme activity during accumulation of sterols in developing rape and tobacco seed. *Planta, 216*(4), 707–715. https://doi.org/10.1007/s00425-002-0913-3.

Harker, M., Holmberg, N., Clayton, J. C., Gibbard, C. L., Wallace, A. D., Rawlins, S., et al. (2003). Enhancement of seed phytosterol levels by expression of an N-terminal truncated Hevea brasiliensis (rubber tree) 3-hydroxy-3-methylglutaryl-CoA reductase. *Plant Biotechnology Journal, 1*(2), 113–121. https://doi.org/10.1046/j.1467-7652.2003.00011.x.

Hase, Y., Fujioka, S., Yoshida, S., Sun, G., Umeda, M., & Tanaka, A. (2005). Ectopic endoreduplication caused by sterol alteration results in serrated petals in Arabidopsis. *Journal of Experimental Botany, 56*(414), 1263–1268. https://doi.org/10.1093/jxb/eri122.

Heinz, E., Dieler, H. P., & Rullkötter, J. (1975). Enzymatic acylation of steryl glycoside. *Zeitschrift Für Pflanzenphysiologie, 75*(1), 78–87. https://doi.org/10.1016/s0044-328x(75)80077-8.

Heinz, P., & Glomb, M. A. (2018). Characterization and quantitation of steryl glycosides in solanum melongena. *Journal of Agricultural and Food Chemistry*. https://doi.org/10.1021/acs.jafc.8b04045.

Hemmerlin, A., Hoeffler, J. F., Meyer, O., Tritsch, D., Kagan, I. A., Grosdemange-Billiard, C., et al. (2003). Cross-talk between the cytosolic mevalonate and the plastidial methylerythritol phosphate pathways in tobacco bright yellow-2 cells. *Journal of Biological Chemistry, 278*(29), 26666–26676. https://doi.org/10.1074/jbc.M302526200.

Hernández-Pinzón, I., Ross, J. H. E., Barnes, K. A., Damant, A. P., & Murphy, D. J. (1999). Composition and role of tapetal lipid bodies in the biogenesis of the pollen coat of Brassica napus. *Planta, 208*(4), 588–598. https://doi.org/10.1007/s004250050597.

Hey, S. J., Powers, S. J., Beale, M. H., Hawkins, N. D., Ward, J. L., & Halford, N. G. (2006). Enhanced seed phytosterol accumulation through expression of a modified HMG-CoA reductase. *Plant Biotechnology Journal, 4*(2), 219–229. https://doi.org/10.1111/j.1467-7652.2005.00174.x.

Hillig, I., Leipelt, M., Ott, C., Zähringer, U., Warnecke, D., & Heinz, E. (2003). Formation of glucosylceramide and sterol glucoside by a UDP-glucose-dependent glucosylceramide synthase from cotton expressed in Pichia pastoris. *FEBS Letters, 553*(3), 365–369. https://doi.org/10.1016/S0014-5793(03)01058-5.

Hirschmann, F., Krause, F., & Papenbrock, J. (2014). The multi-protein family of sulfotransferases in plants: Composition, occurrence, substrate specificity, and functions. *Frontiers in Plant Science, 5*(Oct), 1–13. https://doi.org/10.3389/fpls.2014.00556.

Holmberg, N., Harker, M., Gibbard, C. L., Wallace, A. D., Clayton, J. C., Rawlins, S., et al. (2002). Sterol C-24 methyltransferase type 1 controls the flux of carbon into sterol biosynthesis in tobacco seed. *Plant Physiology, 130*(1), 303–311. https://doi.org/10.1104/pp.004226.

Holmberg, N., Harker, M., Wallace, A. D., Clayton, J. C., Gibbard, C. L., & Safford, R. (2003). Co-expression of N-terminal truncated 3-hydroxy-3-methylglutaryl CoA reductase and C24-sterol methyltransferase type 1 in transgenic tobacco enhances carbon flux towards end-product sterols. *Plant Journal, 36*(1), 12–20. https://doi.org/10.1046/j.1365-313X.2003.01851.x.

Hopp, H. E., Romero, P. A., Daleo, G. R., & Lezica, R. P. (1978). Steryl glucoside biosynthesis in the alga Prototheca zopfii. *Phytochemistry, 17*(6), 1049–1052. https://doi.org/10.1016/S0031-9422(00)94278-X.

Hovenkamp, E., Demonty, I., Plat, J., Lütjohann, D., Mensink, R. P., & Trautwein, E. A. (2008). Biological effects of oxidized phytosterols: A review of the current knowledge. *Progress in Lipid Research, 47*(1), 37–49. https://doi.org/10.1016/j.plipres.2007.10.001.

Ischebeck, T., Krawczyk, H. E., Mullen, R. T., Dyer, J. M., & Chapman, K. D. (2020). Lipid droplets in plants and algae: Distribution, formation, turnover and function. *Seminars in Cell and Developmental Biology*, *108*(November 2019), 82–93. https://doi.org/10.1016/j.semcdb.2020.02.014.

Itkin, M., Heinig, U., Tzfadia, O., Bhide, A. J., Shinde, B., Cardenas, P. D., et al. (2013). Biosynthesis of antinutritional alkaloids in solanaceous crops is mediated by clustered genes. *Science*, *341*(6142), 175–179. https://doi.org/10.1126/science.1240230.

Ito, J., Sawada, K., Ogura, Y., Xinyi, F., Rahmania, H., Mohri, T., et al. (2019). Definitive evidence of the presence of 24-methylenecycloartanyl ferulate and 24-methylenecycloartanyl caffeate in barley. *Scientific Reports*, *9*(1), 1–11. https://doi.org/10.1038/s41598-019-48985-6.

Jan, H. M., Chen, Y. C., Yang, T. C., Ong, L. L., Chang, C. C., Muthusamy, S., et al. (2020). Cholesteryl α-D-glucoside 6-acyltransferase enhances the adhesion of Helicobacter pylori to gastric epithelium. *Communications Biology*, *3*(1). https://doi.org/10.1038/s42003-020-0855-y.

Jaramillo-Madrid, A. C., Abbriano, R., Ashworth, J., Fabris, M., Pernice, M., & Ralph, P. J. (2020). Overexpression of key sterol pathway enzymes in two model marine diatoms alters sterol profiles in phaeodactylum tricornutum. *Pharmaceuticals*, *13*(12), 1–19. https://doi.org/10.3390/ph13120481.

Kajikawa, M., Kinohira, S., Ando, A., Shimoyama, M., Kato, M., & Fukuzawa, H. (2015). Accumulation of squalene in a microalga Chlamydomonas reinhardtii by genetic modification of squalene synthase and squalene epoxidase genes. *PLoS One*, *10*(3), 1–21. https://doi.org/10.1371/journal.pone.0120446.

Kalinowska, M., & Wojciechowski, Z. A. (1983). The occurrence of sterol-ester hydrolase activity in roots of white mustard seedlings. *Phytochemistry*, *22*(1), 59–63. https://doi.org/10.1016/S0031-9422(00)80058-8.

Kalinowska, M., Zimowski, J., Bucholc, M., & Wojciechowski, Z. A. (1989). Reversible enzymatic transesterification between wax esters and sterols in Sinapis alba roots. *Phytochemistry*, *28*(11), 2931–2935. https://doi.org/10.1016/0031-9422(89)80256-0.

Kaneko, R., & Tsuchiya, T. (1955). New compound in rice bran and germ oils. *Chemical Abstracts*, *49*, 4897b.

Kato, T., Matsukawa, T., & Horibata, A. (2017). Quantitative trait loci responsible for the difference in γ-oryzanol content in brown rice between japonica-type and indica-type rice cultivars. *Plant Production Science*, *20*(4), 459–466. https://doi.org/10.1080/1343943X.2017.1372109.

Kellogg, B. A., & Poulter, C. D. (1997). Chain elongation in the isoprenoid biosynthetic pathway. *Current Opinion in Chemical Biology*, *1*(4), 570–578. https://doi.org/10.1016/S1367-5931(97)80054-3.

Kim, Y. J., Lee, O. R., Oh, J. Y., Jang, M. G., & Yang, D. C. (2014). Functional analysis of 3-hydroxy-3-methylglutaryl coenzyme a reductase encoding genes in triterpene saponin-producing ginseng. *Plant Physiology*, *165*(1), 373–387. https://doi.org/10.1104/pp.113.222596.

Klein, J., Horn, E., Ernst, M., Leykauf, T., Leupold, T., Dorfner, M., et al. (2021). RNAi-mediated gene knockdown of progesterone 5β-reductases in Digitalis lanata reduces 5β-cardenolide content. *Plant Cell Reports*, *40*(9), 1631–1646. https://doi.org/10.1007/s00299-021-02707-3.

Knoch, E., Sugawara, S., Mori, T., Poulsen, C., Fukushima, A., Harholt, J., et al. (2018). Third DWF1 paralog in Solanaceae, sterol Δ24-isomerase, branches withanolide biosynthesis from the general phytosterol pathway. *Proceedings of the National Academy of Sciences of the United States of America*, *115*(34), E8096–E8103. https://doi.org/10.1073/pnas.1807482115.

Kodner, R. B., Pearson, A., Summons, R. E., & Knoll, A. H. (2008). Sterols in red and green algae: Quantification, phylogeny, and relevance for the interpretation of geologic steranes. *Geobiology*, *6*(4), 411–420. https://doi.org/10.1111/j.1472-4669.2008.00167.x.
Kojima, M., Ohnishi, M., Ito, S., & Fujino, Y. (1989). Characterization of acylmono-, mono-, di-, tri- and tetraglycosylsterol and saponin in Adzuki bean (Vigna angularis) seeds. *Lipids*, *24*(10), 849–853. https://doi.org/10.1007/BF02535758.
Kovganko, N. V., & Kashkan, Z. N. (1999). Sterol glycosides and acylglycosides. *Chemistry of Natural Compounds*, *35*(5), 479–497. https://doi.org/10.1007/BF02323277.
Ktari, L., Blond, A., & Guyot, M. (2000). 16β-Hydroxy-5α-cholestane-3,6-dione, a novel cytotoxic oxysterol from the red alga Jania rubens. *Bioorganic & Medicinal Chemistry Letters*, *10*(22), 2563–2565. https://doi.org/10.1016/S0960-894X(00)00504-7.
Ktari, L., & Guyot, M. (1999). A cytotoxic oxysterol from the marine alga Padina pavonica (L.) Thivy. *Journal of Applied Phycology*, *11*(6), 511–513. https://doi.org/10.1023/A:1008162624027.
Lange, I., Poirier, B. C., Herron, B. K., & Lange, B. M. (2015). Comprehensive assessment of transcriptional regulation facilitates metabolic engineering of isoprenoid accumulation in Arabidopsis. *Plant Physiology*, *169*(3), 1595–1606. https://doi.org/10.1104/pp.15.00573.
Lara, J. A., Burciaga-Monge, A., Chávez, A., Revés, M., Lavilla, R., Arró, M., et al. (2018). Identification and characterization of sterol acyltransferases responsible for steryl ester biosynthesis in tomato. *Frontiers in Plant Science*, *9*(May), 1–18. https://doi.org/10.3389/fpls.2018.00588.
Largeau, C., Goad, L. J., & Goodwin, T. W. (1977). Biosynthesis of sitosterol in barley seedlings. *Phytochemistry*, *16*(12), 1925–1930. https://doi.org/10.1016/0031-9422(77)80097-6.
Lepage, M. (1964). Isolation and characterization of an esterified form of steryl. *Journal of Lipid Research*, *5*(4), 587–592. https://doi.org/10.1016/S0022-2275(20)40188-9.
Li, J., Nagpal, P., Vitart, V., McMorris, T. C., & Chory, J. (1996). A Role for brassinosteroids in light-dependent development of arabidopsis. *Science*, *272*(5260), 398–401. https://doi.org/10.1126/science.272.5260.398.
Li, X., Xia, T., Huang, J., Guo, K., Liu, X., Chen, T., et al. (2014). Distinct biochemical activities and heat shock responses of two UDP-glucose sterol glucosyltransferases in cotton. *Plant Science*, *219–220*, 1–8. https://doi.org/10.1016/j.plantsci.2013.12.013.
Liao, P., Chen, X., Wang, M., Bach, T. J., & Chye, M. L. (2018). Improved fruit α-tocopherol, carotenoid, squalene and phytosterol contents through manipulation of Brassica juncea 3-HYDROXY-3-METHYLGLUTARYL-COA SYNTHASE1 in transgenic tomato. *Plant Biotechnology Journal*, *16*(3), 784–796. https://doi.org/10.1111/pbi.12828.
Lichtenthaler, H. K. (1999). The 1-deoxy-d-xylulose-5-phosphate pathway of isoprenoid biosynthesis in plants. *Annual Review of Plant Physiology and Plant Molecular Biology*, *50*(1), 47–65. https://doi.org/10.1146/annurev.arplant.50.1.47.
Lindow, T. (2016). *Arboris LCC—The global phytosterol market and cooperation with the steroid hormone business*. https://pt.slideshare.net/TomLindow1/the-global-phytosterol-market-and-cooperation-with-the-steroid-hormone-business.
Lu, Y., Jiang, J., Zhao, H., Han, X., Xiang, Y., & Zhou, W. (2020). Clade-specific sterol metabolites in dinoflagellate endosymbionts are associated with coral bleaching in response to environmental cues. *MSystems*, *5*(5). https://doi.org/10.1128/mSystems.00765-20.
Lupette, J., Jaussaud, A., Seddiki, K., Morabito, C., Brugière, S., Schaller, H., et al. (2019). The architecture of lipid droplets in the diatom Phaeodactylum tricornutum. *Algal Research*, *38*(October 2018), 101415. https://doi.org/10.1016/j.algal.2019.101415.

Lynch, D. V., Criss, A. K., Lehoczky, J. L., & Bui, V. T. (1997). Ceramide glucosylation in bean hypocotyl microsomes: Evidence that steryl glucoside serves as glucose donor. *Archives of Biochemistry and Biophysics, 340*(2), 311–316. https://doi.org/10.1006/abbi.1997.9928.

Lynch, D. V., & Steponkus, P. L. (1987). Plasma membrane lipid alterations associated with cold acclimation of winter rye seedlings (Secale cereale L. cv Puma). *Plant Physiology, 83*(4), 761–767. https://doi.org/10.1104/pp.83.4.761.

MacLachlan, P. L. (1936). Fat metabolism in plants with special reference to sterols. *Journal of Biological Chemistry, 113*(1), 197–204. https://doi.org/10.1016/S0021-9258(18)74906-5.

Maillot-Vernier, P., Gondet, L., Schaller, H., Benveniste, P., & Belliard, G. (1991). Genetic study and further biochemical characterization of a tobacco mutant that overproduces sterols. *MGG Molecular & General Genetics, 231*(1), 33–40. https://doi.org/10.1007/BF00293818.

Mamode Cassim, A., Gouguet, P., Gronnier, J., Laurent, N., Germain, V., Grison, M., et al. (2019). Plant lipids: Key players of plasma membrane organization and function. *Progress in Lipid Research, 73*(July 2018), 1–27. https://doi.org/10.1016/j.plipres.2018.11.002.

Martens, N., Schepers, M., Zhan, N., Leijten, F., Voortman, G., Tiane, A., et al. (2021). 24(S)-Saringosterol prevents cognitive decline in a mouse model for Alzheimer's disease. *Marine Drugs, 19*(4), 190. https://doi.org/10.3390/md19040190.

Merah, O., Langlade, N., Alignan, M., Roche, J., Pouilly, N., Lippi, Y., et al. (2012). Genetic analysis of phytosterol content in sunflower seeds. *Theoretical and Applied Genetics, 125*(8), 1589–1601. https://doi.org/10.1007/s00122-012-1937-0.

Merah, O., & Mouloungui, Z. (2019). Tetraploid wheats: Valuable source of phytosterols and phytostanols. *Agronomy, 9*(4), 10–12. https://doi.org/10.3390/agronomy9040201.

Miller, M. B., Haubrich, B. A., Wang, Q., Snell, W. J., & Nes, W. D. (2012). Evolutionarily conserved Δ25(27)-olefin ergosterol biosynthesis pathway in the alga Chlamydomonas reinhardtii. *Journal of Lipid Research, 53*(8), 1636–1645. https://doi.org/10.1194/jlr.M027482.

Mishra, M. K., Chaturvedi, P., Singh, R., Singh, G., Sharma, L. K., Pandey, V., et al. (2013). Overexpression of WsSGTL1 gene of withania somnifera enhances salt tolerance, heat tolerance and cold acclimation ability in transgenic Arabidopsis plants. *PLoS One, 8*(4), e63064. https://doi.org/10.1371/journal.pone.0063064.

Mishra, M. K., Singh, G., Tiwari, S., Singh, R., Kumari, N., & Misra, P. (2015). Characterization of Arabidopsis sterol glycosyltransferase TTG15/UGT80B1 role during freeze and heat stress. *Plant Signaling & Behavior, 10*(12), e1075682. https://doi.org/10.1080/15592324.2015.1075682.

Mitchell, J. W., Mandava, N., Worley, J. F., Plimmer, J. R., & Smith, M. V. (1970). Brassins--a new family of plant hormones from rape pollen. *Nature, 225*(5237), 1065–1066. https://doi.org/10.1038/2251065a0.

Mohammady, N. G. (2004). Total, free and conjugated sterolic forms in three microalgae used in mariculture. *Zeitschrift Fur Naturforschung—Section C Journal of Biosciences, 59*(9–10), 619–624. https://doi.org/10.1515/znc-2004-9-1002.

Mordor Intelligence—Phytosterols Market—Growth, Trends and Forecast. (2019-2024). https://www.mordorintelligence.com/industry-reports/phytosterols-market. accessed 01/09/2019.

Moreau, R. A., Nyström, L., Whitaker, B. D., Winkler-Moser, J. K., Baer, D. J., Gebauer, S. K., et al. (2018). Phytosterols and their derivatives: Structural diversity, distribution, metabolism, analysis, and health-promoting uses. *Progress in Lipid Research, 70*(January), 35–61. https://doi.org/10.1016/j.plipres.2018.04.001.

Moreau, R. A., Singh, V., & Hicks, K. B. (2001). Comparison of oil and phytosterol levels in germplasm accessions of corn, teosinte, and job's tears. *Journal of Agricultural and Food Chemistry, 49*(8), 3793–3795. https://doi.org/10.1021/jf010280h.

Moreau, R. A., Whitaker, B. D., & Hicks, K. B. (2002). Phytosterols, phytostanols, and their conjugates in foods: Structural diversity, quantitative analysis, and health-promoting uses. *Progress in Lipid Research, 41*(6), 457–500. https://doi.org/10.1016/S0163-7827(02)00006-1.

Morikawa, T., Mizutani, M., Aoki, N., Watanabe, B., Saga, H., Saito, S., et al. (2006). Cytochrome P450 CYP710A encodes the sterol C-22 desaturase in Arabidopsis and tomato. *Plant Cell, 18*(4), 1008–1022. https://doi.org/10.1105/tpc.105.036012.

Morikawa, T., Saga, H., Hashizume, H., & Ohta, D. (2009). CYP710A genes encoding sterol C22-desaturase in Physcomitrella patens as molecular evidence for the evolutionary conservation of a sterol biosynthetic pathway in plants. *Planta, 229*(6), 1311–1322. https://doi.org/10.1007/s00425-009-0916-4.

Moss, G. P. (1989). International union of pure and applied chemistry and international union of biochemistry: Joint commission on biochemical nomenclature: Nomenclature of steroids. *Pure and Applied Chemistry, 61*(10), 1783–1822. https://doi.org/10.1351/pac198961101783.

Msanne, J., Chen, M., Luttgeharm, K. D., Bradley, A. M., Mays, E. S., Paper, J. M., et al. (2015). Glucosylceramides are critical for cell-type differentiation and organogenesis, but not for cell viability in Arabidopsis. *Plant Journal, 84*(1), 188–201. https://doi.org/10.1111/tpj.13000.

Muñoz-Bertomeu, J., Sales, E., Ros, R., Arrillaga, I., & Segura, J. (2007). Up-regulation of an N-terminal truncated 3-hydroxy-3-methylglutaryl CoA reductase enhances production of essential oils and sterols in transgenic Lavandula latifolia. *Plant Biotechnology Journal, 5*(6), 746–758. https://doi.org/10.1111/j.1467-7652.2007.00286.x.

Myeong, J., Park, C. G., Suh, B. C., & Hille, B. (2021). Compartmentalization of phosphatidylinositol 4,5-bisphosphate metabolism into plasma membrane liquid-ordered/raft domains. *Proceedings of the National Academy of Sciences of the United States of America, 118*(9), e2025343118. https://doi.org/10.1073/pnas.2025343118.

Nakamoto, M., Schmit, A. C., Heintz, D., Schaller, H., & Ohta, D. (2015). Diversification of sterol methyltransferase enzymes in plants and a role for β-sitosterol in oriented cell plate formation and polarized growth. *Plant Journal, 84*(5), 860–874. https://doi.org/10.1111/tpj.13043.

Neelakandan, A. K., Chamala, S., Valliyodan, B., Nes, W. D., & Nguyen, H. T. (2012). Metabolic engineering of soybean affords improved phytosterol seed traits. *Plant Biotechnology Journal, 10*(1), 12–19. https://doi.org/10.1111/j.1467-7652.2011.00623.x.

Nes, W. D. (2011). Biosynthesis of cholesterol and other sterols. *Chemical Reviews, 111*(10), 6423–6451. https://doi.org/10.1021/cr200021m.

Nes, W. R., & McKean, M. L. (1977). *Biochemistry of steroids and other isopentenoids*. University Park Press.

Nurmi, T., Nyström, L., Edelmann, M., Lampi, A. M., & Piironen, V. (2008). Phytosterols in wheat genotypes in the healthgrain diversity screen. *Journal of Agricultural and Food Chemistry, 56*(21), 9710–9715. https://doi.org/10.1021/jf8010678.

Nuzzo, G., Gallo, C., D'Ippolito, G., Manzo, E., Ruocco, N., Russo, E., et al. (2018). UPLC–MS/MS identification of sterol sulfates in marine diatoms. *Marine Drugs, 17*(1), 10. https://doi.org/10.3390/md17010010.

Nyström, L., Achrenius, T., Lampi, A. M., Moreau, R. A., & Piironen, V. (2007). A comparison of the antioxidant properties of steryl ferulates with tocopherol at high temperatures. *Food Chemistry, 101*(3), 947–954. https://doi.org/10.1016/j.foodchem.2006.02.046.

Nyström, L., Lampi, A.-M., Andersson, A. A. M., Kamal-Eldin, A., Gebruers, K., Courtin, C. M., et al. (2008). Phytochemicals and dietary fiber components in rye varieties in the HEALTHGRAIN diversity screen. *Journal of Agricultural and Food Chemistry, 56*(21), 9758–9766. https://doi.org/10.1021/jf801065r.

Nyström, L., Moreau, R. A., Lampi, A. M., Hicks, K. B., & Piironen, V. (2008). Enzymatic hydrolysis of steryl ferulates and steryl glycosides. *European Food Research and Technology*, *227*(3), 727–733. https://doi.org/10.1007/s00217-007-0780-z.

Nyström, L., Schär, A., & Lampi, A. M. (2012). Steryl glycosides and acylated steryl glycosides in plant foods reflect unique sterol patterns. *European Journal of Lipid Science and Technology*, *114*(6), 656–669. https://doi.org/10.1002/ejlt.201200033.

Ohyama, K., Suzuki, M., Kikuchi, J., Saito, K., & Muranaka, T. (2009). Dual biosynthetic pathways to phytosterol via cycloartenol and lanosterol in Arabidopsis. *Proceedings of the National Academy of Sciences of the United States of America*, *106*(3), 725–730. https://doi.org/10.1073/pnas.0807675106.

Olkkonen, V. M., Gylling, H., & Ikonen, E. (2017). Plant sterols, cholesterol precursors and oxysterols: Minute concentrations—Major physiological effects. *The Journal of Steroid Biochemistry and Molecular Biology*, *169*, 4–9. https://doi.org/10.1016/j.jsbmb.2015.12.026.

Patterson, G. W., & Krauss, R. W. (1965). Sterols of chlorella. I. The naturally occurring sterols of chlorella vulgaris, C. ellipsoidea, and C. saccharophila. *Plant and Cell Physiology*, *6*(2), 211–220. https://doi.org/10.1093/oxfordjournals.pcp.a079095.

Payne, A. H., & Hales, D. B. (2004). Overview of steroidogenic enzymes in the pathway from cholesterol to active steroid hormones. *Endocrine Reviews*, *25*(6), 947–970. https://doi.org/10.1210/er.2003-0030.

Peng, L., Kawagoe, Y., Hogan, P., & Delmer, D. (2002). Sitosterol-β-glucoside as primer for cellulose synthesis in plants. *Science*, *295*(5552), 147–150. https://doi.org/10.1126/science.1064281.

Pinkwart, K., Schneider, F., Lukoseviciute, M., Sauka-Spengler, T., Lyman, E., Eggeling, C., et al. (2019). Nanoscale dynamics of cholesterol in the cell membrane. *Journal of Biological Chemistry*, *294*(34), 12599–12609. https://doi.org/10.1074/jbc.RA119.009683.

Plat, J., Baumgartner, S., Vanmierlo, T., Lütjohann, D., Calkins, K. L., Burrin, D. G., et al. (2019). Plant-based sterols and stanols in health & disease: "Consequences of human development in a plant-based environment?". *Progress in lipid research*, *74*, 87–102. https://doi.org/10.1016/j.plipres.2019.02.003.

Plat, J, Brzezinka, H., Lütjohann, D., Mensink, R. P., & von Bergmann, K. (2001). Oxidized plant sterols in human serum and lipid infusions as measured by combined gas-liquid chromatography-mass spectrometry. Journal of Lipid Research, 42(12), 2030–2038. https://doi.org/10.1016/S0022-2275(20)31532-7.

Pollier, J., Vancaester, E., Kuzhiumparambil, U., Vickers, C. E., Vandepoele, K., Goossens, A., et al. (2019). A widespread alternative squalene epoxidase participates in eukaryote steroid biosynthesis. *Nature Microbiology*, *4*(2), 226–233. https://doi.org/10.1038/s41564-018-0305-5.

Power, F. B., & Salway, A. H. (1913). XLVII.—The identification of ipuranol and some allied compounds as phytosterol glucosides. *Journal of the Chemical Society, Transactions*, *103*, 399–406. https://doi.org/10.1039/CT9130300399.

Pütter, K. M., van Deenen, N., Unland, K., Prüfer, D., & Schulze Gronover, C. (2017). Isoprenoid biosynthesis in dandelion latex is enhanced by the overexpression of three key enzymes involved in the mevalonate pathway. *BMC Plant Biology*, *17*(1), 1–13. https://doi.org/10.1186/s12870-017-1036-0.

Rahier, A., & Benveniste, P. (1989). Mass spectral identification of phytosterols. *Analysis of Sterols and Other Biologically Significant Steroids*, 223–250. https://doi.org/10.1016/b978-0-12-515445-1.50016-1.

Ramirez-Estrada, K., Castillo, N., Lara, J. A., Arró, M., Boronat, A., Ferrer, A., et al. (2017). Tomato UDP-glucose sterol glycosyltransferases: A family of developmental and stress regulated genes that encode cytosolic and membrane-associated forms of the enzyme. *Frontiers in Plant Science*, *8*(June), 1–21. https://doi.org/10.3389/fpls.2017.00984.

Randhir, A., Laird, D. W., Maker, G., Trengove, R., & Moheimani, N. R. (2020). Microalgae: A potential sustainable commercial source of sterols. *Algal Research, 46*(December 2019), 101772. https://doi.org/10.1016/j.algal.2019.101772.

Ras, R. T., Geleijnse, J. M., & Trautwein, E. A. (2014). LDL-cholesterol-lowering effect of plant sterols and stanols across different dose ranges: A meta-analysis of randomised controlled studies. *British Journal of Nutrition, 112*(2), 214–219. https://doi.org/10.1017/S0007114514000750.

Reed, M. J., Purohit, A., Woo, L. W. L., Newman, S. P., & Potter, B. V. L. (2005). Steroid sulfatase: Molecular biology, regulation, and inhibition. *Endocrine Reviews, 26*(2), 171–202. https://doi.org/10.1210/er.2004-0003.

Renard, D., Perruchon, J., Giera, M., Müller, J., & Bracher, F. (2009). Side chain azasteroids and thiasteroids as sterol methyltransferase inhibitors in ergosterol biosynthesis. *Bioorganic & Medicinal Chemistry, 17*(23), 8123–8137. https://doi.org/10.1016/j.bmc.2009.09.037.

Rieck, C., Geiger, D., Munkert, J., Messerschmidt, K., Petersen, J., Strasser, J., et al. (2019). Biosynthetic approach to combine the first steps of cardenolide formation in Saccharomyces cerevisiae. *MicrobiologyOpen, 8*(12). https://doi.org/10.1002/mbo3.925.

Roche, J., Alignan, M., Bouniols, A., Cerny, M., Mouloungui, Z., Vear, F., et al. (2010). Sterol content in sunflower seeds (Helianthus annuus L.) as affected by genotypes and environmental conditions. *Food Chemistry, 121*(4), 990–995. https://doi.org/10.1016/j.foodchem.2010.01.036.

Rodríguez-Concepción, M., Forés, O., Martínez-García, J. F., González, V., Phillips, M. A., Ferrer, A., et al. (2004). Distinct light-mediated pathways regulate the biosynthesis and exchange of isoprenoid precursors during arabidopsis seedling development. *Plant Cell, 16*(1), 144–156. https://doi.org/10.1105/tpc.016204.

Rohmer, M. (1999). The mevalonate-independent methylerythritol 4-phosphate (MEP) pathway for isoprenoid biosynthesis, including carotenoids. *Pure and Applied Chemistry, 71*(12), 2279–2284. https://doi.org/10.1351/pac199971122279.

Rohmer, M., Kokke, W. C., Fenical, W., & Djerassi, C. (1980). Isolation of two new C30 sterols, (24E)-24-N-propylidenecholesterol and 24 epsilon-N-propylcholesterol, from a cultured marine Chrysophyte. *Steroids, 35*(2), 219–231. https://doi.org/10.1016/0039-128x(80)90104-x.

Sales, E., Muñoz-Bertomeu, J., Arrillaga, I., & Segura, J. (2007). Enhancement of cardenolide and phytosterol levels by expression of an N-terminally truncated 3-hydroxy-3-methylglutaryl CoA reductase in transgenic digitalis minor. *Planta Medica, 73*(6), 605–610. https://doi.org/10.1055/s-2007-967199.

Salimova, E., Boschetti, A., Eichenberger, W., & Lutova, L. (1999). Sterol mutants of Chlamydomonas reinhardtii: Characterisation of three strains deficient in C24(28) reductase. *Plant Physiology and Biochemistry, 37*(4), 241–249. https://doi.org/10.1016/S0981-9428(99)80022-9.

Saravanan, R. S., Slabaugh, E., Singh, V. R., Lapidus, L. J., Haas, T., & Brandizzi, F. (2009). The targeting of the oxysterol-binding protein ORP3a to the endoplasmic reticulum relies on the plant VAP33 homolog PVA12. *The Plant Journal, 58*(5), 817–830. https://doi.org/10.1111/j.1365-313X.2009.03815.x.

Sasso, S., Pohnert, G., Lohr, M., Mittag, M., & Hertweck, C. (2012). Microalgae in the postgenomic era: a blooming reservoir for new natural products. *FEMS Microbiology Reviews, 36*(4), 761–785. https://doi.org/10.1111/j.1574-6976.2011.00304.x.

Sawai, S., Ohyama, K., Yasumoto, S., Seki, H., Sakuma, T., Yamamoto, T., et al. (2014). Sterol side chain reductase 2 is a key enzyme in the biosynthesis of cholesterol, the common precursor of toxic steroidal glycoalkaloids in potato. *Plant Cell, 26*(9), 3763–3774. https://doi.org/10.1105/tpc.114.130096.

Schaeffer, A., Bouvier-Navé, P., Benveniste, P., & Schaller, H. (2000). Plant sterol-C24-methyl transferases: Different profiles of tobacco transformed with SMT1 or SMT2. *Lipids*, *35*(3), 263–269. https://doi.org/10.1007/s11745-000-0522-1.

Schaeffer, A., Bronner, R., Benveniste, P., & Schaller, H. (2001). The ratio of campesterol to sitosterol that modulates growth in Arabidopsis is controlled by STEROL METHYLTRANSFERASE 2;1. *Plant Journal*, *25*(6), 605–615. https://doi.org/10.1046/j.1365-313X.2001.00994.x.

Schaller, H. (2010). Sterol and steroid biosynthesis and metabolism in plants and microorganisms. In *Comprehensive natural products II* (pp. 755–787). Elsevier. https://doi.org/10.1016/B978-008045382-8.00008-3.

Schaller, H., Bouvier-Navé, P., & Benveniste, P. (1998). Overexpression of an Arabidopsis cDNA encoding a sterol-C241-methyltransferase in tobacco modifies the ratio of 24-methyl cholesterol to sitosterol and is associated with growth reduction. *Plant Physiology*, *118*(2), 461–469. https://doi.org/10.1104/pp.118.2.461.

Schaller, H., Grausem, B., Benveniste, P., Chye, M. L., Tan, C. T., Song, Y. H., et al. (1995). Expression of the Hevea brasiliensis (H.B.K.) Müll. Arg. 3-hydroxy-3-methylglutaryl-coenzyme a reductase 1 in tobacco results in sterol overproduction. *Plant Physiology*, *109*(3), 761–770. https://doi.org/10.1104/pp.109.3.761.

Schaller, H., Maillot-Vernier, P., Belliard, G., & Benveniste, P. (1992). Increased sterol biosynthesis in tobacco calli resistant to a triazole herbicide which inhibits demethylation of 14α-methyl sterols. *Planta*, *187*(3), 315–321. https://doi.org/10.1007/BF00195654.

Schaller, H., Maillot-Vernier, P., Gondet, L., Belliard, G., & Benveniste, P. (1993). Biochemical characterization of tobacco mutants resistant to azole fungicides and herbicides. *Biochemical Society Transactions*, *21*(4), 1052–1057. https://doi.org/10.1042/bst0211052.

Schär, A., & Nyström, L. (2016). Enzymatic synthesis of steryl ferulates. *European Journal of Lipid Science and Technology*, *118*(10), 1557–1565. https://doi.org/10.1002/ejlt.201500586.

Schrick, K., Shiva, S., Arpin, J. C., Delimont, N., Isaac, G., Tamura, P., et al. (2012). Steryl glucoside and acyl steryl glucoside analysis of Arabidopsis seeds by electrospray ionization tandem mass spectrometry. *Lipids*, *47*(2), 185–193. https://doi.org/10.1007/s11745-011-3602-9.

Seitz, L. M. (1989). Stanol and sterol esters of ferulic and p-coumaric acids in wheat, corn, rye, and triticale. *Journal of Agricultural and Food Chemistry*, *37*(3), 662–667. https://doi.org/10.1021/jf00087a019.

Sharma, L. K., Madina, B. R., Chaturvedi, P., Sangwan, R. S., & Tuli, R. (2007). Molecular cloning and characterization of one member of 3β-hydroxy sterol glucosyltransferase gene family in Withania somnifera. *Archives of Biochemistry and Biophysics*, *460*(1), 48–55. https://doi.org/10.1016/j.abb.2007.01.024.

Shewry, P. R., Piironen, V., Lampi, A. M., Nyström, L., Li, L., Rakszegi, M., et al. (2008). Phytochemical and fiber components in oat varieties in the healthgrain diversity screen. *Journal of Agricultural and Food Chemistry*, *56*(21), 9777–9784. https://doi.org/10.1021/jf801880d.

Shimada, T. L., Shimada, T., Okazaki, Y., Higashi, Y., Saito, K., Kuwata, K., et al. (2019). HIGH STEROL ESTER 1 is a key factor in plant sterol homeostasis. *Nature Plants*, *5*(11), 1154–1166. https://doi.org/10.1038/s41477-019-0537-2.

Silvestro, D., Andersen, T. G., Schaller, H., & Jensen, P. E. (2013). Plant sterol metabolism. Δ7-Sterol-C5-desaturase (STE1/DWARF7), Δ5,7-sterol-Δ7-reductase (DWARF5) and Δ24-sterol-Δ24-reductase (DIMINUTO/DWARF1) show multiple subcellular localizations in *Arabidopsis thaliana* (Heynh) L. *PLoS One*, *8*(2). https://doi.org/10.1371/journal.pone.0056429.

Silvestro, D., Villette, C., Delecolle, J., Olsen, C. E., Motawia, M. S., Geoffroy, P., et al. (2018). Vitamin D5 in *Arabidopsis thaliana*. *Scientific Reports*, *8*(1), 1–8. https://doi.org/10.1038/s41598-018-34775-z.

Sitbon, F., & Jonsson, L. (2001). Sterol composition and growth of transgenic tobacco plants expressing type-1 and type-2 sterol methyltransferases. *Planta*, *212*(4), 568–572. https://doi.org/10.1007/s004250000417.

Sonawane, P. D., Pollier, J., Panda, S., Szymanski, J., Massalha, H., Yona, M., et al. (2016). Plant cholesterol biosynthetic pathway overlaps with phytosterol metabolism. *Nature Plants*, *3*(December). https://doi.org/10.1038/nplants.2016.205.

Soteriou, C., Kalli, A. C., Connell, S. D., Tyler, A., & Thorne, J. L. (2021). Advances in understanding and in multi-disciplinary methodology used to assess lipid regulation of signalling cascades from the cancer cell plasma membrane. *Progress in Lipid Research*, *81*, 101080. https://doi.org/10.1016/j.plipres.2020.101080.

Stirk, W. A., Bálint, P., Tarkowská, D., Novák, O., Strnad, M., Ördög, V., et al. (2013). Hormone profiles in microalgae: Gibberellins and brassinosteroids. *Plant Physiology and Biochemistry: PPB*, *70*, 348–353. https://doi.org/10.1016/j.plaphy.2013.05.037.

Stirk, W. A., Bálint, P., Tarkowská, D., Strnad, M., van Staden, J., & Ördög, V. (2018). Endogenous brassinosteroids in microalgae exposed to salt and low temperature stress. *European Journal of Phycology*, *53*(3), 273–279. https://doi.org/10.1080/09670262.2018.1441447.

Stonik, V. A., & Stonik, I. V. (2018). Sterol and sphingoid glycoconjugates from microalgae. *Marine Drugs*, *16*(12), 1–20. https://doi.org/10.3390/md16120514.

Strott, C. A., & Higashi, Y. (2003). Cholesterol sulfate in human physiology: What's it all about? *Journal of Lipid Research*, *44*(7), 1268–1278. https://doi.org/10.1194/jlr.R300005-JLR200.

Stucky, D. F., Arpin, J. C., & Schrick, K. (2015). Functional diversification of two UGT80 enzymes required for steryl glucoside synthesis in Arabidopsis. *Journal of Experimental Botany*, *66*(1), 189–201. https://doi.org/10.1093/jxb/eru410.

Suzuki, M., Nakagawa, S., Kamide, Y., Kobayashi, K., Ohyama, K., Hashinokuchi, H., et al. (2009). Complete blockage of the mevalonate pathway results in male gametophyte lethality. *Journal of Experimental Botany*, *60*(7), 2055–2061. https://doi.org/10.1093/jxb/erp073.

Szekeres, M., Németh, K., Koncz-Kálmán, Z., Mathur, J., Kauschmann, A., Altmann, T., et al. (1996). Brassinosteroids rescue the deficiency of CYP90, a cytochrome P450, controlling cell elongation and De-etiolation in Arabidopsis. *Cell*, *85*(2), 171–182. https://doi.org/10.1016/S0092-8674(00)81094-6.

Takahashi, K., Kanerva, K., Vanharanta, L., Almeida-Souza, L., Lietha, D., Olkkonen, V. M., et al. (2021). ORP2 couples LDL-cholesterol transport to FAK activation by endosomal cholesterol/PI(4,5)P 2 exchange. *The EMBO Journal*, *40*(14), 1–20. https://doi.org/10.15252/embj.2020106871.

Tang, J., Ohyama, K., Kawaura, K., Hashinokuchi, H., Kamiya, Y., Suzuki, M., et al. (2011). A new insight into application for barley chromosome addition lines of common wheat: Achievement of stigmasterol accumulation. *Plant Physiology*, *157*(3), 1555–1567. https://doi.org/10.1104/pp.111.183533.

Tarazona, P., Feussner, K., & Feussner, I. (2015). An enhanced plant lipidomics method based on multiplexed liquid chromatography-mass spectrometry reveals additional insights into cold- and drought-induced membrane remodeling. *Plant Journal*, *84*(3), 621–633. https://doi.org/10.1111/tpj.13013.

Tasende, M. G. (2000). Fatty acid and sterol composition of gametophytes and sporophytes of *Chondrus crispus (Gigartinaceae, Rhodophyta)*. *Scientia Marina*, *64*, 421–426. https://doi.org/10.3989/scimar.2000.64n4421.

Teh, L., & Möllers, C. (2016). Genetic variation and inheritance of phytosterol and oil content in a doubled haploid population derived from the winter oilseed rape Sansibar × Oase cross. *Theoretical and Applied Genetics*, *129*(1), 181–199. https://doi.org/10.1007/s00122-015-2621-y.

Thorn, A., Egerer-Sieber, C., Jäger, C. M., Herl, V., Müller-Uri, F., Kreis, W., et al. (2008). The crystal structure of progesterone 5β-reductase from *Digitalis lanata* defines a novel class of short chain dehydrogenases/reductases. *Journal of Biological Chemistry*, *283*(25), 17260–17269. https://doi.org/10.1074/jbc.M706185200.

Tiwari, P., Sangwan, R. S., Asha, Mishra, B. N., Sabir, F., & Sangwan, N. S. (2014). Molecular cloning and biochemical characterization of a recombinant sterol 3-o-glucosyltransferase from *Gymnema* sylvestre R.Br. Catalyzing biosynthesis of steryl glucosides. *BioMed Research International*. https://doi.org/10.1155/2014/934351.

Vanbrabant, K., Van Meel, D., Kerksiek, A., Friedrichs, S., Dubbeldam, M., Schepers, M., et al. (2021). 24(R, S)-Saringosterol—From artefact to a biological medical agent. *The Journal of Steroid Biochemistry and Molecular Biology*, *212*, 105942. https://doi.org/10.1016/j.jsbmb.2021.105942.

Véron, B., Billard, C., Dauguet, J. C., & Hartmann, M. A. (1996). Sterol composition of *Phaeodactylum tricornutum* as influenced by growth temperature and light spectral quality. *Lipids*, *31*(9), 989–994. https://doi.org/10.1007/BF02522694.

Véron, B., Dauguet, J. C., & Billard, C. (1998). Sterolic biomarkers in marine phytoplankton. II. Free and conjugated sterols of seven species used in mariculture. *Journal of Phycology*, *34*(2), 273–279. https://doi.org/10.1046/j.1529-8817.1998.340273.x.

Vert, G., Nemhauser, J. L., Geldner, N., Hong, F., & Chory, J. (2005). Molecular mechanisms of steroid hormone signaling in plants. *Annual Review of Cell and Developmental Biology*, *21*(1), 177–201. https://doi.org/10.1146/annurev.cellbio.21.090704.151241.

Vikström, F., Holmbom, B., & Hamunen, A. (2005). Sterols and triterpenyl alcohols in common pulpwoods and black liquor soaps. *Holz Als Roh-Und Werkstoff*, *63*(4), 303–308. https://doi.org/10.1007/s00107-005-0025-4.

Villette, C., Berna, A., Compagnon, V., & Schaller, H. (2015). Plant sterol diversity in pollen from angiosperms. *Lipids*, *50*(8), 749–760. https://doi.org/10.1007/s11745-015-4008-x.

Vlahakis, C., & Hazebroek, J. (2000). Phytosterol accumulation in canola, sunflower, and soybean oils: Effects of genetics, planting location, and temperature. *JAOCS, Journal of the American Oil Chemists' Society*, 77(1), 49–53. https://doi.org/10.1007/s11746-000-0008-6.

Voshall, A., Christie, N. T. M., Rose, S. L., Khasin, M., Van Etten, J. L., Markham, J. E., et al. (2021). Sterol biosynthesis in four green algae: a bioinformatic analysis of the ergosterol versus phytosterol decision point. *Journal of Phycology*, *57*(4), 1199–1211. https://doi.org/10.1111/jpy.13164.

Wang, H., & Mao, H. (2014). On the origin and evolution of plant brassinosteroid receptor kinases. *Journal of Molecular Evolution*, *78*(2), 118–129. https://doi.org/10.1007/s00239-013-9609-5.

Wang, H., Nagegowda, D. A., Rawat, R., Bouvier-Navé, P., Guo, D., Bach, T. J., et al. (2012). Overexpression of Brassica juncea wild-type and mutant HMG-CoA synthase 1 in Arabidopsis up-regulates genes in sterol biosynthesis and enhances sterol production and stress tolerance. *Plant Biotechnology Journal*, *10*(1), 31–42. https://doi.org/10.1111/j.1467-7652.2011.00631.x.

Warnecke, D. C., Baltrusch, M., Buck, F., Wolter, F. P., & Heinz, E. (1997). UDP-glucose: sterol glucosyltransferase: Cloning and functional expression in Escherichia coli. *Plant Molecular Biology*, *35*(5), 597–603. https://doi.org/10.1023/a:1005806119807.

Warnecke, D. C., & Heinz, E. (1994). Purification of a membrane-bound UDP-glucose: sterol beta-D-glucosyltransferase based on its solubility in diethyl ether. *Plant Physiology*, *105*(4), 1067–1073. https://doi.org/10.1104/pp.105.4.1067.

Wentzinger, L. F., Bach, T. J., & Hartmann, M. A. (2002). Inhibition of squalene synthase and squalene epoxidase in tobacco cells triggers an up-regulation of 3-hydroxy-3-methylglutaryl coenzyme a reductase. *Plant Physiology*, *130*(1), 334–346. https://doi.org/10.1104/pp.004655.

Wewer, V., Dombrink, I., vom Dorp, K., & Dörmann, P. (2011). Quantification of sterol lipids in plants by quadrupole time-of-flight mass spectrometry. *Journal of Lipid Research*, *52*(5), 1039–1054. https://doi.org/10.1194/jlr.D013987.

Whitaker, B. D. (1988). Changes in the steryl lipid content and composition of tomato fruit during ripening. *Phytochemistry*, *27*(11), 3411–3416. https://doi.org/10.1016/0031-9422(88)80740-4.

Wilson, T. A., Nicolosi, R. J., Woolfrey, B., & Kritchevsky, D. (2007). Rice bran oil and oryzanol reduce plasma lipid and lipoprotein cholesterol concentrations and aortic cholesterol ester accumulation to a greater extent than ferulic acid in hypercholesterolemic hamsters. *Journal of Nutritional Biochemistry*, *18*(2), 105–112. https://doi.org/10.1016/j.jnutbio.2006.03.006.

Withers, N. W., Kokke, W. C. M. C., Fenical, W., & Djerassi, C. (1982). Sterol patterns of cultured zooxanthellae isolated from marine invertebrates: Synthesis of gorgosterol and 23-desmethylgorgosterol by aposymbiotic algae. *Proceedings of the National Academy of Sciences*, *79*(12), 3764–3768. https://doi.org/10.1073/pnas.79.12.3764.

World-Heart-Federation. (2019). https://www.world-heart-federation.org/resources/key-facts/. accessed 01/09/2019.

Wu, S. S. H., Moreau, R. A., Whitaker, B. D., & Huang, A. H. C. (1999). Steryl esters in the elaioplasts of the tapetum in developing Brassica anthers and their recovery on the pollen surface. *Lipids*, *34*(5), 517–523. https://doi.org/10.1007/s11745-999-0393-5.

Yang, H., Yan, F., Wu, D., Huo, M., Li, J., Cao, Y., et al. (2010). Recovery of phytosterols from waste residue of soybean oil deodorizer distillate. *Bioresource Technology*, *101*(5), 1471–1476. https://doi.org/10.1016/j.biortech.2009.09.019.

Yokota, T., Arima, M., & Takahashi, N. (1982). Castasterone, a new phytosterol with plant-hormone potency, from chestnut insect gall. *Tetrahedron Letters*, *23*(12), 1275–1278. https://doi.org/10.1016/S0040-4039(00)87081-1.

Yokota, T., Kim, S. K., Fukui, Y., Takahashi, N., Takeuchi, Y., & Takematsu, T. (1987). Brassinosteroids and sterols from a green alga, *Hydrodictyon reticulatum*: Configuration at C-24. *Phytochemistry*, *26*(2), 503–506. https://doi.org/10.1016/S0031-9422(00)81442-9.

Yokota, T., Ohnishi, T., Shibata, K., Asahina, M., Nomura, T., Fujita, T., et al. (2017). Occurrence of brassinosteroids in non-flowering land plants, liverwort, moss, lycophyte and fern. *Phytochemistry*, *136*, 46–55. https://doi.org/10.1016/j.phytochem.2016.12.020.

Zhang, A., Xie, Y., He, Y., Wang, W., Sen, B., & Wang, G. (2019). Bio-based squalene production by Aurantiochytrium sp. through optimization of culture conditions, and elucidation of the putative biosynthetic pathway genes. *Bioresource Technology*, *287*, 121415. https://doi.org/10.1016/j.biortech.2019.121415.

CHAPTER NINE

The biosynthesis and roles of *N*-acylethanolamines in plants

Omar Arias-Gaguancela and Kent D. Chapman*

Department of Biological Sciences, BioDiscovery Institute, University of North Texas, Denton, TX, United States

*Corresponding author: e-mail address: chapman@unt.edu

Contents

Abstract

N-acylethanolamines (NAEs) are a group of lipid signaling molecules derived from the phospholipid precursor *N*-acylphosphatidylethanolamine (NAPE). NAEs can be processed by a wide range of metabolic processes including hydrolysis by fatty acid amide hydrolase (FAAH), peroxidation by lipoxygenases (LOX), and conjugation by glycosyl- and malonyl-transferases. The diversity of NAE metabolites points to participation in multiple downstream pathways for regulation and function. NAEs with acyl chains of 18C are typically the most predominant types in vascular plants. Whereas in nonvascular

Advances in Botanical Research, Volume 101
ISSN 0065-2296
https://doi.org/10.1016/bs.abr.2021.07.002

plants and some algae, the arachidonic acid-containing NAE, anandamide (a functional "endocannabinoid" in animal systems), was recently reported. A signaling role for anandamide and other NAEs is well established in vertebrates, while NAEs and their oxylipin metabolites are recently becoming appreciated for lipid mediator roles in vascular plants. Here, the NAE metabolism and function in plants are overviewed, with particular emphasis on processes described in vascular plants where most attention has been focused.

1. Introduction

N-acylethanolamines (NAEs) are conserved lipophilic acyl amides well-recognized for their signaling roles in animals (Alhouayek, Stafberg, Karlsson, Bergström, & Fowler, 2020; Dlugos, Childs, Stuhr, Hillard, & de Wit, 2012). Significant attention was brought to this group of acyl amides, due to their participation in a wide range of neurological processes. For example, one of the most studied NAEs in mammals, NAE20:4 (anandamide), is an endogenous ligand for cannabinoid (CB) receptors that triggers signaling cascades to regulate processes such as pain perception (Alharthi et al., 2018; Clapper et al., 2010), inflammatory responses (Lowin, Apitz, Anders, & Straub, 2015; Rettori, De Laurentiis, Zorrilla Zubilete, Rettori, & Elverdin, 2012) and anxiety behaviors (Bluett et al., 2014). This has captured the interest of many research groups, which see the therapeutic potential of NAE-like molecules as pharmacological targets (Bruijnzeel et al., 2016).

In plants, the understanding of NAE metabolism and function has gained ground using both biochemical and genetic approaches. Several lines of research have help to reveal the formation and metabolic fates of NAEs. In addition, studies have demonstrated the involvement of NAE and NAE-like metabolites in processes such as seedling establishment and development (Cotter, Teaster, Blancaflor, & Chapman, 2011; Keereetaweep, Blancaflor, Hornung, Feussner, & Chapman, 2013, 2015; Wang et al., 2006), chloroplast development (Cannon & Chapman, 2021; Keereetaweep et al., 2013; Yan et al., 2020), cellular organization (Blancaflor, Hou, & Chapman, 2003; Motes et al., 2005), flowering (Teaster et al., 2012), and biotic responses (Aziz & Chapman, 2020; Kang et al., 2008; Kim et al., 2009; Palmer, Senechal, Mukherjee, Ané, & Blackwell, 2014; Zhang et al., 2019), all of which have expanded our knowledge about these lipids. This information may be useful for future applications in crop improvement or management.

Here, a comprehensive overview of NAE metabolism and functions in plants is provided with the hope to stimulate interest towards this class of signaling molecules.

2. NAE occurence, formation and metabolic fates

2.1 NAE ocurrence in plants

N-acylethanolamines (NAEs) were initially thought to be lipophilic artifacts with minor bioactive effects. However, over the past 30 years, many studies have demonstrated the substantial biological impact associated with these lipid molecules (Blancaflor et al., 2014; Chapman, 2004). The binary designation, namely, NAE(X):(Y), is one of the most common nomenclatures to identity a particular NAE type. In this numeric based designation, X and Y represent the length of the acyl chain linked to ethanolamine and the number of double bonds in the acyl moiety, respectively. The reported NAEs (Chilufya, Devaiah, Sante, & Kilaru, 2015) in plants mostly have acyl chain lengths of 12–18C, and several degrees of unsaturation (up to 3 double bonds) (Fig. 1); however, 20C NAEs also have been identified in some nonvascular plants and algae (Gachet, Schubert, Calarco, Boccard, & Gertsch, 2017).

In vascular plants, NAE content varies with tissue, stage of development and plant species. NAEs are present in the range of nano- to micrograms per gram of fresh weight. Concordantly, many analytical methods needed to be adapted for their detection and quantification (Chapman, Tripathy, Venables, & Desouza, 1998; Gachet et al., 2017; Keereetaweep & Chapman, 2016; Venables, Waggoner, & Chapman, 2005). In crop seeds (e.g., tomato and corn), endogenous NAE18:2, NAE18:1, NAE16:0, NAE12:0, NAE18:0, and NAE14:0 have been identified, with NAE18:2 being the most abundant species. By contrast, in legumes such as alfalfa (*Medicago sativa*) and *Medicago truncatula*, NAE18:3 represents the most abundant NAE species (Table 1). Interestingly, in vegetative tissues, NAE profiles markedly differ from that of seeds (Chapman, Venables, Markovic, Blair, & Bettinger, 1999; Venables et al., 2005). Indeed, in seeds of the model plant *Arabidopsis thaliana* (hereafter referred to as Arabidopsis), the content of 18C polyunsaturated NAEs (e.g., NAE18:2 or NAE18:3) were many times greater than in seedling tissues (Fig. 2) (Wang et al., 2006). This shows that NAE occurrence and composition are influenced by developmental

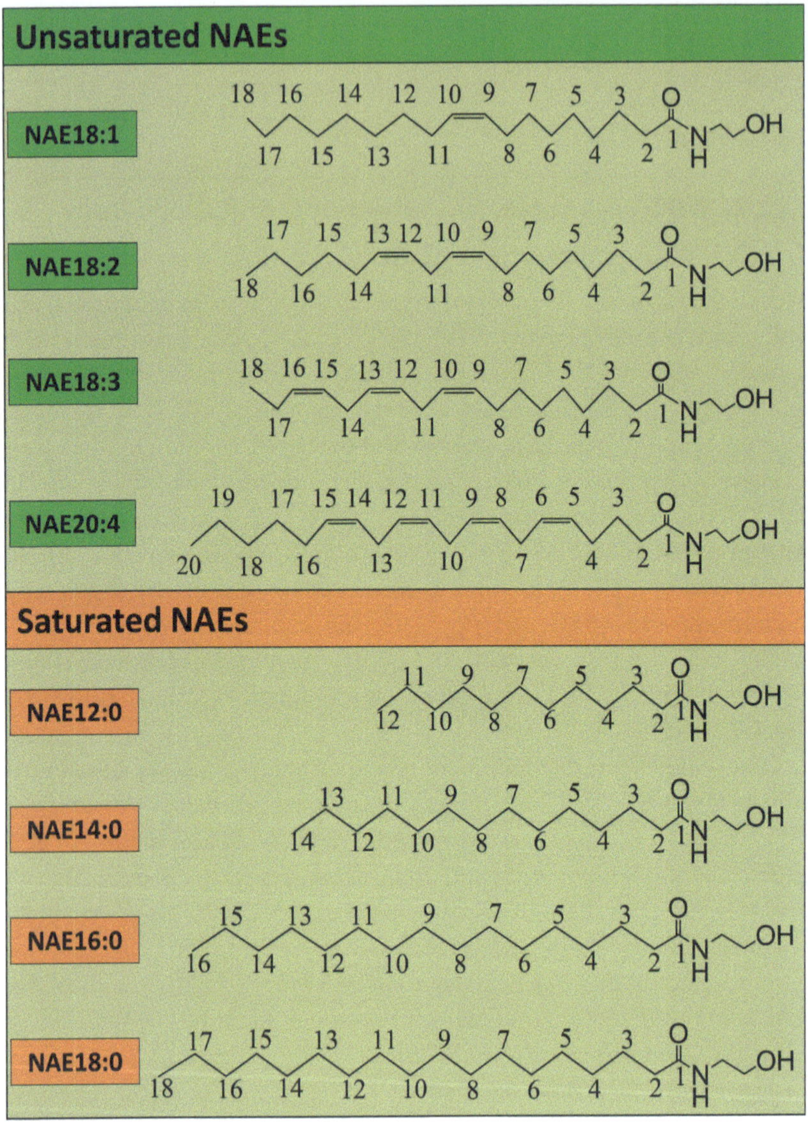

Fig. 1 Unsaturated and saturated *N*-acylethanolamines (NAEs) reported in vascular and nonvascular plants.

changes, and cell or tissue type. Further, these NAE-profiles indicate that 18C NAEs are major contributors of the total NAE pool.

In nonvascular plants, NAE occurrence has been assessed to understand evolution of lipid signaling across multiple taxa. For a long time, it was

Table 1 Endogenous concentrations of *N*-acylethanolamines (NAEs) in seeds of selected angiosperms.

Plant Species	NAE species (μg/g fresh weight)						
	NAE12:0	NAE14:0	NAE16:0	NAE18:0	NAE18:1	NAE18:2	NAE18:3
Solanum lycopersicum	0.15	0.03	0.1	0.03	0.13	0.29	0
Zea mays	0.18	0.02	0.2	0.07	0.33	0.38	0
Gossypium hirsutum	0.15	0.06	0.35	0.06	0.16	0.87	0
Abelmoschus esculentus	0.15	0.04	0.13	0.03	0.12	0.29	0
Medicago sativa	<0.02	<0.02	10.4	4.28	9.4	18.1	25.1
Medicago truncatula cv. A17	<0.02	<0.02	1.7	0.6	2.9	9.1	25.3
Glycine max cv. Dare	2.48	2.72	35.7	8.6	26.02	67.6	18.6

Cells in orange represent the most abundant NAE in a particular plant species.
Data derived from Chapman, K. D., Venables, B., Markovic, R., Blair, R. W., & Bettinger,-C. (1999). N-acylethanolamines in seeds. Quantification of molecular species and their degradation upon imbibition. Plant Physiology, 120, 1157–1164; Venables, B. J., Waggoner, C. A., & Chapman, K. D. (2005). N-acylethanolamines in seeds of selected legumes. *Phytochemistry*, 66, 1913–1918.

thought that the endogenous cannabinoid (NAE20:4) was present exclusively in nonplant systems (Gómez-Boronat et al., 2019; Pacioni et al., 2015). However, recent reports indicate that NAE20:4 alongside other NAE species (e.g., NAE18:2) are present in many nonvascular plants and even some algae. For instance, mosses (e.g., *Physcomitrella patens*), liverworts (e.g., *Conocephalum conicum*), monilophytes (e.g., *Salvinia molesta*), hornworts (e.g., *Anthoceros agrestis*), and even in the algae *Chara vulgaris* (Gachet et al., 2017). Unlike the studies conducted for vascular plants, NAE quantification

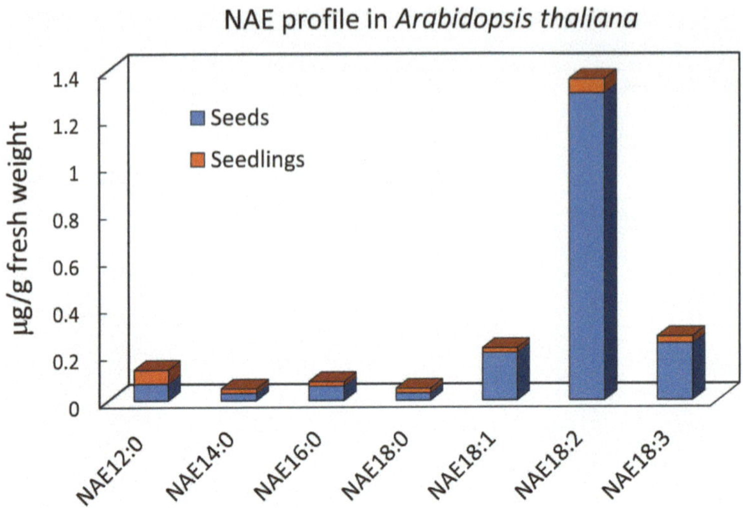

Fig. 2 NAE profile in *Arabidopsis thaliana*. Bar graphs depict the amount of NAEs found in seed and seedling tissues. *Data derived from Wang, Y. S., Shrestha, R., Kilaru, A., Wiant, W., Venables, B. J., Chapman, K. D., et al. (2006). Manipulation of Arabidopsis fatty acid amide hydrolase expression modifies plant growth and sensitivity to N-acylethanolamines.* Proceedings of the National Academy of Sciences of the United States of America, *103, 12197–12202.*

in nonvascular plants was done using whole organisms (sample tissue), and therefore, variations of NAE content associated with different tissues and developmental stages remain to be more fully investigated. Nevertheless, the detection of NAE20:4 in these taxonomic groups point to a previously unrecognized evolutionary plasticity for these lipid-signaling molecules.

2.2 NAE biosynthesis via NAPE

In mammals, multiple metabolic pathways lead to the production of NAEs (Fig. 3). One of the most common paths involves the use of membrane lipid *N*-acylphosphatidylethanolamine (NAPE) as the precursor (Hussain, Uyama, Tsuboi, & Ueda, 2017). NAPEs are found as minor membrane lipids with respect to the total phospholipid pool (Hansen, Kleberg, & Hassing, 2015). Their hydrolysis into NAEs is a conserved mechanism mediated by a membrane-associated enzyme, namely, NAPE-phospholipase D (NAPE-PLD) (Magotti et al., 2015). This phospholipase D type-enzyme cleaves NAPE phosphodiester bond, releasing NAE and phosphatidic acid (Janfelt et al., 2012). Further, NAPE-PLD catalytic activity is independent of nature of the acyl chain, which enables the formation of NAE species of various degrees of saturation and length.

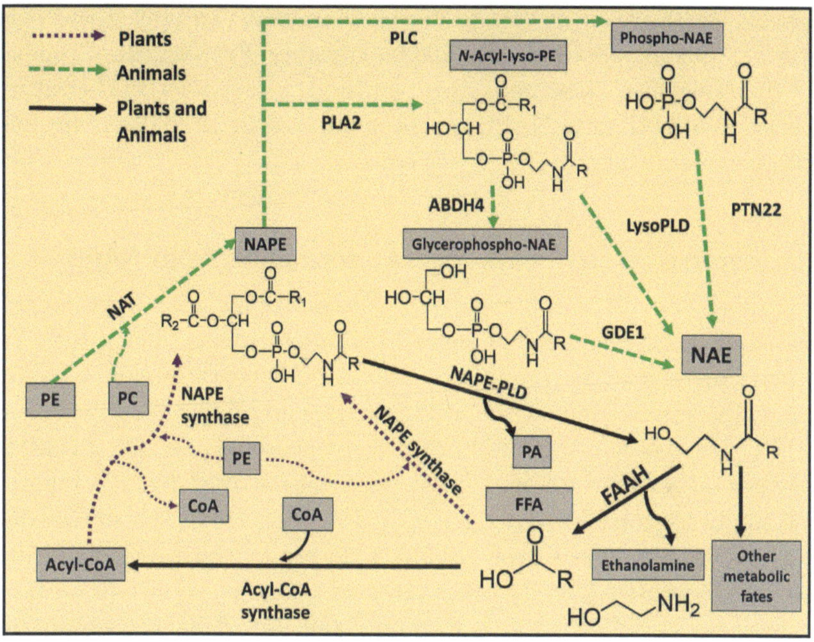

Fig. 3 NAE biosynthesis in plants and animals. Color-coded arrows represent pathways that are exclusive or conserved in plants and animals. Abbreviations: NAE, *N*-acylethanolamines; NAPE, *N*-acylphosphatidylethanolamine; NAPE-PLD, NAPE- phospholipase D; PLA2, phospholipase A2; *N*-acyl-lyso-PE, *N*-acyl-lyso-phosphatidylethanolamine; lysoPLD, lyso phospholipase D; ABDH4, α/β domain-containing hydrolase 4; GDE1, glycerophosphodiesterase 1; PLC, phospholipase C; PTN22, protein tyrosine phosphatase type 22; NAT, *N*-acyltransferase; PC, phosphatidylcholine; PE, phosphatidylethanolamine; FAAH, fatty acid amide hydrolase; FFA, free fatty acids; R, R1, or R2 represent different fatty acid acyl chains linked to the molecule.

In plants, NAE formation from NAPE was first proposed in studies conducted with microsomes of cotton seedlings (Chapman, Lin, & DeSouza, 1995; Chapman & Moore, 1993). Later, many PLD isoforms, namely, α, γ, and β were cloned from Arabidopsis cDNAs (Qin, Pappan, & Wang, 1997) and transformed into *Escherichia coli* for recombinant protein production and biochemical characterization. *In vitro* assays revealed that isoforms β and γ were capable of hydrolyzing NAPE into NAE. Further, ^{14}C radiolabeling experiments with tobacco cell suspensions shown that NAE formation from NAPE was stimulated by exposure with GTPase enhancers (e.g., mastoparan) (Chapman et al., 1998; Pappan, Austin-Brown, Chapman, & Wang, 1998). Similarly, tobacco leaves treated with fungal elicitors (e.g., xylanase) induced the depletion of NAPE and increased NAE 14:0

content (≈10-fold) compared to untreated controls (Tripathy, Kleppinger-Sparace, Dixon, & Chapman, 2003; Tripathy, Venables, & Chapman, 1999). Altogether, these and other similar data support a PLD-mediated formation of NAE from NAPE in a manner that is similar to that of the bon-afide pathway in animals (Fig. 3) (Kilaru et al., 2012).

2.3 Alternative pathways for NAE formation from NAPE

In animals, NAPE also is reported to be converted to *N*-acyl-lyso-PE, glycerophospho-NAE, and phospho-NAE intermediates, all of which ultimately lead to NAE formation (Maccarrone, 2017; Ueda, Tsuboi, & Uyama, 2013). Evidence in animals showed that phospholipase A2 (PLA2) converts NAPE into *N*-acyl-lyso-PE. Then, lyso phospholipase D (LysoPLD) converts *N*-acyl-lyso-PE into NAE or it can be used by α/β domain-containing hydrolase 4 (ABDH4) to generate glycerophospho-NAE (Sun et al., 2004; Simon & Cravatt, 2006), followed by the action of glycerophosphodiesterase 1 (GDE1) which removes the glycerol-3-phosphate group to finally yield NAE (Simon & Cravatt, 2008). An additional pathway uses phospholipase C (PLC) which removes 1,2-diacylglycerol from NAPE, resulting in phospho-NAE, followed by protein tyrosine phosphatase type 22 (PTN22) activity that converts this intermediate into NAE (Liu et al., 2006). These multistep routes constitute alternative pathways that lead to NAE formation, in addition to the one-step canonical PLD-mediated pathway (Fig. 3).

Attempts have been made to identify homologs of some of the mammalian enzymes for these alternative pathways in plants. For instance, the Arabidopsis gene At4g24160 (a putative ABDH4 homolog) was solubilized *in vitro* and shown to have acyltransferase, lipase and hydrolytic activities (Ghosh, Chauhan, Rajakumari, Daum, & Rajasekharan, 2009). The sequence similarity of At4g24160 to that of the mammalian enzyme, in addition to the presence of a GXSXG lipase motif in its amino acid sequence, makes this Arabidopsis gene a potential candidate for further experimental testing using NAPE as substrate (Blancaflor et al., 2014); however, other reports suggest that this gene encodes a homolog of the comparative gene identifier 58, a protein that was shown to facilitate acyl-CoA import into peroxisomes in plant leaves (James et al., 2010; Park et al., 2013), so its role in NAPE metabolism is uncertain. Further, the use of genomics and other similar technologies have made it possible to identify multiple phospholipase A and C gene families in corn, cotton, and *Brassica napus* (Iqbal et al., 2020; Zhang, Wang, & Liu, 2018; Zhu, Zhou, Li, & Li, 2020). Data derived from these studies could be used in future work to identify A or C-type phospholipases that might act upon NAPE to generate *N*-acyl-lyso-PE or phospho-NAE, respectively.

2.4 NAPE formation

The acyltransferase, namely, *N*-acyltransferase (NAT) is responsible of NAPE formation in multiple eukaryotic species (Fig. 3). NAT transfers *O*-acyl groups of a phospholipid donor (e.g., phosphatidylcholine (PC)) to the amine head group (*N*-acyl position) of phosphatidylethanolamine (PE) in a calcium dependent manner (Dosoky, Guo, Chen, Feigley, & Davies, 2018; Ogura, Parsons, Kamat, & Cravatt, 2016). In fact, rat neuron cells incubated with PC and nanomolar concentrations of Ca_2^+ produced NAPE, whereas no evident synthesis was observed otherwise (Wellner, Diep, Janfelt, & Hansen, 2013). The same experiment also revealed that acyl moieties in glycerophospholipids in the sn-1 position, rather than free fatty acids (e.g., oleic acid), were the preferred substrates used for NAPE synthesis.

In plants, a candidate NAPE synthase gene was identified and biochemically characterized in Arabidopsis (At1g78690p) (Faure et al., 2009). Recombinant protein and membrane extracts were used to test NAPE synthase activity from ^{14}C-labeled lipid substrates. In both experiments, NAPE synthase was capable of using acyl-CoA rather than free fatty acids (Faure et al., 2009). Later work has suggested that this protein might also synthesize other acyl lipids (Bulat & Garrett, 2011), so further work is required to fully reconcile the role of this protein in NAPE formation. Alternatively, an NAPE synthase activity associated with cottonseed microsomes was capable of producing NAPE from ^{14}C-labeled free fatty acids much more readily than from ^{14}C-acyl CoA or ^{14}C-labeled PC (Chapman & Moore, 1993). Later, three distinct bioactive isozymes were separated from the original protein extract. Transient expression of YFP-tagged At1g78690p in tobacco leaves showed exclusive localization to the plasma membrane whereas NAPE synthase activity in cotton microsomes was detected in three membrane compartments: plasma membrane, Golgi apparatus and the endoplasmic reticulum (Chapman & Moore, 1993). Further, spatiotemporal analysis of At1g78690p in transgenic Arabidopsis lines, revealed preferential localization and expression of the synthase gene in roots and young tissues of embryos (Faure et al., 2009). Similar findings also support the notion that NAPE production may differ depending on the plant species, and tissue analyzed (Kotel'nikova, 2011).

Environmental cues also influence NAPE formation. For instance, evidence has shown that chilling stress, and hydration/dehydration cycles impact NAPE formation (Chapman & Sprinkle, 1996; Sandoval, Huang, Garrett, Gage, & Chapman, 1995; Zhang & Tian, 2010). Similarly, increasing NAPE content was detected in potato cells challenged under artificial anoxic environments or treated with respiratory inhibitors (e.g., salicylhydroxamate). These abiotic factors are associated with the release of free fatty acids, and as a result, the cell counteracts the membrane destabilizing effects by promoting NAPE formation from PE and free fatty acids (Rawyler & Braendle, 2001). Overall, NAPE metabolic machineries adapt to physiological and homeostatic conditions for subsequent NAE modulation, but much remains to be learned about NAPE biosynthesis in plants.

2.5 NAE signal termination by fatty acid amide hydrolase

NAEs can be processed by fatty acid amide hydrolase (FAAH) into ethanolamine and free fatty acids (Aziz, Wang, Tripathi, Bankaitis, & Chapman, 2019; Balsevich et al., 2018; Lin, Metherel, Jones, & Bazinet, 2017). A cDNA encoding FAAH was first identified in rat, and following studies in multiple nonplant organisms have reported NAE hydrolysis by FAAH. For example, mouse, rat, and human FAAHs hydrolyzed NAE20:4 into arachidonic acid and ethanolamine (Long, LaCava, Jin, & Cravatt, 2011; Palermo et al., 2015; Strittmatter et al., 2012). Similarly, FAAH from the slime mold, *Dictyostelium discoideum*, was capable of hydrolyzing NAE20:4, along with other saturated (e.g., NAE16:0) and unsaturated NAEs (e.g., NAE18:2) (Neelamegan, Schoenhofen, Richards, & Cox, 2012). These data suggested that FAAH catalytic capabilities might be more broadly distributed in other biological systems.

In plants, the molecular basis for NAE hydrolysis by FAAH was initially revealed following identification of the Arabidopsis homolog to rat FAAH (37% amino acid similarity within the amidase signature domain) (Kim et al., 2009; Shrestha, Dixon, & Chapman, 2003). Experiments with recombinant Arabidopsis FAAH (AtFAAH) have shown its capacity to hydrolyze a wide range of NAEs, including NAE18:2, NAE16:0, NAE14:0, and NAE12:0. Similar work in other plant systems showed that recombinant rice and *Medicago truncatula* FAAHs, namely, OsFAAH and MtFAAH, also metabolized NAEs in a manner that resembles that of the Arabidopsis FAAH

(Shrestha, Kim, Dyer, Dixon, & Chapman, 2006). Enzyme kinetics studies also have shown that AtFAAH is capable of hydrolyzing NAE18:2-derived oxylipin, 9-hydroxy octadecadienoyl ethanolamide (9-NAE-HOD) with a similar catalytic efficiency to that of the unsubstituted NAE18:2 parent molecule (Aziz et al., 2019), which was not the case for rat FAAH. Further, AtFAAH has also been shown to irreversibly cleave the amide bond of *N*-acyl-L-homoserine lactones (e.g., 3-oxo-C12-HSL), bacteria-derived lipids known to participate in quorum signaling (Palmer et al., 2014).

Recent studies in the moss, *Physcomitrella patens,* identified nine FAAHs (PpFAAHs). PpFAAH1–4 were closely related to AtFAAH whereas PpFAAH5–9 were more similar to rat FAAH (Haq & Kilaru, 2020). Biochemical characterization of one of these enzymes (PpFAAH1) revealed amidohydrolase activity toward endogenous NAE20:4 and other NAEs (e.g., NAE16:0). Collectively, this information shows an expanded role for FAAHs to hydrolyze a wide range of NAE and NAE-like structures, and further supports a conserved catalytic mechanism across multiple species.

2.6 FAAH structural features accommodate NAEs

FAAH is a widely conserved amidase in eukaryotes which is characterized by a Ser-Ser-Lys catalytic triad of amino acid residues (Haq & Kilaru, 2020; McKinney & Cravatt, 2003, 2005). FAAH is a homodimer protein with each monomer consisting of a core of twisted beta sheets surrounded by alpha helices (Dainese & Oddi, 2020). The resolved crystal structures of AtFAAH (Aziz et al., 2019) and rat FAAH (Bracey, Hanson, Masuda, Stevens, & Cravatt, 2002) have made it possible to gain additional insights about their structural conformations and enzymatic behavior. Analysis of their structures has helped to recognize some key structural similarities and differences between plant and mammalian FAAHs.

Unlike rat FAAH, AtFAAH has an N-terminal region that is postulated to anchor the protein to *endo*- and plasma membranes and facilitate its interaction with lipophilic substrates (Fig. 4). Additional structural differences were revealed in the acyl binding and membrane access channels, namely, ABC and MAC, respectively (Aziz & Chapman, 2020; Aziz et al., 2019). Rat FAAH has two cavities for acyl binding and substrate access, separated by a region known as the "dynamic paddle." In contrast, FAAH from Arabidopsis and the FAAH modeled from *P. patens*, have a more open cavity for substrate access and binding (Aziz et al., 2019; Haq & Kilaru, 2020). The absence of a "dynamic paddle" seems to be a key conserved evolutionary

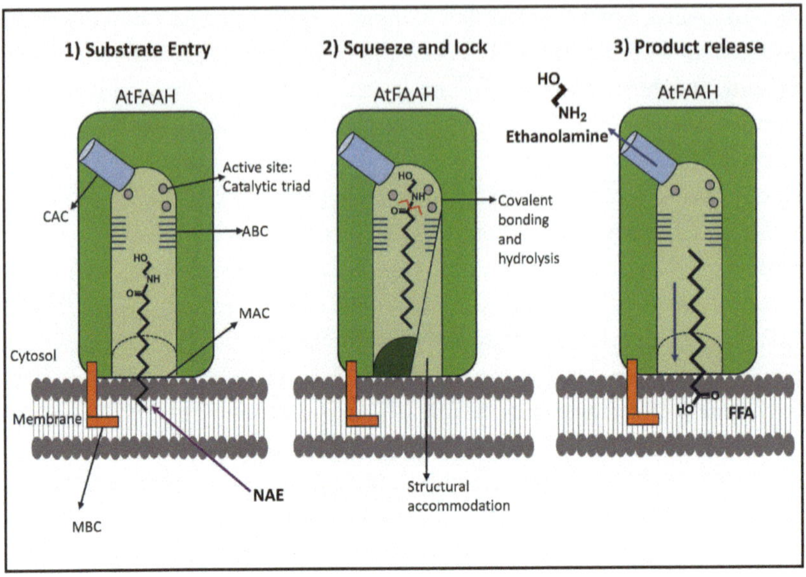

Fig. 4 Arabidopsis fatty acid amide hydrolase (AtFAAH) uses a "squeeze and lock" mechanism to accommodate and hydrolyze NAEs. Abbreviations: CAC, cytosolic access channel; ABC, acyl-binding channel; MAC, membrane access channel; MBC, membrane binding cap; FFA, free fatty acids. *Model derived from Aziz, M., Wang, X., Tripathi, A., Bankaitis, V. A., & Chapman, K. D. (2019). Structural analysis of a plant fatty acid amide hydrolase provides insights into the evolutionary diversity of bioactive acylethanolamides.* The Journal of Biological Chemistry, *294, 7419–7432.*

trait in vascular and nonvascular plants, as observed in sequence homology and studies of their 3D structures. Such differences alter the means by which plant FAAHs can accommodate NAEs and other lipid substrates into their active site. Based on X-Ray crystallographic data, AtFAAH is proposed to undergo a "squeeze and lock" mechanism upon substrate binding. When the ligand has reached the active site of the ABC, there is a conformational change of the helix region conformed by amino acids 531 to 537—"squeeze" the substrate in the binding site. In parallel, residues 25–28 in the MAC undergo movement along with Leu55 to secure the substrate-ligand complex—"lock." Upon reaction completion, the water-soluble product ethanolamine is released into the cytosol via the cytoplasmic access channel, while the fatty acid product is released back to the hydrophobic environment of the membrane (Fig. 4). In contrast, rat FAAH-substrate accommodation relies on a "dynamic paddle" formed by Phe432 and Trp531, to partially close its short MAC, and to further secure the ligand complex into its wider ABC (Aziz & Chapman, 2020; Aziz et al., 2019). This suggests that in animals, NAE-processing mediated by FAAH may have evolved substrate preferences that are more restrictive than FAAHs found in plants.

2.7 A second group of FAAH enzymes in plants

Recently, an analysis of FAAH amino acid sequences with special reference to amino acid residues in the substrate binding pocket have made it possible to categorize FAAH amino acid sequences from vascular plants into two major phylogenetic clusters, namely, group I and II FAAH (Aziz & Chapman, 2020). Angiosperm plants such as Arabidopsis and *Ricinus communis* have only group I FAAH whereas most angiosperms have genes that encode for multiple group I and II FAAHs, as predicted for cotton plants (*Gossypium arboretum*, *G. raimondii*, and *G. hirsutum*), Solanaceous plants (e.g., *S. lycopersicum*, *S. tuberosum*) and legumes (e.g., *Glycine* max, *Medicago truncatula*). This in silico analysis highlights the possibility that through evolution plants may have adapted to utilize one or both FAAH groups, perhaps providing for an expanded substrate repertoire beyond processing NAEs.

Homology modeling of group I and II FAAH proteins from soybean (*G. max*) based on the AtFAAH three-dimensional crystal structure (PDB: 6DHV), revealed some similarities and differences that were conserved among members of each FAAH group (Aziz & Chapman, 2020). Like group I FAAH, the group II FAAH protein from soybean maintained a substrate binding pocket with a MAC, an ABC, and a cytosolic access channel. Also, they both have the characteristic N-terminal region for membrane anchoring, and a conserved catalytic triad. In contrast, the main differences between group I and II FAAHs reside in the substitution of key conserved amino acids surrounding the substrate binding pocket. For example; the cytosolic access channel of group I FAAH was highly hydrophilic and had a more polar substrate binding pocket, whereas group II FAAH had a more hydrophobic cytosolic access channel as well as a more hydrophobic substrate binding pocket, characterized by the presence of more aromatic amino acids than group I FAAHs. Based on these differences it is reasonable to think that that group II FAAHs might process substrates with less polarity compared to group I FAAHs. In fact, computational docking shown that GmFAAH II can accommodate alkamides (molecules structurally similar to NAEs), and *N*-acyl L-homoserine lactones (AHLs) in a more efficient manner compared to GmFAAH I (Aziz & Chapman, 2020; Aziz et al., 2019). However, these differences are yet to be experimentally tested.

Further, such experiments to examine substrate promiscuity also should include known FAAH substrates (e.g., unsaturated, saturated or hydroxylated NAEs).

2.8 NAE oxidation via lipoxygenase pathways

Alternative to the traditional NAE signaling termination mediated by FAAH, polyunsaturated 18C NAE species like NAE18:2 or NAE18:3 also can be oxidized by lipoxygenase enzymes (LOX) to generate hydroperoxides that can be reduced nonenzymatically to hydroxides (Fig. 5) (Keereetaweep et al., 2013, 2015; Kilaru et al., 2011). There are two broad

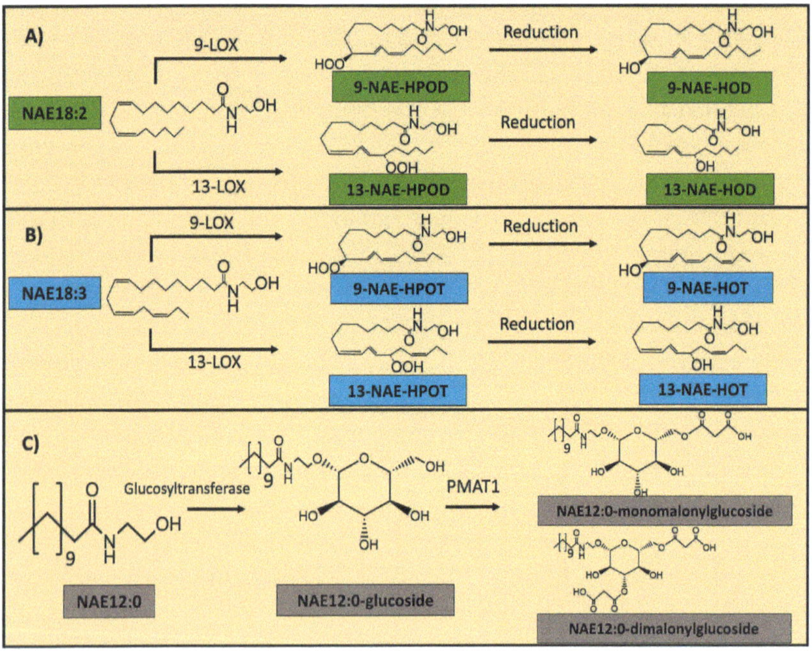

Fig. 5 Summary of selected NAE metabolic fates. Formation of hydroperoxides and hydroxides from (A) NAE18:2, (B) NAE18:3 and (C) NAE12:0-conjugates. Abbreviations: 9-NAE-HPOD, (9S,12Z,10E)-9-hydroperoxy-10,12-octadecadienoylethanolamide; 13-NAE-HPOD, (13S,9Z,11E)-13-hydroperoxy-9,11-octadecadienoylethanolamide; 9-NAE-HOD, (9S,12Z,10E)-9-hydroxy-10,12-octadecadienoylethanolamide; 13-NAE-HOD, (13S,9Z,11E)-13-hydroxy-9,11-octadecadienoylethanolamide; 9-NAE-HPOT, (9S,12Z,10E,15Z)-9-hydroperoxy-10,12,15-octadecatrienoylethanolamide; 13-NAE-HPOT, (13S,9Z,11E,15Z)-13-hydroperoxy-9,11,15-octadecatrienoylethanolamide; 9-NAE-HOT, (9S,12Z,10E,15Z)-9-hydroxy-10,12,15-octadecatrienoylethanolamide; 13-NAE-HOT, (13S,9Z,11E,15Z)-13-hydroxy-9,11,15-octadecatrienoylethanolamide; PMAT, glucoside malonyltransferase 1; 9-LOX, 9-lipoxygenase; 13-LOX, 13-lipoxygenase.

families of LOX enzymes in plants based on their position of peroxidation, namely, 9-LOX and 13-LOX (Hayward, Cilliers, & Swart, 2017). They introduce molecular oxygens at the C9 or C13 positions of the acyl chain (Andreou & Feussner, 2009). These same LOX enzymes can act on polyunsaturated NAEs, free fatty acids or acyl groups of glycerolipids (Chechetkin, Osipova, Antsygina, Gogolev, & Grechkin, 2011; Gabbs & Leng, 2015; Kilaru et al., 2011).

Evidence of oxylipin formation from NAEs in plants was first described in experiments with imbibed cottonseed extracts, where a 13-LOX activity mediated the oxidation of NAE18:2 (Shrestha, Noordermeer, van der Stelt, Veldink, & Chapman, 2002). Experiments with recombinant Arabidopsis LOX enzymes also demonstrated activity toward NAEs, where two 9-LOX genes (AtLOX1 and AtLOX5) and four 13-LOX genes (AtLOX2, AtLOX3, AtLOX4, and AtLOX6) have been reported (Bannenberg, Martínez, Hamberg, & Castresana, 2009). 9-LOX or 13-LOX enzymes act upon NAE18:2 to generate (9S,12Z,10E)-9-hydroperoxy-10,12-octadecadienoylethanolamide (9-NAE-HPOD) or (13S,9Z,11E)-13-hydroperoxy-9,11-octadecadienoylethanolamide (13-NAE-HPOD) (Kilaru et al., 2011). Subsequently, these hydroperoxides may be reduced to (9S,12Z,10E)-9-hydroxy-10,12-octadecadienoylethanolamide (9-NAE-HOD) or (13S,9Z,11E)-13-hydroxy-9,11-octadecadienoylethanolamide (13-NAE-HOD). Interestingly, 9-NAE-HOD rather than 13-NAE-HOD acts as a potent bioactive inhibitor of seedling growth (Keereetaweep et al., 2015), pointing to a selective lipid mediator function for specific NAE metabolites. Furthermore, the action of 9-NAE-HOD requires an intact abscisic acid (ABA) signaling pathway and works synergistically with ABA to modulate growth in Arabidopsis seedlings (Keereetaweep et al., 2015).

Similarly, NAE18:3 also can be used as lipophilic substrate by 9-LOX or 13-LOX to form (9S,12Z,10E,15Z)-9-hydroperoxy-10,12,15-octadecatrienoylethanolamide (9-NAE-HPOT) or (13S,9Z,11E,15Z)-13-hydroperoxy-9,11,15-octadecatrienoylethanolamide (13-NAE-HPOT) (Keereetaweep et al., 2013). These intermediates are then reduced to (9S,12Z,10E,15Z)-9-hydroxy-10,12,15-octadecatrienoylethanolamide (9-NAE-HOT) or (13S, 9Z,11E,15Z)-13-hydroxy-9,11,15-octadecatrienoylethanolamide (13-NAE-HOT), and both exert negative regulatory effects in seedling growth. Alternatively, 13-NAE-HPOT could undergo additional metabolic steps that lead to (13S,9Z,10E,15Z)-12-oxo-10,15-phytodienoic acid ethanolamide (NAE-OPDA) formation. Overall, these 9-LOX and 13-LOX mediated routes indicate a previously unrecognized versatility in NAE oxidative metabolism.

Moreover, the interplay between LOX and FAAH enzymes is key in determining NAE oxylipin production. Side by side experiments comparing the NAE-derived oxylipin profiles of Arabidopsis wild-type vs *atfaah*-knockout or *AtFAAH* overexpressing lines fed with micromolar concentrations of NAE18:2 or NAE18:3, demonstrated that *atfaah*-knockouts produced an increase of NAE-HODs or NAE-HOTs contents, whereas, NAE18:2- or NAE18:3-oxylipin levels in *FAAH* overexpressing lines were much lower than in wild-type controls (Keereetaweep et al., 2013, 2015). It is reasonable to assume that the suppression of FAAH leads to an increase of the NAE pool content (especially of 18C polyunsaturated NAEs), which can be further metabolized by LOX enzymes (hence the elevated NAE-oxylipin content). Conversely, in the overexpressing lines, an increase of *FAAH* expression and hydrolytic activity reduces the bioavailability of polyunsaturated NAE substrates that could be used by LOX enzymes to form NAE-HODs or NAE-HOTs (Keereetaweep et al., 2013, 2015; Kilaru et al., 2011). Overall, LOX and FAAH interactions for NAE processing provide new layers of complexity and fine tuning for these lipid molecules that represent opportunities for further research in other plant systems and physiological conditions.

2.9 NAE12:0 glyco-conjugates

The study of NAE-metabolic fates led to somewhat of a surprise when using untargeted analytical tools. Khan et al. demonstrated the presence of NAE12:0 conjugates in Arabidopsis seedlings treated with micromolar concentrations of NAE12:0 (Fig. 5) (Khan et al., 2016). Data from mass spectrometry (MS) and nuclear magnetic resonance (NMR) spectroscopy identified detectable levels of NAE12:0-glucoside. In this reaction, a yet-to-be determined glycosyltransferase is responsible for adding a sugar moiety to the free hydroxyl group of NAE12:0. In addition, NAE12:0-monomalonylglucoside and NAE12:0-dimalonylglucoside were identified, representing two additional previously unknown NAE conjugates. *In vivo* and *in vitro* experiments showed that the enzyme phenolic glucoside malonyltransferase 1 (PMAT1) attaches malonic acid to the carbon 3 or/and 6 of NAE12:0-glucoside conjugates (Khan et al., 2016). The identification of these novel NAE derivatives suggest additional metabolic pathways that plants might use to process and transport NAEs to different cellular compartments, thus expanding the potential routes for NAE metabolism (Fig. 5). However, the biological effects or bioactivity of these NAE12:0-glyco-conjugates remain to be elucidated in future research.

2.10 Some NAEs untilize G-proteins for signal transduction

The propagation of NAE signal(s) in animal systems depends upon G protein-coupled receptors (GPCRs) and associated G proteins. This is the topic of intense investigations in vertebrates where these receptors are the same as those that are activated by cannabinoids. As such, specific NAEs are considered endogenous, or "endocannabinoids," and regulate numerous physiological processes as part of the endocannabinoid signaling system (Kilaru & Chapman, 2020). For example, in mammals, postsynaptic neurons produce NAE20:4 which binds to CB receptors, and this leads to the activation of heterotrimeric G proteins (Howlett, Blume, & Dalton, 2010). Activated trimeric G-proteins undergo dismantling of the G$\alpha\beta\gamma$ trimeric state, which results in the release of Gα-GTP and G$\beta\gamma$ dimer, and these units are free to interact with downstream factors to regulate multiple neurological processes, thus, propagating NAE signaling (Mahavadi, Sriwai, Huang, Grider, & Murthy, 2014; Samson et al., 2003).

In Arabidopsis, NAE18:3-induced chloroplast responses have been linked to G-protein activity (Cannon & Chapman, 2021; Cannon et al., 2020; Yan et al., 2020). For instance, the de-greening of chloroplasts induced by NAE 18:3 application to seedlings, was inhibited in G-protein mutants. Exogenous applications of NAE18:3 also led to differential expression of many genes associated with G-protein activity, stress responses, senescence, autophagy, and chlorophyll anabolic and catabolic pathways (Cannon et al., 2020; Yan et al., 2020). This and other similar experiments support a model in which G-proteins are required, at least in part, for NAE18:3 action in plants. Altogether, studies from animals and plants suggest a partially-conserved mechanism for NAE signal transduction requiring G-proteins, and also suggest different components that may have arisen during the course of evolutionary modification of NAE-signaling.

3. Functions of NAEs

3.1 Plant growth

Pharmacological approaches have revealed multiple effects of NAEs on plant growth (Blancaflor et al., 2014; Wang et al., 2006). For example, NAE12:0-treated Arabidopsis seedlings showed a dose-dependent, reversible inhibition of primary root elongation, abnormal root apices with disrupted cell walls, and reduced root hair length and numbers compared to controls (Blancaflor et al., 2003). Also, NAE18:2- and NAE18:3-derived oxylipins (but not the parent NAEs) inhibited seedling growth (Keereetaweep et al., 2013, 2015). ABA seems to be required for the action of NAE18: 2-derived oxylipin (9-NAE-HOD) on overall seedling growth, thus providing evidence of cross talk between NAE and ABA pathways (Keereetaweep et al., 2015). These examples together have suggested a link between NAE metabolism and seedling development.

To gain insights of the physiological role of FAAH-mediated NAE hydrolysis in Arabidopsis, several genetic approaches also have been taken. Different FAAH overexpression and knockout lines were compared to wild-type plants to assess phenotypic changes. Studies showed that overexpression of Arabidopsis FAAH results in an increase in plant growth, a notion derived from the enhanced seedling growth and early flowering phenotypes found in the *FAAH* overexpressors compared with wild-type seedlings (Teaster et al., 2012, 2007; Wang et al., 2006). Interestingly, the quantities of endogenous NAE levels were reduced significantly in seedlings of *FAAH* overexpressing lines compared to wild-type. When genetic and pharmacological approaches were combined, it was evident that the growth effects were even more pronounced when seedlings were challenged with NAEs. For instance, exogenous applications of NAE12:0 or NAE18:2 did not inhibited growth in FAAH overexpressing plants to the same extent as wild-type or knockouts. By contrast, *atfaah*-knockouts showed only a modest reduction in seedling growth and minor discernible impacts on endogenous NAE levels during seedling growth, but these growth effects were much more pronounced in the presence of added NAEs than those of wild-type seedlings (Cotter et al., 2011; Teaster et al., 2007; Wang et al., 2006). Altogether, work derived from these studies support an important function(s) for NAEs in plant development.

Alkamides are a group of NAE-like structures known to be hydrolyzed by Arabidopsis FAAH (Faure et al., 2015). Experiments have shown that alkamides are negative regulators of plant development (Ramírez-Chávez, López-Bucio, Herrera-Estrella, & Molina-Torres, 2004). Exogenous applications of affinin (one type of alkamide) greatly impacted Arabidopsis seedling growth, and such effects were more noticeable in the context of FAAH altered lines. For instance, *atfaah*-knockout plants exposed to exogenous alkamides showed reduced root and seedling growth compared with wild-type seedlings, whereas *AtFAAH* overexpressing lines were more tolerant to the exposure (Faure et al., 2015; Ramírez-Chávez et al., 2004). Similarly, *N*-acyl L-homoserine lactones (another type of NAE-like structures) have been associated with the alteration of root architecture, and overall seedling growth. Like NAEs and alkamides, *N*-acyl L-homoserine

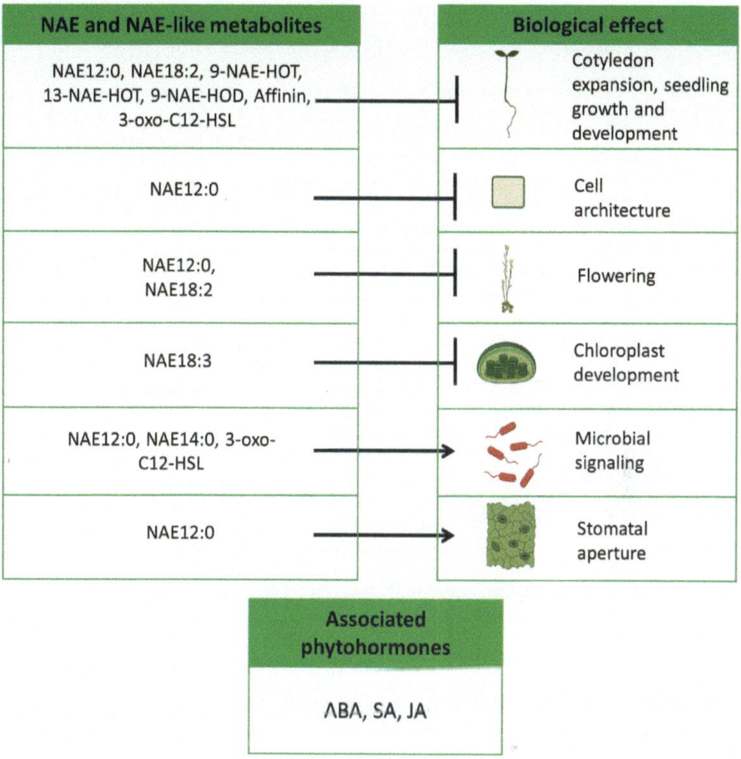

Fig. 6 Selected plant NAE, and NAE-like metabolites and their corresponding biological effects, and associated phytohormones. *Figure created with BioRender.com and Power Point 2016.*

lactone-induced effects are influenced in a FAAH-depended manner (Ortíz-Castro, Martínez-Trujillo, & López-Bucio, 2008; Palmer et al., 2014). These examples support the notion that NAE and NAE-like molecules can influence plant development in a FAAH-dependent manner (Fig. 6).

3.2 Flowering

NAEs have been implicated in the modulation of flowering time in Arabidopsis plants (Teaster et al., 2012). *FAAH* overexpressing lines had the tendency to flower up to 10 days earlier than wild-type and *atfaah*-knockouts plants under noninductive short-day conditions. This phenotype coincided with 30% reduction of NAE12:0 and NAE18:2 contents, and elevated expression of genes associated with flowering time. For example, *FLOWERING LOCUS* (*FT*) and *SEPTELLA3* (*SEP3*) genes were elevated in *FAAH* overexpressing plants. *FT* is an important transcription factor which activates or represses flowering gene regulators (e.g., *SEP3*). Given that NAE12:0 was one of the NAEs reduced in *FAAH* overexpressing plants, micromolar concentrations of NAE12:0 were irrigated to the base of wild-type plants every 3 days to assess phenotypes. Data showed a decrease in the number of rosette leaves, shorter inflorescent stems, and flowering delay of 6 days compared to DMSO controls. Consistent with these findings, NAE12:0-treated plants had reduced *FT* and *SEP3* expression levels (Teaster et al., 2012). These pieces of evidence point to NAE12:0 capacity to delay the onset of flowering in plants (Fig. 6), maybe in association with flowering gene networks, and that this can be modulated by FAAH.

3.3 Chloroplast degradation

Destabilization of chlorophyll components are also an effect associated with NAEs. Exogenous applications of NAE18:3 to Arabidopsis wild-type seedlings led to reduced growth and an evident cotyledon bleaching phenotype. The de-greening effect was even more severe in *atfaah*-knockout seedlings whereas *AtFAAH* overexpressing lines were unaffected and did not undergo de-greening in the presence of NAE (Keereetaweep et al., 2013). In this context, a FAAH-suppressed plant would be unable to efficiently hydrolyze NAEs which renders it more susceptible to NAE18:3. By contrast, FAAH-overexpressing plants tolerate NAE negative effect (s) due to an enhanced FAAH activity and capacity for NAE18:3 turnover. Further, analysis at the transcriptome level, revealed that exogenous applications of NAE18:3 to Arabidopsis wild-type seedlings result in down-regulation of multiple genes associated with chlorophyll biosynthesis and development (Cannon et al., 2020). These data show that exogenous application of NAE18:3 triggers detrimental effect (s) through molecular networks associated with chlorophyll metabolism, and retards the normal chloroplast development and cotyledon expansion associated with seedling growth.

Further observations of NAE18:3-treated seedlings with confocal laser scanning microscopy and by transmission electron microscopy revealed a disruption of thylakoid membranes and an overall dismantling of chloroplasts over a 3-day time course (Keereetaweep et al., 2013). The same study also showed that the loss of chlorophyll observed upon NAE18:3-treatment was reversible, and stage-specific. Experiments showed that removal of NAE18:3 resulted in the restoration of chloroplast architecture and a re-greening of cotyledons over a 4-day time course. Interestingly, addition of micromolar concentrations of NAE12:0 prevented NAE18:3-seedling bleaching effect (Keereetaweep et al., 2013; Keereetaweep, Kilaru, Feussner, Venables, & Chapman, 2010). Thus, this may suggest that certain NAEs can interact with other NAE types to influence the ultimate physiological outcome (Khan, Chapman, & Blancaflor, 2021), perhaps by competing for the active site of FAAH or LOX enzymes. Overall, these experiments suggest that levels of NAE18:3 or metabolites thereof can impact chloroplast homeostasis in plants (Fig. 6).

3.4 Stomatal closure

As previously mentioned, there appears to be crosstalk between NAE and ABA signaling pathways (Cotter et al., 2011). Another piece of information that supports such a notion is the finding that NAE12:0 blocks PLDα activity and its ability to modulate ABA-induced closure of stomata. Indeed, tobacco guard cells treated with ABA alone led to reduction of the diameters of the stomatal pores, whereas treatments at low micromolar concentrations of both NAE12:0 and ABA led to greater pore diameters over a 60-min course (Austin-Brown & Chapman, 2002). These results indicated that NAE12:0 can modulate stomatal aperture in a dose- and time-dependent manner (Fig. 6).

3.5 Cellular reorganization

The link between cell architecture and NAEs is an area that has received attention (Blancaflor et al., 2014). Pharmacological applications of NAE 12:0 to different types of hypocotyl and cotyledon cells caused a dramatic impact on cellular organization and integrity. Data showed that long-term exposures of NAE 12:0 led to transverse alignment of the microtubules and an aggregation of F-actin bundles (Blancaflor et al., 2003; Motes et al., 2005). Furthermore, these changes compromise other cellular processes such as endomembrane trafficking and cell expansion. Together these support the notion that NAE12:0 interferes with regular cell expansion, which may be responsible, in part, for the overall reduced growth phenotypes observed in NAE-treated seedlings.

3.6 Microbial interactions

One of the first indications that NAE could participate in plant–pathogen interactions came from studies in tobacco cells and leaves, where NAE14:0 levels were dramatically increased in the presence of fungal elicitors (Tripathy et al., 2003, 1999). This NAE-profile coincided with increased expression of defense genes (e.g., phenylalanine ammonia lyase) and the inhibition of extracellular alkalinization in a time- and dose-dependent manner (Tripathy et al., 1999). The influence of NAEs on plant–fungal interactions also has been examined in cotton (Zhang et al., 2019). Research revealed that defoliating strains (DS) of the plant pathogenic fungus *Verticillium dahliae* were capable of producing NAE12:0 during infection of cotton seedlings. Furthermore, the leaf defoliation phenotype seemed to depend on ABA induction. Notably, in these studies, neither *FAAH* overexpressors nor *atfaah*-knockouts were tested for sensitivity to *V. dahliae* (Zhang et al., 2019). Nevertheless, the study found that many *FAAH* transcripts increased upon *V. dahliae* infection or NAE 12:0 treatment. These findings support a model in which FAAHs and NAEs modulate leaf defoliation phenotypes, suggesting a role for NAEs in plant–microbe interactions (Fig. 6).

NAE modulation in plant–bacterial interactions has been addressed as well. *AtFAAH* overexpressing or knockout lines with altered NAE profiles have been used to understand their link to plant–pathogen interactions (Kang et al., 2008; Kim et al., 2009). *FAAH* overexpressing plants were found to be more susceptible to different strains of the bacterial plant pathogen *Pseudomonas syringae* compared to wild-type. The enhanced bacterial growth rate in Arabidopsis leaves also was associated with a reduction of multiple phytohormone levels. Free salicylic acid (SA) and conjugated forms were decreased by several fold compared to wild-type controls. Concordantly, SA-associated genes were also significantly downregulated in *FAAH* overexpressors. Similarly, jasmonic acid (JA) and several of its biosynthetic associated genes were reduced in *FAAH* overexpressing tissues. Abscisic acid (ABA) was another phytohormone that was found to be reduced in *FAAH* overexpressing lines. These indicate a potential interaction of NAE signaling in plant hormone regulation during biotic stress and that NAE metabolism by FAAH may participate in the modulation of plant immunity (Kang et al., 2008; Kim et al., 2009).

Although, the role(s) of NAEs in the influence of endogenous plant microbial populations remains unknown, promising evidence for NAE modulation of microbiomes has been recently described in mammals

(Fornelos & Franzosa, 2020). Tissues from inflammatory bowel disease patients had elevated amounts of NAE18:2 and NAE20:4, and this mis-regulation was shown to shift the composition of bacterial communities in humans. Further, metagenomic analysis revealed that *ex vivo* treatments with NAE18:2, NAE20:4, or a combination of both NAEs enhanced the growth of Proteobacteria (especially of Enterobacteriaceae) while depleting Bacteroidetes. While the exact mechanism by which bacteria process these host-derived NAEs is still unknown, transcriptomic analysis in Enterobacteriaceae communities revealed the need for elements of the respiratory and electron transfer chains for NAE metabolism (Fornelos & Franzosa, 2020). In this context, it is reasonable to speculate that an altered NAE content or composition may be associated with host-microbe communications in plants.

4. Future prospectives

While the advances in NAE-related research have produced meaningful insights regarding NAE-metabolism and regulation in plants, many open questions are yet to be addressed. For instance, does the discovery of structurally and phylogenetically different group I and II FAAHs in vascular plants greatly expand the types of lipophilic acyl amides (NAE-like structures) that could be potential substrates for FAAHs? Efforts to characterize these two distinct FAAH groups should be studied in systems beyond Arabidopsis. Extending the scope of research toward crops of agricultural impotence could lead to the discovery of previously unrecognized functions. Further, the notion that NAEs may also participate in endogenous microbial communications is an intriguing possibility that remains to be addressed. Certainly, the ability of AtFAAH to hydrolyze *N*-acyl-L-homoserine lactones (a quorum sensing molecule) supports a link between NAE and NAE-like structures and the potential for endogenous microbial signaling. Understanding the connection between plant-microbial communication and endogenous plant growth regulation is likely to provide much needed insights into the functional roles of NAEs, acylamides and their potential regulatory hydrolases—the FAAHs. Finally, the evolutionary conservation of NAE signaling in nonvascular plants and algae compared to vascular plants and nonplant systems seems to be an area open for exploration, especially with recent studies in *P. patens* for the occurrence of mammalian type NAE and FAAHs (Haq & Kilaru, 2020).

Acknowledgments

Work in the senior author's laboratory was supported initially by the US Dept of Energy, Office of Science under DE-FG02-05ER15647, and more recently by the National Science Foundation—Division of Integrative and Organismal Sciences (IOS-1656263).

References

Alharthi, N., Christensen, P., Hourani, W., Ortori, C., Barrett, D. A., Bennett, A. J., et al. (2018). n−3 polyunsaturated N-acylethanolamines are CB2 cannabinoid receptor-preferring endocannabinoids. *Biochimica et Biophysica Acta (BBA)—Molecular and Cell Biology of Lipids*, *1863*, 1433–1440.

Alhouayek, M., Stafberg, L., Karlsson, J., Bergström, S. H., & Fowler, C. J. (2020). Effects of orthotopic implantation of rat prostate tumour cells upon components of the N-acylethanolamine and monoacylglycerol signalling systems: An mRNA study. *Scientific Reports*, *10*, 6314.

Andreou, A., & Feussner, I. (2009). Lipoxygenases—Structure and reaction mechanism. *Phytochemistry*, *70*, 1504–1510.

Austin-Brown, S. L., & Chapman, K. D. (2002). Inhibition of phospholipase D alpha by N-acylethanolamines. *Plant Physiology*, *129*, 1892–1898.

Aziz, M., & Chapman, K. D. (2020). Fatty acid amide hydrolases: An expanded capacity for chemical communication? *Trends in Plant Science*, *25*, 236–249.

Aziz, M., Wang, X., Tripathi, A., Bankaitis, V. A., & Chapman, K. D. (2019). Structural analysis of a plant fatty acid amide hydrolase provides insights into the evolutionary diversity of bioactive acylethanolamides. *The Journal of Biological Chemistry*, *294*, 7419–7432.

Balsevich, G., Sticht, M., Bowles, N. P., Singh, A., Lee, T. T. Y., Li, Z., et al. (2018). Role for fatty acid amide hydrolase (FAAH) in the leptin-mediated effects on feeding and energy balance. *Proceedings of the National Academy of Sciences of the United States of America*, *115*, 7605–7610.

Bannenberg, G., Martínez, M., Hamberg, M., & Castresana, C. (2009). Diversity of the enzymatic activity in the lipoxygenase gene family of Arabidopsis thaliana. *Lipids*, *44*, 85.

Blancaflor, E. B., Hou, G., & Chapman, K. D. (2003). Elevated levels of N-lauroylethanolamine, an endogenous constituent of desiccated seeds, disrupt normal root development in Arabidopsis thaliana seedlings. *Planta*, *217*, 206–217.

Blancaflor, E. B., Kilaru, A., Keereetaweep, J., Khan, B. R., Faure, L., & Chapman, K. D. (2014). N-Acylethanolamines: Lipid metabolites with functions in plant growth and development. *The Plant Journal*, *79*, 568–583.

Bluett, R. J., Gamble-George, J. C., Hermanson, D. J., Hartley, N. D., Marnett, L. J., & Patel, S. (2014). Central anandamide deficiency predicts stress-induced anxiety: Behavioral reversal through endocannabinoid augmentation. *Translational Psychiatry*, *4*, e408.

Bracey, M. H., Hanson, M. A., Masuda, K. R., Stevens, R. C., & Cravatt, B. F. (2002). Structural adaptations in a membrane enzyme that terminates endocannabinoid signaling. *Science*, *298*, 1793–1796.

Bruijnzeel, A. W., Qi, X., Guzhva, L. V., Wall, S., Deng, J. V., Gold, M. S., et al. (2016). Behavioral characterization of the effects of cannabis smoke and anandamide in rats. *PLoS One*, *11*, e0153327.

Bulat, E., & Garrett, T. A. (2011). Putative N-acylphosphatidylethanolamine synthase from Arabidopsis thaliana is a lysoglycerophospholipid acyltransferase. *The Journal of Biological Chemistry*, *286*, 33819–33831.

Cannon, A. E., & Chapman, K. D. (2021). Lipid signaling through G proteins. *Trends in Plant Science*, *26*, 720–728.

Cannon, A. E., Yan, C., Burks, D. J., Rao, X., Azad, R. K., & Chapman, K. D. (2020). Lipophilic signals lead to organ-specific gene expression changes in Arabidopsis seedlings. *Plant Direct*, *4*, e00242.

Chapman, K. D. (2004). Occurrence, metabolism, and prospective functions of N-acylethanolamines in plants. *Progress in Lipid Research*, *43*, 302–327.

Chapman, K. D., Lin, I., & DeSouza, A. D. (1995). Metabolism of cottonseed microsomal N-acylphosphatidylethanolamine. *Archives of Biochemistry and Biophysics*, *318*, 401–407.

Chapman, K. D., & Moore, T. S., Jr. (1993). Catalytic properties of a newly discovered acyltransferase that synthesizes N-acylphosphatidylethanolamine in cottonseed (Gossypium hirsutum L.) microsomes. *Plant Physiology*, *102*, 761.

Chapman, K. D., & Sprinkle, W. B. (1996). Developmental, tissue-specific, and environmental factors regulate the biosynthesis of N-acylphosphatidyl-ethanolamine in cotton (Gossypium hirsutum L.). *Journal of Plant Physiology*, *149*, 277–284.

Chapman, K. D., Tripathy, S., Venables, B., & Desouza, A. D. (1998). N-acylethanolamines: Formation and molecular composition of a new class of plant lipids. *Plant Physiology*, *116*, 1163–1168.

Chapman, K. D., Venables, B., Markovic, R., Blair, R. W., Jr., & Bettinger, C. (1999). N-acylethanolamines in seeds. Quantification of molecular species and their degradation upon imbibition. *Plant Physiology*, *120*, 1157–1164.

Chechetkin, I. R., Osipova, E. V., Antsygina, L. L., Gogolev, Y. V., & Grechkin, A. N. (2011). Oxidation of glycerolipids by maize 9-lipoxygenase and its A562G mutant. *Chemistry and Physics of Lipids*, *164*, 216–220.

Chilufya, J., Devaiah, S., Sante, R. R., & Kilaru, A. (2015). *Endocannabinoid-like lipids in plants*. John Wiley & Sons, Ltd.

Clapper, J. R., Moreno-Sanz, G., Russo, R., Guijarro, A., Vacondio, F., Duranti, A., et al. (2010). Anandamide suppresses pain initiation through a peripheral endocannabinoid mechanism. *Nature Neuroscience*, *13*, 1265–1270.

Cotter, M. Q., Teaster, N. D., Blancaflor, E. B., & Chapman, K. D. (2011). N-acylethanolamine (NAE) inhibits growth in Arabidopsis thaliana seedlings via ABI3-dependent and -independent pathways. *Plant Signaling & Behavior*, *6*, 671–679.

Dainese, E., & Oddi, S. (2020). The endocannabinoid hydrolase FAAH is an allosteric enzyme. *Scientific Reports*, *10*, 2292.

Dlugos, A., Childs, E., Stuhr, K. L., Hillard, C. J., & de Wit, H. (2012). Acute stress increases circulating anandamide and other N-acylethanolamines in healthy humans. *Neuropsychopharmacology: Official Publication of the American College of Neuropsychopharmacology*, *3-7*, 2416–2427.

Dosoky, N. S., Guo, L., Chen, Z., Feigley, A. V., & Davies, S. S. (2018). Dietary fatty acids control the species of N-acyl-phosphatidylethanolamines synthesized by therapeutically modified Bacteria in the intestinal tract. *ACS Infectious Diseases*, *4*, 3–13.

Faure, L., Cavazos, R., Khan, B. R., Petros, R. A., Koulen, P., Blancaflor, E. B., et al. (2015). Effects of synthetic alkamides on Arabidopsis fatty acid amide hydrolase activity and plant development. *Phytochemistry*, *110*, 58–71.

Faure, L., Coulon, D., Laroche-Traineau, J., Le Guedard, M., Schmitter, J. M., Testet, E., et al. (2009). Discovery and characterization of an Arabidopsis thaliana N-acylphosphatidylethanolamine synthase. *The Journal of Biological Chemistry*, *284*, 18734–18741.

Fornelos, N., & Franzosa, E. A. (2020). Growth effects of N-acylethanolamines on gut bacteria reflect altered bacterial abundances in inflammatory bowel disease. *Nature Microbiology*, *5*, 486–497.

Gabbs, M., & Leng, S. (2015). Advances in our understanding of oxylipins derived from dietary PUFAs. *Advances in Nutrition*, *6*, 513–540.

Gachet, M. S., Schubert, A., Calarco, S., Boccard, J., & Gertsch, J. (2017). Targeted metabolomics shows plasticity in the evolution of signaling lipids and uncovers old and new endocannabinoids in the plant kingdom. *Scientific Reports*, *7*, 41177.

Ghosh, A. K., Chauhan, N., Rajakumari, S., Daum, G., & Rajasekharan, R. (2009). At4g24160, a soluble acyl-coenzyme A-dependent lysophosphatidic acid acyltransferase. *Plant Physiology*, *151*, 869–881.

Gómez-Boronat, M., Isorna, E., Armirotti, A., Delgado, M. J., Piomelli, D., & de Pedro, N. (2019). Diurnal profiles of N-acylethanolamines in goldfish brain and gastrointestinal tract: Possible role of feeding. *Frontiers in Neuroscience*, *13*, 450.

Hansen, H. S., Kleberg, K., & Hassing, H. A. (2015). Chapter 1—Non-endocannabinoid N-acylethanolamines and monoacylglycerols: Old molecules new targets. In V. Di Marzo, & J. Wang (Eds.), *The endocannabinoidome* (pp. 1–13). Boston: Academic Press.

Haq, I., & Kilaru, A. (2020). An endocannabinoid catabolic enzyme FAAH and its paralogs in an early land plant reveal evolutionary and functional relationship with eukaryotic orthologs. *Scientific Reports*, *10*, 3115.

Hayward, S., Cilliers, T., & Swart, P. (2017). Lipoxygenases: From isolation to application. *Comprehensive Reviews in Food Science and Food Safety*, *16*, 199–211.

Howlett, A. C., Blume, L. C., & Dalton, G. D. (2010). CB(1) cannabinoid receptors and their associated proteins. *Current Medicinal Chemistry*, *17*, 1382–1393.

Hussain, Z., Uyama, T., Tsuboi, K., & Ueda, N. (2017). Mammalian enzymes responsible for the biosynthesis of N-acylethanolamines. *Biochimica et Biophysica Acta - Molecular and Cell Biology of Lipids*, *1862*, 1546–1561.

Iqbal, S., Ali, U., Fadlalla, T., Li, Q., Liu, H., Lu, S., et al. (2020). Genome wide characterization of phospholipase A & C families and pattern of lysolipids and diacylglycerol changes under abiotic stresses in Brassica napus L. *Plant Physiology and Biochemistry*, *147*, 101–112.

James, C. N., Horn, P. J., Case, C. R., Gidda, S. K., Zhang, D., Mullen, R. T., et al. (2010). Disruption of the Arabidopsis CGI-58 homologue produces Chanarin-Dorfman-like lipid droplet accumulation in plants. *Proceedings of the National Academy of Sciences of the United States of America*, *107*, 17833–17838.

Janfelt, C., Wellner, N., Leger, P. L., Kokesch-Himmelreich, J., Hansen, S. H., Charriaut-Marlangue, C., et al. (2012). Visualization by mass spectrometry of 2-dimensional changes in rat brain lipids, including N-acylphosphatidylethanolamines, during neonatal brain ischemia. *The FASEB Journal*, *26*, 2667–2673.

Kang, L., Wang, Y. S., Uppalapati, S. R., Wang, K., Tang, Y., Vadapalli, V., et al. (2008). Overexpression of a fatty acid amide hydrolase compromises innate immunity in Arabidopsis. *The Plant Journal*, *56*, 336–349.

Keereetaweep, J., Blancaflor, E. B., Hornung, E., Feussner, I., & Chapman, K. D. (2013). Ethanolamide oxylipins of linolenic acid can negatively regulate Arabidopsis seedling development. *Plant Cell*, *25*, 3824–3840.

Keereetaweep, J., Blancaflor, E. B., Hornung, E., Feussner, I., & Chapman, K. D. (2015). Lipoxygenase-derived 9-hydro(pero)xides of linoleoylethanolamide interact with ABA signaling to arrest root development during Arabidopsis seedling establishment. *The Plant Journal*, *82*, 315–327.

Keereetaweep, J., & Chapman, K. D. (2016). Lipidomic analysis of endocannabinoid signaling: Targeted metabolite identification and quantification. *Neural Plasticity*, *2016*, 2426398.

Keereetaweep, J., Kilaru, A., Feussner, I., Venables, B. J., & Chapman, K. D. (2010). Lauroylethanolamide is a potent competitive inhibitor of lipoxygenase activity. *FEBS Letters*, *584*, 3215–3222.

Khan, B. R., Chapman, K. D., & Blancaflor, E. B. (2021). Chemical genetics to uncover mechanisms underlying lipid-mediated signaling events in plants. *Methods in Molecular Biology*, *2213*, 3–16.

Khan, B. R., Wherritt, D. J., Huhman, D., Sumner, L. W., Chapman, K. D., & Blancaflor, E. B. (2016). Malonylation of glucosylated N-lauroylethanolamine: A new pathway that determines N-acylethanolamine metabolic fate in plants. *The Journal of Biological Chemistry*, *291*, 27112–27121.

Kilaru, A., & Chapman, K. D. (2020). The endocannabinoid system. *Essays in Biochemistry*, *64*, 485–499.

Kilaru, A., Herrfurth, C., Keereetaweep, J., Hornung, E., Venables, B. J., Feussner, I., et al. (2011). Lipoxygenase-mediated oxidation of polyunsaturated N-acylethanolamines in Arabidopsis. *The Journal of Biological Chemistry*, *286*, 15205–15214.

Kilaru, A., Tamura, P., Isaac, G., Welti, R., Venables, B. J., Seier, E., et al. (2012). Lipidomic analysis of N-acylphosphatidylethanolamine molecular species in Arabidopsis suggests feedback regulation by N-acylethanolamines. *Planta*, *236*, 809–824.

Kim, S. C., Kang, L., Nagaraj, S., Blancaflor, E. B., Mysore, K. S., & Chapman, K. D. (2009). Mutations in Arabidopsis fatty acid amide hydrolase reveal that catalytic activity influences growth but not sensitivity to abscisic acid or pathogens. *The Journal of Biological Chemistry*, *284*, 34065–34074.

Kotel'nikova, I. M. (2011). The content of N-acylphosphatidylethanolamines in the seeds of cultivated plants and grain products. *Russian Journal of Plant Physiology*, *58*, 364–369.

Lin, L., Metherel, A. H., Jones, P. J., & Bazinet, R. P. (2017). Fatty acid amide hydrolase (FAAH) regulates hypercapnia/ischemia-induced increases in n-acylethanolamines in mouse brain. *Journal of Neurochemistry*, *142*, 662–671.

Liu, J., Wang, L., Harvey-White, J., Osei-Hyiaman, D., Razdan, R., Gong, Q., et al. (2006). A biosynthetic pathway for anandamide. *Proceedings of the National Academy of Sciences of the United States of America*, *103*, 13345–13350.

Long, J. Z., LaCava, M., Jin, X., & Cravatt, B. F. (2011). An anatomical and temporal portrait of physiological substrates for fatty acid amide hydrolase. *Journal of Lipid Research*, *52*, 337–344.

Lowin, T., Apitz, M., Anders, S., & Straub, R. H. (2015). Anti-inflammatory effects of N-acylethanolamines in rheumatoid arthritis synovial cells are mediated by TRPV1 and TRPA1 in a COX-2 dependent manner. *Arthritis Research & Therapy*, *17*, 321.

Maccarrone, M. (2017). Metabolism of the endocannabinoid anandamide: Open questions after 25 years. *Frontiers in Molecular Neuroscience*, *10*, 166.

Magotti, P., Bauer, I., Igarashi, M., Babagoli, M., Marotta, R., Piomelli, D., et al. (2015). Structure of human N-acylphosphatidylethanolamine-hydrolyzing phospholipase D: Regulation of fatty acid ethanolamide biosynthesis by bile acids. *Structure*, *23*, 598–604.

Mahavadi, S., Sriwai, W., Huang, J., Grider, J. R., & Murthy, K. S. (2014). Inhibitory signaling by CB1 receptors in smooth muscle mediated by GRK5/β-arrestin activation of ERK1/2 and Src kinase. *American Journal of Physiology. Gastrointestinal and Liver Physiology*, *306*, G535–G545.

McKinney, M. K., & Cravatt, B. F. (2003). Evidence for distinct roles in catalysis for residues of the serine-serine-lysine catalytic triad of fatty acid amide hydrolase. *The Journal of Biological Chemistry*, *278*, 37393–37399.

McKinney, M. K., & Cravatt, B. F. (2005). Structure and function of fatty acid amide hydrolase. *Annual Review of Biochemistry*, *74*, 411–432.

Motes, C. M., Pechter, P., Yoo, C. M., Wang, Y. S., Chapman, K. D., & Blancaflor, E. B. (2005). Differential effects of two phospholipase D inhibitors, 1-butanol and N-acylethanolamine, on in vivo cytoskeletal organization and Arabidopsis seedling growth. *Protoplasma*, *226*, 109–123.

Neelamegan, D., Schoenhofen, I. C., Richards, J. C., & Cox, A. D. (2012). Identification and recombinant expression of anandamide hydrolyzing enzyme from Dictyostelium discoideum. *BMC Microbiology*, *12*, 124.

Ogura, Y., Parsons, W. H., Kamat, S. S., & Cravatt, B. F. (2016). A calcium-dependent acyltransferase that produces N-acyl phosphatidylethanolamines. *Nature Chemical Biology*, *12*, 669–671.

Ortíz-Castro, R., Martínez-Trujillo, M., & López-Bucio, J. (2008). N-acyl-L-homoserine lactones: A class of bacterial quorum-sensing signals alter post-embryonic root development in Arabidopsis thaliana. *Plant, Cell & Environment*, *31*, 1497–1509.

Pacioni, G., Rapino, C., Zarivi, O., Falconi, A., Leonardi, M., Battista, N., et al. (2015). Truffles contain endocannabinoid metabolic enzymes and anandamide. *Phytochemistry*, *110*, 104–110.

Palermo, G., Bauer, I., Campomanes, P., Cavalli, A., Armirotti, A., Girotto, S., et al. (2015). Keys to lipid selection in fatty acid amide hydrolase catalysis: Structural flexibility, gating residues and multiple binding pockets. *PLoS Computational Biology*, *11*, e1004231.

Palmer, A. G., Senechal, A. C., Mukherjee, A., Ané, J. M., & Blackwell, H. E. (2014). Plant responses to bacterial N-acyl L-homoserine lactones are dependent on enzymatic degradation to L-homoserine. *ACS Chemical Biology*, *9*, 1834–1845.

Pappan, K., Austin-Brown, S., Chapman, K. D., & Wang, X. (1998). Substrate selectivities and lipid modulation of plant phospholipase Dα, −β, and -γ. *Archives of Biochemistry and Biophysics*, *353*, 131–140.

Park, S., Gidda, S. K., James, C. N., Horn, P. J., Khuu, N., Seay, D. C., et al. (2013). The α/β hydrolase CGI-58 and peroxisomal transport protein PXA1 coregulate lipid homeostasis and signaling in Arabidopsis. *Plant Cell*, *25*, 1726–1739.

Qin, W., Pappan, K., & Wang, X. (1997). Molecular heterogeneity of phospholipase D (PLD). Cloning of PLDgamma and regulation of plant PLDgamma, −beta, and -alpha by polyphosphoinositides and calcium. *The Journal of Biological Chemistry*, *272*, 28267–28273.

Ramírez-Chávez, E., López-Bucio, J., Herrera-Estrella, L., & Molina-Torres, J. (2004). Alkamides isolated from plants promote growth and alter root development in Arabidopsis. *Plant Physiology*, *134*, 1058–1068.

Rawyler, A. J., & Braendle, R. A. (2001). N-Acylphosphatidylethanolamine accumulation in potato cells upon energy shortage caused by anoxia or respiratory inhibitors. *Plant Physiology*, *127*, 240–251.

Rettori, E., De Laurentiis, A., Zorrilla Zubilete, M., Rettori, V., & Elverdin, J. C. (2012). Anti-inflammatory effect of the endocannabinoid anandamide in experimental periodontitis and stress in the rat. *Neuroimmunomodulation*, *19*, 293–303.

Samson, M.-T., Small-Howard, A., Shimoda, L. M. N., Koblan-Huberson, M., Stokes, A. J., & Turner, H. (2003). Differential roles of CB1 and CB2 cannabinoid receptors in mast cells. *The Journal of Immunology*, *170*, 4953–4962.

Sandoval, J. A., Huang, Z. H., Garrett, D. C., Gage, D. A., & Chapman, K. D. (1995). N-acylphosphatidylethanolamine in dry and imbibing cottonseeds (amounts, molecular species, and enzymatic synthesis). *Plant Physiology*, *109*, 269–275.

Shrestha, R., Dixon, R. A., & Chapman, K. D. (2003). Molecular identification of a functional homologue of the mammalian fatty acid amide hydrolase in Arabidopsis thaliana. *The Journal of Biological Chemistry*, *278*, 34990–34997.

Shrestha, R., Kim, S. C., Dyer, J. M., Dixon, R. A., & Chapman, K. D. (2006). Plant fatty acid (ethanol) amide hydrolases. *Biochimica et Biophysica Acta*, *1761*, 324–334.

Shrestha, R., Noordermeer, M. A., van der Stelt, M., Veldink, G. A., & Chapman, K. D. (2002). N-acylethanolamines are metabolized by lipoxygenase and amidohydrolase in competing pathways during cottonseed imbibition. *Plant Physiology*, *130*, 391–401.

Simon, G. M., & Cravatt, B. F. (2006). Endocannabinoid biosynthesis proceeding through glycerophospho-N-acyl ethanolamine and a role for alpha/beta-hydrolase 4 in this pathway. *The Journal of Biological Chemistry*, *281*, 26465–26472.

Simon, G. M., & Cravatt, B. F. (2008). Anandamide biosynthesis catalyzed by the phosphodiesterase GDE1 and detection of glycerophospho-N-acyl ethanolamine precursors in mouse brain. *The Journal of Biological Chemistry, 283*, 9341–9349.

Strittmatter, F., Gandaglia, G., Benigni, F., Bettiga, A., Rigatti, P., Montorsi, F., et al. (2012). Expression of fatty acid amide hydrolase (FAAH) in human, mouse, and rat urinary bladder and effects of FAAH inhibition on bladder function in awake rats. *European Urology, 61*, 98–106.

Sun, Y. X., Tsuboi, K., Okamoto, Y., Tonai, T., Murakami, M., Kudo, I., et al. (2004). Biosynthesis of anandamide and N-palmitoylethanolamine by sequential actions of phospholipase A2 and lysophospholipase D. *The Biochemical Journal, 380*, 749–756.

Teaster, N. D., Keereetaweep, J., Kilaru, A., Wang, Y. S., Tang, Y., Tran, C. N., et al. (2012). Overexpression of fatty acid amide hydrolase induces early flowering in Arabidopsis thaliana. *Frontiers in Plant Science, 3*, 32.

Teaster, N. D., Motes, C. M., Tang, Y., Wiant, W. C., Cotter, M. Q., Wang, Y. S., et al. (2007). N-Acylethanolamine metabolism interacts with abscisic acid signaling in Arabidopsis thaliana seedlings. *Plant Cell, 19*, 2454–2469.

Tripathy, S., Kleppinger-Sparace, K., Dixon, R. A., & Chapman, K. D. (2003). N-Acylethanolamine signaling in tobacco is mediated by a membrane-associated, high-affinity binding protein. *Plant Physiology, 131*, 1781.

Tripathy, S., Venables, B. J., & Chapman, K. D. (1999). N-Acylethanolamines in signal transduction of elicitor perception. Attenuation of alkalinization response and activation of defense gene expression. *Plant Physiology, 121*, 1299–1308.

Ueda, N., Tsuboi, K., & Uyama, T. (2013). Metabolism of endocannabinoids and related N-acylethanolamines: Canonical and alternative pathways. *The FEBS Journal, 280*, 1874–1894.

Venables, B. J., Waggoner, C. A., & Chapman, K. D. (2005). N-acylethanolamines in seeds of selected legumes. *Phytochemistry, 66*, 1913–1918.

Wang, Y. S., Shrestha, R., Kilaru, A., Wiant, W., Venables, B. J., Chapman, K. D., et al. (2006). Manipulation of Arabidopsis fatty acid amide hydrolase expression modifies plant growth and sensitivity to N-acylethanolamines. *Proceedings of the National Academy of Sciences of the United States of America, 103*, 12197–12202.

Wellner, N., Diep, T. A., Janfelt, C., & Hansen, H. S. (2013). N-acylation of phosphatidylethanolamine and its biological functions in mammals. *Biochimica et Biophysica Acta (BBA)—Molecular and Cell Biology of Lipids, 1831*, 652–662.

Yan, C., Cannon, A., Watkins, J., Keereetaweep, J., Khan, B. R., Jones, A. M., et al. (2020). Seedling chloroplast responses induced by N-linolenoylethanolamine require intact G-protein complexes. *Plant Physiology, 184*, 01552.02019.

Zhang, C., & Tian, S. (2010). Peach fruit acquired tolerance to low temperature stress by accumulation of linolenic acid and N-acylphosphatidylethanolamine in plasma membrane. *Food Chemistry, 120*, 864–872.

Zhang, B., Wang, Y., & Liu, J. Y. (2018). Genome-wide identification and characterization of phospholipase C gene family in cotton (Gossypium spp.). *Science China. Life Sciences, 61*, 88–99.

Zhang, D. D., Wang, J., Wang, D., Kong, Z. Q., Zhou, L., Zhang, G. Y., et al. (2019). Population genomics demystifies the defoliation phenotype in the plant pathogen Verticillium dahliae. *The New Phytologist, 222*, 1012–1029.

Zhu, J., Zhou, Y., Li, J., & Li, H. (2020). Genome-wide investigation of the phospholipase C gene family in Zea mays. *Frontiers in Genetics, 11*, 611414.

CHAPTER TEN

The biosynthesis and metabolism of prostaglandins in microalgae

Ida Orefice, Giovanna Romano, and Valeria Di Dato*
Marine Biotechnology Department, Stazione Zoologica Anton Dohrn Napoli, Villa Comunale, Napoli, Italy
*Corresponding author: e-mail address: valeria.didato@szn.it

Contents

Abstract

Prostaglandins (Pgs) are hormone-like mediators in many physiological and pathological processes that are present in all vertebrates, in some terrestrial and aquatic invertebrates, and have also been identified in some micro and macroalgae. In mammals, where they were mostly studied, they are key players in a great variety of physiological and pathological processes, playing similar roles in lower organisms, including marine invertebrates. Very recently the Pgs pathway has been identified in some species of marine microalgae, particularly in diatoms and dinoflagellates, suggesting a broad diffusion and conservation of these important pathway, strengthening Pgs importance in living organisms. The expression of the enzymes composing the pathway has been demonstrated in these microalgae, as well as the synthesis and release of the corresponding molecules outside the cells.

The phylogenetic analysis of the limiting enzyme, the cyclooxygenase (COX), in diatoms indicated the presence of additional enzyme types related to the oxidative metabolism of fatty acids belonging to the peroxidase-cyclooxygenase superfamily. These findings suggest a more complex evolution and diversity of metabolic pathways leading to the synthesis of lipid mediators in diatoms.

Advances in Botanical Research, Volume 101
ISSN 0065-2296
https://doi.org/10.1016/bs.abr.2021.09.004

The occurrence of prostaglandins in unicellular photosynthetic eukaryotes opens new intriguing perspectives in the evolution and role of these molecules in the marine environment as possible mediators in cell-to-cell signaling, eventually influencing population dynamics in the plankton.

1. Introduction

1.1 Overview of prostaglandin structure, synthesis and function

Prostaglandins (Pgs) are hormone-like mediators in many physiological and pathological processes that are present in all vertebrates, in some terrestrial and aquatic invertebrates, and have also been identified in some micro and macroalgae. Due to their discovery and deep characterization just in animals, their presence in microalgae was surprising and let questioned about their origin and evolutionary history. However, based on the strong similarity among the molecules identified in animals and in microalgae it is possible to consider Pg in photosynthetic organisms as true oxylipins and their metabolism as an eukaryotic metabolism that radiated at one point in the different algal and animal branches.

Prostaglandins (Pgs) have been investigated principally in animals, especially in mammals, where they have a broad range of biological activities, both in physiological and pathological processes (Fig. 1A). They were first discovered in human seminal fluid as a factor able to induce the contraction of smooth muscle, which is reason for the Pgs name, considering their origin from the prostate gland. However, years later their provenience was established from the seminal vesicles and afterwards their presence was discovered in almost all tissues, each producing a specific Pg or expressing a specific Pg receptor (Clària, 2003; Crofford, 2001; Markovič, Jakopin, Dolenc, & Mlinarič-Raščan, 2017; Versteeg, van Bergen en Henegouwen, van Deventer, & Peppelenbosch, 1999). Functioning as autocrine or paracrine factors, they behave as hormones-like molecules carrying their signaling to tissue very far from the site of synthesis, through the blood circulation system in which they enter after being released outside the cells (Biringer, 2020; Hara, 2017).

Pgs belong to the eicosanoids, a class of molecules derived from ω3 and ω6 fatty acids, composed of 20 carbon atoms, a cyclic ring, a C 13/C 14 *trans* double bond and a hydroxyl group at C-15 (Fig. 1B) (Versteeg et al., 1999). Depending on the type of fatty acid functioning as precursor, three series of Pgs, 1 to 3, arise from di-homo-γ-linolenic (DGLA), arachidonic (AA) and

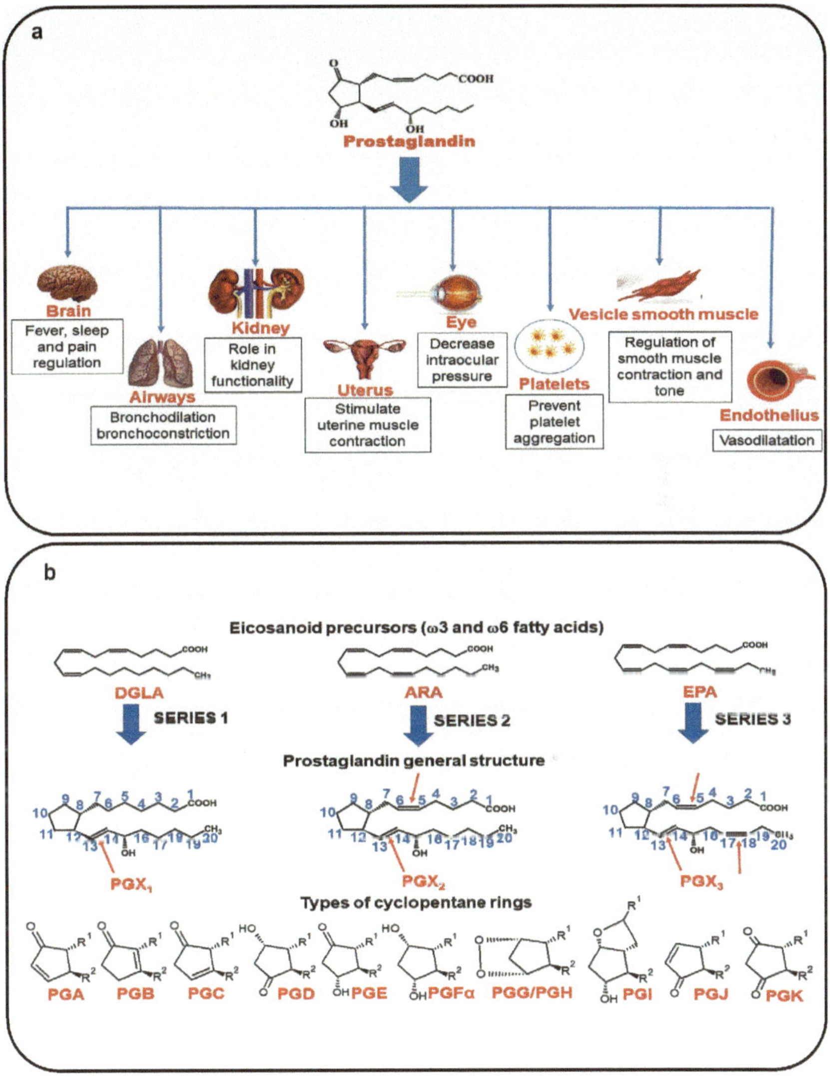

Fig. 1 Prostaglandin functions, series and nomenclature. (A) Prostaglandin tissue/organ localization and corresponding function. (B) Prostaglandin series derivation from the specific precursor: PGX_1, PGX_2 and PGX_3 from di-homo-γ-linolenic acid (DGLA), arachidonic acid (AA) and eicosapentaenoic acid (EPA), respectively. Subscript number indicate the number of double bonds in the carbon chains and the X indicates the type of cyclopentane rings. Based on the position of the substituents on the cyclopentane ring, the X is named with a letter from A to K. In the chemical structure, R_1 and R_2 represent the carboxylic and the hydroxyl groups, respectively.

eicosapentaenoic (EPA) acids, respectively (Fig. 1B). Each Pg is identified by a subscribed number, indicating the series, near a letter, from A to J, depending on the functional groups attached to the cyclopentane ring (Fig. 1B) (Piper, 1977; Versteeg et al., 1999). PGE_2, PGD_2, PGI_2 and $PGF_{2\alpha}$, derived from the arachidonic acid are the best characterized Pgs (Clària, 2003; Versteeg et al., 1999). Each Pg interacts with a specific receptor, which is cell and tissue specific (Table 1), whose binding activates the specific function. These receptors belong to the seven transmembrane spanning G protein coupled receptor (GPCR) of which at least 9 are known, and are identified by a number and two letters, the first of which corresponding to the binder Pg (Fig. 2A). The receptor subtypes EP_1-EP_4 bind PGE_2, DP_1 and DP_2 bind PGD_2, FP and IP are single receptors for $PGF_{2\alpha}$ and PGI_2. IP, DP_1, EP_2 and EP_4 receptors are coupled with the G_s type proteins and linked to the increase in intracellular cAMP, whereas EP_1, FP and TP are coupled with the G_q type proteins that mediate the increase of intracellular calcium concentration. The EP_3 receptor is the only one coupled with the G_i-type protein inducing a decrease in cAMP formation (Biringer, 2020; Clària, 2003; Crofford, 2001; Hara, 2017; Markovič et al., 2017).

Table 1 Most studied prostaglandins derived from arachidonic acid: specific receptors, corresponding prostaglandin terminal synthase enzymes, tissue distribution and subcellular localization in animals.

Prostaglandin	Receptor	Prostaglandin terminal synthase	Tissue distribution	Subcellular localization
PGE_2	EP_1	$mPGES_1$	Brain, Kidneys, VSMCs, Platelets	Perinuclear membrane, Cytosol, Golgi
	EP_2	$mPGES_2$		
	EP_3	cPGES		
	EP_4			
PGD_2	DP_1	H-PGDS	Mast cells, Brain, Airways	Cytosol, Secreted
	DP_2	L-PGDS		
PGI_2	IP	PGIS	Endothelium, VSMCs, Platelets, Kidneys, Brain	Endoplasmic reticulum
$PGF_{2\alpha}$	FP	PGFS	Uterus, Airways, VSMCs, Eyes	Secreted
		AKR		

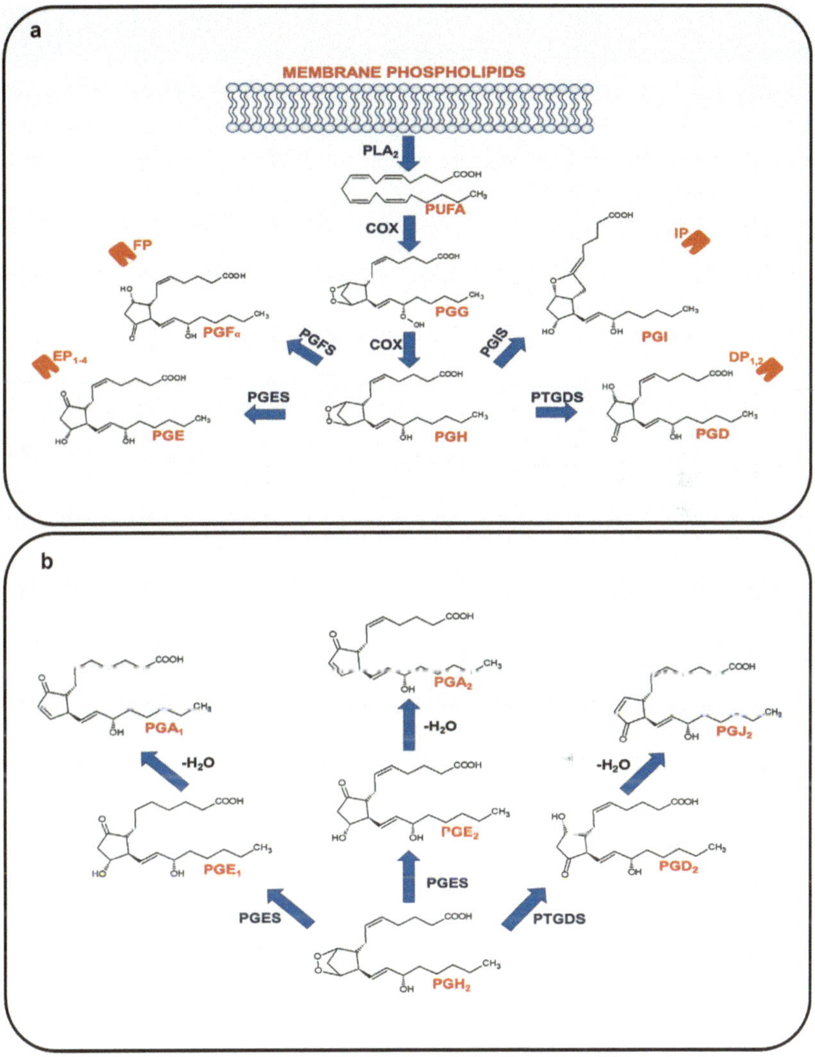

Fig. 2 Prostaglandin biosynthetic pathways. (A) Enzyme mediated prostaglandin synthesis. Enzymes involved in the pathway are reported next to the arrows in blue; receptors are reported in orange. (B) Prostaglandin synthesis by spontaneous dehydration. Abbreviation: PLA_2, phospholipase A2; PUFA, polyunsaturated fatty acid; COX, cyclooxygenase; PGG, prostaglandin G; PGH, prostaglandin H; PGES, prostaglandin E synthase; PGE, prostaglandin E; PGFS, prostaglandin F synthase; PGFα, prostaglandin Fα; PTGDS, prostaglandin D synthase; PGD, prostaglandin D; PGIS, prostaglandin I synthase; PGI, prostaglandin I; PGA, prostaglandin A; PGJ, prostaglandin J.

Pgs synthesis requires three classes of enzymes, phospholipases (PLA2), prostaglandin endoperoxide synthases or cyclooxygenases (PGHS or COX), and different Pg terminal synthases, each one acting sequentially (Fig. 2A): Prostaglandin E synthase (PGES), Prostaglandin D synthase or H isomerase (PGDS or PGHi), Prostaglandin F synthase or $PGF_{2\alpha}$ synthetase or PGD_2 11-Ketoreductase (PGFS), Prostaglandin I synthase (PGIS).

Following a stimulus, PLA_2 act first to release the fatty acids at the *sn*-2 position of the membrane glycerophospholipids (Biringer, 2020; Crofford, 2001). The released fatty acid becomes the target of the COX isozymes that soon isomerize it to the very unstable PGG_2, soon converted in PGH_2, and in turn converted to each of the different Pgs by the specific relative Pg terminal synthases (Table 1). In alternative, spontaneous isomerization or dehydration can form Pgs, as it is the case for the cyclopentenone prostaglandin PGA_1, PGA_2 and PGJ_2, formed by dehydration of the PGE_1, PGE_2 and PGD_2 ring respectively (Fig. 2B) (Biringer, 2020; Hara, 2017; Straus & Glass, 2001).

Studies on animals, highlighted a wide range of biological effects of eicosanoids on inflammatory responses, intensity and duration of pain and fever, and on reproductive function. They also play important roles in inhibiting gastric acid secretion, regulating blood pressure through vasodilation or constriction, and inhibiting or activating platelet aggregation and thrombosis (Fig. 1A) (Giera & Galano, 2019; Wolfe, 1982).

The major action of the series 1 is anti-inflammatory, vasodilation and antiplatelet aggregation agent. The series 2 are pro-inflammatory and the series 3 tend to have anti-proliferative and anti-inflammatory activities potentially antagonizing the tumor promoting effect of PGE_2 in tumorigenic cells (Wolfe, 1982; Yang, Jiang, & Fischer, 2014).

1.1.1 Prostaglandin-endoperoxide G/H synthase (PTGS) or cyclooxygenase (COX)

In animals three isoforms of the COX enzymes, -1, -2 and -3 have been identified. They are integral membrane proteins located at the inner leaflet of the nuclear and endoplasmic reticulum membrane lipid bilayer to which the cytosolic PLA type 2 translocate in response to external stimuli (Crofford, 2001). After activation by stimuli, PLA_2 free membrane fatty acids that can become COX substrate (DGLA, AA, EPA and α and γ linolenic acid). DGLA is converted to PGH_1, AA to PGH_2 and EPA to PGH_3 (Figs. 1B and 2A) whereas the 18-carbon fatty acids are converted into monohydroxy acids. Each of them has preferences for one of the

COX isoforms, especially 1 or 2, suggesting that their role could be related to the one of the relative COX to which they have preference for (Clària, 2003).

COX-1, is constitutively expressed in almost all tissue (Korbecki, Baranowska-Bosiacka, Gutowska, & Chlubek, 2014; Rajakariar, Yaqoob, & Gilroy, 2006; Simmons, Botting, & Hla, 2004; Tanabe & Tohnai, 2002) while COX-2 is instead an inducible isoform similar to an early response gene. It has been associated to the synthesis of Pgs in inflammatory processes and is inhibited by PGJ_2 (Korbecki et al., 2014; Rajakariar et al., 2006; Simmons et al., 2004; Tanabe & Tohnai, 2002). COX-2 metabolizing AA to PGH_2 from lower levels than those required for COX-1, justify its preferential activation in the delayed response (Murakami et al., 2000).

COX-1 and COX-2 share 60% of homology and the genes differ, other than for their gene expression regulation, type of expression (constitutive or inducible) and tissue localization, as well as structurally for the loss in the COX-2 gene of the first COX-1 intron and for shorter COX-2 intron with respect to the corresponding one in the COX-1 gene (Tanabe & Tohnai, 2002). Although COX-1 and COX-2 oxygenate arachidonate with almost identical kinetics, in general, COX-2 is much more efficient with alternative substrates utilizing both fatty acids and endocannabinoids. The Pgs generated by the COXs protein mediate their signal through G-protein linked receptors following alternative way depending from the COX sub-type that have generated them. For example, Pgs derived from the COX-2 located in the nuclear envelope mediate their signal through a nuclear pathway while the COX-1 derived Pgs can act only through cell surface receptors (Clària, 2003).

COX-3 is a splice isoform of COX-1 and, similarly to other spliced isoforms, it's characterization is not yet clear (Rajakariar et al., 2006; Simmons et al., 2004).

In few specie of worms, bacteria and plants, pathogen-inducible oxygenases (PIOXs) have been identified whose members have been included together with the COX enzymes in a large fatty acid oxygenase family. PIOXs sequences share 30% of homology with COXs sequences and the critical residues needed for fatty acid oxygenation. Similarity in critical residues have been found also in the linoleate diol synthase (LDS). The presence of PIOX and LDS in plant, bacteria, fungus and lower animals suggest the existence of a progenitor of this enzyme very early in the evolution and the development of an ancient cell signaling mechanism that gave rise to oxygenated fatty acids by COXs enzymes. In addition, the possibility to produce PGE_2 from arachidonic acid have been discovered in *Entamoeba histolytica* by a cyclooxygenase lacking any similarity with COX, PIOX or LDS (Simmons et al., 2004).

1.1.2 Prostaglandin E synthase (PTGES)

PTGES is one of the fundamental enzymes for the synthesis of prostaglandin, catalyzing the synthesis of PGE_2 from PGH_2 through an intramolecular rearrangement. In humans, PTGES activity is present in both cytosolic and membrane-associated fractions of different cells and tissues, requiring in most cases glutathione (GSH) for catalytic activity. The GSH-dependent cytosolic PTGES (cPGES/p23) are constitutively expressed in many type of cells and tissues functioning preferentially coupled to COX-1. On the other side, during the delayed response induced by pro-inflammatory stimuli, PTGES seems to be selectively linked to COX-2 and indeed in many cells, during that phase, PGE_2 is the main Pg produced and PTGES activity is increased. In addition, COX-2 inhibitors principally affect the synthesis of PGE_2 in respect to the other Pgs (Biringer, 2020; Murakami et al., 2000).

Similarly to cyclooxygenase, three isoforms of Prostaglandin E Synthase have been identified, two of them are membrane-bound and have been designated as mPGES-1 and mPGES-2, and the third is a cytosolic enzyme referred to as cytosolic PGES (cPGES).

mPGES-1 is a membrane spanning and hometrimeric protein of 152 amino acids and is a member of the membrane associated proteins in eicosanoid and glutathione metabolism family of proteins (MAPEG). As other members of this family, it requires glutathione (GSH) as a cofactor to isomerize PGH_2 to PGE_2 other than to convert PGG_2 to 15-hydroperoxy-PGE_2. Its expression is induced by pro-inflammatory stimuli with a concomitant expression of COX-2 and their localization at the perinuclear membrane could allow a more efficient transfer of the unstable substrate PGH_2 from COX-2 to mPGES-1.

mPGES-2 is a 378–385 amino acid protein synthesized as a Golgi membrane-associated protein that becomes cytosolic after the cutting of the N-terminal hydrophobic domain. Its expression is constitutive, can be coupled with both COX-1 and COX-2 and is not induced by inflammation. It has a catalytic glutaredoxin/thioredoxin-like domain and is activated by various thiol reagents (Biringer, 2020).

cPGES, also is a GSH-requiring enzyme and is constitutively expressed in a wide variety of cells. Being located in the cytosol, it is easily coupled with COX-1 at the endoplasmic reticulum (ER) rather than with

COX-2, which is located more distantly, at the perinuclear envelope. Indeed, studies revealed that it is involved in the conversion of the COX 1-derived PGH_2 to PGE_2 during the immediate PGE_2 response (Biringer, 2020; Hara, 2017; Hara et al., 2010; Kudo & Murakami, 2005; Murakami, Nakatani, Tanioka, & Kudo, 2002; Park, Pillinger, & Abramson, 2006; Sampey, Monrad, & Crofford, 2005).

1.1.3 Prostaglandin D2 synthase (PTGDS) or prostaglandin-H2 D-isomerase (HPGDS)

Prostaglandin D synthase (PTGDS) or Prostaglandin-H2 D-isomerase (HPGDS) catalyzes the isomerization of the PGH precursor to PGD. There are three types of PTGDS: the lipocalin-type PGDS (L-PTGDS) GSH independent, the hematopoietic PGDS (HPGDS) GSH dependent and a GSH S-transferase. In addition, albumin also is known to catalyze the conversion of PGH_2 into PGD_2 without requiring any cofactor for the catalysis (Urade & Hayaishi, 2000; Urade, Watanabe, & Hayaishi, 1995).

L-PTGDS is a member of the lipocalin (lipophilic ligand-carrier protein) superfamily and is an N-glycosylated monomeric protein, composed of 180 190 amino acid residues, that requires free sulfhydryl compounds for the reaction. The signal peptide of about 20 amino acid residues is present at the N terminus and is removed to form the mature enzyme. Asn51 and78 are the two N-glycosylation sites conserved among mammalian species, but not in amphibian homologs. The Cys65, 89 e 186, instead, are conserved among all species. The residues Cys89 and Cys186 form a disulphide bridge while the Cys65 is the key residue for the catalytic activity of the enzyme (Urade et al., 1995; Urade & Hayaishi, 2000).

HPGDS is a cytosolic, 199 amino acid long, GSH requiring protein inactive with other thiol compounds, without signal peptide or N-glycosylation sites (Urade & Hayaishi, 2000).

1.1.4 Prostaglandin F synthase or $PGF_{2\alpha}$ synthetase or PGD_2 11-ketoreductase

PGF synthase is the name that has been attributed to PGD 11-ketoreductase when it was discovered that the latter catalyzed the 11-ketoreduction of PGD_2 to PGF_2. It is a member of the aldo-keto reductase family and an NADPH requiring enzyme for which it possesses one binding site. PGFS have two active functional sites, one catalyzing the reduction of PGD_2 to 9α, 11β-PGF_2, a stereoisomer of $PGF_{2\alpha}$, and the other catalyzing the conversion of PGH_2 to $PGF_{2\alpha}$ (Urade et al., 1995; Watanabe, 2002). There are two isozymes of PGD 11-ketoreductase in mammals with different Km values for PGD_2, named PGFS I and PGFS II (Watanabe, 2002).

$PGF_{2\alpha}$ is also synthesized by other two enzymes. PGE 9-ketoreductase is one of them and has 15-hydroxy dehydrogenase activity toward Pgs through which it catalyzes the conversion of PGE_2 to $PGF_{2\alpha}$, and its reverse reaction, in presence of NADH or NADPH. Also this enzyme is a member of the aldo-keto reductase (AKR) superfamily (Watanabe, 2002).

The third enzyme to catalyze the conversion of PGH_2 to $PGF_{2\alpha}$ is the PGH 9-, 11- endoperoxide reductase. Also this enzyme needs NADPH or NADH as cofactor (Watanabe, 2002).

1.2 Prostaglandin occurrence in plant and macroalgae

In plants, arachidonic acid is absent and enzymes involved in Pg biosynthesis have not been identified. However, prostaglandin-like molecules, called phytoprostanes (PP) in plant, can be formed by oxidation of the linolenic acid (C18:3, ω–3), very abundant in these organisms, followed by reduction or regioisomeric rearrangements (Imbusch & Mueller, 2000a).

The search for Pgs in plant started in the 70s when PGA_1 was detected in *Allium cepa*. Meanwhile Russian studies identified PGA_2, PGB_2, PGE_2, and $PGF_{2\alpha}$ in *Populus balsamifera* and *Larix sibirica*. Later on, Pgs were also identified in other plants such as Aloe, soya bean, *Kalanchoe blossfeldiana v. Poelln*, sugar cane, coconut and banana plants (Table 2) (Bosisio, 1988). Pgs were found abundant in the buds and cambial zone indicating a relation with biochemical processes connected with the different phases of the annual cycle, linked to the reproductive processes. Pgs content in these plants was completely abolished by treatments with the cyclooxygenase inhibitor indomethacin suggesting the presence of COX-like enzymes, in plant even if they have never been identified. Similar results were obtained in *Aloe vera*, studying the synthesis of the identified Pgs 6-keto-$PGF_{1\alpha}$, $PGF_{2\alpha}$, PGE_2, and PGD_2 (Groenewald & van der Westhuizen, 1997). However, the α-linolenic acid (C18:3), one of the major plant fatty acid, can be released from the membrane and be converted by autoxidation reactions in Pg-like molecules, starting through a non-enzymatic formation of an H_2-isoprostane precursor (Fig. 3) (Bosisio, 1988; Groenewald & van der Westhuizen, 1997; Imbusch & Mueller, 2000a, 2000b; Panossian, 1987).

Differently from higher plants, lower plants such as green, red and brown/stramenopile macroalgae, have all the three Pg precursors, AA, DGLA and EPA, and thus it was not surprising to find prostaglandins in their

Table 2 Prostaglandins in plants and macroalgae.

Specie	**Prostaglandin**
Plants	
Plants	
Allium cepa	PGA_1
Populus balsamifera	PGA_2, PGB_2, PGE_2, $PGF_{2\alpha}$
Larix sibirica	PGA_2, PGB_2, PGE_2, $PGF_{2\alpha}$
Aloe vera	6-keto-$PGF_{1\alpha}$, $PGF_{2\alpha}$, PGE_2, PGD_2
Macroalgae	
Laminaria digitata (Phaeophyceae)	PGA_2, PGB_2, PGD_1, PGE_1, PGE_2, PGJ_2, 15-keto-PGE_2, 15-keto-$PGF_{2\alpha}$
Gracilaria sp. (Rhodophyta)	PGA_2, PGE_2, $PGF_{2\alpha}$, 15-keto-PGE_2,
Chondrus crispus (Rhodophyta)	PGB_1, PGB_2, PGA_2, 15-keto-PGE_2

extract even if their biosynthetic route and function still await to be clarified (Table 2) (Groenewald & van der Westhuizen, 1997).

In the '70s, Pgs were isolated from macroalgae of the genus *Gracilaria* that seems to be a good reservoir of prostaglandins as indeed PGA_2, PGE_2, PGF_2 and 15-keto-PGE_2 have been found in different species of the genus, but without a defined associated role (Imbs et al., 2012; Sajiki & Kakimi, 1998). In the year 1979, PGE_2 and $PGF_{2\alpha}$ were identified in *G. lichenoides*. Soon after, in 1983, Fusetani and Hashimoto demonstrated the occurrence of elevated levels of PGA_2 and PGE_2 in *Gracilaria verrucosa* when analyzed to search for compounds responsible for the so called "oogonori" poisoning. Elevated levels of PGA_2, PGE_2, PGF_2 and 15-keto-PGE_2 were also detected in the same species as a function of the solar irradiance seasonality (Fusetani & Hashimoto, 1984; Imbs, Vologodskaya, Nevshupova, Khotimchenko, & Titlyanov, 2001; Sajiki, 1997; Sajiki & Kakimi, 1998) and predation. It was in fact observed that PGA_2 inhibited predators grazing activity while PGE_2 and 15-keto-PGE_2 increased the alga palatability (Imbs et al., 2001; Nylund, Weinberger, Rempt, & Pohnert, 2011).

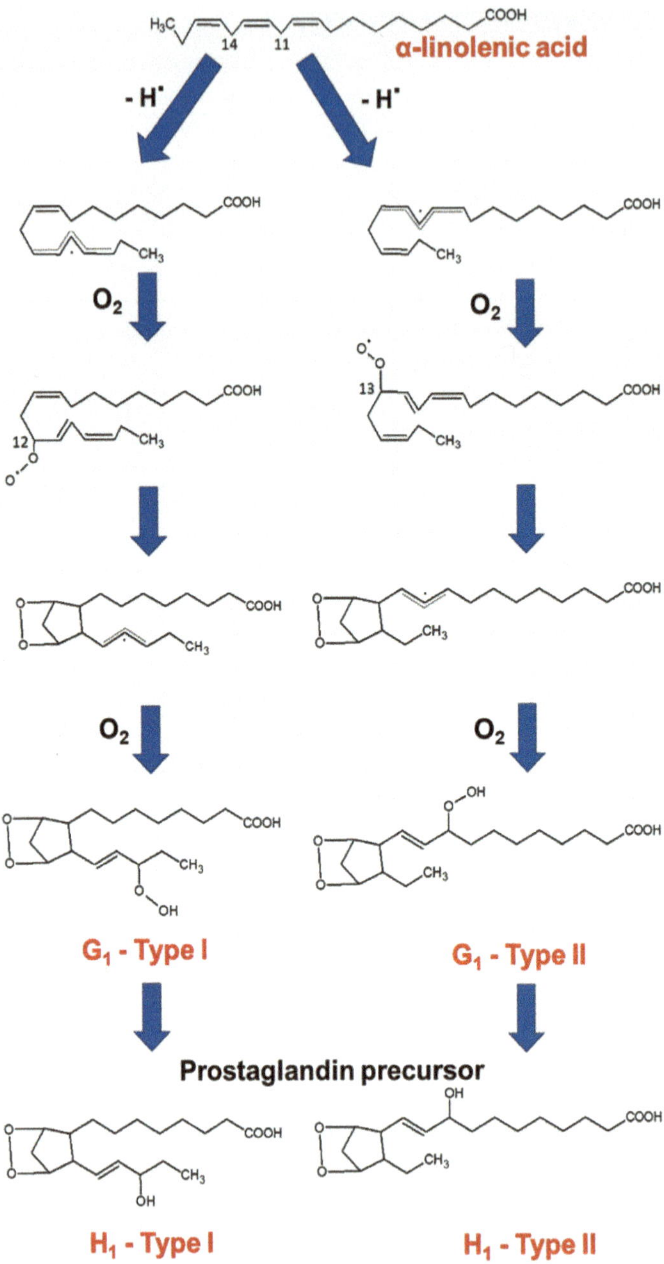

Fig. 3 Production of H1-isoprostane prostaglandin-like precursor in plants. LA released from the membrane can be converted, by autoxidation reactions, in H1-isoprostane prostaglandin-like precursor successively then converted in Pg-like molecules.

Extract of another specie of the red algae group, *Chondrus crispus*, contained PGB_1 and PGB_2 when challenged in vitro with a pathogen extract incubated with a fatty acid precursor. On the other side, the same alga, after a stimulus with a plant hormone, the methyljasmonate, released in vivo 15-keto-PGE_2 and PGA_2, but the exact role of these Pgs was not clarified (Bouarab et al., 2004; Gaquerel et al., 2007).

Later, PGE_1, PGD_1, 15-keto-$PGF_{2\alpha}$, PGE_2, 15-keto-PGE_2, PGA_2, PGB_2 and PGJ_2 were detected in another macroalga, a brown type of the genus *Laminaria*, following a lipid peroxidation induced by copper (Table 2) (Ritter et al., 2008).

2. Prostaglandin pathway metabolism in Bacillariophyceae and Dinoflagellata

In 1984, Levin et al reported the synthesis of 6-keto-$PGF_{1\alpha}$, $PGF_{2\alpha}$, PGE_2, PGA_2 and other kind of eicosanoids in the green unicellular microalga *Euglena gracilis* Klebs, var. bacillaris, especially when grown in the dark (Levine, Sneiders, Kobayashi, & Schiff, 1984). The presence of PGA, PGE, and PGF series and their esters have also been documented in the cyanobacteria of the taxa *Oscillatoria* and *Microcystidaceae*, and $PGF_2$was identified in *Microcystis aeruginosa* (Kützing) Kützing, 1846 (Table 3) (Kafanova, Busarova, Isai, & Zvyagintseva, 1996; Krüger, Groenewald, de Wet, & Botes, 1990).

Few years later, in 1990, two isomeric compound resembling leukotrienes, named Bacillariolides I and II, were described in extracts of the diatom *Nitzschia pungens*. Both compounds had formula $C_{20}H_{28}O_3$ and represented a cyclopentane eicosanoid deriving from cyclization at C-2 ~ C-6 instead of the normal C-8 ~ C-12 whose biosynthesis seemed closely related to eicosanoids (Wang & Shimizu, 1990). In addition, other cyclopentenone-like molecules (CPPGs) have been identified in cyanobacteria (Kafanova et al., 1996).

Table 3 Prostaglandins in microalgae.

Specie	**Class**	**Prostaglandin**
Euglena gracilis	Green alga	$PGF_{2\alpha}$, PGE_2
Microcystis aeruginosa	Cyanobacteria	$PGF_{2\alpha}$, CPPGs
Skeletonema marinoi	Diatom	PGE_2, PGD_2, PGA_2, PGB_2, PGJ_2, PGK_2, $PGF_{2\alpha}$
Thalassiosira rotula	Diatom	PGE_1, PGE_2, PGD_1, PGD_3, PGA_2, PGB_2

Despite these quite old studies describing Pgs or Pg-like molecules synthesis in microalgae, only in 2017 the existence of the canonical "animal-like" Pg biosynthetic pathway has started to be described in some species of the Bacillariophyceae and Dinoflagellata classes of marine eukaryotic unicellular microalgae (Di Dato et al., 2017, 2019, 2020; Di Dato, Ianora, & Romano, 2020).

The studies by Di Dato et al, revealed the expression and enzymatic activity of the enzymes known in animals for being necessary for the prostaglandin synthesis and catabolism, i.e. the COX, PGES, PGHi or PGDS, PGFS, 15-hydroxyprostaglandin dehydrogenase, Prostaglandin-E(2) 9-reductase, Prostaglandin reductases. The expression of these enzymes was experimentally studied in two clones of the diatom *Skeletonema marinoi* and in one clone of another diatom species, *Thalassiosira rotula*, isolated in the Adriatic Sea and in the Gulf of Naples, respectively. The enzymatic activity was demonstrated by the intracellular measurement and mass spectrometry identification of a complete panel of Pgs deriving from each of the three precursors, DGLA, AA and EPA (Table 3). On the other side, diverse dinoflagellate species have been explored in silico for expression of the Pg biosynthetic enzymes, by mining the corresponding transcriptomes sequenced from cultures growing in different conditions. The in silico study revealed the expression of the complete set of enzymes needed for Pg metabolism, both the catabolic and the anabolic ones, and revealed differences in the expression at the species and clone level, and based on the growth conditions. It is interesting to observe the absence of COX in several species, which should be the key enzyme for Pg enzymatic synthesis, while other enzymes of the pathway are more frequently present. However, a lack of expression does not necessarily correspond to a real loss of the corresponding gene, but it could correspond to a very low expression level not sufficient to produce a signal over the detection limit of the sequencing technology used. In addition, COX gene absence not always corresponds to an incapability to synthesize prostaglandin molecules, as isoprostanes were identified by mass spectrometry in another diatom species, *Phaeodactylum tricornutum*, whose genome does not have genes coding for Pgs biosynthetic enzymes (Lupette et al., 2018).

2.1 Prostaglandin-endoperoxide G/H synthase (PTGS) or cyclooxygenase (COX)

In 2017 COXs was identified for the first time in some diatom and dinoflagellates, by transcriptome analysis of selected species. Based on the functional annotation, COX was found annotated as prostaglandin G/H synthase 1 or prostaglandin G/H synthase 2 suggesting the presence of only one of the two mammalian isoforms homologs, most likely the isoform 2 (Di Dato et al., 2017, 2019, Di Dato, Ianora, & Romano, 2020).

2.1.1 Phylogenetic analysis

To ascertain that the COX sequence identified in diatoms was actually a cyclooxygenase, two phylogenetic analyses have been conducted on diatom sequences retrieved from the "Marine Microbial Eukaryote Transcriptome Sequencing Project (MMETSP)" (https://www.imicrobe.us) together with canonical representative COX sequences from other organisms. In the first analysis, diatom sequences were obtained by blasting the sequence of the *Skeletonema marinoi* clone named FE7 (Gerecht et al., 2011) against a database containing the peptide sequences of all the diatom species sequenced. The 32 sequences obtained were aligned with characterized COX sequences from the animal kingdom, i.e. *Caprella sp*, *Ciona intestinalis* both isoforms (COXa and COXb), *Gammarus sp*, *Gersemia fruticosa* both isoforms (COXa and COXb), *Homo sapiens* both isoforms (COX-1 and COX-2), *Oncorhynchus mykiss*, *Ovis aries* and *Plexaura homomalla* both 15S-specific-cyclooxygenase and COX-1, corresponding to shrimp, ascidia, soft coral, humans, rainbow trout, sheep and gorgonian type coral, plus a red alga, *Coccotylus truncatus,* and a cyanobacteria, *Nostoc sp* (Di Dato et al., 2017).

The sequences alignment showed the conservation of the amino acidic residues involved in peroxidase and cyclooxygenase activities while the Bayesian phylogeny separated the considered sequences in two main clades, one of which constituted of two sister clades (Fig. 4A). The latter clustered most of the diatom species and included also the characterized COX sequence from *Nostoc sp.* (cyanobacteria) (Brash, Niraula, Boeglin, & Mashhadi, 2014) and *C. truncates* (red alga) (Varvas et al., 2013) confirming the diatom sequences identified as putative COX protein be actually a cyclo-oxygenase enzyme. This was supported by the clustering in its sister clade of the animal sequences (crustaceans, tunicates and mammals) strengthening the identity of the diatom cyclooxygenase.

The other small basal clade grouped only a small portion of diatom sequences suggesting that in this group of microalgae at least two different proteins can function as COX, one of which found only in four diatom genera, possibly representing a different protein sharing common domains with the first one. In support to this hypothesis, two genera, *Skeletonema* and *Pseudo-nitzschia,* were also present in the first clade, revealing that the two proteins may coexist (Di Dato et al., 2017).

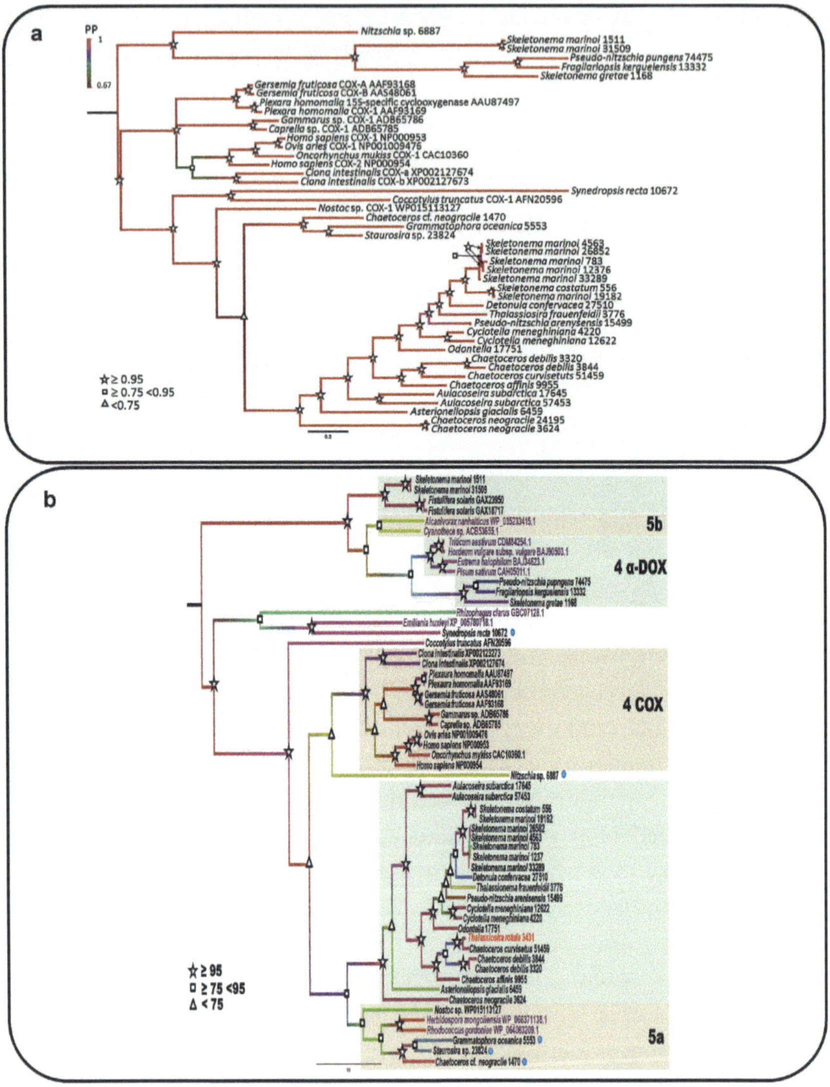

Fig. 4 Phylogenetic analysis of the diatom COX sequences. (A) Bayesian mid-point routed phylogenetic trees based on *Skeletonema marinoi* COX-1 (see the text for details). Sequences from other organisms are also included as outgroup. The tree was built on scale in the branch length (scale bars reported). Sequences are identified by the species name and annotation (where applicable), followed by their MMETSP ID deprived of the taxon ID for simplicity or GenBank accession number. Posterior probability (PP) values are represented with symbols at the nodes: star represents PP above or equal to 0.95; square PP between 0.75 and 0.95; triangle below 0.75. Branch color refers to PP, color code is reported in figure (2,000,000 generations, split frequency standard deviation = 0.002). (B) Bayesian mid-point routed phylogenetic trees based on *Thalassiosira rotula* COX. Legend: *Trot*COX is indicated in red; cyan dot: diatom

(Continued)

A second analysis to confirm the diatom COX sequences as real cyclooxygenases was done last year on the COX sequence retrieved from the diatom specie *Thalassiosira rotula*, named *Trot*COX. A multiple alignment developed by the Phyre2 (Protein Homology/analogY Recognition Engine V 2.0) program retrieved as best hits 13 structures of proteins belonging to the peroxidase-cyclooxygenase family, all sharing 100% confidence and 94–99% coverage (Table 4). The best hits corresponded to a *Bos taurus* lactoperoxidase (*Bt*LPO) (19% identity) already reported as representative of the peroxidase-cyclooxygenase superfamily, to an *Ovis aries* prostaglandin G/H synthase (*Oa*COX1) (28% identity), cyclooxygenases, myeloperoxidase or oxidoreductase protein structures (26% to 19% identity) from *Homo sapiens*, *Mus musculus*, *Arabidopsis thaliana* and *Oryza sativa*, respectively, confirming the cyclooxygenase structure of the *Trot*COX protein (Di Dato, Barbarinaldi, et al., 2020).

The comparison among *Trot*COX predicted structure and known models of the peroxidase-cyclooxygenase family, alias the *Bt*LPO (3BXI.pdb, 10.2210/pdb3bxi/pdb) and *Oa*COX1 structures, resulted in an almost perfect overlap of the main alpha helices that characterize the cyclooxygenases and accommodate the heme prosthetic group (Fig. 5).

The in silico predicted model constructed on the BtLPO showed the conservation in the diatom sequences of the catalytic arginine (R275) in the distal heme side, embedded in the characteristic motif XRXXEX, while the glutamate, usually conserved, was mutated to leucine (L278) in *Trot*COX. The latter seems to be a characteristic of diatoms as none of the diatom sequences analyzed showed a glutamate in that position (Fig. 6). The proximal heme sides appeared relatively conserved in diatom COX as the highly conserved arginine (R371), histidine (H373), isoleucine (I376) and glutamic acid (E454) were conserved in most of the diatom sequences (Fig. 6).

Fig. 4—Cont'd sequences not clustering the diatoms-specific clade; cyan-shaded clade: diatom specific clade; 5a: 5a-family in the diatom/bacterial clade; 4 COX: animal COX clade (family 4, cyclooxygenases); 4 α-DOX: plant α-DOX proteins (family 4, cyclooxygenases); 5b: 5b-family cyanobacteria. The nomenclature used for the family classification follows Zámocký et al. (2015). *Part A: From Di Dato, V., Orefice, I., Amato, A., Fontanarosa, C., Amoresano, A., Cutignano, A., Ianora, A., & Romano, G. (2017). Animal-like prostaglandins in marine microalgae.* The ISME Journal, 11*(7), 1722–1726. https://doi.org/10.1038/ismej.2017.27. Part B: From Di Dato, V., Barbarinaldi, R., Amato, A., Di Costanzo, F., Fontanarosa, C., Perna, A., Amoresano, A., Esposito, F., Cutignano, A., Ianora, A., & Romano, G. (2020). Variation in prostaglandin metabolism during growth of the diatom Thalassiosira rotula.* Scientific Reports, 10*(1), 5374. https://doi.org/10.1038/s41598-020-61967-3.*

Table 4 Results of *Trot*COX protein sequence multi alignment with available cyclooxygenase/peroxidase protein sequences.

Template alignment	Template Id/DOI/web link	Coverage	Confidence	% of identity
Buffalo lactoperoxidase	c2gjmA 10.2210/pdb2GJM/pdb	97	100	19
Sheep prostaglandin g/h synthase 1	c2oyup 10.2210/pdb2OYU/pdb	99	100	28
Sheep prostaglandin g/h synthase 1	c1pggB 10.2210/pdb1PGG/pdb	99	100	28
Sheep prostaglandin g/h synthase 1	c1ht8B 10.2210/pdb1HT8/pdb	99	100	28
Sheep prostaglandin g/h synthase 2	d1q4ga1 http://scop.berkeley.edu/search/?oldURL=1&key=%22d1q4ga1%22&key=d1q4ga1	99	100	28
Dictyostelium discoideum Peroxidase	c6ercA 10.2210/pdb6ERC/pdb	97	100	26
Human promyeloperoxidase	c5mfaA 10.2210/pdb5MFA/pdb	97	100	20
Mouse Cycloxygenase 2	c1ddxA 10.2210/pdb1DDX/pdb	99	100	27
Mouse Cycloxygenase 2	c3pghD 10.2210/pdb3PGH/pdb	99	100	27
Mouse Cycloxygenase 2	d1cvua1 http://scop.berkeley.edu/search/?oldURL=1&key=%22d1cvua1%22&key=d1cvua1	99	100	27
Human Myeloperoxidase isoform C	c1d2vD 10.2210/pdb1D2V/pdb	94	100	20
Arabidopsis fatty acid alphadioxygenase (arabidopsis2 thaliana)	c4hhsA 10.2210/pdb4HHS/pdb	98	100	19
Oryza sativa fatty acid alphadioxygenase	c4kvjA 10.2210/pdb4KVJ/pdb	99	100	19

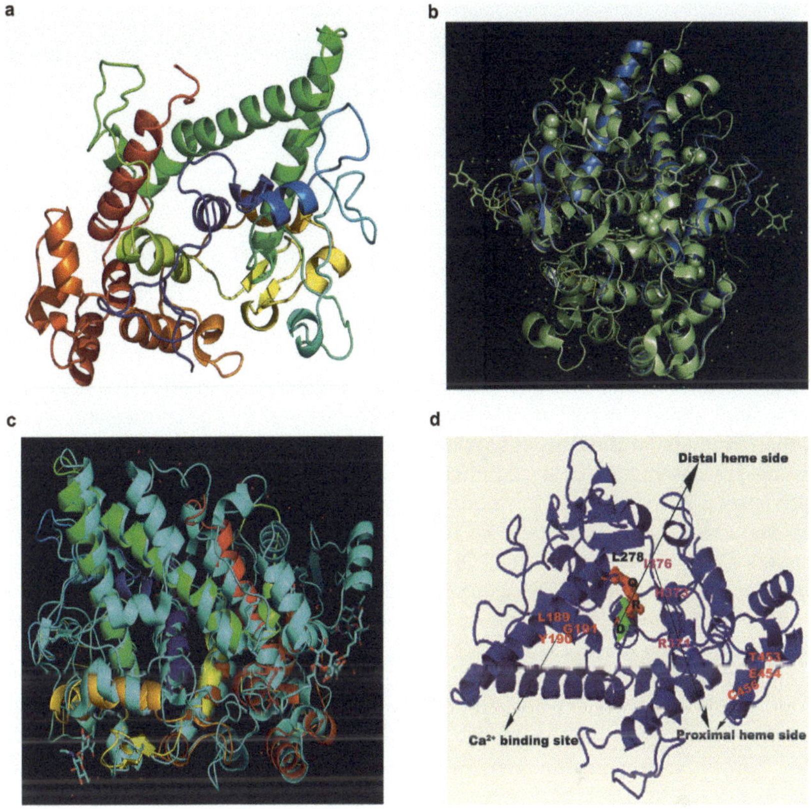

Fig. 5 In silico reconstruction of the *Trot*COX protein structure. Comparison with representative protein structures of the peroxidase-cyclooxygenase superfamily. (A) *Trot*COX structure based on the 2.75 Å resolved X-ray diffraction of *Bt*LPO. (B) Comparison of the in silico structure of the *Trot*COX protein and the 2.75 Å resolved X-ray diffraction of *Bt*LPO showing the perfect overlap of the alpha-helices embedding the heme sides. In blue the *Trot*COX, in light green the *Bt*LPO. (C) Comparison of the in silico structure of the *Trot*COX protein and the 2.75 Å resolved X-ray diffraction of *Oa*COX1 showing the perfect overlap of the alpha-helices embedding the heme sides. Rainbow colored helices correspond to *Trot*COX, the light-blue colored helices correspond to *Oa*COX1. (D) Illustration of the conserved catalytic sites described in the text. Q, R, D and L278 black letters refer to the conserved amino acids of the two distal heme sides with L278 being the mutated original (E). Pink and light red letters indicate the conserved amino acids in the two proximal heme sides. The red letters indicate the amino acids forming the calcium binding site *Bt*LPO = *Bos taurus* LPO; *Trot*COX = *Thalassiosira rotula COX; Oa*COX1: *Ovis aries* Cox1. *From Di Dato, V., Barbarinaldi, R., Amato, A., Di Costanzo, F., Fontanarosa, C., Perna, A., Amoresano, A., Esposito, F., Cutignano, A., Ianora, A., & Romano, G. (2020). Variation in prostaglandin metabolism during growth of the diatom Thalassiosira rotula.* Scientific Reports, 10*(1), 5374. https://doi.org/10.1038/s41598-020-61967-3.*

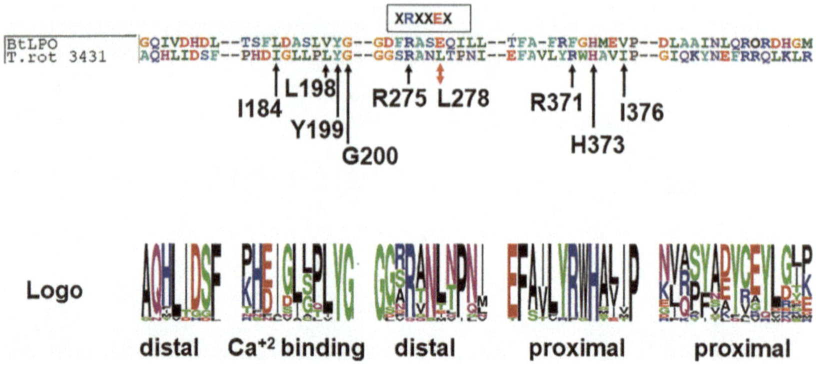

Fig. 6 *Trot*COX vs *Bt*LPO amino acid sequence alignment at the distal and proximal heme sides and at the calcium binding sites. The sequence Logo of the same portion of the alignment for the diatom specific clade is shown below. The sequence Logo is a resumed representation of an alignment among two or more sequences. The height of each letter is proportional to its frequency and the most common one is placed on the top. Important amino acidic residues reported in the text are indicated on the *T. rotula* sequence. *Bt*LPO = *Bos taurus* LPO; *T.rot* = *Thalassiosira rotula. From Di Dato, V., Barbarinaldi, R., Amato, A., Di Costanzo, F., Fontanarosa, C., Perna, A., Amoresano, A., Esposito, F., Cutignano, A., Ianora, A., & Romano, G. (2020). Variation in prostaglandin metabolism during growth of the diatom Thalassiosira rotula.* Scientific Reports, 10*(1), 5374. https://doi.org/10.1038/s41598-020-61967-3.*

The conserved asparagine and arginine were instead substituted with a threonine (T453) and a cysteine (C456), respectively, and the calcium binding site conserved the Isoleucine (I184) and LYG motif (L198, Y199, G200) (Di Dato, Barbarinaldi, et al., 2020).

A refined phylogenetic analysis comparing the COX sequence from *T. rotula* with that of other organisms retrieved from a NCBI (National Center for Biotechnology Information) pBLAST resulted in a large well-supported diatom clade (Maximum Likelihood (ML) bootstrap value 100) closely related to the animal COX clade (family 4, cyclooxygenases), although the phylogenetic relationship was not well resolved (ML 35) (Fig. 4B).

This large diatom clade was divided into a main well-supported clade including the vast majority of diatom sequences and in an end-clade (ML 94) grouping both prokaryotic (*Nostoc sp.*, a cyanobacterium, the actinomycetales *Herbidospora mongoliensis* and *Rhodococcus gordoniae*) and diatom (*Grammatophora oceanica*, *Staurosira sp.*, and *Chaetoceros cf. neogracile*) sequences. This diatom/bacterial clade can be hypothesized as being the 5a family within the peroxidase-cyclooxygenase super family (Zámocký et al., 2015). Noteworthy, the sequence from the pennate biraphid diatom

Nitzschia sp. clustered basally to the animal COX clade (ML 56). In addition to that large clade, a second very well resolved (ML 100) super clade was composed of two clades. One of these forks included only diatom sequences (*Fistulifera solaris* and *S. marinoi* sequence number 31509 and 1511), whereas the other bifurcates in two sister clades. One of these sister clades grouped only two prokaryotic sequences, and the other bifurcates further in two branches. The first branch groups plant α-DOX proteins (family 4 cyclooxygenase) and the other groups diatoms (*Skeletonema gretae*, *Pseudo-nitzschia pungens* and *Fragilariopsis kerguelensis*) (Fig. 4B).

In this analysis, with the addition of new bacterial sequences, the clade grouping the three diatoms and *Nostoc* sequences was identified as a bacterial-like clade. Moreover, this phylogenetic analysis highlighted the existence of more than one peroxidase-cyclooxygenase enzyme related to the oxidative metabolism of fatty acids in diatoms. The four sub-families in the peroxidase-cyclooxygenase superfamily (family 4: cyclooxygenase and α-DOX; 5a and 5b, bacterial peroxidase-cyclooxygenase families) did not evolved from one single protein, but rather appeared independently during evolution. α-DOX belongs to the same peroxidase–cyclooxygenase superfamily as COX and some diatoms (e.g. *S. marinoi*) possess this protein along with COX. This questions the origin and evolution of this class of enzymes in diatoms and shows that this group of microalgae is capable of manipulating fatty acids in different ways: diatom α-DOX enzymes may act on medium chain saturated fatty acids, such as 16:0, while COX on long chain-PUFAs (e.g. AA and EPA). The close proximity of bacterial and diatom sequences opened the questions whether diatoms acquired the genes coding for COX via lateral gene transfer, which occurred during diatom evolution, or whether these genes were acquired during endosymbiotic events, at least for species like *S. marinoi* possessing both α-DOX-like and COX-like enzymes. In the case of *T. rotula*, α-DOX-like is absent, while the pennate model diatom *Phaeodactylum tricornutum* completely lacks the enzymatic pathway leading to prostaglandins though it can produce isoprostanoids non-enzymatically. Indeed, the ability to synthesize eicosanoids, including lipoxygenase-derived oxylipins and aldehydes, is not robustly conserved in diatoms where genes involved in these metabolisms might have been lost during evolution (Di Dato, Barbarinaldi, et al., 2020).

2.1.2 Gene expression analysis

COX gene expression have been experimentally analyzed by quantitative Real Time PCR (qPCR) during the growth of only two diatom species, *Skeletonema marinoi* and *Thalassiosira rotula* (Di Dato et al., 2017, 2019; Di

Dato, Barbarinaldi, et al., 2020). During diatom culture growth, it is possible to identify different phases based on the cell division rate: an exponential phase that last in these species about 3 days, during which cells divide fast; a late exponential-early stationary phase, lasting one or two days during which the division rate starts slowing down; a stationary phase in which the culture division rate is slower, lasting few days; a senescent phase during which the cell division stops, the population reach a maximum and cells start to die, mostly for nutrient starvation. Regarding *S. marinoi*, COX gene expression has been measured in two strains, named FE7 and FE60, known to differ for the production of oxylipins, at three time points: exponential (4th growing day), stationary (7th growing day) and senescent (11th growing day). Similarly, the *T. rotula* COX gene expression have been measured along the growth of the corresponding culture considering six time-points from the third to the seventh and the tenth day.

On the other side, a wider in silico gene expression analysis was extended to all diatom and dinoflagellates species whose transcriptomes were available in the MMETSP database (MMETSP, www-imicrobe.us) (Keeling et al., 2014).

2.1.2.1 Experimental evaluation of gene expression levels

qPCR results normalized with respect to the exponential phase showed a down regulation of COX expression in *S. marinoi* FE7 (FE7) at the senescent phase while in *S. marinoi* FE60 (FE60) at the stationary phase. In addition, the FE60 COX gene expression level was lower than the FE7 COX (Fig. 7A) (Di Dato et al., 2017).

T. rotula qPCR results (Fig. 7B) showed a bell like trend of expression with a statistically significant expression peak around the late exponential/early stationary phase, i.e. day 5. In addition, COX-day 6 expression persisted with significant difference versus COX-day 7 and COX-day 10 expressions. However, the expression peak was only transient because a quick decrease in the expression level occurred soon afterwards, reaching very low levels at the end of the stationary (day 7) and senescent (day 10) phases (Di Dato, Barbarinaldi, et al., 2020). In addition, COX gene expression was susceptible to stress conditions such as nutrient depletion of silica, a very important nutrient for the diatom physiology. In this condition indeed, COX expression was significantly downregulated (Fig. 7C) (Di Dato et al., 2019).

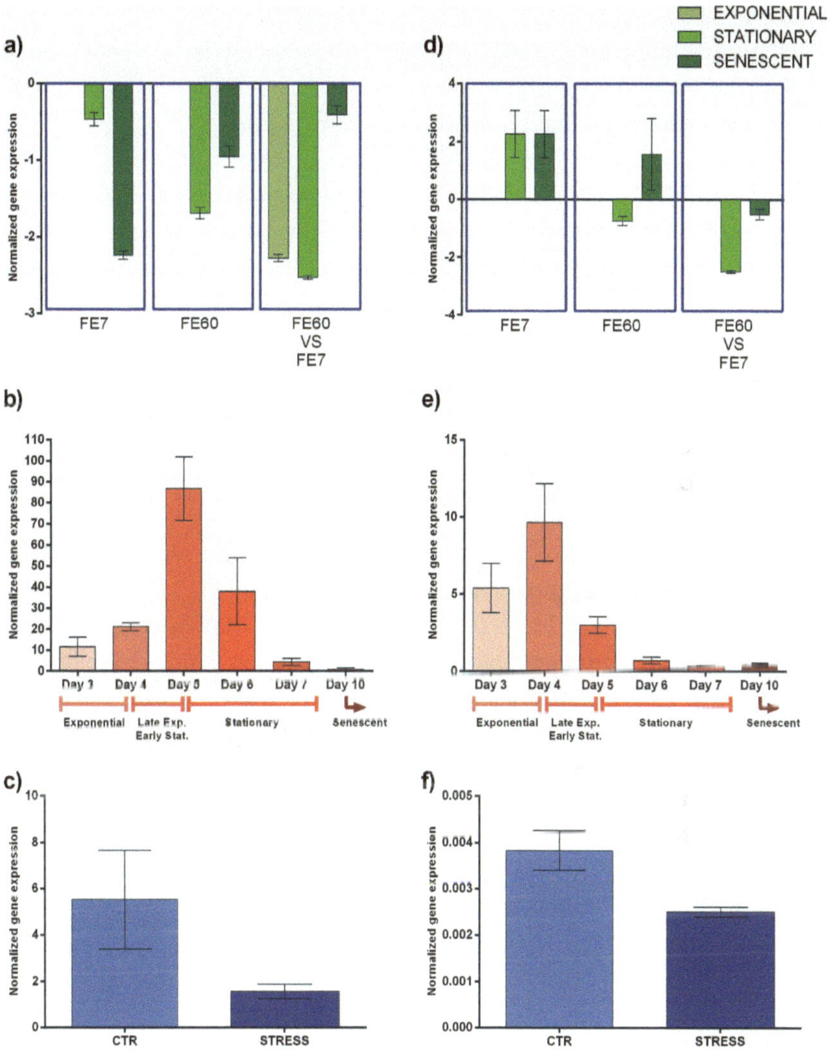

Fig. 7 Gene expression analysis of COX and PGES by Real-time qPCR. (A and D) *Skeletonema marinoi* FE7 and FE60 clones. COX (A) and PGES (D) gene expression levels were measured at the exponential, stationary and senescent phases of the algae cultures growth. Results were analyzed with the REST software and reported as 2 log expression ratio using the exponential phase as reference condition. Comparison between the two strain is also reported as 2 log expression ratio of FE60 with respect to FE7. Relative expression ratios above two folds were considered significant. (B, C, E and F) *Thalassiosira rotula*. COX (B) and PGES (E) gene expression levels were measured along *T. rotula* growth from day 3 to day 10. The Y axis reports the mean normalized expression (MNE) of three biological and technical replicates on a housekeeping gene *(Continued)*

2.1.2.2 In silico evaluation of gene expression levels

Data mining approach of transcriptome sequencing have been used to extend the COX expression analysis on additional microalgal species and to identify conditions that could affect its expression.

Results revealed a huge variability among species and among different culture conditions with only a few species expressing the COX gene. Moreover, different treatments (nutrient deprivation, different growth temperature, CO_2 concentrations) did not had the same effect on different species showing high genetic and metabolic variability among species and clones as already observed previously (Gerecht et al., 2011; Lamari et al., 2013).

Among diatoms, COX has been annotated in *Asterionellopsis glacialis*, different species of *Chaetoceros*, i.e. *affinis*, *debilis*, *neogracile*, different clones of *S. marinoi*, named SkelA, FE7, FE 60, *Thalassiosira gravida* (Table 5) (Di Dato et al., 2017).

Asterionellopsis glacialis COX appeared expressed in almost similar amount when the cultures were grown in nitrogen or silica depleted media. *C. affinis* COX expression, instead, was two times more expressed in nitrogen depleted media containing current CO_2 concentration compared to cultures grown in standard f/2 media and 5 and 17 time more expressed compared to cultures grown in phosphorus depleted media and media containing CO_2 concentration mimicking future scenarios, respectively. In addition, *C. debilis* COX expression in iron depleted medium, was half of the control medium while the one of *C. neogracile* was almost doubled expressed, with respect to control media or nitrogen depleted media, when the growth media was depleted of silica or contained nocodazole, but there were no differences between control and nitrogen depleted media. Also in *S. marinoi*,

Fig. 7—Cont'd (TBP), calculated with the qGene tool for qPCR analysis. COX (C) and PGES (F) gene expression levels measured in *T. rotula* cultures grown in normal condition (CTR) versus cultures grown in silica depletion (STRESS). *From Di Dato, V., Barbarinaldi, R., Amato, A., Di Costanzo, F., Fontanarosa, C., Perna, A., Amoresano, A., Esposito, F., Cutignano, A., Ianora, A., & Romano, G. (2020). Variation in prostaglandin metabolism during growth of the diatom Thalassiosira rotula.* Scientific Reports, 10*(1), 5374. https://doi.org/10.1038/s41598-020-61967-3; Di Dato, V., Orefice, I., Amato, A., Fontanarosa, C., Amoresano, A., Cutignano, A., Ianora, A., & Romano, G. (2017). Animal-like prostaglandins in marine microalgae.* The ISME Journal, 11*(7), 1722–1726. https://doi.org/10.1038/ismej.2017.27, Di Dato, V., Di Costanzo, F., Barbarinaldi, R., Perna, A., Ianora, A., & Romano, G. (2019). Unveiling the presence of biosynthetic pathways for bioactive compounds in the Thalassiosira rotula transcriptome.* Scientific Reports, 9*(1), 9893. https://doi.org/10.1038/s41598-019-46276-8.*

Table 5 List of diatom species whose transcriptome annotation retrieved a transcript with a functional assignment for a Prostaglandin endoperoxide G/H synthase 1/2 (PGHS or COX) (iMicrobe database, www.imicrobe.us).

Genus	Species	Clone	MMETSP code	Treatment	COX FPKM
Asterionellopsis	*glacialis*	CCMP134	0707	N-limited	7305.03
			0706	Si-limited	8052.7
Chaetoceros	*affinis*	CCMP159	0088	Replete-ModernCO$_2$	22383
			0090	N-ModernCO$_2$	43066
			0091	-P-ModernCO$_2$	7994.5
			0092	Replete-FutureCO$_2$	2526.8
Chaetoceros	*debilis*	MM31A-1	0149	CTR	21207.5
			0150	-Fe	10718.8
Chaetoceros	*neogracile*	CCMP1317	0751	Exponential	2887
			0753	NLim	2311.38
			0754	Nocodazole	5420.1
			0752	SiLim	4620
Skeletonema	*marinoi*	SkelA	0918	Replete	10270
			0920	SiLim	24515
Thalassiosira	*gravida*	GMp14c1	0492	Replete	5168.2
			0493	NLim	2093
			0494	PLim	5953.6

For each specie, are reported the clone name, the MMETSP code, the growth conditions and the Fragments Per Kilobase of transcript per Million mapped reads (FPKM) value relative to the corresponding COX sequence.

clone SkelA, the condition depleted of silica induced an over expression of the COX with respect to the control (Table 5) (Di Dato et al., 2017). The *T. rotula* clones, sequenced in the frame of the MMETSP did not retrieved any annotation for the COX function, however the one out of the MMETSP, from the Mediterranean sea, yes, showing an in silico expression values threefold lower when growing in silica depleted media compared to cultures growing in complete media (Table 6) (Di Dato et al., 2019).

Table 6 In silico expression analysis of the prostaglandin pathway in *Thalassiosira rotula* strains from the MMETSP project, grown in different conditions, in comparison with the *T. rotula* strain from the Mediterranean Sea.

	FPKM							
	Light cycle	Dark cycle	PO_4 limited	Fe limited	PO_4 and Fe limited	Replete	Complete medium	SiO_4 limited
	T. rotula MMETSP data						Mediterranean *T. rotula*	
COX	–	–	–	–	–	–	132.2	45
PGES-2	4.86	0	0	0	0	0	4.3	3.2
PGFS I	15.5	9.24	27.6	58.3	54.65	46.92	46	24
PGFS II	1.4	25	12.8	0	10.2	0	0	0

Expression levels are reported in Fragments Per Kilobase of transcript per Million mapped reads (FPKM).

Considering another major group of marine eukaryotic microalgae, the Dinoflagellata, the COX gene function has been found annotated only in two species, over a total of 42 transcriptomes analyzed, i.e. *Amphidinium massartii* and *Polarella glacialis*, with a highest expression in *P. glacialis* compared to *A. massartii* in which it was very low. The two species differed not only for the sampling site, the Pacific Ocean and the MacMurdo Sound, respectively, but also for the growing conditions. For *Polarella glacialis*, conditions were very high light (3,800 lumen) with long day length (14h), and 0.3 μmol/kg phosphate, 3 μmol nitrate, 35% of salinity, whereas they were 80 lumen, 12h of day length, 25 μmol/kg phosphate, 645 μmol nitrate, 35% of salinity for *Amphidinium massartii* (Fig. 8) (Di Dato, Ianora, & Romano, 2020).

2.2 Prostaglandin E synthase (PGES)

Differently from COXs, the PGES have been studied only in mammals, so there is no information about similar enzymes in other type of organisms.

Only in 2017, together with COX, PGES starts to be described in microalgae. Based on transcriptomic functional annotations, PGES was annotated as microsomal prostaglandin E synthase 1 in *S. marinoi*, as Prostaglandin E synthase 2-like (mPGES-2) in *T. rotula* and as Prostaglandin E synthase 2 in dinoflagellates transcriptomes (Di Dato et al., 2017, 2019; Di Dato, Ianora, & Romano, 2020).

2.2.1 Phylogenetic analysis

Similarly to the COX protein, PGES phylogenetic analysis has been conducted on diatom sequences retrieved by blasting the *S. marinoi* sequence

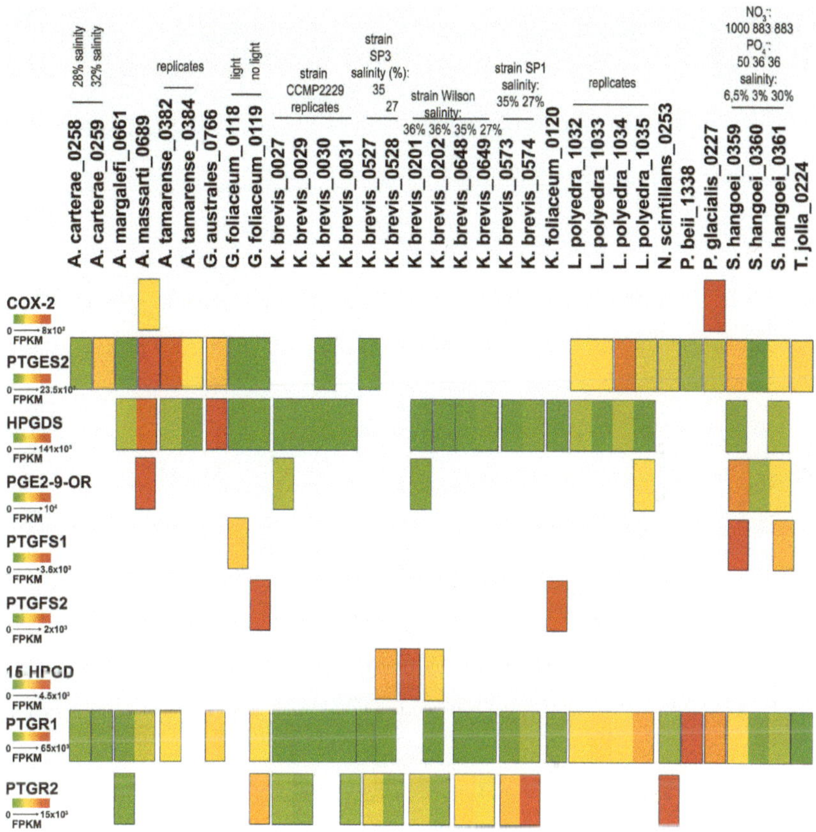

Fig. 8 In silico expression of the transcripts associated to prostaglandin-related functions in dinoflagellates. Heat map representation of the expression levels, expressed in Fragments per Kilobase per Million Mapped Reads (FPKM), of the Pgs metabolism related enzymes identified in each transcriptome of the different dinoflagellate species. *From Di Dato, V., Ianora, A., & Romano, G. (2020). Identification of prostaglandin pathway in dinoflagellates by transcriptome data mining.* Marine Drugs, 18*(2). https://doi.org/10.3390/md18020109.*

against a database containing the peptide sequences of all the diatom species sequenced in the MMETSP (https://www.imicrobe.us). In addition, were also added to the analysis the sequences retrieved by the *S. marinoi*-FE7 PGES blastp against the NCBI protein database. This blast search retrieved plant sequences annotated as "glutathione S-transferase" or "predicted prostaglandin E synthase 2-like" obtained by an automated computational analysis on deposited genomic sequences.

The phylogeny analysis resulted in a forked tree showing two sister clades separating diatoms and plant/macroalgae sequences (Fig. 9A). The diatom

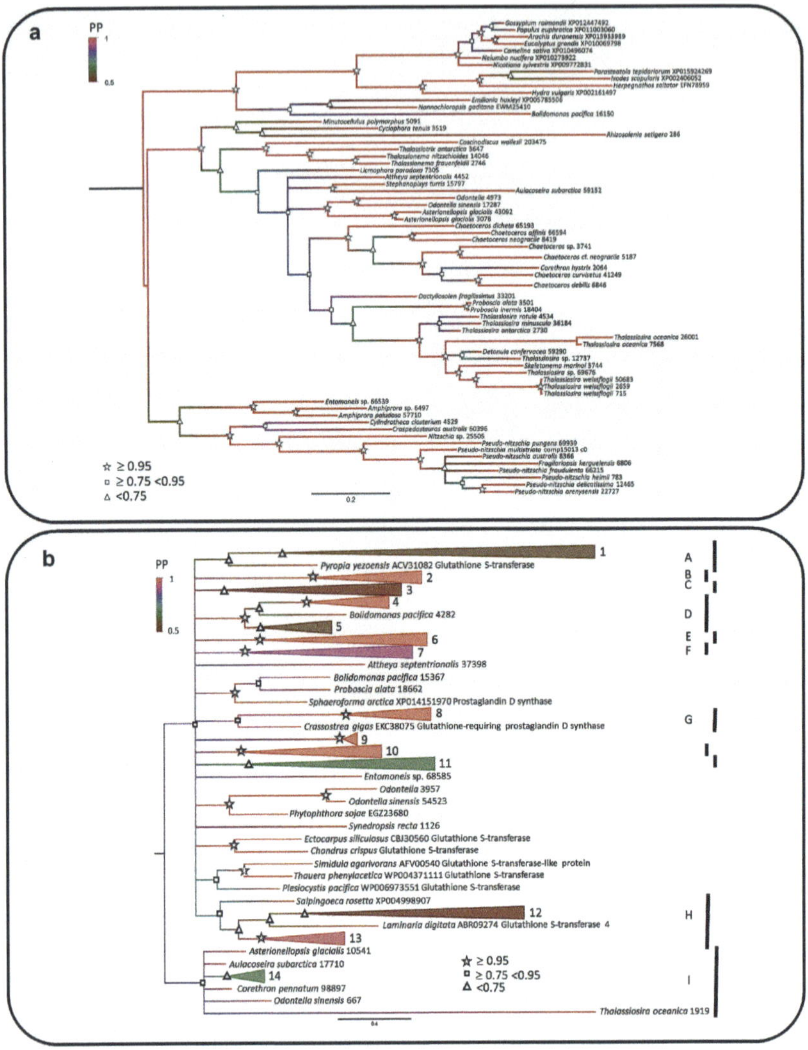

Fig. 9 Phylogenetic analysis of diatom's PGES and PTGDS sequences. Bayesian mid-point routed phylogenetic trees based on *Skeletonema marinoi* PGES (A) and PTGDS (B) proteins from pennate and centric diatoms. Sequences from other organisms are also included as outgroup. The trees were built on scale in the branch length (scale bars reported). Sequences are identified by the species name and annotation (where applicable), followed by their MMETSP ID deprived of the taxon ID for simplicity or GenBank accession number. Posterior probability (PP) values are represented with symbols at the nodes: star represents PP above or equal to 0.95; square PP between 0.75 and 0.95; triangle below 0.75. Branch color refers to PP, color code is reported in figure (5,000,000 generations, split frequency standard deviation = 0.01). For simplicity, for PTGDS tree, large sequence clusters were condensed and identified by numbers and the nine main clades were identified by letters highlighted with vertical bars on the right side of the picture. *From Di Dato, V., Orefice, I., Amato, A., Fontanarosa, C., Amoresano, A., Cutignano, A., Ianora, A., & Romano, G. (2017). Animal-like prostaglandins in marine microalgae.* The ISME Journal, 11*(7), 1722–1726. https://doi.org/10.1038/ismej.2017.27.*

clade forked again creating a separation between the two diatom subgroups, centric (radial frustule symmetry) and pennate (bilateral frustule symmetry). The two subgroups clustered separately in two sisters with only two exceptions represented by the two pennates *Cyclophora tenuis* and *Licmophora paradoxa* that clustered with centric diatoms (Di Dato et al., 2017).

2.2.2 Gene expression analysis

Similarly to COX, PGES gene expression have been experimentally analyzed by qPCR, during the growth of *S. marinoi* FE7 and FE60 and *T. rotula* (Di Dato et al., 2017, 2019; Di Dato, Barbarinaldi, et al., 2020), while a wider in silico gene expression analysis was extended to all diatom and dinoflagellata species whose transcriptomes were available in the MMETSP database (www.imicrobe.us) (Keeling et al., 2014) (see Section 2.1.2).

2.2.2.1 Experimental evaluation of gene expression levels

qPCR results normalized with respect to the exponential phase showed an almost two-fold upregulation of FE7 PGES expression at both stationary and senescent phase (Fig. 7D). On the contrary, FE60 PGES expression showed a tendency to be downregulated at the stationary phase without varying at the senescent phase (Fig. 7D). Interestingly, comparing FE60 PGES gene expression level with the FE7 one, it was downregulated by twofold and half at the stationary phase while showing no differences at the senescent phase (Fig. 7D) (Di Dato et al., 2017).

On the other side, *T. rotula* PGES gene expression, by qPCR, showed a statistically significant expression peak around the late exponential phase, one day before COX peak (Fig. 7E). PGES expression at day 3 was statistically significantly higher than days 6, 7 and 10 but was lower than the one at day 4. The day 4 expression peak was transient as a quick decrease in the expression level occurred soon afterwards. At the edge of the exponential/stationary phase, when cells start to feel the nutrient depletion and overpopulation stresses, the expression crashes by two and a half fold (day5). Soon afterward, at the full stationary and up to senescent phase, the expression seems to decline reaching values near zero (day 6–10) (Fig. 7E) (Di Dato, Barbarinaldi, et al., 2020). This sensitivity to nutrient depletion was confirmed by measuring PGES expression at day 7 in culture growing in complete or in silica depleted medium, the latter being the condition in which the gene expression levels were lower than the complete medium (Fig. 7F) (Di Dato et al., 2019).

2.2.2.2 In silico evaluation of gene expression levels

Data mining analysis revealed a huge variability for the presence of PTGES among species and also as a function of the culture conditions (Table 7) (Di Dato et al., 2017). However, the transcript expression levels based on the reads value were not calculated. The diatom species in which it was possible to identify the PTGES expression are listed in Table 7.

The species *Amphiprora, Asterionellopsis glacialis, Corethron hystrix* and *Chaetoceros* species, *Fragilariopsis kerguelensis, Leptocylindrus* species, *Minutocellus polymorphus, Nitzschia punctata, Proboscia* species, *Pseudonitzschia* species, *Skeletonema* species, *Thalassionema, Thalassiosira* and *Thalassiothrix antarctica* showed differences among clones and treatments, being PTGES expression detectable only in certain clones or treatments of the same species (Di Dato et al., 2017). However, in silico comparison of PTGES expression in different *T. rotula* clones and in different growing conditions showed no difference in mPGES-2 expression calculated in FPKM, when detectable (Table 6) (Di Dato et al., 2019).

Considering the dinoflagellates, the PTGES gene function is annotated in twelve species and 20 transcriptomes, over a total of 42 transcriptomes analyzed (Table 8), i.e. *Alexandrium tamarense, Amphidinium carterae, Amphidinium massartii, Gambierdiscus australes, Glenodium foliaceum, Karenia brevis, Lingulodinium polyedra, Noctiluca scintillans, Pelagodinium beii, Polarella glacialis, Scrippsiella hangoei, Togula jolla.* Among these species, PGES expression in FPKM was very low, and the species with the highest expression was *A. massartii*, followed by *A. tamarense, G. australes* and *L. polyedra* (Fig. 8). PTGES expression seems to be sensitive to salinity percentage. In *S. hangoei* it showed a higher expression when the cells were growing at 6.5% of salinity with respect to 3% but also to 30% of salinity. In addition, in *K. brevis* SP3, Wilson and SP1 clones, PTGES showed higher expression at 35 % of salinity, compared to 27%. Similarly, 32% of salinity induced PTGES expression in *A. carterae* in comparison to 28% (Di Dato, Ianora, & Romano, 2020).

2.3 Prostaglandin D2 synthase (PTGDS) or prostaglandin-H2 D-isomerase (HPGDS)

As for PTGES, also PTGDS have not been studied in the vegetal reign until our studies in microalgae started in 2017. The functional annotation identified this gene in only one of the diatom species used as model, *Skeletonema marinoi* but not in *Thalassiosira rotula*. However, the in silico

Table 7 Presence of the prostaglandin pathway enzymes in diatom transcriptomes (www.imicrobe.us).

Genus	Species	Clone	MMETSP code	Treatment	PGES	PTGDS	PGT
Amphiprora	*paludosa*	CCMP125	1065	–	+	+	+
Amphiprora	*sp*	CCMP467	0724	Exponential	+	+	+
Amphiprora	*sp*	CCMP467	0727	Nocodazole	-	-	+
Amphiprora	*sp*	CCMP467	0725	SiLim	+	+	+
Amphora	*coffeaeformis*	CCMP127	0316	f/2 – silicate, adjusted to 1 mM nitrate and amended with 200 μM acetate	-	-	+
Amphora	*coffeaeformis*	CCMP127	0317	f/2 – silicate, but without nitrate	-	-	-
Amphora	*coffeaeformis*	CCMP127	0318	f/2 – silicate, adjusted to 1 mM nitrate	-	+	-
Asterionellopsis	*glacialis*	CCMP1581	1394	–	+	+	+
Asterionellopsis	*glacialis*	–	0713	–	+	+	+
Asterionellopsis	*glacialis*	CCMP134	0705	Exponential	-	-	-
Asterionellopsis	*glacialis*	CCMP134	0708	Nocodazole	-	-	-
Asterionellopsis	*glacialis*	CCMP134	0707	N-limited	-	+	+
Asterionellopsis	*glacialis*	CCMP134	0706	Si-limited	-	+	-
Astrosyne	*radiata*	13vi08-1A	0418	–	-	+	+
Attheya	*septentrionalis*	CCMP2084	1449	–	+	+	+
Aulacoseira	*subarctica*	CCAP1002/5	1064	–	+	+	+
Bolidomonas	*pacifica*	CCMP1866	0785	–	+	+	-
Chaetoceros	*affinis*	CCMP159	0088	Replete-ModernCO_2	-	+	-
Chaetoceros	*affinis*	CCMP159	0090	-N-ModernCO_2	-	+	+
Chaetoceros	*affinis*	CCMP159	0091	-P-ModernCO_2	+	+	+

Continued

Table 7 Presence of the prostaglandin pathway enzymes in diatom transcriptomes (www.iMicrobe.us).—cont'd

Genus	Species	Clone	MMETSP code	Treatment	PGES	PTGDS	PGT
Chaetoceros	*affinis*	CCMP159	0092	Replete-FutureCO_2	-	+	-
Chaetoceros	*brevis*	CCMP164	1435	–	-	+	-
Chaetoceros	*curvisetus*		0716	Exponential	-	-	-
Chaetoceros	*curvisetus*		0717	Exponential	+	+	+
Chaetoceros	*curvisetus*		0718	Exponential	+	+	+
Chaetoceros	*curvisetus*		0719	Exponential	-	-	-
Chaetoceros	*debilis*	MM31A-1	0149	CTR	-	+	+
Chaetoceros	*debilis*	MM31A-1	0150	-Fe	+	+	+
Chaetoceros	*dichaeta*	CCMP1751	1447	–	+	+	+
Chaetoceros	*neogracile*	RCC1993	1336	–	+	+	+
Chaetoceros	*neogracile*	CCMP1317	0751	Exponential	-	+	+
Chaetoceros	*neogracile*	CCMP1317	0753	NLim	+	-	-
Chaetoceros	*neogracile*	CCMP1317	0754	Nocodazole	-	+	+
Chaetoceros	*neogracile*	CCMP1317	0752	SiLim	-	+	+
Chaetoceros	*sp.*	clone GSL56	0200	–	+	+	+
Chaetoceros	*sp.*	clone UNC1202	1429	–	+	+	+
Corethron	*hystrix*	308	0010	–	+	+	+
Corethron	*pennatum*	L29A3	0169	CTR	-	+	+
Corethron	*pennatum*	L29A3	0171	-Fe	-	+	+
Coscinodiscus	*wailesii*	CCMP2513	1066	–	+	+	+

Craspedostauros	*australis*	CCMP3328	1442	–	+	+	+
Cyclophora	*tenuis*	ECT3854	0397	–	+	+	+
Cyclotella	*meneghiniana*	CCMP 338	1057	–	-	+	+
Cylindrotheca	*closterium*	KMMCC:B-181	0017	–	+	+	+
Dactyliosolen	*fragilissimus*		0580	–	+	+	+
Detonula	*confervacea*	CCMP 353	1058	–	+	+	+
Entomoneis	*sp*	CCMP2396	1443	–	+	+	-
Eucampia	*antarctica*	CCMP1452	1437	–	+	+	+
Fragilariopsis	*kerguelensis*	L26-C5	0733	Fe LL	-	+	+
Fragilariopsis	*kerguelensis*	L26-C5	0734	Fe SL	-	+	+
Fragilariopsis	*kerguelensis*	L26-C5	0735	-Fe LL	+	+	-
Fragilariopsis	*kerguelensis*	L26-C5	0736	-Fe SL	-	+	+
Fragilariopsis	*kerguelensis*	L2-C3	0906	Freezing temperature (T = -2 °C), light = 45	-	-	-
Fragilariopsis	*kerguelensis*	L2-C3	0907	Control growth, T = 5 °C, light = 100	-	-	+
Fragilariopsis	*kerguelensis*	L2-C3	0908	Iron limitation, T = 4 °C, light = 80	-	+	+
Fragilariopsis	*kerguelensis*	L2-C3	0909	Darkness, T = 5 °C	-	-	-
Grammatophora	*oceanica*	CCMP 410	0009	–	-	+	+
Leptocylindrus	*apora*	B651	0322	–	+	+	+
Leptocylindrus	*danicus*	B650	0321	–	-	-	-
Leptocylindrus	*danicus*	CCMP1856	1362	–	+	+	+
Licmophora	*paradoxa*	CCMP2313	1360	–	+	+	+

Continued

Table 7 Presence of the prostaglandin pathway enzymes in diatom transcriptomes (www.iMicrobe.us).—cont'd

Genus	Species	Clone	MMETSP code	Treatment	PGES	PTGDS	PGT
Minutocellus	*polymorphus*	NH13	1070	–	-	+	+
Minutocellus	*polymorphus*	CCMP3303	1434	–	-	-	+
Minutocellus	*polymorphus*	RCC2270	1322	–	+	+	+
Nitzschia	*punctata*	CCMP561	0744	Exponential	-	+	+
Nitzschia	*puncatata*	CCMP561	0745	SiLim	+	-	+
Nitzschia	*puncatata*	CCMP561	0746	N-limit	-	-	-
Nitzschia	*punctata*	CCMP561	0747	Nocodazole	-	+	+
Nitzschia	*sp*	RCC80	0014	–	+	+	+
Odontella	*aurita*	1302-5	0015	–	+	+	+
Odontella	*sinensis*	Grunow1884	0160	–	+	+	+
Proboscia	*alata*	PI-D3	0174	CTR	-	+	+
Proboscia	*alata*	PI-D3	0176	-Fe	+	+	+
Proboscia	*inermis*	CCAP1064/1	0816	–	+	+	+
Pseudonitzschia	*arenisensys*	B593	0329	–	+	+	+
Pseudonitzschia	*australis*	10AB	0142_2	Glutamate as N source	-	+	+
Pseudonitzschia	*australis*	10AB	0140	NH_4 as N source	-	+	-
Pseudonitzschia	*australis*	10AB	0139	NO_3 as N source	-	-	+
Pseudonitzschia	*australis*	10AB	0141	UREA as N source	+	+	+
Pseudonitzschia	*delicatissima*	B596	0327	–	+	+	+
Pseudonitzschia	*fraudolenta*	WWA7	0850	Si Replete 390 ppm	-	-	+

Pseudonitzschia	*fraudolenta*	WWA7	0851	Si Lim 390 ppm	–	+	+
Pseudonitzschia	*fraudolenta*	WWA7	0852	Si Replete 750 ppm	+	+	–
Pseudonitzschia	*fraudolenta*	WWA7	0853	Si Lim 750 ppm	–	–	–
Pseudonitzschia	*heimii*	UNC1101	1423	–	+	+	+
Pseudonitzschia	*multistriata*	Female at little size	–		+	+	+
Pseudonitzschia	*pungens*		1061	–	–	+	–
Pseudonitzschia	*pungens f cingulata*		1060	–	+	–	+
Rhizosolenia	*setigera*	CCMP 1694	0789	–	+	+	+
Skeletonema	*costatum*	1716	0013_2	–	–	+	+
Skeletonema	*marinoi*	SkelB	0562	Replete	–	–	+
Skeletonema	*marinoi*	SkelB	0563	N limit	–	–	+
Skeletonema	*grethea*	CCMP 1804	0578	–	–	+	+
Skeletonema	*japonicum*	CCMP2506	0593	–	–	+	+
Skeletonema	*marinoi*	FE60	1040	–	+	+	+
Skeletonema	*marinoi*	FE7	1039	–	+	+	+
Skeletonema	*marinoi*	SkelA	0918	Replete	–	–	–
Skeletonema	*marinoi*	SkelA	0920	Si Lim	–	–	–
Skeletonema	*marinoi*	SM1012Den-03	0320	No Cu, No Si	+	+	+
Skeletonema	*marinoi*	SM1012Hels-07	0319	No Cu, No Si	–	–	+
Skeletonema	*marinoi*	UNC1201	1428	–	–	–	–

Continued

Table 7 Presence of the prostaglandin pathway enzymes in diatom transcriptomes (www.iMicrobe.us).—cont'd

Genus	Species	Clone	MMETSP code	Treatment	PGES	PTGDS	PGT
Skeletonema	*menzelii*	CCMP793	0603	Replete	-	+	+
Skeletonema	*menzelii*	CCMP793	0604	N Lim	-	-	+
Staurosira	*complex*	CCMP2646	1361	–	-	-	+
Stephanophyxis	*turris*	CCMP 815	0794	–	+	+	+
Striatella	*unipunctata*	CCMP2910	0800	–	-	+	+
Synedropsis	*recta*	CCMP1620	1176	–	-	+	-
Thalassionema	*frauenfeldii*	CCMP 1798	0786	–	+	+	+
Thalassionema	*nitzschioides*	L26-B	0156	CTR 1	+	+	+
Thalassionema	*nitzschioides*	L26-B	0158	-Fe	-	+	+
Thalassiosira	*antarctica*	CCMP982	0905	Control growth	+	-	+
Thalassiosira	*antarctica*	CCMP982	0902	Freezing temperature	-	+	+
Thalassiosira	*antarctica*	CCMP982	0903	Iron limitation	-	+	-
Thalassiosira	*antarctica*	CCMP982	0904	Silica limitation	-	+	-
Thalassiosira	*gravida*	GMp14c1	0492	Replete	-	+	-
Thalassiosira	*gravida*	GMp14c1	0493	N Lim	-	-	+
Thalassiosira	*gravida*	GMp14c1	0494	P Lim	-	-	+
Thalassiosira	*minuscola*	CCMP1093	0737	Exponential	-	+	+
Thalassiosira	*minuscola*	CCMP1093	0738	Si Lim	-	-	-
Thalassiosira	*minuscola*	CCMP1093	0739	N Lim	+	+	+
Thalassiosira	*minuscola*	CCMP1093	0740	Nocodazole	-	-	-

Thalassiosira	*oceanica*	CCMP1005	0970	Replete	+	+	+
Thalassiosira	*oceanica*	CCMP1005	0971	Iron limitation	+	-	+
Thalassiosira	*oceanica*	CCMP1005	0972	Silicate limitation	-	-	+
Thalassiosira	*oceanica*	CCMP1005	0973	Nitrate limitation	-	-	-
Thalassiosira	*puntigera*	Tpunct2005C2	1067	–	-	+	+
Thalassiosira	*rotula*	CCMP3096	0404	Dark cycle	-	+	-
Thalassiosira	*rotula*	CCMP3096	0403	Light cycle	+	-	+
Thalassiosira	*rotula*	GSO102	0910	P Lim	-	+	+
Thalassiosira	*rotula*	GSO102	0911	Fe Lim	-	-	-
Thalassiosira	*rotula*	GSO102	0912	Fe Lim_2	-	-	-
Thalassiosira	*rotula*	GSO102	0910	Replete	-	-	-
Thalassiosira	*sp.*	FreshWater	1059	–	+	+	+
Thalassiosira	*sp.*	NH16	1071	–	+	+	+
Thalassiosira	*weissflogii*	CCMP1336	0873	20 nM Fe	-	-	-
Thalassiosira	*weissflogii*	CCMP1336	0879	60 nM Fe	-	-	-
Thalassiosira	*weissflogii*	CCMP1336	0880	20 nM Fe_2	-	-	-
Thalassiosira	*weissflogii*	CCMP1336	0881	60 nM Fe_2	-	-	+
Thalassiosira	*weissflogii*	CCMP1010	0898	Replete	-	-	-
Thalassiosira	*weissflogii*	CCMP1010	0899	Fe Lim	+	+	-
Thalassiosira	*weissflogii*	CCMP1010	0900	Si Lim	-	-	-
Thalassiosira	*weissflogii*	CCMP1010	0901	N Lim	+	+	-
Thalassiosira	*weissflogii*	CCMP1010	1405	Replete A	+	-	-

Continued

Table 7 Presence of the prostaglandin pathway enzymes in diatom transcriptomes (www.iMicrobe.us).—cont'd

Genus	Species	Clone	MMETSP code	Treatment	PGES	PTGDS	PGT
Thalassiosira	*weissflogii*	CCMP1010	1406	Replete B	–	–	–
Thalassiosira	*weissflogii*	CCMP1010	1407	Replete C	–	–	+
Thalassiosira	*weissflogii*	CCMP1010	1408	Si Lim A	–	–	–
Thalassiosira	*weissflogii*	CCMP1010	1409	Si Lim B	–	–	–
Thalassiosira	*weissflogii*	CCMP1010	1410	Si Lim C	–	–	–
Thalassiosira	*weissflogii*	CCMP1010	1411	Fe Lim A	–	–	–
Thalassiosira	*weissflogii*	CCMP1010	1412	Fe Lim B	–	–	+
Thalassiosira	*weissflogii*	CCMP1010	1413	Fe Lim C	–	–	+
Thalassiosira	*weissflogii*	CCMP1010	1414	+2HrFeAddA	–	–	–
Thalassiosira	*weissflogii*	CCMP1010	1415	+2HrFeAddB	–	+	–
Thalassiosira	*weissflogii*	CCMP1010	1416	+2HrFeAddC	–	–	+
Thalassiosira	*weissflogii*	CCMP1010	1417	+8HrFeAddA	–	–	–
Thalassiosira	*weissflogii*	CCMP1010	1418	+8HrFeAddB	–	–	–
Thalassiosira	*weissflogii*	CCMP1010	1419	+8HrFeAddC	–	–	+
Thalassiosira	*weissflogii*	CCMP1010	1420	+24HrFeAddA	–	–	–
Thalassiosira	*weissflogii*	CCMP1010	1421	+24HrFeAddA	–	–	–
Thalassiosira	*weissflogii*	CCMP1010	1422	+24HrFeAddB	–	+	+
Thalassiothrix	*antarctica*	L6-D1	0152	CTR1	–	+	+
Thalassiothrix	*antarctica*	L6-D1	0154	Fe1	+	+	+

For each specie is reported the clone name, the MMETSP code, the growth conditions and the presence/absence of transcripts associated to the prostaglandin E synthase (PGES), the prostaglandin D synthase (PTGDS) and the prostaglandin transporter (PGT) functions.

Table 8 List of dinoflagellate species, in the iMicrobe database, whose transcriptomes contained a transcript associated to the Prostaglandin E synthase function.

Genus	Species	MMETSP code	Treatment
Alexandrium	*margalefi*	0661	GSe; Light 60; day portion 12 h; exp salinity 35
Alexandrium	*tamarense*	0384	F/20 no Silica; Nitrate 88.2; Light 150; day portion 16 h; phosphate 3.62 μmol/kg; exp salinity 30
Alexandrium	*tamarense*	0382	F/20 no Silica; Nitrate 88.2; Light 150; day portion 16 h; phosphate 3.62 μmol/kg; exp salinity 30
Amphidinium	*carterae*	0258	Light 60; day portion 12 h; nitrate 60; phosphate 60 μmol/kg; exp salinity 28
Amphidinium	*carterae*	0259	Light 60; day portion 12 h; L1–N; phosphate 60 μmol/kg; exp salinity 32
Amphidinium	*massartii*	0689	No silica; Nitrate 645; Light 80; day portion 12 h; phosphate 25 μmol/kg; exp salinity 35
Gambierdiscus	*australes*	0766	No silica; Nitrate 645; Light 80; day portion 14 h; phosphate 25 μmol/kg; exp salinity 35
Glenodinium	*foliaceum*	0118	F/2-Si amended sea water; 48 h before harvest antibiotics were added to medium: 500 mg/L penicillin; 200 mg/L ampicillin; 100 mg/L streptomycin; 100 mg/L neomycin; Nitrate 882; Light 550; day portion 12 h; phosphate 36.2 μmol/kg; exp salinity 21
Glenodinium	*foliaceum*	0119	F/2-Si amended sea water; 48 h before harvest antibiotics were added to medium: 500 mg/L penicillin; 200 mg/L ampicillin; 100 mg/L streptomycin; 100 mg/L neomycin; Nitrate 882; night portion 48 h; phosphate 36.2 μmol/kg; exp salinity 21
Karenia	*brevis*	0030	F/2; Nitrate 60; Light 60; day portion 12 h; phosphate 60 μmol/kg; exp salinity 32
Karenia	*brevis*	0527	F/2; Nitrate 882; Light 70; day portion 12 h; phosphate 36.2 μmol/kg; exp salinity 35
Lingulodinium	*polyedra*	1034	No silica; Nitrate 882; Light 150; day portion 12 h; phosphate 36.2 μmol/kg; exp salinity 28
Lingulodinium	*polyedra*	1033	No silica; Nitrate 882; Light 150; day portion 12 h; phosphate 36.2 μmol/kg; exp salinity 28
Lingulodinium	*polyedra*	1032	No silica; Nitrate 882; Light 150; day portion 12 h; phosphate 36.2 μmol/kg; exp salinity 28
Lingulodinium	*polyedra*	1035	No silica; Nitrate 882; Light 150; day portion 12 h; phosphate 36.2 μmol/kg; exp salinity 28

Continued

Table 8 List of dinoflagellate species, in the iMicrobe database, whose transcriptomes contained a transcript associated to the Prostaglandin E synthase function.—cont'd

Genus	Species	MMETSP code	Treatment
Noctiluca	*scintillans*	0253	HESNW no silicate; Nitrate 3; Light 3800; day portion 14h; phosphate 0.3 μmol/kg; exp salinity 30
Pelagodinium	*beii*	1338	K/2; Nitrate 882; Light 40; day portion 12h; phosphate 36.2 μmol/kg; exp salinity 35
Polarella	*glacialis*	0227	HESNW no silicate; Nitrate 3; Light 3800; day portion 14h; phosphate 0.3 μmol/kg; exp salinity 30
Scrippsiella	*hangoei*	0359	MWC; Light 20; day portion 12h; Selenium added to a final concentration of 4.55 nmol/L; nitrate 1000; silicate 100 μmol/kg; phosphate 50 μmol/kg; env salinity 6.5
Scrippsiella	*hangoei*	0361	f/2 amended sea water without silica; Light 20; day portion 12h; nitrate 883; silicate 100 μmol/kg; phosphate 36 μmol/kg; env salinity 6.5; exp salinity 30
Scrippsiella	*hangoei*	0360	f/2 no silica; Light 20; day portion 12h; silicate 100 μm/kg; nitrate 883; phosphate 36 μmol/kg; env salinity 6.5; exp salinity 3
Togula	*jolla*	0224	Light 3800; day portion 14h; no silica; nitrate 3; phosphate 0.3 μmol/kg; exp salinity 30

For each specie is reported the MMETSP code and the growth conditions.

search in many different diatoms and dinoflagellates transcriptomes revealed its presence in other *T. rotula* strains (Di Dato et al., 2017, 2019; Di Dato, Ianora, & Romano, 2020).

2.3.1 Phylogenetic analysis

The Bayesian phylogeny of PTGDS was performed with the alignment of 230 sequences from diatoms and other organisms. The protein sequences were very variable and the analysis resulted in a complicated tree that showed at least nine clades, five of which containing only diatom sequences, with nodes that were variably supported statistically (PP from 0.53 to1.00) (Fig. 9B). The remaining four clades contained at least one outgroup sequence, some of which annotated as Glutathione S-transferase 4, Glutathione S-transferase-like protein, Glutathione-requiring prostaglandin D synthase or Glutathione S-transferase. The other outgroup sequences clustered in unresolved branches or in supported but not resolved small clades. All but one clade contained both centric and pennate diatoms, suggesting that the nine clades cluster different proteins present in both diatom classes. This most likely indicates that in diatoms different proteins can have the same function or can be expressed only in specific conditions. This suggests that the sequences share a certain identity and contain similar domains, but may belong to different protein families. Also, PTGDS can have evolved quite differently in different organisms and also in diatoms different proteins can serve the same function or be involved in different conditions (Di Dato et al., 2017).

2.3.2 Gene expression

Similarly to COX and PTGES, PTGDS gene expression have been experimentally analyzed, by qPCR, only during the growth of *S. marinoi* FE7 and FE60 (Di Dato et al., 2017). In the other diatom model species *T. rotula*, the transcriptome did not reveal any functional annotation for PTGDs. A wider only in silico gene expression analysis extended to all diatom and dinoflagellates species whose transcriptomes were available in the MMETSP database detected its expression in different dinoflagellata and diatom species (www-imicrobe.us) (Keeling et al., 2014)

2.3.2.1 Experimental evaluation of gene expression levels

The PTGDS gene expression measured by qPCR revealed no difference of expression between the exponential and stationary phase of the growth curve of the *S. marinoi* FE7 while showing a tendency to be upregulated in the senescent phase (Fig. 10A). On the contrary, with respect to the

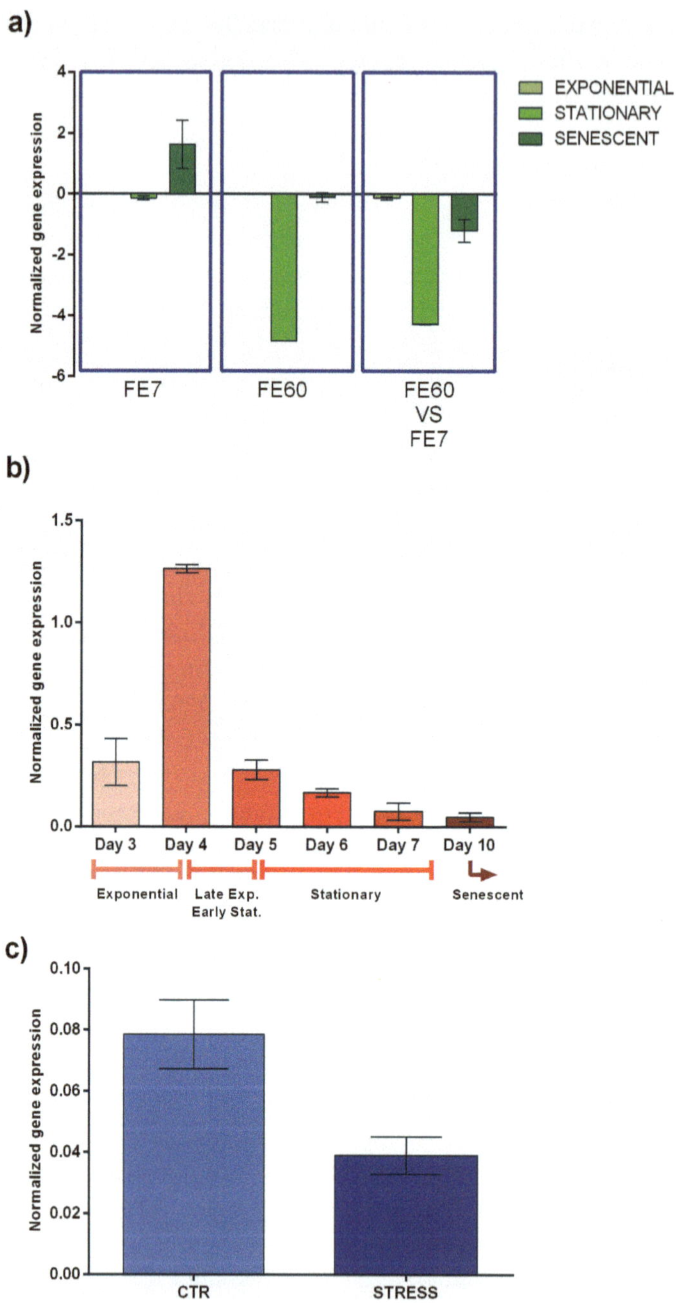

Fig. 10 Gene expression analysis of *S. marinoi* PTGDS and *T. rotula* PGFS by Real-time qPCR. (A) PTGDS gene expression levels were measured at the exponential, stationary and senescent phases of the FE7 or FE60 cultures growth. Results were analyzed with *(Continued)*

exponential phase, the *S. marinoi* FE60 PTGDS was almost five folds down-regulated at the stationary phase while returning to the initial value (exponential phase) at the senescent phase. Comparing the expression levels between the two *S. marinoi* clones, FE7 and FE60, PTGDS expression appeared similar in both strains at the exponential phase while being four and a half times less expressed in FE60 at the stationary phase, and almost similar at the senescent phase.

2.3.2.2 In silico evaluation of gene expression levels

Similarly to the other enzymes, PTGDS annotation has been searched in all available diatom and dinoflagellate transcriptome datasets. Also in this case, the presence/absence of the functional annotation revealed an intra-genera and intra-clonal variability in diatoms, other than a correlation with the condition in which the algae have been cultivated, such as temperatures levels, salinity concentrations and nutrient concentration (Table 7) (Di Dato et al., 2017).

Data mining of 42 transcriptomes corresponding to different species and growth conditions of Dinoflagellata, identified the expression of PTGDS in 9 species and 18 transcriptomes, and it was one of the most widespread function of the Pg pathway among the species considered. However, when comparing all the species together, the in silico determined gene expression was very low and at similar levels in all species except for *G. australes* and *A. massartii*. When the comparison was made among different transcriptomes and different growth conditions for each species, PTGDS

Fig. 10—Cont'd the REST software and reported as 2 log expression ratio using the exponential phase as reference condition. Comparison between the two strain is also reported as 2 log expression ratio of FE60 with respect to FE7 strain. Relative expression ratios above two folds were considered significant. (B) PGFS gene expression levels were measured along the algae culture growth from day 3 to day 10. The Y axis reports the Mean Normalized Expression (MNE) of three biological and technical replicates on a housekeeping gene (TBP), calculated with the qGene tool for qPCR analysis. (C) PGFS gene expression levels measured in cultures grown in normal condition (CTR) versus cultures grown in silica depletion (STRESS). *From Di Dato, V., Barbarinaldi, R., Amato, A., Di Costanzo, F., Fontanarosa, C., Perna, A., Amoresano, A., Esposito, F., Cutignano, A., Ianora, A., & Romano, G. (2020). Variation in prostaglandin metabolism during growth of the diatom Thalassiosira rotula.* Scientific Reports, 10*(1), 5374. https://doi.org/10.1038/s41598-020-61967-3; Di Dato, V., Di Costanzo, F., Barbarinaldi, R., Perna, A., Ianora, A., & Romano, G. (2019). Unveiling the presence of biosynthetic pathways for bioactive compounds in the Thalassiosira rotula transcriptome.* Scientific Reports, 9*(1), 9893. https://doi.org/10.1038/s41598-019-46276-8.*

was the most expressed compared to the expression levels of all the identified Pg pathway functions, and seemed to be regulated by different salinity percentage (Table 9; Fig. 8) (Di Dato, Ianora, & Romano, 2020).

2.4 Prostaglandin F synthase or $PGF_{2\alpha}$ synthetase or PGD_2 11-ketoreductase

Until now, these enzymes have not been identified in plants nor in macroalgae. In microalgae they were identified in the transcriptome of *T. rotula* and in few dinoflagellates species. In *T. rotula* a Prostaglandin F Synthase was identified by functional annotation of its de novo transcriptome. Similarly, in Dinoflagellata, transcriptome functional annotations identified PGFS I and II (Di Dato et al., 2019; Di Dato, Barbarinaldi, et al., 2020; Di Dato, Ianora, & Romano, 2020).

2.4.1 Gene expression

PGFS gene expression have been experimentally analyzed by qPCR only during the growth of *T. rotula*. Differently, in the *S. marinoi* transcriptome no functional annotation of PGFSs was detected (Di Dato et al., 2019). In silico analysis for these enzymes have been performed widely on dinoflagellates transcriptomes and on different clones of *T. rotula*.

2.4.1.1 Experimental evaluation of gene expression levels

qPCR expression analysis of PGFS during 10-day *T. rotula* growth showed a statistically significant difference ($0.0001 < P$ value < 0.0002) among the six time-points considered along the period (Fig. 10B). PGFS gene expression showed a statistically significant peak at day 4, around the late exponential/early stationary phase. Day 3 expression was statistically significantly lower than day 4 but higher than the expression levels at days 7 and 10. Analogously, expression at day 5 was still statistically significantly higher than the expression values at days 7 and 10. The peak of expression was only transient because a quick decrease in the expression level occurred soon afterwards, reaching very low levels at the end of the stationary (day 7) and senescent (day 10) phases (Fig. 10B) (Di Dato, Barbarinaldi, et al., 2020).

In addition, at stationary phase, i.e. day 7, a stress as silica depletion determined a downregulation of PGFS expression levels compared to the culture growing without stress (Fig. 10C) (Di Dato et al., 2019).

Table 9 List of dinoflagellate species, in the iMicrobe database, whose transcriptomes contained a transcript associated to the prostaglandin D synthase function.

Genus	Species	MMETSP code	Treatment
Alexandrium	*margalefi*	0661	GSe; Light 60; day portion 12 h; exp salinity 35
Alexandrium	*tamarense*	0384	F/20 no silica; Nitrate 88.2; Light 150; day portion 16 h; phosphate 3.62 μmol/kg; exp salinity 30
Alexandrium	*tamarense*	0382	F/20 no silica; Nitrate 88.2; Light 150; day portion 16 h; phosphate 3.62 μmol/kg; exp salinity 30
Amphidinium	*massartii*	0689	No silica; Nitrate 645; Light 80; day portion 12 h; phosphate 25 μmol/kg; exp salinity 35
Gambierdiscus	*australes*	0766	No silica; Nitrate 645; Light 80; day portion 14 h; phosphate 25 μmol/kg; exp salinity 35
Glenodinium	*foliaceum*	0118	F/2-Si amended sea water; 48 h before harvest antibiotics were added to medium: 500 mg/L penicillin; 200 mg/L ampicillin; 100 mg/L streptomycin; 100 mg/L neomycin; Nitrate 882; Light 550; day portion 12 h; phosphate 36.2 μmol/kg; exp salinity 21
Glenodinium	*foliaceum*	0119	F/2-Si amended sea water; 48 h before harvest antibiotics were added to medium: 500 mg/L penicillin; 200 mg/L ampicillin; 100 mg/L streptomycin; 100 mg/L neomycin; Nitrate 882; night portion 48 h; phosphate 36.2 μmol/kg; exp salinity 21
Karenia	*brevis*	0201	f/2; Nitrate 883; Light 50; day portion 16 h; phosphate 36 μmol/kg; exp salinity 36
Karenia	*brevis*	0202	f/2; Nitrate 883; Light 50; day portion 12 h; phosphate 36 μmol/kg; exp salinity 36
Karenia	*brevis*	0648	f/2; Nitrate 882; Light 70; day portion 16 h; phosphate 36.2 μmol/kg; exp salinity 35
Karenia	*brevis*	0649	f/2; Nitrate 882; Light 70; day portion 16 h; phosphate 36.2 μmol/kg; exp salinity 27
Karenia	*brevis*	0573	f/2; Nitrate 882; Light 70; day portion 12 h; phosphate 36.2 μmol/kg; exp salinity 35
Karenia	*brevis*	0574	f/2; Nitrate 882; Light 70; day portion 12 h; phosphate 36.2 μmol/kg; exp salinity 27

Continued

Table 9 List of dinoflagellate species, in the iMicrobe database, whose transcriptomes contained a transcript associated to the prostaglandin D synthase function.—cont'd

Genus	Species	MMETSP code	Treatment
Karenia	*brevis*	0027	F/2; Nitrate 60; Light 60; day portion 12h; phosphate 60 μmol/kg; exp salinity 32
Karenia	*brevis*	0029	F/2; Nitrate N.I.; Light 60; day portion 12h; phosphate 60 μmol/kg; exp salinity 32
Karenia	*brevis*	0030	F/2; Nitrate 60; Light 60; day portion 12h; phosphate 60 μmol/kg; exp salinity 32
Karenia	*brevis*	0031	F/2; Nitrate 60; Light 600; day portion 12h; phosphate 60 μmol/kg; exp salinity 32
Kryptoperidinium	*foliaceum*	0120	F/2-Si amended sea water; 48h before harvest antibiotics were added to medium: 500mg/L penicillin; 200mg/L ampicillin; 100mg/L streptomycin; 100mg/L neomycin; Nitrate 882; Light 550; day portion 12h; phosphate 36.2 μmol/kg; exp salinity 21
Lingulodinium	*polyedra*	1034	No silica; Nitrate 882; Light 150; day portion 12h; phosphate 36.2 μmol/kg; exp salinity 28
Lingulodinium	*polyedra*	1033	No silica; Nitrate 882; Light 150; day portion 12h; phosphate 36.2 μmol/kg; exp salinity 28
Lingulodinium	*polyedra*	1032	No silica; Nitrate 882; Light 150; day portion 12h; phosphate 36.2 μmol/kg; exp salinity 28
Lingulodinium	*polyedra*	1035	No silica; Nitrate 882; Light 150; day portion 12h; phosphate 36.2 μmol/kg; exp salinity 28
Scrippsiella	*hangoei*	0359	MWC; Light 20; day portion 12h; Selenium added to a final concentration of 4.55nmol/L; nitrate 1000; silicate 100μmol/kg; phosphate 50 μmol/kg; env salinity 6.5
Scrippsiella	*hangoei*	0361	f/2 amended sea water without silica; Light 20; day portion 12h; nitrate 883; silicate 100μmol/kg; phosphate 36 μmol/kg; env salinity 6.5; exp salinity 30

For each specie is reported the MMETSP code and the growth conditions.

2.4.1.2 In silico evaluation of gene expression levels

The in silico analysis of *T. rotula* transcriptomes revealed three putative peptides and two transcripts associated to the "Prostaglandin F synthase" function. In addition, six peptides were annotated as "Carbonyl reductase (NADPH)" or "Prostaglandin E2 9-reductase" that reduces PGE_2 to $PGF_{2\alpha}$ differently from PGFS that produces $PGF_{2\alpha}$ starting from PGD_2. Comparing different clones using the transcriptomes from the MMETSP database (Keeling et al., 2014), we observed that PGFS was present with two isoforms in some of the strains and the depletion of iron or iron and phosphate together induced an higher expression of PGFS I while PGFS II was stimulated by a longer dark cycle (Table 6) (Di Dato et al., 2019).

In the dinoflagellate available transcriptomes, PGFS have been identified as PGFS I, PGFS II and Prostaglandin-E(2) 9-reductase (PGE_2-9-OR) (Di Dato, Ianora, & Romano, 2020). PGFS I has been detected in three species and four transcriptomes, PGFS II just in one species, and PGE_2-9-OR in four species and six transcriptomes (Table 10). Comparing the FPKM expression among all the species, PGE_2-9-OR was highly expressed in *A. massartii*, which also showed the highest expression compared to other species such as *K. brevis* CCMP2229 and Wilson clones, *L. polyedra* and *S. hangoei*. PTGFS1 and PTGFS2 were found expressed only in three species, *G. foliaceum*, *K. foliaceum* and *S. hangoei* with highest values of PTGFS1 in S. *hangoei* and highest values of PTGFS2 in *G. foliaceum*. Comparing the expression values of these enzymes with the others in the pathway, for each species, clones and condition, the PGF synthase enzymes were the ones with the lowest value in each condition tested except for *K. brevis* clone CCMP2229 for which PGE_2-9-OR expression was slightly induced under low light regimen (Fig. 8) (Di Dato, Ianora, & Romano, 2020).

2.5 Enzymes involved in prostaglandin catabolism

Similarly to the synthetizing processes, fine-tuned mechanisms of Pgs catalysis are needed to rapidly stop Pgs action involved in inflammatory and other pathological processes. Prostaglandin reductases (PGRs or PTGRs) are key enzymes in the irreversible inactivation of these eicosanoids and are represented by three members named PGR1, PGR2 or ZADH1, and PGR3 or ZADH2 (Mesa et al., 2015). They are zinc-independent members of the medium-chain dehydrogenase/reductase (MDR) superfamily and in animals they are able to reduce 15-keto-PGE_2 to the inactive 13,14-dehydro-15-keto-PGE_2, with different tissue specificities (Mesa et al., 2015).

Table 10 List of dinoflagellate species, in the iMicrobe database, whose transcriptomes contained a transcript associated to the Prostaglandin F synthase.

Genus	Species	MMETSP code	Treatment	PTGFS type
Amphidinium	*massartii*	0689	No silica; Nitrate 645; Light 80; day portion 12 h; phosphate 25 μmol/kg; exp salinity 35	PGE_2-9-OR
Glenodinium	*foliaceum*	0118	F/2-Si amended sea water; 48 h before harvest antibiotics were added to medium: 500 mg/L penicillin; 200 mg/L ampicillin; 100 mg/L streptomycin; 100 mg/L neomycin; Nitrate 882; Light 550; day portion 12 h; phosphate 36.2 μmol/kg; exp salinity 21	$PTGFS_2$
Glenodinium	*foliaceum*	0119	F/2-Si amended sea water; 48 h before harvest antibiotics were added to medium: 500 mg/L penicillin; 200 mg/L ampicillin; 100 mg/L streptomycin; 100 mg/L neomycin; Nitrate 882; night portion 48 h; phosphate 36.2 μmol/kg; exp salinity 21	$PTGFS_1$
Karenia	*brevis*	0201	f/2; Nitrate 883; Light 50; day portion 16 h; phosphate 36 μmol/kg; exp salinity 36	PGE_2-9-OR
Karenia	*brevis*	0027	F/2; Nitrate 60; Light 60; day portion 12 h; phosphate 60 μmol/kg; exp salinity 32	PGE_2-9-OR
Kryptoperidinium	*foliaceum*	0121	F/2-Si amended sea water; 48 h before harvest antibiotics were added to medium: 500 mg/L penicillin; 200 mg/L ampicillin; 100 mg/L streptomycin; 100 mg/L neomycin; Nitrate 882; night portion 48 h; phosphate 36.2 μmol/kg; exp salinity 21	$PTGFS_1$
Lingulodinium	*polyedra*	1035	No silica; Nitrate 882; Light 150; day portion 12 h; phosphate 36.2 μmol/kg; exp salinity 28	PGE_2-9-OR
Scrippsiella	*hangoei*	0359	MWC; Light 20; day portion 12 h; Selenium added to a final concentration of 4.55 nmol/L; nitrate 1000; silicate 100 μmol/kg; phosphate 50 μmol/kg; env salinity 6.5	PGE_2-9-OR $PTGFS_1$
Scrippsiella	*hangoei*	0361	f/2 amended sea water without silica; Light 20; day portion 12 h; nitrate 883; silicate 100 μmol/kg; phosphate 36 μmol/kg; env salinity 6.5; exp salinity 30	PGE_2-9-OR $PTGFS_1$
Scrippsiella	*hangoei*	0360	f/2 no silica; Light 20; day portion 12 h; silicate 100 μm/kg; nitrate 883; phosphate 36 μmol/kg; env salinity 6.5; exp salinity 3	PGE_2-9-OR

For each specie is reported the MMETSP code, the growth conditions and the annotation type.

Besides PGRs, 15-hydroxyprostaglandin dehydrogenases (15-PGDHs) type I and II, being able to oxidise the Pgs' 15(S)-hydroxyl group, are considered key enzymes for the biological inactivation of Pgs and related eicosanoids. Indeed, the 16-ketoprostaglandins has a greatly reduced biological activities. The two isoforms need NAD+ as cofactor but differ by the possibility of isoform II to utilize also NADP+ and by their specificity of action. The isoform II has a broader substrate specificity, while the isoform I, which is more specific for prostaglandins and eicosanoids, is also involved in the Pgs/eicosanoid activity control (Tai, Ensor, Tong, Zhou, & Yan, 2002).

These enzymes have not been found in diatoms, while they were found in dinoflagellate transcriptomes. PGR I is the one with a wider distribution being expressed in 24 transcriptomes corresponding to 14 species. On the contrary, PGR II was found to be expressed only in 7 transcriptomes corresponding to 4 species, and 15-PGDH was expressed only in 2 transcriptomes corresponding to one species (Di Dato, Ianora, & Romano, 2020). 15-HPGD was expressed in two clones of *K. brevis*, i.e. Wilson and SP3, with a major expression in the clone Wilson with respect to the clone SP3. PTGR2 was expressed in *A. margalefi*, *G. foliaceum*, *N. scintillans* and some clones of *K. brevis*, one of which showed the highest expression value. *P. beii* was the species with the highest in silico expression of PTGR1 among the 24 species tested (Fig. 8).

Low salinity seems to stimulate the expression of PTGR1 and PTGR2 in the clone SP1 of *K. brevis*, as well as in *S. hangoei* for PTGR1 only. On the other side, in another *K. brevis* clone, the Wilson clone, PTGR2 was more expressed compared to PGR1, especially at normal salinities percentage (35ppm), similarly to the SP3 clone of *K. brevis* where PTGR2 was more expressed than PTGR1 and 15-HPGD at normal salinity. At lower salinities PTGR1 was the one most expressed (Fig. 8).

2.6 Prostaglandin transporter

The analysis of the *S. marinoi* transcriptome identified a coding sequence whose functional annotation corresponded to "Prostaglandin transporter" (Di Dato et al., 2017).

The clearance of Pgs outside the cell represents a crucial step to end their signaling activity. In 1995, it was isolated a protein with 12 hydrophobic domains, defined as a prostaglandin transporter (PGT) that could mediate the release of newly synthesized Pgs from cells, the transepithelial transport of Pgs, and the clearance of Pgs from the circulation (Kanai et al., 1995).

Subsequent studies, identified PGT as a lactate/Pg exchanger co-ordinately regulated with COX, being expressed in cyclooxygenase positive cells. The role of PGT seems to be the regulation of Pg levels in the pericellular environment via re-uptake and re-direction towards and/or away from Pg receptors. Further studies are still needed to better clarify the complete Pg metabolisms (Chi & Schuster, 2010; Schuster, 1998, 2002).

Studies regarding PGT have been performed only in mammals and similarly to the other proteins involved in Pg metabolism there is no information for its presence in plants.

The corresponding transcript was analyzed for its expression by qPCR in the two *S. marinoi* clone, FE7 and FE60, similarly to the Pgs synthetic enzymes (Fig. 11). Interestingly its expression did not change along the growth curve of the clone FE7 but was downregulated by twofold at the senescent and stationary phase in the clone FE60, compared to the exponential phase (Di Dato et al., 2017).

The search in the transcriptome of the *T. rotula* clone isolated from the Gulf of Naples, did not detect any transcript associated to the "Prostaglandin Transporter" function (Di Dato et al., 2019). However, depending on the

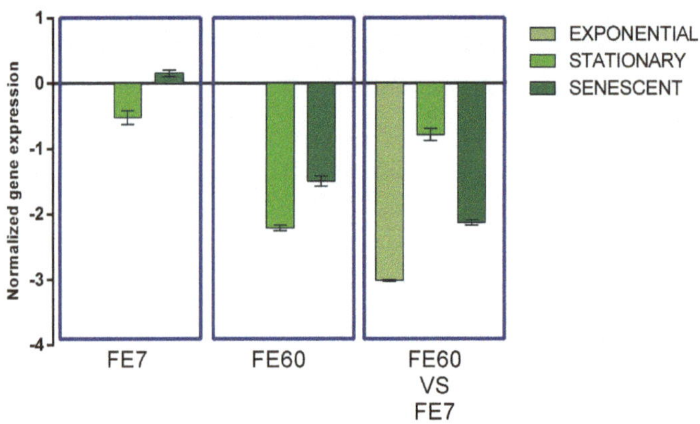

Fig. 11 S. *marinoi* PGT Gene expression analysis by Real-time qPCR. PGT gene expression levels were measured at the exponential, stationary and senescent phases of the FE7 or FE60 *Skeletonema marinoi* clone cultures growth. Results were analyzed with the REST software and reported as 2 log expression ratio using the exponential phase as reference condition. Comparison between the two strain is also reported as 2 log expression ratio of FE60 with respect to FE7 strain. Relative expression ratios above two folds were considered significant. *From Di Dato, V., Orefice, I., Amato, A., Fontanarosa, C., Amoresano, A., Cutignano, A., Ianora, A., & Romano, G. (2017). Animal-like prostaglandins in marine microalgae.* The ISME Journal, 11*(7), 1722–1726. https://doi.org/10.1038/ismej.2017.27.*

growing conditions, this transcript was found in the annotation table of other *T. rotula* clones. Similarly to the other enzymes in the pathway, many species showed the expression of PGT to vary among different clones of a same species and depending on the growing conditions (Table 7) (Di Dato et al., 2017). Also the search in the dinoflagellates transcriptome didn't retrieve any "Prostaglandin transporter" annotated function (Di Dato, Ianora, & Romano, 2020).

Blastp of *S. marinoi* FE7 PGT protein sequence in the NCBI database retrieved similarity with the broad ABC transporter family which includes less specific transporters. For this reason, Bayesian analysis of the PTGT sequence was restricted only to diatom sequences, retrieved by a blastp in a database containing only diatom transcriptomes and showing a low e-value (indicating high rate of similarity). The analyses showed four main clades with some diatom genera represented in all clades, while some others appearing in only one (*Cyclotella, Amphora, Coscinodiscus, Astrosyne, Cyclophora, Nitzschia, Attheya, Striatella*), two (*Staurosira, Grammatophora, Dactyliosolen, Odontella, Aulacoseira, Thalassiotrix, Minutocellulus*) or three clades (*Amphiprora, Rhizosolenia, Asterionellopsis, Licmophora, Thalassionema, Stephanopyxis, Proboscia, Letpocylindrus* (Di Dato et al., 2017).

3. Occurrence of prostaglandin molecules in Bacillariophyceae

In animals, were Pgs have been mostly studied, PGD_2, PGE_2, PGI_2, $PGF_{2\alpha}$ are the principal identified prostanoids, each one induced by a specific stimulus and synthesized from a specific cell type (Versteeg et al., 1999).

In microalgae, Pgs occurrence started to be investigated only in 2017 and have been analyzed in only two diatom species, *S. marinoi* and *T. rotula* (Di Dato et al., 2017; Di Dato, Barbarinaldi, et al., 2020). In the same period, in *Phaeodactylum tricornutum*, it was discovered a non-enzymatic formation of isomers of prostanoids, also called isoprostanoids, such as PhytoPs, F_{2t}-isoprostanes (F_{2t}-IsoPs), F_{3t}-isoprostanes (F_{3t}-IsoPs) and F_{4t}-neuroprostanes (Lupette et al., 2018).

In *S. marinoi* the Pg content was measured inside the cells during the growth of the algae. The identified metabolites derived not only from the most abundant PUFA precursor in diatoms, EPA (Stonik & Stonik, 2015), but also from eicosatrienoic (ETE) and AA acids, both found in very low amounts in diatoms (Table 11) (d'Ippolito et al., 2004). Quantitative

Table 11 List of the prostaglandin molecules identified intracellularly in the FE7 and FE60 clones of *Skeletonema marinoi* and their concentration measured during the exponential, stationary and senescence phases of the cultures growth.

Pgs (ng μg prot^{-1})	FE7 exponential	FE7 stationary	FE7 senescence	FE60 exponential	FE60 stationary	FE60 senescence
PGE_1	1.09 ± 1.89	2.49 ± 0.60	–	3.18 ± 0.82	2.11 ± 0.01	–
PGD_1	2.82 ± 1.45	–	–	1.06 ± 0.27	–	–
$PGF_{1\alpha}$	–	1.13 ± 0.27	–	–	1.34 ± 0.03	–
6k PGE_1	0.37 ± 0.64	0.75 ± 0.17	–	–	–	–
PGE_2	–	11.62 ± 2.73	8.14 ± 0.79	–	7.64 ± 0.13	–
PGD_2	–	2.81 ± 2.15	3.53 ± 0.30	2.18 ± 0.59	3.32 ± 0.02	–
$PGF_{2\alpha}$	7.37 ± 2.08	3.8 ± 0.87	–	2.68 ± 0.68	1.44 ± 0.03	1.41 ± 0.11
PGA_2	–	6.08 ± 1.42	1.71 ± 0.17	–	4.99 ± 0.05	–
PGB_2	–	11.70 ± 2.75	4.7 ± 0.40	–	9.44 ± 0.18	–
PGJ_2	–	1.44 ± 0.51	–	1.77 ± 0.56	1.65 ± 0.02	–
PGK_2	5.18 ± 2.01	–	–	–	0.52 ± 0.06	–
11b PGF_2	–	–	–	2.06 ± 1.02	–	–
15d PGA_2	–	5.29 ± 1.25	5.95 ± 0.47	–	5.46 ± 0.07	–
15d PGD_2	–	0.67 ± 0.15	–	–	–	–
15d PGJ_2	3.20 ± 0.73	1.01 ± 0.23	2.51 ± 0.21	–	1.36 ± 0.03	–
15k PGD_2	–	0.97 ± 0.99	–	2.36 ± 0.35	–	–

15k PGE_2	–	6.62 ± 1.54	2.09 ± 0.18	–	3.57 ± 0.03	–
15k $PGF_{2\alpha}$	–	8.02 ± 1.80	1.73 ± 0.14	–	6.90 ± 0.11	–
dhk PGE_2	–	1.32 ± 0.30	30.90 ± 2.70	–	1.80 ± 0.02	–
dhk-PGD_2	–	0.77 ± 0.17	–	–	–	–
dh-$PGF_{2\alpha}$	1.73 ± 1.39	11.66 ± 2.59	–	–	6.41 ± 0.10	–
dhomo-$PGF_{2\alpha}$	–	2.57 ± 0.57	–	–	2.65 ± 0.007	–
dihomo-PGJ_2	11.35 ± 3.18	3.78 ± 1.68	–	4.10 ± 1.02	2.57 ± 0.001	1.51 ± 0.09
PGE_3	17.14 ± 5,01	4.8 ± 1.08	9.33 ± 0.76	6.19 ± 1.67	6.84 ± 0.11	1.56 ± 0.11
PGD_3	–	1.15 ± 0.51	3.76 ± 0.30	1.55 ± 0.42	–	–
$PGF_{3\alpha}$	–	–	2.47 ± 0.20	–	–	–

analysis revealed a wide variability among replicates and an overall lower production of prostaglandins in the clone FE60 compared to FE7. Molecule quantity decreased during the stationary and senescent phases of growth, especially in the FE7 strain, coherently with the downregulation of cyclooxygenase expression. Interestingly, the major EPA metabolite expected, PGE_3, was at least two and six times more abundant in the FE7 with respect to FE60 during both exponential and senescent phases while in the stationary phase the amounts were comparable (Fig. 12 and Table 11).

In *T. rotula*, the Pg content was measured outside the cells, all along the culture growth from the 3rd day to the 10th, demonstrating also the actual release of the molecules outside the cell. In the culture medium were detected PGE_2, PGB_2 and 15-d-PGJ_2 in addition to PGA_2, PGD_1, PGD_3, PGE_1, PGE_2, PGE_3, PGEM, PGFM, 15-d-PGD_2 and 2,3-dinor-11b-PGF_2 that were classified from the authors as "putative" since they were identified through comparison of experimental mass fragmentation data obtained with those reported in the literature (Table 12). Overall, it was observed a high variability among replicates with an up-and-down trend of each measured Pg concentration from days 3 to 5. Measurements from day 6, corresponding to the full stationary phase, up to day 10, showed a linear increasing trend of Pg concentration, except for PGE_2, PGE_3, 15-d-PGJ_2, and PGFM. PGE_2 had a constant low level reaching zero on day 10, while PGEM showed an exponential trend starting from day 4, i.e. late exponential phase of the growth, reaching the highest value compared to all the other Pgs identified (Fig. 13 and Table 12) (Di Dato, Barbarinaldi, et al., 2020).

These data demonstrate that *Trot*COX is active and function like the animal COX, being able to synthesize the same Pgs produced by animals. The presence of PGE_3, a specific derivative of EPA, is in line with the abundance of this PUFA precursor, which represents a hallmark of diatoms. Interestingly, the general increase of Pgs released outside the cells started from day 5 onwards, i.e., when cells were entering the stationary phase that coincides with the maximum expression of COX. This stage is generally associated with an increase in the naturally associated bacterial population. During the transition between the stationary and senescent phases, nutrients decrease while dying algal cells generate organic matter that can feed associated bacteria. Bacteria are able to assimilate phosphate better than algae, especially at low concentrations, and compete with algae for inorganic nutrients. This increased pressure may stimulate Pgs synthesis in diatoms as already shown in animals, where COX expression is stimulated in the

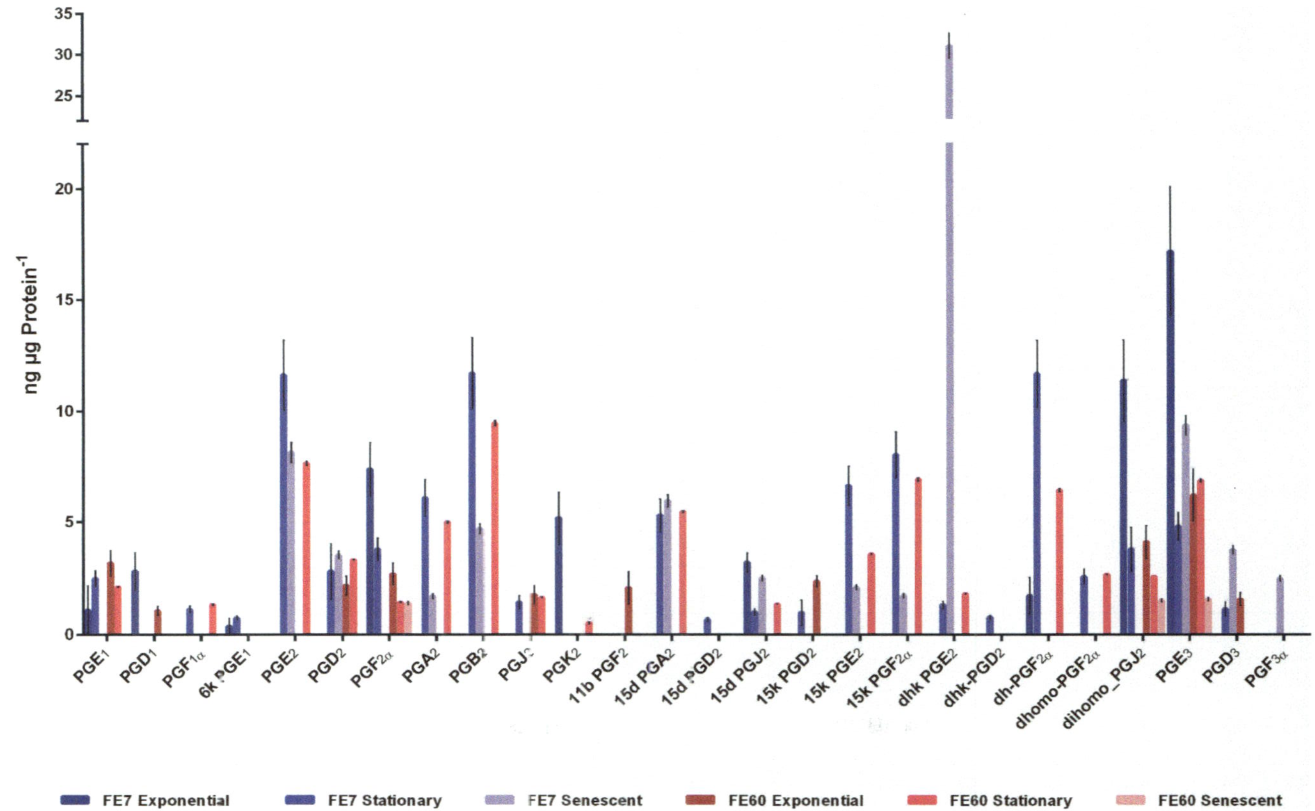

Fig. 12 Prostaglandin molecules in *S. marinoi*. Identification and abundance of the prostaglandin metabolites synthesized by *Skeletonema marinoi* FE7 and FE60 clones in the three different phases of growth (N = 3), quantified with LC/MSMS. Results are means of three technical replicates for each biological replicate (N = 3). *From Di Dato, V., Orefice, I., Amato, A., Fontanarosa, C., Amoresano, A., Cutignano, A., Ianora, A., & Romano, G. (2017). Animal-like prostaglandins in marine microalgae.* The ISME Journal, 11*(7), 1722–1726. https://doi.org/10.1038/ismej.2017.27.*

Table 12 List of *T. rotula* identified prostaglandin molecules and their concentration measured in the spent medium along the cultures growth.

Pgs (pg mL^{-1} cell^{-1})	*T. rotula* Day 3	*T. rotula* Day 4	*T. rotula* Day 5	*T. rotula* Day 6	*T. rotula* Day 7	*T. rotula* Day 10
PGA_2	5.99 ± 7.04	–	4.23 ± 0.48	2.33 ± 2.25	4.11 ± 1.88	10.85 ± 2.87
PGB_2	3.23 ± 3.43	–	2.43 ± 0.06	1.33 ± 1.25	2.16 ± 1.01	6.53 ± 1.71
PGD_1	1.85 ± 1.67	0.94 ± 0.43	0.587 ± 0.51	0.85 ± 0.472	1.08 ± 0.38	2.93 ± 0.75
PGD_3	2.94 ± 3.37	–	1.99 ± 0.22	0.56 ± 0.15	2.11 ± 1.30	5.72 ± 0.88
PGE_1	1.88 ± 1.71	0.96 ± 0.45	0.41 ± 0.41	0.86 ± 0.47	1.07 ± 0.37	2.99 ± 0.68
PGE_2	5.92 ± 5.52	0.50 ± 0.48	0.93 ± 0.82	0.47 ± 0.05	0.82 ± 0.76	–
PGE_3	2.81 ± 3.25	–	2.08 ± 0.23	0.54 ± 0.13	2.22 ± 1.22	4.20 ± 3.66
PGEM	6.80 ± 4.63	0.43 ± 0.45	1.27 ± 1.27	2.18 ± 1.04	11.95 ± 6.74	31.82 ± 4.87
PGFM	20.85 ± 21.46	6.10 ± 4.88	8.04 ± 1.68	5.06 ± 4.41	6.11 ± 3.12	10.94 ± 5.63
15-d-PGD_2	5.18 ± 6.10	–	2.80 ± 0.77	1.96 ± 1.88	3.34 ± 1.27	9.78 ± 2.56
15-d-PGJ_2	6.78 ± 7.49	0.29 ± 0.50	3.08 ± 1.82	1.65 ± 1.53	2.68 ± 1.79	9.67 ± 3.95
2,3-dinor-11b-PGF_2	2.18 ± 2.34	0.096 ± 0.17	0.43 ± 0.74	0.38 ± 0.36	1.76 ± 0.59	3.65 ± 1.62

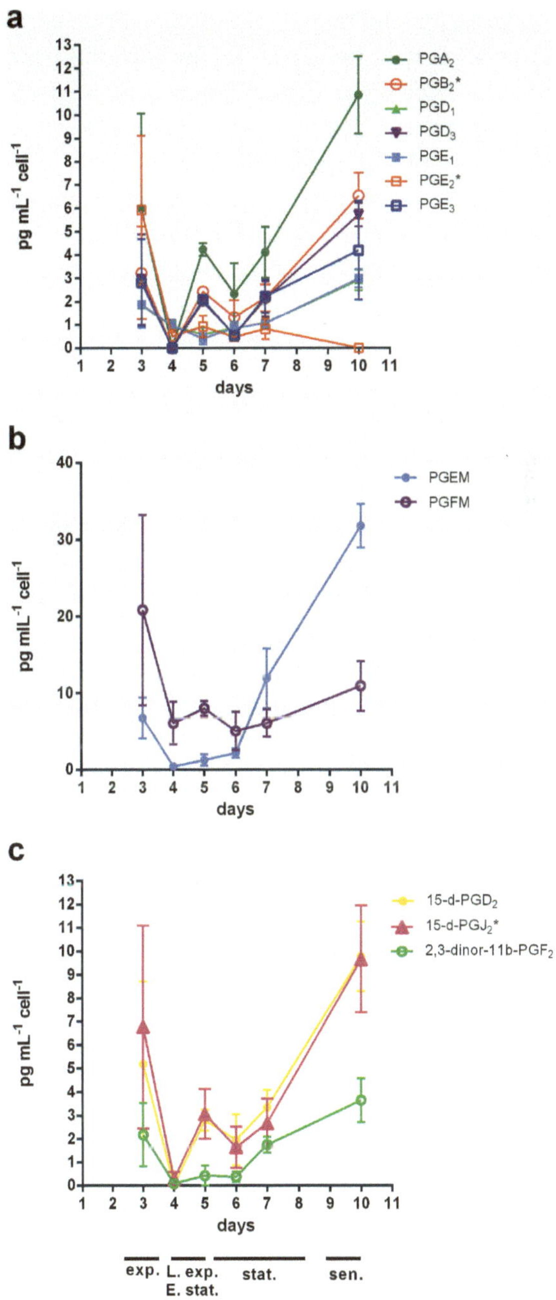

Fig. 13 Prostaglandin molecules in *T. rotula*. Identification and abundance of the prostaglandin metabolites, by LC/MSMS, released by *T. rotula* in the spent medium during the culture growth. Results are means of three technical replicates (N = 3). *From Di Dato, V., Barbarinaldi, R., Amato, A., Di Costanzo, F., Fontanarosa, C., Perna, A., Amoresano, A., Esposito, F., Cutignano, A., Ianora, A., & Romano, G. (2020). Variation in prostaglandin metabolism during growth of the diatom Thalassiosira rotula.* Scientific Reports, 10*(1), 5374. https://doi.org/10.1038/s41598-020-61967-3.*

presence of bacteria, and in humans PGE_2 has been correlated to viral load and infection severity in influenza. In line with this hypothesis, similar compounds, such as hydroxylated eicosapentaenoic acid 15-HEPE, were shown to be up-regulated and released from cells when the diatom *S. marinoi* was exposed to pathogenic bacteria.

PGE_2 is the most abundant prostanoid in the human body, but it is also very unstable since it is rapidly converted into PGE-M that is considered to mirror the systemic levels of PGE_2 formation. PGE_2 was present at low levels in the medium during the exponential growth phase and was almost absent in the late stationary phase while PGE-M levels increased exponentially up to the senescent phase. The peak in the expression of the related enzyme, PTGES, at the onset of the stationary phase and the sustained release of Pgs afterwards suggested a possible role of prostaglandins in communication and cell signaling. Their release outside the cells, in addition to their sustained presence in a saline environment, is quite striking and has never been observed before in diatoms. Pgs are known to exert a wide range of effects in different organisms, including the induction of inflammatory processes, injury and pain in humans where they have been best studied. The fact that they have been identified in organisms ranging from unicellular diatoms, corals and jellyfish to arthropods, mollusks and mammals denotes that they regulate important physiological processes that have been conserved through evolution (Di Dato, Barbarinaldi, et al., 2020).

References

Biringer, R. G. (2020). The enzymology of the human prostanoid pathway. *Molecular Biology Reports*, *47*(6), 4569–4586. https://doi.org/10.1007/s11033-020-05526-z.

Bosisio, E. (1988). Prostaglandins and prostaglandin-like compounds in plants. *Pharmacological Research Communications*, *20*, 99–103. https://doi.org/10.1016/S0031-6989(88)80851-8.

Bouarab, K., Adas, F., Gaquerel, E., Kloareg, B., Salaün, J.-P., & Potin, P. (2004). The innate immunity of a marine red alga involves oxylipins from both the eicosanoid and octadecanoid pathways. *Plant Physiology*, *135*(3), 1838–1848. https://doi.org/10.1104/pp.103.037622.

Brash, A. R., Niraula, N. P., Boeglin, W. E., & Mashhadi, Z. (2014). An ancient relative of cyclooxygenase in cyanobacteria is a linoleate 10S-dioxygenase that works in tandem with a catalase-related protein with specific 10S-hydroperoxide lyase activity. *The Journal of Biological Chemistry*, *289*(19), 13101–13111. https://doi.org/10.1074/jbc.M114.555904.

Chi, Y., & Schuster, V. L. (2010). The prostaglandin transporter PGT transports PGH(2). *Biochemical and Biophysical Research Communications*, *395*(2), 168–172. https://doi.org/10.1016/j.bbrc.2010.03.108.

Clària, J. (2003). Cyclooxygenase-2 biology. *Current Pharmaceutical Design*, *9*(27), 2177–2190.

Crofford, L. J. (2001). Prostaglandin biology. *Gastroenterology Clinics of North America*, *30*(4), 863–876. https://doi.org/10.1016/s0889-8553(05)70217-x.

d'Ippolito, G., Tucci, S., Cutignano, A., Romano, G., Cimino, G., Miralto, A., et al. (2004). The role of complex lipids in the synthesis of bioactive aldehydes of the marine diatom *Skeletonema costatum*. *Biochimica Et Biophysica Acta*, *1686*(1–2), 100–107. https://doi.org/10.1016/j.bbalip.2004.09.002.
Di Dato, V., Barbarinaldi, R., Amato, A., Di Costanzo, F., Fontanarosa, C., Perna, A., et al. (2020). Variation in prostaglandin metabolism during growth of the diatom *Thalassiosira rotula*. *Scientific Reports*, *10*(1), 5374. https://doi.org/10.1038/s41598-020-61967-3.
Di Dato, V., Di Costanzo, F., Barbarinaldi, R., Perna, A., Ianora, A., & Romano, G. (2019). Unveiling the presence of biosynthetic pathways for bioactive compounds in the *Thalassiosira rotula* transcriptome. *Scientific Reports*, *9*(1), 9893. https://doi.org/10.1038/s41598-019-46276-8.
Di Dato, V., Ianora, A., & Romano, G. (2020). Identification of prostaglandin pathway in dinoflagellates by transcriptome data mining. *Marine Drugs*, *18*(2). https://doi.org/10.3390/md18020109.
Di Dato, V., Orefice, I., Amato, A., Fontanarosa, C., Amoresano, A., Cutignano, A., et al. (2017). Animal-like prostaglandins in marine microalgae. *The ISME Journal*, *11*(7), 1722–1726. https://doi.org/10.1038/ismej.2017.27.
Fusetani, N., & Hashimoto, K. (1984). Prostaglandin E2: A candidate for causative agent of "ogonori" poisoning. *Nippon Suisan Gakkaishi*, *50*(3), 465–469. https://doi.org/10.2331/suisan.50.465.
Gaquerel, E., Hervé, C., Labrière, C., Boyen, C., Potin, P., & Salaün, J.-P. (2007). Evidence for oxylipin synthesis and induction of a new polyunsaturated fatty acid hydroxylase activity in *Chondrus crispus* in response to methyljasmonate. *Biochimica et Biophysica Acta (BBA) - Molecular and Cell Biology of Lipids*, *1771*(5), 565–575. https://doi.org/10.1016/j.bbalip.2007.02.007.
Gerecht, A., Romano, G., Ianora, A., d'Ippolito, G., Cutignano, A., & Fontana, A. (2011). Plasticity of oxylipin metabolism among clones of the marine diatom *skeletonema marinoi* (bacillariophyceae). *Journal of Phycology*, *47*(5), 1050–1056. https://doi.org/10.1111/j.1529-8817.2011.01030.x.
Giera, M., & Galano, J. M. (2019). Eicosanoids. In P. Worsfold, C. Poole, A. Townshend, & M. Miró (Eds.), *Encyclopedia of analytical science* (3rd ed., pp. 259–263). Academic Press. https://doi.org/10.1016/B978-0-12-409547-2.13984-8.
Groenewald, E. G., & van der Westhuizen, A. J. (1997). Prostaglandins and related substances in plants. *The Botanical Review*, *63*(3), 199–220. https://doi.org/10.1007/BF02857948.
Hara, S. (2017). Prostaglandin terminal synthases as novel therapeutic targets. *Proceedings of the Japan Academy. Series B, Physical and Biological Sciences*, *93*(9), 703–723. https://doi.org/10.2183/pjab.93.044.
Hara, S., Kamei, D., Sasaki, Y., Tanemoto, A., Nakatani, Y., & Murakami, M. (2010). Prostaglandin E synthases: Understanding their pathophysiological roles through mouse genetic models. *Biochimie*, *92*(6), 651–659. https://doi.org/10.1016/j.biochi.2010.02.007.
Imbs, A. B., Latyshev, N. A., Svetashev, V. I., Skriptsova, A. V., Le, T. T., Pham, M. Q., et al. (2012). Distribution of polyunsaturated fatty acids in red algae of the genus *Gracilaria*, a promising source of prostaglandins. *Russian Journal of Marine Biology*, *38*(4), 339–345. https://doi.org/10.1134/S1063074012040049.
Imbs, A. B., Vologodskaya, A. V., Nevshupova, N. V., Khotimchenko, S. V., & Titlyanov, E. A. (2001). Response of prostaglandin content in the red alga *Gracilaria verrucosa* to season and solar irradiance. *Phytochemistry*, *58*(7), 1067–1072.
Imbusch, R., & Mueller, M. J. (2000a). Formation of isoprostane F(2)-like compounds (phytoprostanes F(1)) from alpha-linolenic acid in plants. *Free Radical Biology & Medicine*, *28*(5), 720–726. https://doi.org/10.1016/s0891-5849(00)00154-4.

Imbusch, R., & Mueller, M. J. (2000b). Analysis of oxidative stress and wound-inducible dinor isoprostanes F(1) (phytoprostanes F(1)) in plants. *Plant Physiology*, *124*(3), 1293–1304. https://doi.org/10.1104/pp.124.3.1293.

Kafanova, T. V., Busarova, N. G., Isai, S. V., & Zvyagintseva, T. Y. (1996). Fatty acids and prostaglandins of thermal cyanobacteria. *Chemistry of Natural Compounds*, *32*(6), 861–865. https://doi.org/10.1007/BF01374015.

Kanai, N., Lu, R., Satriano, J. A., Bao, Y., Wolkoff, A. W., & Schuster, V. L. (1995). Identification and characterization of a prostaglandin transporter. *Science*, *268*(5212), 866–869. https://doi.org/10.1126/science.7754369.

Keeling, P. J., Burki, F., Wilcox, H. M., Allam, B., Allen, E. E., Amaral-Zettler, L. A., et al. (2014). The marine microbial eukaryote transcriptome sequencing project (MMETSP): Illuminating the functional diversity of eukaryotic life in the oceans through transcriptome sequencing. *PLoS Biology*, *12*(6), e1001889. https://doi.org/10.1371/journal.pbio.1001889.

Korbecki, J., Baranowska-Bosiacka, I., Gutowska, I., & Chlubek, D. (2014). Cyclooxygenase pathways. *Acta Biochimica Polonica*, *61*(4), 639–649.

Krüger, G. H. J., Groenewald, E. G., de Wet, H., & Botes, P. J. (1990). The occurrence of prostaglandin F2α in procaryotic organisms. *South African Journal of Botany*, *56*(2), 150–153. https://doi.org/10.1016/S0254-6299(16)31081-X.

Kudo, I., & Murakami, M. (2005). Prostaglandin E synthase, a terminal enzyme for prostaglandin E2 biosynthesis. *Journal of Biochemistry and Molecular Biology*, *38*(6), 633–638. https://doi.org/10.5483/bmbrep.2005.38.6.633.

Lamari, N., Ruggiero, M. V., d'Ippolito, G., Kooistra, W. H. C. F., Fontana, A., & Montresor, M. (2013). Specificity of lipoxygenase pathways supports species delineation in the marine diatom genus *Pseudo-nitzschia*. *PLoS One*, *8*(8), e73281. https://doi.org/10.1371/journal.pone.0073281.

Levine, L., Sneiders, A., Kobayashi, T., & Schiff, J. A. (1984). Serologic and immunochromatographic detection of oxygenated polyenoic acids in *Euglena gracilis var. Bacillaris*. *Biochemical and Biophysical Research Communications*, *120*(1), 278–285.

Lupette, J., Jaussaud, A., Vigor, C., Oger, C., Galano, J.-M., Réversat, G., et al. (2018). Non-enzymatic synthesis of bioactive isoprostanoids in the diatom *Phaeodactylum* following oxidative stress. *Plant Physiology*, *178*(3), 1344–1357. https://doi.org/10.1104/pp.18.00925.

Markovič, T., Jakopin, Ž., Dolenc, M. S., & Mlinarič-Raščan, I. (2017). Structural features of subtype-selective EP receptor modulators. *Drug Discovery Today*, *22*(1), 57–71. https://doi.org/10.1016/j.drudis.2016.08.003.

Mesa, J., Alsina, C., Oppermann, U., Parés, X., Farrés, J., & Porté, S. (2015). Human prostaglandin reductase 1 (PGR1): Substrate specificity, inhibitor analysis and site-directed mutagenesis. *Chemico-Biological Interactions*, *234*, 105–113. https://doi.org/10.1016/j.cbi.2015.01.021.

Murakami, M., Nakatani, Y., Tanioka, T., & Kudo, I. (2002). Prostaglandin E synthase. *Prostaglandins & Other Lipid Mediators*, *68–69*, 383–399. https://doi.org/10.1016/s0090-6980(02)00043-6.

Murakami, M., Naraba, H., Tanioka, T., Semmyo, N., Nakatani, Y., Kojima, F., et al. (2000). Regulation of prostaglandin E2 biosynthesis by inducible membrane-associated prostaglandin E2 synthase that acts in concert with cyclooxygenase-2. *The Journal of Biological Chemistry*, *275*(42), 32783–32792. https://doi.org/10.1074/jbc.M003505200.

Nylund, G. M., Weinberger, F., Rempt, M., & Pohnert, G. (2011). Metabolomic assessment of induced and activated chemical defence in the invasive red alga *Gracilaria vermiculophylla*. *PLoS One*, *6*(12), e29359. https://doi.org/10.1371/journal.pone.0029359.

Panossian, A. G. (1987). Search for prostaglandins and related compounds in plants: A review of the occurrence of prostaglandins and prostaglandin-like compounds in plants. *Prostaglandins*, *33*(3), 363–381. https://doi.org/10.1016/0090-6980(87)90019-0.

Park, J. Y., Pillinger, M. H., & Abramson, S. B. (2006). Prostaglandin E2 synthesis and secretion: The role of PGE2 synthases. *Clinical Immunology (Orlando, FL)*, *119*(3), 229–240. https://doi.org/10.1016/j.clim.2006.01.016.

Piper, P. J. (1977). Introduction to the biosynthesis and metabolism of prostaglandins. *Postgraduate Medical Journal*, *53*(625), 643–646.

Rajakariar, R., Yaqoob, M. M., & Gilroy, D. W. (2006). COX-2 in inflammation and resolution. *Molecular Interventions*, *6*(4), 199–207. https://doi.org/10.1124/mi.6.4.6.

Ritter, A., Goulitquer, S., Salaün, J.-P., Tonon, T., Correa, J. A., & Potin, P. (2008). Copper stress induces biosynthesis of octadecanoid and eicosanoid oxygenated derivatives in the brown algal kelp *Laminaria digitata*. *The New Phytologist*, *180*(4), 809–821. https://doi.org/10.1111/j.1469-8137.2008.02626.x.

Sajiki, J. (1997). Effects of acetic acid treatment on the concentrations of arachidonic acid and prostaglandin E_2 in the red algae, *Gracilaria asiatica* and *G. rhodocaudata*. *Fisheries Science*, *63*(1), 128–131. https://doi.org/10.2331/fishsci.63.128.

Sajiki, J., & Kakimi, H. (1998). Identification of eicosanoids in the red algae, *Gracilaria asiatica*, using high-performance liquid chromatography and electrospray ionization mass spectrometry. *Journal of Chromatography. A*, *795*(2), 227–237.

Sampey, A. V., Monrad, S., & Crofford, L. J. (2005). Microsomal prostaglandin E synthase-1: The inducible synthase for prostaglandin E2. *Arthritis Research & Therapy*, 7(3), 114–117. https://doi.org/10.1186/ar1748.

Schuster, V. L. (1998). Molecular mechanisms of prostaglandin transport. *Annual Review of Physiology*, *60*, 221–242. https://doi.org/10.1146/annurev.physiol.60.1.221.

Schuster, V. L. (2002). Prostaglandin transport. *Prostaglandins & Other Lipid Mediators*, *68–69*, 633–647.

Simmons, D. L., Botting, R. M., & Hla, T. (2004). Cyclooxygenase isozymes: The biology of prostaglandin synthesis and inhibition. *Pharmacological Reviews*, *56*(3), 387–437. https://doi.org/10.1124/pr.56.3.3.

Stonik, V., & Stonik, I. (2015). Low-molecular-weight metabolites from diatoms: Structures, biological roles and biosynthesis. *Marine Drugs*, *13*(6), 3672–3709. https://doi.org/10.3390/md13063672.

Straus, D. S., & Glass, C. K. (2001). Cyclopentenone prostaglandins: New insights on biological activities and cellular targets. *Medicinal Research Reviews*, *21*(3), 185–210.

Tai, H.-H., Ensor, C. M., Tong, M., Zhou, H., & Yan, F. (2002). Prostaglandin catabolizing enzymes. *Prostaglandins & Other Lipid Mediators*, *68–69*, 483–493.

Tanabe, T., & Tohnai, N. (2002). Cyclooxygenase isozymes and their gene structures and expression. *Prostaglandins & Other Lipid Mediators*, *68–69*, 95–114. https://doi.org/10.1016/s0090-6980(02)00024-2.

Urade, Y., & Hayaishi, O. (2000). Prostaglandin D synthase: Structure and function. *Vitamins and Hormones*, *58*, 89–120. https://doi.org/10.1016/s0083-6729(00)58022-4.

Urade, Y., Watanabe, K., & Hayaishi, O. (1995). Prostaglandin D, E, and F synthases. *Journal of Lipid Mediators and Cell Signalling*, *12*(2–3), 257–273. https://doi.org/10.1016/0929-7855(95)00032-l.

Varvas, K., Kasvandik, S., Hansen, K., Järving, I., Morell, I., & Samel, N. (2013). Structural and catalytic insights into the algal prostaglandin H synthase reveal atypical features of the first non-animal cyclooxygenase. *Biochimica Et Biophysica Acta*, *1831*(4), 863–871. https://doi.org/10.1016/j.bbalip.2012.11.010.

Versteeg, H. H., van Bergen en Henegouwen, P. M., van Deventer, S. J., & Peppelenbosch, M. P. (1999). Cyclooxygenase-dependent signalling: Molecular events and consequences. *FEBS Letters*, *445*(1), 1–5. https://doi.org/10.1016/s0014-5793(99)00105-2.

Wang, R., & Shimizu, Y. (1990). Bacillariolides I and II, a new type of cyclopentane eicosanoids from the diatom *Nitzschia pungens*. *Journal of the Chemical Society, Chemical Communications*, *5*, 413–414. https://doi.org/10.1039/C39900000413.

Watanabe, K. (2002). Prostaglandin F synthase. *Prostaglandins & Other Lipid Mediators*, *68–69*, 401–407. https://doi.org/10.1016/s0090-6980(02)00044-8.

Wolfe, L. S. (1982). Eicosanoids: Prostaglandins, thromboxanes, leukotrienes, and other derivatives of carbon-20 unsaturated fatty acids. *Journal of Neurochemistry*, *38*(1), 1–14. https://doi.org/10.1111/j.1471-4159.1982.tb10847.x.

Yang, P., Jiang, Y., & Fischer. (2014). Prostaglandin E3 metabolism and cancer. *Cancer Letters*, *348*(1–2), 1–11. https://doi.org/10.1016/j.canlet.2014.03.010. 2014.

Zámocký, M., Hofbauer, S., Schaffner, I., Gasselhuber, B., Nicolussi, A., Soudi, M., et al. (2015). Independent evolution of four heme peroxidase superfamilies. *Archives of Biochemistry and Biophysics*, *574*, 108–119. https://doi.org/10.1016/j.abb.2014.12.025.

CHAPTER ELEVEN

Alpha-linolenic acid, phytoprostanes and phytofurans in plant, algae and food

Kin Sum Leung[a], Camille Oger[b], Alexandre Guy[b], Valérie Bultel-Poncé[b], Claire Vigor[b], Thierry Durand[b], Angel Gil-Izquierdo[c], Sonia Medina[c], Jean-Marie Galano[b], and Jetty Chung-Yung Lee[a,*]

[a]School of Biological Sciences, The University of Hong Kong, Pokfulam, Hong Kong SAR
[b]Institut des Biomolécules Max Mousseron, IBMM, Université de Montpellier, CNRS, ENSCM Batiment Balard, Montpellier, France
[c]Research Group on Quality, Safety and Bioactivity of Plant Foods, Department of Food Science and Technology, CEBAS (CSIC), 30100 Campus University, Murcia, Spain
*Corresponding author: e-mail address: jettylee@hku.hk

Contents

Abstract

Alpha-linolenic acid (ALA) is abundant in plants and essential in human diet. Due to the occurrence of skipped diene in the structure, it can steadily become oxidized non-enzymatically in plants and human, and generate phytoprostanes (PhytoPs) and phytofurans (PhytoFs). Insofar, there are Type A-, B-, D-, E-, F-, J- and L-PhytoPs that comprises of different isomers, and two types of PhytoFs; the alkenyl and enediol sub-structures identified by analysts and synthesized by organic chemists. In agricultural products such as olives and almonds, these compounds are valuable biomarkers for

ISSN 0065-2296
https://doi.org/10.1016/bs.abr.2021.09.005

adverse environmental change and quality assurance. In plants they regulate immunological functions and take part in gene interaction for growth, while in human, they retain antioxidative and anti-inflammatory properties. This chapter introduces the types of PhytoPs and PhytoFs found in ALA rich plants and algae, and overlooks at some potential biological functions in plants and human.

1. Introduction

Alpha-linolenic acid (ALA) is probably the most abundant and the richest source of polyunsaturated fatty acid with chemical diversity in the plant kingdom, especially as the precursor of plant oxylipins produced by aerobic oxidation (Gerwick, Moghaddam, & Hamberg, 1991). Enzymatically produced oxylipins from this omega-3 polyunsaturated fatty acid (C18:3, n-3 PUFA), which is not in scope of this chapter, are known to be implicated in the development and defenses of plants. The understanding of the non-enzymatic ALA-derived oxylipins is still in progress, but their widespread in plant is surely as important as their enzymatic congeners, while their biological relevance in plant or human food is accumulating. Hereby, we are reviewing the main knowledge of such non-enzymatic oxylipins studied in our groups, *i.e.*, the phytoprostanes (PhytoPs) discovered by Mueller and coworker in 1998 (Parchmann & Mueller, 1998) and the phytofurans (PhytoFs) discovered by us in 2015 (Cuyamendous et al., 2015).

2. Chemical structure and total synthesis of phytoprostanes and phytofurans

The biosynthesis of PhytoPs and PhytoFs originates by the abstraction of H-atom at one of the bisallylic position of ALA in its esterified form. The resulting isomeric pentadienyl radicals (only 2 out of the 4 possible can lead to PhytoP or PhytoF) react with molecular oxygen generating linolenic acid hydroperoxyl radical, which subsequently attack the adjacent double bond following a 5-*exo*-trig cyclization. (Mueller and coworkers initially demonstrated this pathway *via* LOX-induced formation of the corresponding hydroperoxide; Parchmann & Mueller, 1998). The transient 1,2-dioxolanylcarbinyl radicals can then lead to several pathways and the generation of specific type of metabolites. PhytoPs are subsequently formed by following a second 5-exo trig cyclization to produce the well-known bicyclic endoperoxide; thereafter a last oxygen trapping and further reduction and hydrolysis occurs to give access to F-PhytoP. PhytoPs are produced as racemates and in theory all isomers are produced as *cis* configuration of the lateral chains in the biological system. Although it is the backbone of the stereo structure, 1,2-trans configuration can parallelly take place. This has been recently reaffirmed and highlighted by Smrcek et al. (2021). However, for PhytoFs several biosynthetic possibilities still coexist (Fessel, Porter, Moore, Sheller, & Roberts, 2002) while a unified route is based on the known formation of bisepoxide and thus explains the formation of both alkenyl and the enediol PhytoFs (Scheme 1) (Jahn, Galano, & Durand, 2008).

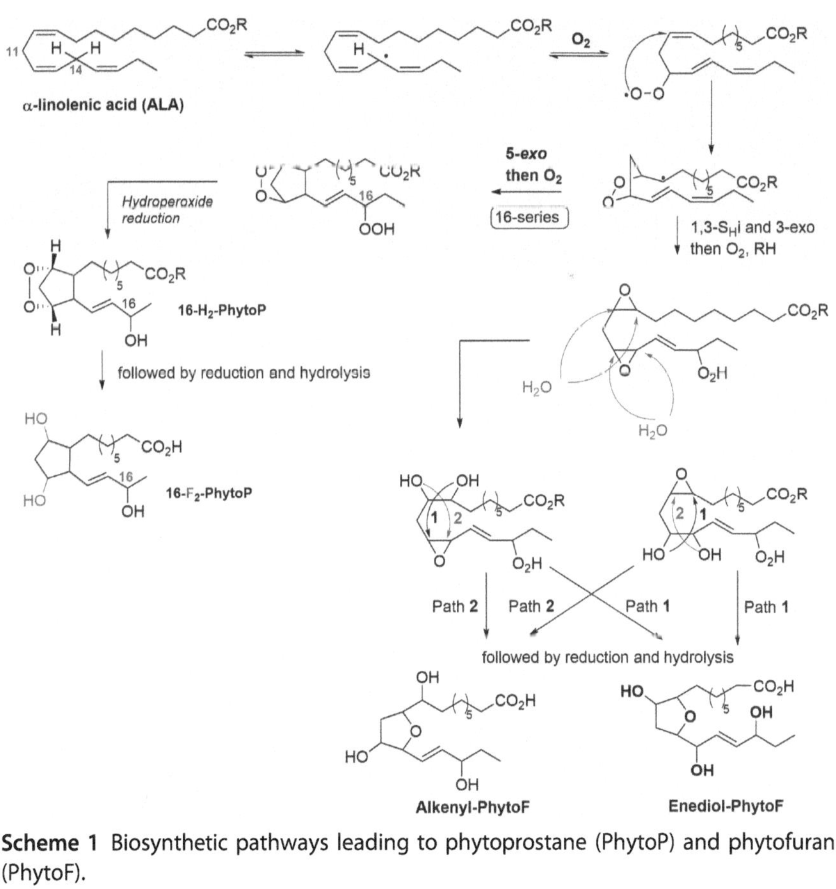

Scheme 1 Biosynthetic pathways leading to phytoprostane (PhytoP) and phytofuran (PhytoF).

2.1 Phytoprostanes

Several types of phytoprostanes have been identified and isolated. They encompass the extent of prostaglandin like structure (Fig. 1) where two series exist per type, the 16th and 9th series. However, a nomenclature update should be consulted to better familiarize with their naming (Jahn, Galano, & Durand, 2010).

2.1.1 F-type

The F-type is one of the most synthesized phytoprostane (Fig. 2), and all isomers of the *syn*-anti-syn stereostructures have been made as racemic or enantiomerically enriched. Two series exist for phytoprostane the 9th and the 16th (Fig. 3) (El Fangour et al., 2004; El Fangour, Guy, Vidal, Rossi, & Durand, 2003).

Recently, the group of Jahn made the prostaglandin version of the 16 series of PhytoP, and they confirmed their presence in edible oils (Smrček et al., 2021).

2.1.2 E- and D-types

As noted in Fig. 4, the 9-D_1- and 16-E_1-PhytoPs made by Durand's and Spur's groups (Jahn et al., 2008) are the less susceptible ketohydroxy cyclopentane to dehydrate or to epimerize, and this is probably the reason why the 16-D_1- and the 9-E_1-PhytoPs have not been synthesized so far.

2.1.3 A- and J-types

The A- and J-type (Fig. 5) have been obtained by the group of Vidari and Zanoni, the 9-J_1-type being obtained as the free acid (Porta et al., 2014).

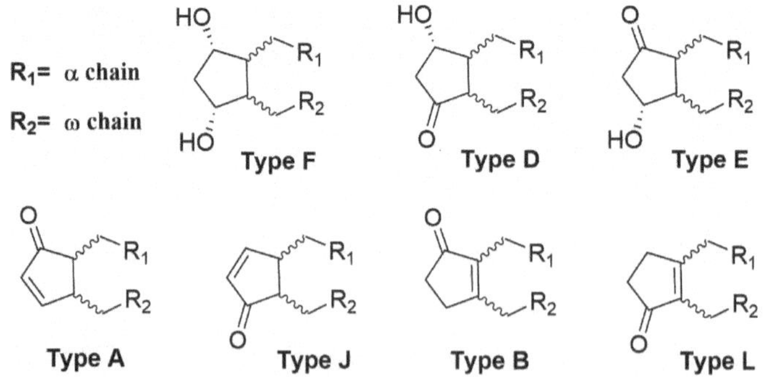

Fig. 1 Representation of different oxidation states of the cyclopentane ring in phytoprostanes.

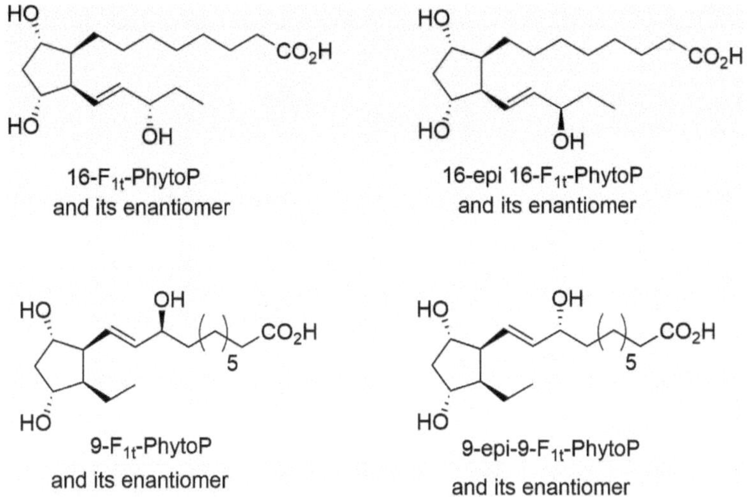

Fig. 2 Available F-type phytoprostane (PhytoP) from Durand's group.

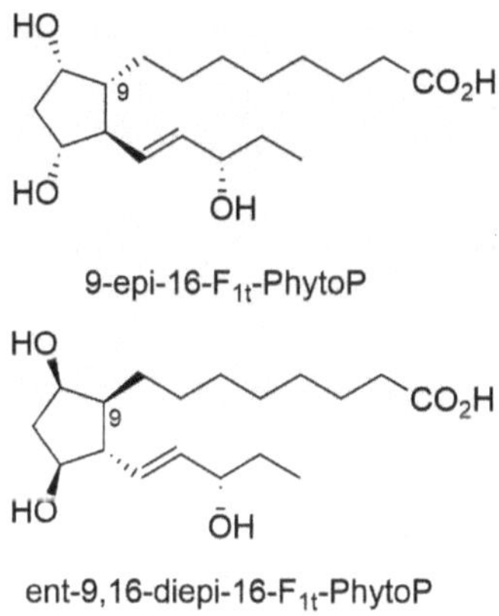

Fig. 3 Prostaglandin isomers of phytoprostane (PhytoP).

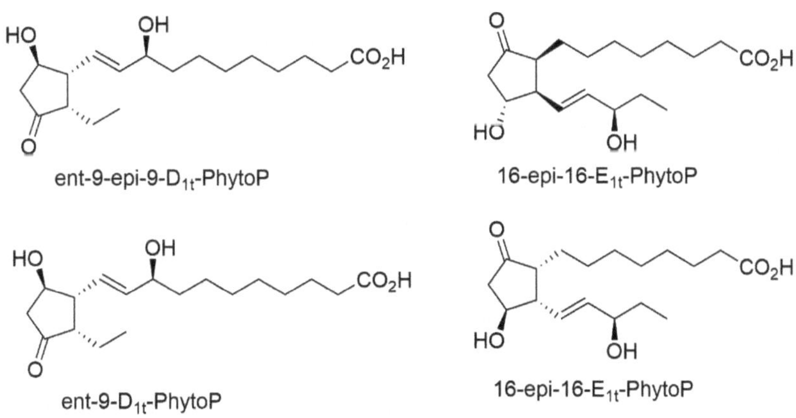

Fig. 4 D_1-Phytoprostane (PhytoP) available from the Durand's group and synthesized E_1-PhytoP from the Spur's group (Jahn et al., 2008).

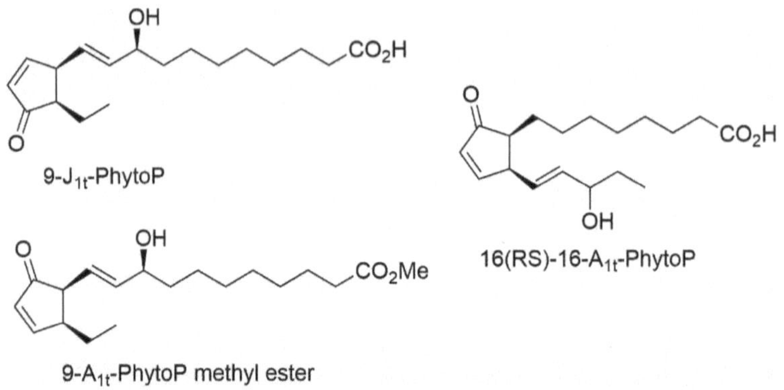

Fig. 5 A_1- and J_1-phytoprostane (PhytoP) synthesized by Vidari and Zanoni's group (Porta et al., 2014), and available at Cayman Chemical®.

Nonetheless, it was problematic to obtain the 9-A_1-PhytoP as a free acid because of decomposition reflecting the high lability of the H-atom at the 12th position. The 16-A_1-PhytoP is now available commercially by Cayman Chemical® (Item No. 9000593).

2.1.4 Deoxy-PhytoP-type

Deoxy phytoprostanes also called dPPJ1 type I and II are produced from dehydration of the 9-A_1-PhytoP and the 16-J_1-PhytoP (Fig. 6), and have been synthesized in 2005 (Iqbal et al., 2005) as their methyl ester.

2.1.5 B- and L-types

Such PhytoP cyclopentenones are the easiest to synthesize (Beretta et al., 2015; Guy, Flanagan, Durand, Oger, & Galano, 2015; Schmidt & Boland, 2007; Vázquez-Romero, Verdaguer, & Riera, 2013) and three out of the four possible were made to date. Durand's group made the 16-B_1-PhytoP and the 9-L_1-PhytoP and their enantiomers. Vidari's team (Beretta et al., 2015) added the 16-L_1-PhytoP, together with the Riera's group (Vázquez-Romero et al., 2013) (Fig. 7).

2.2 Phytofurans

The only available phytofurans were made by the group of Durand during the PhD thesis of Dr. Claire Cuyamendous (Cuyamendous et al., 2015). Three structures were obtained and have been investigated in lipidomic studies (Fig. 8).

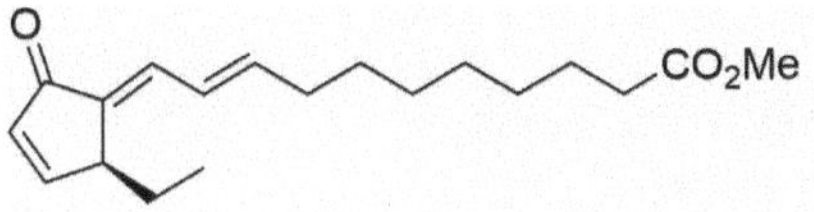

dPPJ$_1$-II
9-deoxy-$\Delta^{9,11}$-A$_{1t}$-PhytoP methyl ester

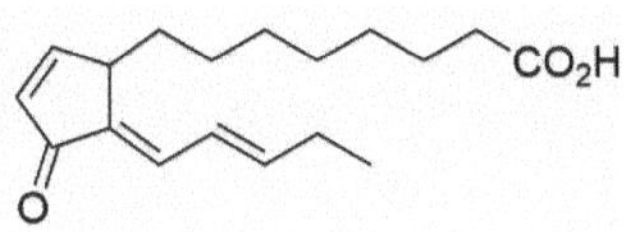

dPPJ$_1$-I
16-deoxy-$\Delta^{13,15}$-J$_1$-PhytoP-

Fig. 6 Deoxy phytoprostane (PhytoP) synthesized (Iqbal et al., 2005).

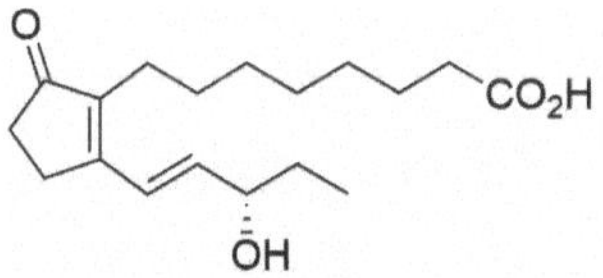

16-B$_1$-PhytoP
and its enantiomer

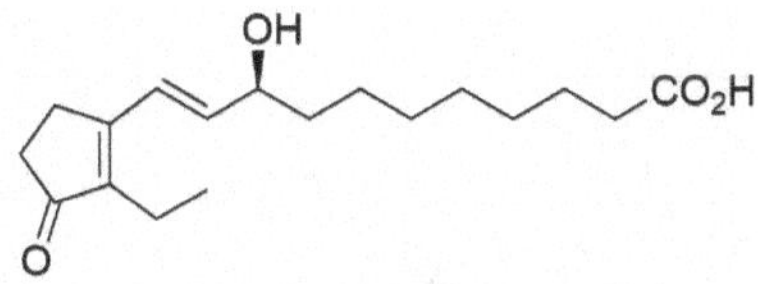

9-L$_1$-PhytoP
and its enantiomer

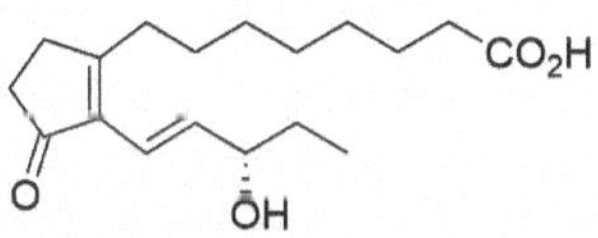

16-L$_1$-PhytoP

16-L$_1$-PhytoP not available

Fig. 7 Available cyclopentenone phytoprostane (PhytoP) (Beretta et al., 2015; Guy et al., 2015; Schmidt & Boland, 2007; Vázquez-Romero et al., 2013).

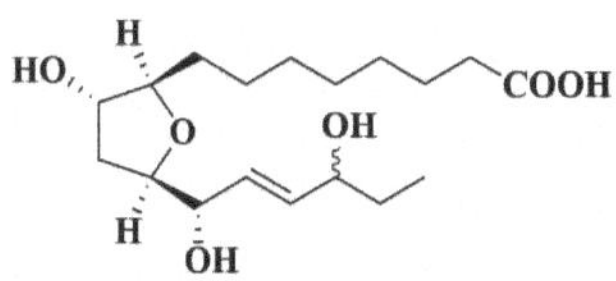

ent-16(RS)-13-epi- ST–Δ^{14}-9-PhytoF

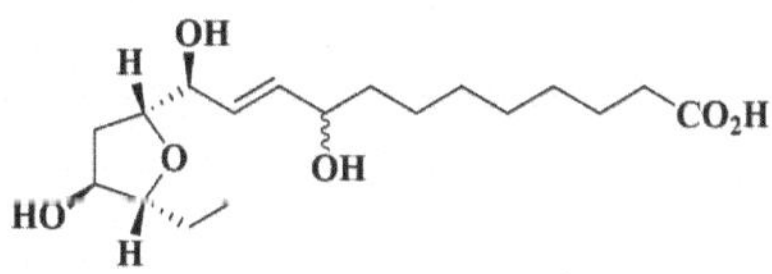

ent-9-(RS)-12-epi-ST-Δ^{10}-13-PhytoF

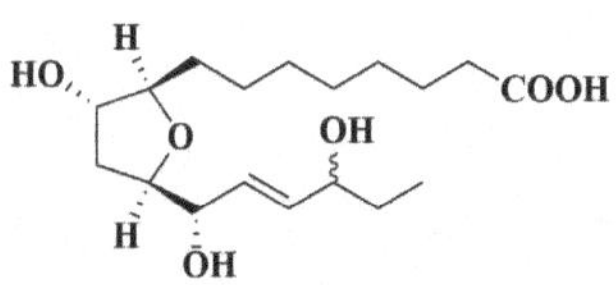

ent-16(RS)-13-epi- ST–Δ^{14}-9-PhytoF

Fig. 8 Available phytofuran (PhytoF) from Durand's group (Cuyamendous et al., 2015).

3. Biological functions and metabolism of phytoprostanes and phytofurans

As noted, PhytoPs are mainly found in higher plants. Although they have been suggested as possible biomarkers of oxidative stress in plants, PhytoPs also display a wide range of biological functions (Collado-González et al., 2015; Imbusch & Mueller, 2000b). Among all the PhytoPs molecules, A_1- and B_1-PhytoPs were the most studied. PhytoPs were shown to up-regulate genes related to detoxification of xenobiotics including phase I enzymes (CYP450 enzyme, aldehyde dehydrogenase, OPDA reductase), phase II conjugating enzymes (glutathione-S-transferase and glycosyl transferase) and putative phase III xenobiological transporter (ABC transporter) (Loeffler et al., 2005; Mueller et al., 2008; Stotz, Mueller, Zoeller, Mueller, & Berger, 2013). They also stimulate genes related to stress, such as heat shock proteins and heat shock factors, improving cell metabolism under stress conditions (Muench, Hsin, Ferber, Berger, & Mueller, 2016). Furthermore, Stotz et al. (2013) and Mueller et al. (2008) reported that A_1-PhytoPs inhibited root growth in seedlings and cell cycle progression in Tobacco cells when stressed by water deficit and heat. Loeffler et al. (2005) also found that B_1-PhytoPs protected Tobacco cells against cell death caused by severe $CuSO_4$ poisoning, further supporting that PhytoPs have potential protective effects for plant cells. In addition, Lupette et al. (2018) discovered PhytoPs are possible oxidative stress biomarker in diatom *Phaeodactylum*. Treatment of millimolar range of B_1-PhytoPs and L_1-PhytoPs can induce triglyceride accumulation and growth reduction in diatoms, which are common responses when they are expose to environmental stress. Despite the number of studies showing PhytoPs to protect plant tissues against stress conditions and prevent cell death triggered by oxidative damage, the protective role of PhytoFs in plants is still unclear, as it is relatively new in the archive, and they are generally less concentrated in most plant tissues compared to PhytoPs.

Apart from biological functions exerted in plants, there are also some studies regarding the bioavailability of PhytoPs and PhytoFs in animal models and human. Probably due to instability of free PhytoPs and PhytoFs during metabolism and lack of commercially available standards, all studies showing *in vitro* and *in vivo* bioavailability data of PhytoPs and PhytoFs are collective and derived from natural sources such as plant materials and vegetable oils instead of pure individual isomers (Barden et al., 2009; Karg et al., 2007; Lee et al., 2018; Leung et al., 2020; Leung, Galano,

Oger, Durand, & Lee, 2021). The earliest study was published in 2007, where Karg et al. (2007) discovered in an *in vitro* model that PhytoPs, including F_1-PhytoPs, E_1-PhytoPs, A_1-PhytoPs and B_1-PhytoPs, can withstand gastric juice and be released in the intestine from triglycerides by the hydrolytic action of pancreatic lipase in the digestive system. In their follow up study, they showed that F_1-PhytoPs in vegetable oils (olive oil and soybean oil) were absorbed in the intestine, circulated in plasma in esterified form, and were excreted in free form in human urine. Barden et al. further confirmed the elevation of F_1-PhytoPs (free + esterified) in plasma and urine (free) after consumption of vegetable oils (flaxseed oil and olive oil) (Barden et al., 2009). It should be noted that flaxseed oil is a rich source of ALA (ALA representing 45–55% of total fatty acids), while there is less in soybean oil (5–10%), suggesting that even at low ALA concentration, it is efficiently metabolized into PhytoPs and excreted (Calder, 2013).

Notwithstanding, the most recent simulated digestion (gastric, intestinal, and gastrointestinal) study showed that PhytoPs and PhytoFs from microalgae (*Gracilaria longissimi*) were significantly absorbed. Moreover, they passed through the Caco-2 cell monolayer (up to 88.2%) indicating PhytoPs and PhytoFs are readily absorbed in the gut system with high bioavailability (Martinez Sanchez et al., 2020). Furthermore, the elevation of PhytoP in the bloodstream is not only the result of direct intestinal absorption from vegetable oils, but also of *in vivo* autoxidation of the ALA present in these oils. Previous speculations suggest that PhytoPs and PhytoFs may be generated *in vivo* and accumulate in different target organs and cells such as adipose tissue and skin (Medina, Gil-Izquierdo, Durand, Ferreres, & Dominguez-Perles, 2018; Sinclair, Attar-Bashi, & Li, 2002). A recent animal study by Lee's group showed that PhytoPs and PhytoFs were up-regulated in the liver and heart tissues of rodents when the diets were enriched with high ALA-containing plant food such as oat bran (Lee et al., 2018), walnut (Leung et al., 2020), flaxseed and flaxseed oil (Leung et al., 2021). This indicates that PhytoPs and PhytoFs other than F_1-PhytoPs may have similar absorption and excretion pathways. In addition, the same group suggested that autoxidation of ALA may have taken place *in vivo* since PhytoPs and PhytoFs levels were more elevated in the rat tissues after consuming ALA enriched diet than in the diet itself (Leung et al., 2021). From a quantitative point of view, the methods used so far in bioavailability studies to discriminate between the free and esterified conjugated forms can be improved by the use of enzymatic hydrolysis with bacteria esterase, as reported by Gil-Izquierdo group (Dominguez-Perles et al., 2020).

Biological functions of PhytoPs and PhytoFs were also investigated in both animal and cell models. The most studied biological effect in human is the immunomodulatory capacity of pollen associated PhytoPs. Traidl-Hoffmann et al. (2005) found that E_1-PhytoPs from aqueous birch pollen modulated the immune system in human dendritic cells (DC) by inhibiting IL-12 production and promoting TH2 pro-allergenic activity. Subsequent studies showed that E_1-PhytoPs were key regulatory mediator in pollen modulating DC by favoring TH2 cell polarization (Gilles et al., 2009; Gutermuth et al., 2007; Mariani et al., 2007; Traidl-Hoffmann et al., 2005). Aside from E_1-PhytoPs, F_1-PhytoPs are also concentrated in pollen, and Gutermuth et al. (2007) reported that intranasal E_1- and F_1-PhytoPs were immunoregulatory in human DCs by reducing TH1 and TH2 cytokine production *in vivo*. Recently, Gonzalez Roldan et al. (2019) found that pollens contain not only E_1- and F_1-PhytoPs but also 16(*RS*)-9-*epi*-ST-Δ^{14}-10-PhytoF, all of which contributing to the immediate effector phase of allergic inflammation through the activation of mast cell chemotaxis, IL-6 generation, and IgE-dependent degranulation. Interestingly, they have also reported that only glycerol-bound PhytoPs (E_1-PhytoPs) were involved in the mast cell chemotaxis initiation, but not free PhytoPs or PhytoF.

On a side note in 2017, Collado-González and colleagues (Collado-Gonzalez et al., 2017) showed in an *in vitro* study that extra virgin olive oils extract, which is rich in PhytoPs (9 series of the F_{1t} class), had potent hypoglycemic effects by strongly inhibiting the α-glucosidase and the α-amylase activities.

PhytoPs and PhytoFs also possess anti-inflammatory and anti-oxidative properties. Minghetti et al. (2014) investigated the effect of PhytoPs in human neuroblastoma (SH-SY5Y) subjected to H_2O_2 induced oxidative stress. They revealed that B_1-PhytoPs enhanced cell metabolic activity and prevented cell death against oxidative damage and that 16-B_1-PhytoP protected SH-SY5Y cells against oxidative injury by activating PPAR-γ, a ligand-dependent transcription factor that regulates several antioxidant enzymes (Minghetti et al., 2014). In addition, PhytoPs and PhytoFs extracted from raw algae were found to contain anti-inflammatory properties in epithelial cells by reducing the expression of pro-inflammatory ICAM-1 and IL-6 inflammation markers (Campillo et al., 2021). Furthermore, 9-F_{1t}-PhytoP, ent-9-D_{1t}-PhytoP, 16-B_1-PhytoP, 9-L_1-PhytoP, and ent-9(*RS*)-12-*epi*-ST-Δ^{10}-13-PhytoF exerted anti-inflammatory property by inhibiting the production of prostaglandins in THP-1 monocytic cells stimulated with lipopolysaccharides.

All these biological activities could be related to their structural similarities with prostanoids and isoprostanes. For example, as described above, E_1-PhytoP may mimic the biological activity of E_2-prostaglandin (PGE_2) (inhibition of the production IL-12). The structural and biological similarities of these molecules could be related to the key positions of the hydroxyl and keto groups on the cyclopentenone ring and their electrophilic features. The cyclopentenone ring of the plant oxylipins are similar to those found in human prostanoids and isoprostanes, and could be the key structure for their biological actions since it can react with proteins (Medina et al., 2018). However, research on the organ-specific biological effect of PhytoPs and PhytoFs in animal models or clinical trial in humans are quite limited, probably due to the lack of commercial standards to measure these compounds in bloodstream or tissues, and/or to test their molecular mechanism in cells. Only one *in vitro* study indicated that 16-F_{1t}-PhytoP could act like TXA_2 to trigger reversible platelet aggregation (Gutermuth et al., 2007). Few other studies suggested that PhytoPs and PhytoFs elevation in heart tissues resulting from plant-based diet with high ALA content might be responsible for the cardio-protective effects observed in rodent models (Lee et al., 2018; Leung et al., 2020, 2021). From these small numbers of studies, it is plausible that PhytoPs and PhytoFs retain bioactive properties *in vivo*. Nonetheless, more investigations are required in terms of toxicity, immunological functions, and gene interaction in different biological models (Fig. 9).

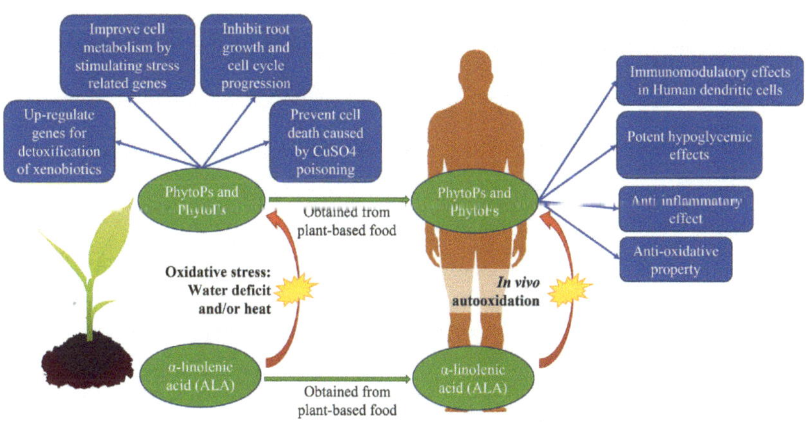

Fig. 9 Summarized biological effects of phytoprostanes (PhytoPs) and phytofurans (PhytoFs) in plants and human.

4. Phytoprostanes and phytofurans in plants and algae

4.1 Phytoprostanes and phytofurans in plants

As noted, phytoprostanes were discovered by Parchmann and Mueller in 1998 (Parchmann & Mueller, 1998). They deciphered the autoxidative radical induced pathway necessary for this transformation. Their investigation highlighted the production of all types of PhytoPs (see Fig. 2) in herbs (intensively studied in *Arabidopsis thaliana cv. Columbia, Lycopersicon esculentum cv. Moneymaker and Nicotiana tabacum cv. Samson*) and pollen birch (Durand et al., 2009). Such early investigations relied on the use of GC–MS and derivatization techniques, with a detection range from low $ng\,g^{-1}$ of dry leaves to more than $1600\,ng\,g^{-1}$ for deoxy-PhytoPs. Recent investigations with modern mass spectrometry (LC–MS/MS without derivatization) showed comparable values for Arabidopsis (Rac et al., 2021) and pollen birch (Gonzalez Roldan et al., 2019), but not in melon leaves where Vigor and coworkers (Yonny et al., 2016) observed 10 times more of the F-type, with values in the range $600–750\,ng\,g^{-1}$. Today, most of the studies are focused on edible oils and food-based plants.

The PhytoFs were also quantified in melon leaves submitted to thermal stress: the levels were 2.5 times higher when the leaves were under thermal stress compared to control (control level range from 250 to $4500\,ng\,g^{-1}$). This study also encompassed the potential use of such compounds as an indicator of oxidative stress since the spraying of bioactive compounds (antioxidants) showed a diminution of PhytoFs and PhytoPs compared to the control. Levels of PhytoFs were also quantified in Arabidopsis plants and showed values of approximately $10\,ng\,g^{-1}$ per type of PhytoF (ent-16(*RS*)-9-*epi*-ST-Δ^{14}-10-PhytoF, ent-9(*RS*)-12-*epi*-ST-Δ^{10}-13-PhytoF, ent-16(*RS*)-13-*epi* -ST-Δ^{14}-9-PhytoF) (Rac et al., 2021).

4.2 Phytoprostanes and phytofurans in algae

Oxylipins have been quantified and identified in macroalgae some time ago (Barbosa, Valentão, & Andrade, 2016), but PhytoPs were observed for the first time in 2015 by Ángel Gil-Izquierdo and coworkers (Barbosa et al., 2015). The free PhytoP composition of 24 macroalga species belonging to Chlorophyta, Phaeophyta and Rhodophyta were determined using ultra-high-performance liquid chromatography coupled with triple quadrupole mass spectrometry (UHPLC-QqQ-MS/MS). The PhytoP profiles varied greatly among all samples, where F_{1t}-PhytoPs and L_1-PhytoPs being the predominant and minor classes respectively. Interestingly, the data could point toward levels of PhytoPs being species-specific or resulting from interspecific interactions. In 2018, Vigor and coworkers (Vigor et al., 2018) further investigated the levels of PhytoFs in red and brown macroalgae as well as PhytoPs and other isoprostanoids derived from arachidonic acid under copper stress conditions. Particularly the ent-16 (*RS*)-9-*epi*-ST-Δ^{14}-10-PhytoF was ubiquitous and the most abundant of all isoprostanoids identified and quantified. Furthermore, they also established a link between the isoprostanoid content of macroalgae and heavy metal stress. For instance, the total isoprostanoid concentrations in *Ectocarpus siliculosus* were 945.8 and 3957.8 $ng\,g^{-1}$ of fresh weight before and after cupric treatment, respectively.

In 2018, Eric Maréchal's team and Durand's group (Lupette et al., 2018) focused on diatom *Phaeodactylum tricornutum,* looking at the entire spectrum of isoprostanoids (derived from ALA, AA, EPA and DHA) in response to oxidative stress. It was shown that, although ALA is present in minor amount, PhytoPs constitute the main family of non-enzymatic oxylipins in *P. tricornutum* (up to 300 pmol of total PhytoP per 1 million cells). This is in the same concentration range of ALA (up to 400 pmol per 1 million cells) (Lupette et al., 2018), with a large preference for the F-type PhytoPs compared to B_1-PhytoPs and L_1-PhytoPs type. A significant increase in the F_{1t}-PhytoP concentration occurred after treatment with 0.25 μM H_2O_2, whereas with 0.75 μM H_2O_2 (lethal dose) the level was close to control *P. tricornutum* cells.

In 2020, a unique study on four marine microalgae (the cryptophyte *Rhodomonas salina*, rich in ALA, the haptophyte *Tisochrysis lutea* and the diatoms *P. tricornutum* and *Chaetoceros gracilis*) by Vigor and coworkers showed a good correlation between the most abundant PUFAs and their respective isoprostanoid and isofuranoid profiles (up to 38 isoprostanoids and isofuranoids were detected) (Vigor et al., 2020). This study also showed no significant variation in the content of oxidized metabolites in *R. salina* and *C. gracilis* under copper stress, while an increase in the production of C18-, C20- and C22-derived isoprostanoids were monitored in *T. lutea* and *P. tricornutum*. In the presence of hydrogen peroxide, no significant changes were observed for *C. gracilis* and *T. lutea*, while it varied for the other two algae (*R. salina* and *P. tricornutum*).

5. Phytoprostanes and phytofurans in food

In the early stage of PhytoPs research, scientists mainly focused on studying PhytoPs in plant tissues (Imbusch & Mueller, 2000a; Loeffler et al., 2005; Traidl-Hoffmann et al., 2005) and plant cell cultures (Thoma et al., 2003), as PhytoPs generated by non-enzymatic oxidation of ALA were first discovered in higher plants (Parchmann & Mueller, 1998). The study of PhytoPs in food (Table 1) was first reported in 2007 by Karg and colleagues (Karg et al., 2007) who found high levels of PhytoPs in fresh vegetable oils. They also noted that PhytoPs in plant food might have potential biological activity in human body. Thereafter, the number of PhytoPs studies on different plant-based food increased.

Of note, only Karg et al. measured PhytoPs in food using GC–MS method, while majority of the newer PhytoPs studies instead used UHPLC-QqQ-MS/MS or liquid chromatography tandem mass spectrometry (LC-MS/MS) thanks to the method developed by Collado-González and co-workers (Collado-Gonzalez, Medina, et al., 2015). The use of LC-MS/MS has more advantages compared to GC–MS when measuring PhytoPs in food samples. First, it is obvious that LC-MS/MS is more compound specific compared to GC–MS. In Karg et al. study, the measured PhytoPs were reported in groups such as F_1-PhytoPs, A_1-PhytoPs and B_1-PhytoPs (Karg et al., 2007), while in Collado-González method (LC-MS/MS), PhytoPs single molecules such as 9-F_{1t}-PhytoP and 16-B_1-PhytoP, and even stereoisomers (*i.e.*, 16-B_1-PhytoP and ent-16-B_{1t}-PhytoP, and 9-L_{1t}-PhytoP and ent-9-L_{1t}-PhytoP) were distinguished using multiple reaction monitoring technique (Collado-Gonzalez, Medina, et al., 2015). Therefore, LC-MS/MS is a more specific measurement method to quantify PhytoPs in biological samples. Second, even though GC–MS is more sensitive, the sample requires stringent extraction and purification process plus derivatization steps before analysis (Karg et al., 2007). On the other hand, LC-MS/MS does not require derivatization process (Collado-Gonzalez, Medina, et al., 2015) hence lesser meticulous techniques. Finally, the LC-MS/MS is time saving since a typical analytical time is one-third to a half shorter than GC–MS.

Measurement and investigations of PhytoFs in food (Table 2) appeared much later than PhytoPs. PhytoFs were first synthesized and characterized by Cuyamendous et al. (2015), and thereafter the group has measured significant amount of PhytoFs in nuts and seeds using LC-MS/MS.

Table 1 Phytoprostanes determined in various food samples.

	PhytoP 1	PhytoP 2	PhytoP 3	PhytoP 4	PhytoP 5	PhytoP 6	PhytoP 7
Nuts and seeds							
Almond (*Vairo Cultivar*) (Lipan et al., 2020) ($ng\,g^{-1}$ DW)	<LOQ-24.32	<LOQ-0.238	<LOQ-30.61	<LOQ-1.086	<LOQ-0.666	<LOQ-15.007	<LOQ-14.258
Pistachio (Collado-Gonzalez et al., 2020) ($ng\,g^{-1}$ FW)	1 16–6.13	0.911–1.31	0.936–4.39	0.239–0.384	0.26–0.293	0.04–0.122	0.039–0.077
Hazelnut (*Gevuina avellana*) (Pino Ramos et al., 2019) ($ng\,g^{-1}$ FW)	<LOQ-167.772	<LOQ-39.628	<LOQ-347.256	<LOQ-12.602	<LOQ-12163	<LOQ-6.80	<LOQ-7.609
Walnut (Carrasco-Del Amor et al., 2017) ($ng\,g^{-1}$ FW)	3 768–3.907	N/A	1.542–1.743	N/A	N/A	0.024–0.035$^{\#}$	0.023–0.036$^{\#}$
Macadamia (Carrasco-Del Amor et al., 2017) ($ng\,g^{-1}$ FW)	3 439–21.241	N/A	2.055–13.405	<LOQ-0.362	<LOQ-0.055	0.137–0.608$^{\#}$	0.127–0.504$^{\#}$
Pecan (Carrasco-Del Amor et al., 2017) ($ng\,g^{-1}$ FW)	4 088–13.704	N/A	2.588–6.851	<LOQ	<LOQ	0.034–0.266$^{\#}$	0.028–0.214$^{\#}$
Almond kernels (Carrasco-Del Amor et al., 2016) ($ng\,g^{-1}$ FW)	1 99–111.98	0.42–1.46	1.49–132.14	0.17–0.74	0.08–0.42	0.22–19.22$^{\#}$	0.25–10.48$^{\#}$
Green table olives (Collado-González, 2015) ($ng\,g^{-1}$ FW)	3 64–37.30	N/A	1.55–50.00	N/A	N/A	0.30–0.83	0.12–0.35
Almond cultivars (Carrasco-Del Amor et al., 2015) ($ng\,g^{-1}$ FW)	10.11–97.597	<LOQ-1.758	12.968–101.22	<LOQ-1.762	<LOQ-0.492	1.065–30.936	0.278–12.141
Edible oils							
Aged rice oil (Smrček et al., 2021) ($ng\,mg^{-1}$)	0.308*	0.462*	0.077*	N/A	N/A	N/A	N/A

Continued

Table 1 Phytoprostanes determined in various food samples.—cont'd

	PhytoP 1	PhytoP 2	PhytoP 3	PhytoP 4	PhytoP 5	PhytoP 6	PhytoP 7
Aged sunflower oil (Smrček et al., 2021) ($ng\,mg^{-1}$)	0.462*	0.769*	0.308*	N/A	N/A	N/A	N/A
Aged pumpkin seed oil (Smrček et al., 2021) ($ng\,mg^{-1}$)	0.192*	0.308*	0.154*	N/A	N/A	N/A	N/A
Aged olive oil (Smrček et al., 2021) ($ng\,mg^{-1}$)	0.846*	2.0*	0.769*	N/A	N/A	N/A	N/A
Aged walnut oil (Smrček et al., 2021) ($ng\,mg^{-1}$)	11.538*	33.846*	13.846*	N/A	N/A	N/A	N/A
Aged hemp oil (Smrček et al., 2021) ($ng\,mg^{-1}$)	23.846*	51.538*	20.00*	N/A	N/A	N/A	N/A
Hemp oil (Smrček et al., 2021) ($ng\,mg^{-1}$)	2.692*	5.923*	2.846*	N/A	N/A	N/A	N/A
Extra virgin olive oil (Dominguez-Perles et al., 2018) ($\mu g\,mL^{-1}$)	4.6–13.6*	<LOQ	3.1–14.0*	<LOQ	<LOQ	1.0–5.0*	1.0–3.0*
Bioorganic virgin flax oil (Dominguez-Perles et al., 2018) ($\mu g\,mL^{-1}$)	14.0–18.0*	14.0–17.0*	10.0–12.0*	7.0–10.0*	28.0–32.0*	28.0–32.0*	20.0–22.0*
Virgin argan oil (Dominguez-Perles et al., 2018) ($\mu g\,mL^{-1}$)	<1.0*	<LOQ	<1.0*	<LOQ	<LOQ	<1.0*	<1.0*
Sesame seed oil (Dominguez-Perles et al., 2018) ($\mu g\,mL^{-1}$)	<1.0*	18.0–19.0*	<LOQ	<LOQ	<LOQ	<1.0*	<1.0*
Grapeseed oil (Dominguez-Perles et al., 2018) ($\mu g\,mL^{-1}$)	<2.0*	<LOQ	<1.0*	<LOQ	<LOQ	<1.0*	<LOQ

Safflower seed oil (Dominguez-Perles et al., 2018) (μg mL^{-1})	<2.0*	19.0–21.0*	<LOQ	<LOQ	<LOQ	<1.0*	<1.0*
Palm oil (Dominguez-Perles et al., 2018) (μg mL^{-1})	<LOQ	<LOQ	<LOQ	<LOQ	<LOQ	<1.0*	<1.0*
Extra virgin olive oil (Collado-Gonzalez et al., 2017) (ng mL^{-1})	6.02–33.20	<LOQ-0.15	5.25–29.14	0.07–0.20	0.04–0.08	0.19–0.62$^{\#}$	0.10–0.35$^{\#}$
Extra virgin olive oil (Collado-Gonzalez et al., 2016) (ng mL^{-1})	5.24–39.19	N/A	2.83–27.28	0.13–0.26	<LOQ-0.45	0.32–0.75	0.25–2.88
Extra virgin olive oil (Collado-Gonzalez et al., 2015) (ng mL^{-1})	5.03–12.02	N/A	2.62–6.19	0.13–0.34	<LOQ-0.56	0.30–0.91$^{\#}$	0.54–3.76$^{\#}$
Extra virgin olive oil (Collado-Gonzalez et al., 2015) (ng mL^{-1})	<LOQ	1.63–2.63	<LOQ	1.13–1.27	10.72–11.86	0.33–0.37$^{\#}$	<LOQ$^{\#}$
Olive oil (Collado-Gonzalez, Medina, et al., 2015) (ng mL^{-1})	<LOQ	3.21–4.19	<LOQ	7.17–11.49	18.58–31.14	<LOQ$^{\#}$	<LOQ$^{\#}$
Refined sunflower oil (Collado-Gonzalez, Medina, et al., 2015) (ng mL^{-1})	24.19–26.09	21.51–27.29	19.02–19.66	54.83–59.43	39.86–53.08	62.73–72.4$^{\#}$	274.02–320.88$^{\#}$
Hoki liver oil (Phung et al., 2020) (pg g^{-1})	1200–1400*	1600–2000*	1000–1200*	N/A	N/A	100*	100*
Anchovy oil (Phung et al., 2020) (pg g^{-1})	190–210*	190–210*	380–410*	N/A	N/A	1000*	1000*

Continued

Table 1 Phytoprostanes determined in various food samples.—cont'd

	PhytoP 1	PhytoP 2	PhytoP 3	PhytoP 4	PhytoP 5	PhytoP 6	PhytoP 7
Other food items							
Sea buckthorn juices (Tkacz et al., 2021) ($ng\,g^{-1}$ DW)	<LOQ-2.26	0.233–0.886	<LOQ-7.569	<LOQ-2.019	<LOQ-1.614	<LOQ-1.561	<LOQ-1.808
Dark chocolate (León-Perez et al., 2020) ($ng\,g^{-1}$ FW)	4.95–252.69	<LOQ-12.20*	<LOQ-595.00*	3.02–8.82	0.72–322.20	7.87 to 26.60	0.57–16.08
Date palm (*Phoenix dactylifera*) (Medina et al., 2020) ($ng\,g^{-1}$ FW)	2.01–72.23	<LOQ-9.00	2.09–72.97	0.02–0.80*	0.02–1.20*	0.075–26.71	0.032–12.99
Mangetout (García-García et al., 2019) ($ng\,g^{-1}$ DW)	66.67–1533.33*	10–2050*	160–1040*	<LOQ-120*	<LOQ-250*	<1–60*	<1–44*
French bean (García-García et al., 2019) ($ng\,g^{-1}$ DW)	66.67–600.00*	50–1500*	40–640*	<LOQ-40*	<LOQ-200*	<1–200*	<1–120*
Pea (García-García et al., 2019) ($ng\,g^{-1}$ DW)	66.67–666.67*	10–1500*	20–840*	<LOQ-260*	<LOQ-100*	<1–100*	<1–76*
Goldenberry (*Physalis peruviana*) (Medina et al., 2019) ($ng\,g^{-1}$ DW)	7.055–9.179	0.0018–0.0022	18.162–25.156	0.0019–0.0025	0.0013–0.0019	0.0127–0.0211#	0.0096–0.0132#
Cocoa beans (Leon-Perez et al., 2019) ($ng\,g^{-1}$ FW)	112.34–1346.78	3.56–74.32	23.67–819.44	1.43–12.42	1.08–17.18	2.55–48.85#	1.88–27.43#
Coffee pulp (Ruesgas-Ramon et al., 2019) ($ng\,g^{-1}$ DW)	139.4–151.8	204.6–215.8	166.6–176.2	N/A	N/A	58.5–62.9	63.7–69.7
Cocoa husk (Ruesgas-Ramon et al., 2019) ($ng\,g^{-1}$ DW)	73.5–79.5	75.5–86.1	66.3–69.1	N/A	N/A	101–120	126.3–151.3

Rice (Pinciroli et al., 2019) ($ng g^{-1}$ DW)	33 98–216.6	94.24–371.31	2.4–7.32	1.69–6.63	13.1–42.11	4.73–18.15	5.86–14.6
Rice (Pinciroli et al., 2018) ($ng g^{-1}$ DW)	27 1–386.6	50.8–871.1	26.8–518.6	2.2–7.0	10.3–84.9	2.1–19.0	1.5–12.6
Banana passion fruit shell (Medina et al., 2017a) ($ng g^{-1}$ DW)	3051.1–3326.5	2.1–2.5	20,487.6–21,890.8	3.1–3.9	2.6–3.8	108.2–131.4#	75.0–90.4#
Gulupa shell (Medina et al., 2017b) ($ng g^{-1}$ DW)	63–79*	N/A	32–48*	1.3–1.6*	1.0–1.5*	22–24*#	15–17*#
Rice flour (Pinciroli et al., 2017) ($ng g^{-1}$ DW)	<LOQ	<LOQ	<LOQ	<1.0–18.75*	<0.1–42.5*	<0.1–50.0*	1.25–33.75*
Must (Marhuenda et al., 2015) ($ng mL^{-1}$)	13.4–151.6	4.42–50.18	N/A	0.04–23.6	1.8–211.5	2.4–2.6#	6.1–6.7#
Wine (Marhuenda et al., 2015) ($ng mL^{-1}$)	71.1–90.94	19.6–136	N/A	0.08–0.42	1.19–17.8	0.299–0.301#	3.6–4.0#

PhytoP 1: 9-F_{1t}-PhytoP; PhytoP 2: *ent*-16-F_{1t}-PhytoP + *ent*-16-*epi*-16-F_{1t}-PhytoP; PhytoP 3: 9-*epi*-9-F_{1t}-PhytoP; PhytoP 4: 9-*epi*-9-D_{1t}-PhytoP; PhytoP 5: 9-D_{1t}-PhytoP; PhytoP 6: 16-B_1-PhytoP; PhytoP 7: 9-L_1-PhytoP. LOQ: limit of quantification; N/A: PhytoP not measured; *Values are estimated from graphs reported; #Regioisomers (ent-16-B_1-PhytoP/ent-9-L_1-PhytoP) were also measured together in the report.

Table 2 Phytofurans measured in various food samples

	PhytoF 1	PhytoF 2	PhytoF 3
Nuts and seeds			
Almond (Vairo Cultivar) (Lipan et al., 2020) (DW)	<LOQ	<LOQ-0.615	<LOQ-0.01
Pistachio (Collado-Gonzalez et al., 2020) (FW)	0.052–0.153	<LOQ-0.121	N/A
Hazelnut (*Gevuina avellana*) (Pino Ramos et al., 2019) ($ng\,g^{-1}$ FW)	<LOQ-43.672	<LOQ-2.095	<LOQ-1.654
Pine nut (Cuyamendous et al., 2015) ($ng\,g^{-1}$ FW)	N/A	N/A	0.12–0.73
Walnut (Cuyamendous et al., 2015) ($ng\,g^{-1}$ FW)	N/A	N/A	7.75–19.28
Chia seed (Cuyamendous et al., 2015) ($ng\,g^{-1}$ FW)	N/A	N/A	1.73–3.03
Flaxseed (Cuyamendous et al., 2015) ($ng\,g^{-1}$ FW)	N/A	N/A	0.50–1.24
Edible oils			
Extra virgin olive oil (Dominguez-Perles et al., 2018) ($\mu g\,mL^{-1}$)	0.33–2.67*	1.0–7.16*	3.0–8.36*
Bioorganic virgin flax oil (Dominguez-Perles et al., 2018) ($\mu g\,mL^{-1}$)	21.3–25.0*	4.16–4.50*	12.0–16.3*
Virgin argan oil (Dominguez-Perles et al., 2018) ($\mu g\,mL^{-1}$)	<0.1*	<LOQ	<LOQ
Sesame seed oil (Dominguez-Perles et al., 2018) ($\mu g\,mL^{-1}$)	<0.1*	<LOQ	<LOQ
Grapeseed oil (Dominguez-Perles et al., 2018) ($\mu g\,mL^{-1}$)	<0.1*	<LOQ	<LOQ
Safflower oil (Dominguez-Perles et al., 2018) ($\mu g\,mL^{-1}$)	<0.1*	<0.1*	<LOQ
Palm oil (Dominguez-Perles et al., 2018) ($\mu g\,mL^{-1}$)	<0.1*	<0.1*	<LOQ
Hoki liver oil (Phung et al., 2020) ($pg\,g^{-1}$)	2600–2800*	N/A	N/A
Anchovy oil (Phung et al., 2020) ($pg\,g^{-1}$)	220–240*	N/A	N/A

Other food items			
Sea buckthorn juices (Tkacz et al., 2021) ($ng\,g^{-1}$ DW)	N/A	<LOQ-1.027	N/A
Dark chocolate (León-Perez et al., 2020) ($ng\,g^{-1}$ FW)	2.01–3.73	N/A	1.80–5.66
Date palm (*Phoenix dactylifera*) (Medina et al., 2020) ($ng\,g^{-1}$ FW)	0.046–1.91	0.008–0.81	0.031–1.50
Mangetout (García-García et al., 2019) ($ng\,g^{-1}$ DW)	3.2–6.0*	<LOQ	<LOQ
French bean (García-García et al., 2019) ($ng\,g^{-1}$ DW)	2.0–4.2*	<LOQ	<LOQ
Pea (García-García et al., 2019) ($ng\,g^{-1}$ DW)	0.4–7.6*	1.8–5.1*	25.0–40.0*
Cacao beans (Leon-Perez et al., 2019) ($ng\,g^{-1}$ FW)	<LOQ-3.09	<LOQ-3.15	<LOQ-7.79
Coffee pulp (Ruesgas-Ramon et al., 2019) ($ng\,g^{-1}$ DW)	157.4–169	176.3–190.5	196–197.2
Cocoa husk (Ruesgas-Ramon et al., 2019) ($ng\,g^{-1}$ DW)	57.2–68.4	88.3–89.7	119.6–132.8
Rice (Pinciroli et al., 2019) ($ng\,g^{-1}$ DW)	6.96–26.91	2.09–4.44	6.15–14.9
Rice (Pinciroli et al., 2018) ($ng\,g^{-1}$ DW)	4.1–24.2	1.2–5.7	2.0–13.3
Rice flour (Pinciroli et al., 2017) ($ng\,g^{-1}$ DW)	<0.1–15.71*	<0.1–5.33*	0.15–17.85*

PhytoF 1: ent-16(*RS*)-9-*epi*-ST-Δ_{14}-10-PhytoF; PhytoF 2: ent-9(*RS*)-12-*epi*-ST-Δ_{10}-13-PhytoF; PhytoF 3: ent-16(*RS*)-13-*epi*-ST-Δ_{14}-9-PhytoF. LOQ indicates the limit of quantification. N/A indicates the PhytoP was not measured in samples. *Indicates values are estimated from graphs found in literatures as no actual values were provided.

Initially, only one type of PhytoF, ent-16-(*RS*)-13-*epi*-*ST*-Δ^{14}-9-PhytoF, was measured successfully in nut and seed samples, but in later studies two more types of PhytoFs, ent-16-(*RS*)-13-*epi*-*ST*-Δ^{14}-9-PhytoF and ent-16(*RS*)-9-*epi*-ST-Δ^{14}-10-PhytoF, were identified in nuts, seeds and vegetable oils (Collado-Gonzalez et al., 2020; Dominguez-Perles et al., 2018; Lipan et al., 2020; Pino Ramos et al., 2019).

5.1 Phytoprostanes and phytofurans in nuts and seeds

Carrasco-Del Amor and colleagues were the first to show the presence of PhytoPs in almonds kernels from different cultivars (0.04×10^3 to $0.24 \times 10^3\,\mathrm{ng\,g^{-1}}$) and to study the relationship between PhytoPs and oxidative stress in almond kernels under different environmental conditions (Table 1) (Carrasco-Del Amor et al., 2015, 2016). Most of the individual PhytoPs levels increased with storage time. The total PhytoPs was significantly elevated when the almond kernels were stored in air atmosphere compared to those stored in vacuum. Temperature was also a key to PhytoPs production in almonds in which the levels increased when stored at 24 °C. Intriguingly, if the temperature is too high, such as roasting and frying process, the concentration of PhytoPs in almond kernels were low due to thermal degradation. Thus, the group successfully demonstrated that PhytoPs can be used as quality assurance biomarkers for oxidative degradation of almonds during storage.

The first report on the levels of PhytoFs in particular ent-16-(*RS*)-13-*epi*-*ST*-Δ^{14}-9-PhytoF was made by Cuyamendous et al. (2015) in nuts and seeds (Table 2). They revealed that PhytoFs are naturally present in common nuts and seeds, and ALA levels were not the only factor that affects the concentration in the samples.

Probably inspired by Collado-Gonzalez and co-worker's studies on EVOO (Collado-Gonzalez et al., 2016; Collado-Gonzalez, Perez-Lopez, et al., 2015), Lipan et al. (2020) continued the almond study to explore the effect of water deficit stress on the PhytoPs and PhytoFs profile in almonds. Similar to what was found in the EVOO studies, almonds obtained from trees maintained under a deficit of irrigation had a higher concentration of PhytoPs and PhytoFs compared to those obtained from trees with sufficient irrigation. More importantly, it was discovered that almonds with sufficient irrigation had PhytoPs and PhytoFs levels below limit of detection while those from trees maintained under deficit irrigation conditions had levels above limit of detection. Therefore, PhytoPs and PhytoFs are ideal biomarkers to estimate if the almonds were harvested from almond trees with drought stress (Tables 1 and 2).

Besides almonds, Carrasco-Del Amor et al. (2017) also studied the effects of processing on PhytoPs levels in walnut, macadamia nuts and pecan. It was reported that nuts subjected to frying process increased the types and concentration of PhytoPs. PhytoPs and PhytoFs in pistachio kernels were also determined by the same group (Collado-Gonzalez et al., 2020). As for almonds, pistachio kernels obtained from trees maintained under a deficit of irrigation had generally higher levels of PhytoPs and PhytoFs, especially 9-series F_1-PhytoPs and *ent*-16(*RS*)-9-*epi*-ST-Δ^{14}-10-PhytoF. In addition, high temperature treatment like roasting caused thermal degradation of PhytoPs in pistachio. This study further suggests that nut and seed cultivars, together with the growing conditions of the plants, indeed influenced the PhytoPs and PhytoFs levels and consequently the quality and levels of ALA (Tables 1 and 2).

Pino Ramos et al. (2019) showed the effect of roasting on the levels of PhytoPs and PhytoFs in Chilean hazelnut. They identified all eight PhytoPs and three PhytoFs in Chilean hazelnuts. However, unlike what was reported in almonds (Carrasco-Del Amor et al., 2016; Collado-Gonzalez et al., 2020), there was no significant reduction of both PhytoPs and PhytoF in Chilean hazelnut after roasting. In their study, they realized that PhytoPs, and PhytoFs in Chilean hazelnut were mostly esterified and not in the free form. This suggests that the esterified PhytoPs and PhytoFs were protected against degradation during high thermal treatment.

5.2 Phytoprostanes and phytofurans in edible oils

As mentioned Karg et al. (2007) measured several types of PhytoPs in vegetable oils including linseed, soybean, olive, rapeseed, walnut, and grapeseed oils (Table 1). More importantly, they reported that PhytoPs were able to be absorbed when consumed and remain bioactive in the human body. This greatly increased researchers' interest in investigating the potential biological effects of PhytoPs, not only in plant tissues but also in animal cells.

The study of PhytoPs in vegetable oils continued 8 years after Karg et al. (2007) study. In 2015, Collado-González and co-workers measured 10 different types of PhytoPs (9-F_{1t}-PhytoP, ent-16-*epi*-16F_{1t}-PhytoP, ent-16-F_{1t}-PhytoP, 9-*epi*-9-F_{1t}-PhytoP, 9-*epi*-9-D_t-PhytoP, 9-D_1-PhytoP,

16-B_1-PhytoP, ent-16-B_{1t}-PhytoP, 9-L_{1t}-PhytoP and ent-9-L_{1t}-PhytoP) in three commercial vegetable oils including extra virgin olive oil (EVOO), olive oil and refined sunflower oil (Collado-Gonzalez, Medina, et al., 2015). They found that refined sunflower oil had more types of PhytoPs and contained a 20- and 8-fold higher level of total PhytoPs compared to EVOO and olive oil. It is plausible that during manufacturing process, the refining treatment increased oxidation in vegetable oils leading to elevated PhytoPs levels. In the same year, it was found that water deficit in agriculture increased PhytoPs concentration in EVOO and suggested PhytoPs as potential oxidative stress biomarkers for olive tree (Collado-Gonzalez, Perez-Lopez, et al., 2015). In addition, they further studied the effect of season (Collado-Gonzalez et al., 2016) and the effect of water stress on PhytoPs content in EVOO and green table olives from olives grown on regulated deficit irrigation of the trees (Collado-González, 2015; Collado-Gonzalez et al., 2016). It was confirmed that water deficiency increased the amount of total PhytoPs in EVOO, and the processing treatment of green table olives was an also complementary action of this enhancing effect. However, not all individual types of PhytoPs proportionally responded to water stress.

In 2018, Dominguez-Perles et al. (2018) were the first to measure PhytoPs and PhytoFs in a variety of vegetable oils including EVOO, bioorganic virgin flax oil, argan oil, sesame oil, grapeseed oil, safflower oil and palm oil (Tables 1 and 2). They revealed that not all vegetable oils contain all types of PhytoPs and PhytoFs. The group showed that out of the vegetable oils tested, only flax oil contained all types of PhytoPs and PhytoFs, which is not surprising as it has high levels of ALA. Furthermore, certain oils like grapeseed oil and palm oil were considered as "negative control" since only 2–3 types of PhytoPs were found and the total PhytoPs and PhytoFs levels were very low (below $0.15\,\mu g\,mL^{-1}$).

Recently, Smrček et al. has successfully found prostaglandin-like PhytoPs including 9-*epi*-16-F_{1t}-PhytoP and *ent*-9,16-*diepi*-16-F_{1t}-PhytoP in aged commercial rice, sunflower, roasted pumpkin seed, olive and hemp oils and fresh hemp oils (Smrček et al., 2021). The report evidently showed that prostaglandin-like metabolites could be formed from PhytoPs in plant-derived food through autoxidation during storage at room temperature.

Interestingly, PhytoPs and PhytoFs were not only found in plant-based oil. Phung et al. (2020) successfully measured PhytoPs and PhytoFs in over-oxidized fish oil. They discovered significant elevation of PhytoPs and PhytoFs levels in over-oxidized Hoki liver oil but not in in anchovy oil. Moreover, a new type of PhytoP, 16(*RS*)-16-A_{1t}-PhytoP was found in both oils in significant amount (4800 pg g^{-1} in Hoki liver oil oxidized for 30 days and 400 pg g^{-1} in anchovy oil). It can be speculated that the PhytoPs and PhytoFs might originate from the fish diet (ALA and/or PhytoPs and PhytoFs) and were taken up and stored in the liver.

5.3 Phytoprostanes and phytofurans in other food items

Aside from oil rich food, including vegetable oils, nuts, and seeds, PhytoPs and PhytoFs were investigated in other food items especially grains (Tables 1 and 2). PhytoPs and PhytoFs were identified and quantified in rice bran, white and brown grain flours (Pinciroli et al., 2017), and they were also shown to be good biomarkers for assessing stress experienced by rice crop (Pinciroli et al., 2019). When foliar fertilization treatment or salicylic acid was applied to reduce the nutrition stress in rice crop, the PhytoPs and PhytoFs levels were significantly reduced. This demonstrates that these two types of agronomical procedures were able to modulate the redox balance by lowering the PUFA oxidation and consequently the synthesis of oxylipins (Pinciroli et al., 2018).

Surprisingly, and to the best of our knowledge, the highest concentrations of PhytoPs and PhytoFs were found in non-oily product, particularly in legume foods. The PhytoPs concentration was low (2.2 mg 100 g^{-1} d.w.) in French beans, high in mangetout (*P. sativum ssp* arvense, 26.3 mg 100 g^{-1} d.w.) and intermediate in pea (13.3 mg 100 g^{-1} d.w.), while PhytoFs ranged from 20 to 500 mg 100 g^{-1} d.w. (García-García et al., 2019). In sea buckthorn juices, the PhytoPs and PhytoFs ranged from 32.31 to 1523.51 ng 100 g^{-1} d.w. and 34.19 to 101.47 μg 100 g^{-1} d.w., respectively (Tkacz et al., 2021).

Researchers also studied the levels of PhytoPs and PhytoFs in fermented foods including musts and wines (Marhuenda et al., 2015), cocoa (Leon-Perez et al., 2019) and its manufactured product, dark chocolate (León-Perez et al., 2020). It was found that PhytoPs levels in musts and wines were affected by both the agriculture conditions of the grapes, and the wine making and aging process (Marhuenda et al., 2015). Generally, if the grapes come from stressed vineyards like old and low yielding ones, the PhytoPs levels were higher. In addition, wine contained higher levels of PhytoPs compared to the musts, probably due to increased level of oxidation during wine making and aging process. PhytoPs provide great information in determining the quality of vineyards, wine making process and wine. Furthermore, PhytoPs and PhytoFs were also identified and quantified in 31 cocoa clones (Leon-Perez et al., 2019). The total PhytoPs and PhytoFs in cocoa ranged from 221.46 to 1589.83 $ng\,g^{-1}$ and from 1.18 to 13.13 $ng\,g^{-1}$, respectively. It was concluded that the levels of PhytoPs and PhytoFs depended on the levels of PUFAs in cocoa. However, when processed to make chocolate, the PhytoPs and PhytoFs levels ranged from 2.45 to 1192.43 $ng\,g^{-1}$ d.w. and from 144.57 to 840.19 $ng\,g^{-1}$ d.w. respectively in different dark chocolates (León-Perez et al., 2020), and did not correlate with the initial levels of PUFAs nor with the initial cocoa percentage.

Recently, PhytoPs and PhytoFs were also measured in food wastes, including shells from banana passion fruit shell (*Passiflora ripartite* var. mollisima) (Medina et al., 2017a), gulupa fruits (*Passiflora edulis* Sims f. edulis) (Medina et al., 2017b), calyces of goldenberry (*Physalis peruviana*) (Medina et al., 2019), byproducts of date palm (*Phoenix dactylifera*) including pulp, skin, pits, leaves, clusters, and pollen (Medina et al., 2020), coffee pulp, coffee husk and pod husk (Ruesgas-Ramon et al., 2019). These parts are inedible and can only be discarded. Researchers identified new alternatives for handling these food wastes in agro-food industry. A noticeable amount of PhytoPs (1.30–67.60 $ng\,g^{-1}$ d.w.) were found in shells from gulupa fruits, while banana passion fruit shell, calyces of goldenberry, date palm by products, coffee pulp, coffee husk and pod husk had extremely high levels of PhytoPs (19,786, 21,659.80, 1,137,514, 654.6, 179.9 and 278 $ng\,g^{-1}$ d.w. respectively) compared to oil rich food items like EVOO or nuts. High PhytoFs levels were also found in date palm skin (32,915 $ng\,g^{-1}$ d.w.) and coffee husk (543.2 $ng\,g^{-1}$ d.w.). These findings suggest that food wastes can be potential source of PhytoPs and PhytoFs, which can be further utilized for pharmaceutical or cosmetic products due to the bioactive properties noted earlier.

6. Concluding remarks

In this chapter, we have described the chemical formation of PhytoPs and PhytoFs from ALA and the different types identified so far (2021). Furthermore, we have shown the significance of these compounds in algae, plants and foods, and potential applications in the agricultural industry. Not only as an index of plant lipid oxidation but also as a valuable biomarker for quality assurance and quality control of edible oils, plant foodstuffs and food products. Moving forward, the action at the molecular level of PhytoPs and PhytoFs in human needs to be further explored. It is plausible that the anti-inflammatory and antioxidant properties associated with the Mediterranean diet may be partly attributed to the richness of phenolic compounds and to PhytoPs generated *in vivo* from ALA (both found in edible oils, nuts and seeds of the diet); perhaps this may be one of the several reasons for the health benefits.

References

Barbosa, M., Collado-Gonzalez, J., Andrade, P. B., Ferreres, F., Valentao, P., Galano, J. M., et al. (2015). Nonenzymatic alpha-linolenic acid derivatives from the sea: Macroalgae as novel sources of phytoprostanes. *Journal of Agricultural and Food Chemistry, 63*(28), 6466–6474. https://doi.org/10.1021/acs.jafc.5b01904.

Barbosa, M., Valentão, P., & Andrade, P. B. (2016). Biologically active oxylipins from enzymatic and nonenzymatic routes in macroalgae. *Marine Drugs, 14*(1), 23.

Barden, A. E., Croft, K. D., Durand, T., Guy, A., Mueller, M. J., & Mori, T. A. (2009). Flaxseed oil supplementation increases plasma F1-phytoprostanes in healthy men. *The Journal of Nutrition, 139*(10), 1890–1895. https://doi.org/10.3945/jn.109.108316.

Beretta, R., Giambelli Gallotti, M., Penne, U., Porta, A., Gil Romero, J. F., Zanoni, G., et al. (2015). General strategy for the synthesis of B1 and L1 prostanoids: Synthesis of phytoprostanes (RS)-9-L1-PhytoP, (R)-9-L1-PhytoP, (RS)-16-B1-PhytoP, and (RS)-16-L1-PhytoP. *The Journal of Organic Chemistry, 80*(3), 1601–1609. https://doi.org/10.1021/jo502538b.

Calder, P. C. (2013). Nutritional benefits of omega-3 fatty acids. In N. S. N. Charlotte Jacobsen, F. H. Anna, & A.-D. M. Sorensen (Eds.), *Food enrichment with omega-3 fatty acids* (pp. 3–26). Woodhead Publishing.

Campillo, M., Medina, S., Fanti, F., Gallego-Gomez, J. I., Simonelli-Munoz, A., Bultel-Ponce, V., et al. (2021). Phytoprostanes and phytofurans modulate COX-2-linked inflammation markers in LPS-stimulated THP-1 monocytes by lipidomics workflow. *Free Radical Biology & Medicine, 167*, 335–347. https://doi.org/10.1016/j.freeradbiomed.2021.03.002.

Carrasco-Del Amor, A. M., Aguayo, E., Collado-Gonzalez, J., Guy, A., Galano, J. M., Durand, T., et al. (2016). Impact of packaging atmosphere, storage and processing conditions on the generation of phytoprostanes as quality processing compounds in almond kernels. *Food Chemistry, 211*, 869–875. https://doi.org/10.1016/j.foodchem.2016.05.132.

Carrasco-Del Amor, A. M., Aguayo, E., Collado-Gonzalez, J., Guy, A., Galano, J. M., Durand, T., et al. (2017). Impact of processing conditions on the phytoprostanes profile of three types of nut kernels. *Free Radical Research, 51*(2), 141–147. https://doi.org/10.1080/10715762.2017.1288909.

Carrasco-Del Amor, A. M., Collado-Gonzalez, J., Aguayo, E., Guy, A., Galano, J.-M., & Durand, T. (2015). Phytoprostanes in almonds: Identification, quantification, and impact of cultivar and type of cultivation. *RSC Advances, 5*, 51233–51241. https://doi.org/10.1039/C5RA07803B.

Collado-González, J. (2015). The phytoprostane content in green table olives is influenced by Spanish-style processing and regulated deficit irrigation. *LWT, 64*(2), 997–1003. v. 1064 no. 1002. https://doi.org/10.1016/j.lwt.2015.07.005.

Collado-Gonzalez, J., Cano-Lamadrid, M., Perez-Lopez, D., Carbonell-Barrachina, A. A., Centeno, A., Medina, S., et al. (2020). Effects of deficit irrigation, rootstock, and roasting on the contents of fatty acids, phytoprostanes, and phytofurans in pistachio kernels. *Journal of Agricultural and Food Chemistry, 68*(33), 8915–8924. https://doi.org/10.1021/acs.jafc.0c02901.

Collado-González, J., Durand, T., Ferreres, F., Medina, S., Torrecillas, A., & Gil-Izquierdo, Á. (2015). Phytoprostanes. *Lipid Technology, 27*(6), 127–130. https://doi.org/10.1002/lite.201500020.

Collado-Gonzalez, J., Grosso, C., Valentao, P., Andrade, P. B., Ferreres, F., Durand, T., et al. (2017). Inhibition of alpha-glucosidase and alpha-amylase by Spanish extra virgin olive oils: The involvement of bioactive compounds other than oleuropein and hydroxytyrosol. *Food Chemistry, 235*, 298–307. https://doi.org/10.1016/j.foodchem.2017.04.171.

Collado-Gonzalez, J., Medina, S., Durand, T., Guy, A., Galano, J. M., Torrecillas, A., et al. (2015). New UHPLC-QqQ-MS/MS method for quantitative and qualitative determination of free phytoprostanes in foodstuffs of commercial olive and sunflower oils. *Food Chemistry, 178*, 212–220. https://doi.org/10.1016/j.foodchem.2015.01.097.

Collado-Gonzalez, J., Perez-Lopez, D., Memmi, H., Gijon, M. C., Medina, S., Durand, T., et al. (2015). Water deficit during pit hardening enhances phytoprostanes content, a plant biomarker of oxidative stress, in extra virgin olive oil. *Journal of Agricultural and Food Chemistry, 63*(14), 3784–3792. https://doi.org/10.1021/acs.jafc.5b00805.

Collado-Gonzalez, J., Perez-Lopez, D., Memmi, H., Gijon, M. C., Medina, S., Durand, T., et al. (2016). Effect of the season on the free phytoprostane content in Cornicabra extra virgin olive oil from deficit-irrigated olive trees. *Journal of the Science of Food and Agriculture, 96*(5), 1585–1592. https://doi.org/10.1002/jsfa.7259.

Cuyamendous, C., Leung, K. S., Durand, T., Lee, J. C., Oger, C., & Galano, J. M. (2015). Synthesis and discovery of phytofurans: Metabolites of alpha-linolenic acid peroxidation. *Chemical Communication (Cambridge), 51*(86), 15696–15699. https://doi.org/10.1039/c5cc05736a.

Dominguez-Perles, R., Abellan, A., Leon, D., Ferreres, F., Guy, A., Oger, C., et al. (2018). Sorting out the phytoprostane and phytofuran profile in vegetable oils. *Food Research International, 107*, 619–628. https://doi.org/10.1016/j.foodres.2018.03.013.

Dominguez-Perles, R., Sanchez-Martinez, I., Rodriguez-Hernandez, M. D., Lopez-Gonzalez, I., Oger, C., Guy, A., et al. (2020). Optimization of free phytoprostane and phytofuran production by enzymatic hydrolysis of pea extracts using esterases. *Journal of Agricultural and Food Chemistry, 68*(11), 3445–3455. https://doi.org/10.1021/acs.jafc.9b06624.

Durand, T., Bultel-Ponce, V., Guy, A., Berger, S., Mueller, M. J., & Galano, J. M. (2009). New bioactive oxylipins formed by non-enzymatic free-radical-catalyzed pathways: The phytoprostanes. *Lipids, 44*(10), 875–888. https://doi.org/10.1007/s11745-009-3351-1.

El Fangour, S., Guy, A., Despres, V., Vidal, J. P., Rossi, J. C., & Durand, T. (2004). Total synthesis of the eight diastereomers of the syn-anti-syn phytoprostanes F1 types I and II. *The Journal of Organic Chemistry, 69*(7), 2498–2503. https://doi.org/10.1021/jo035638i.

El Fangour, S., Guy, A., Vidal, J.-P., Rossi, J.-C., & Durand, T. (2003). Total synthesis of phytoprostane F1 and its 16 epimer. *Tetrahedron Letters, 44*(10), 2105–2108. https://doi.org/10.1016/S0040-4039(03)00171-0.

Fessel, J. P., Porter, N. A., Moore, K. P., Sheller, J. R., & Roberts, L. J., 2nd. (2002). Discovery of lipid peroxidation products formed in vivo with a substituted tetrahydrofuran ring (isofurans) that are favored by increased oxygen tension. *Proceedings of the National Academy of Sciences of the United States of America, 99*(26), 16713–16718. https://doi.org/10.1073/pnas.252649099.

García-García, M. C., del Río Celestino, M., Gil-Izquierdo, Á., Egea-Gilabert, C., Galano, J. M., Durand, T., et al. (2019). The value of legume foods as a dietary source of phytoprostanes and phytofurans is dependent on species, variety, and growing conditions. *European Journal of Lipid Science and Technology*, *121*(8), 1800484. https://doi.org/10.1002/ejlt.201800484.

Gerwick, W. H., Moghaddam, M., & Hamberg, M. (1991). Oxylipin metabolism in the red alga Gracilariopsis lemaneiformis: Mechanism of formation of vicinal dihydroxy fatty acids. *Archives of Biochemistry and Biophysics*, *290*(2), 436–444. https://doi.org/10.1016/0003-9861(91)90563-x.

Gilles, S., Mariani, V., Bryce, M., Mueller, M. J., Ring, J., Jakob, T., et al. (2009). Pollen-derived E1-phytoprostanes signal via PPAR-gamma and NF-kappaB-dependent mechanisms. *Journal of Immunology*, *182*(11), 6653–6658. https://doi.org/10.4049/jimmunol.0802613.

Gonzalez Roldan, N., Engel, R., Dupow, S., Jakob, K., Koops, F., Orinska, Z., et al. (2019). Lipid mediators from Timothy grass pollen contribute to the effector phase of allergy and prime dendritic cells for glycolipid presentation. *Frontiers in Immunology*, *10*, 974. https://doi.org/10.3389/fimmu.2019.00974.

Gutermuth, J., Bewersdorff, M., Traidl-Hoffmann, C., Ring, J., Mueller, M. J., Behrendt, H., et al. (2007). Immunomodulatory effects of aqueous birch pollen extracts and phytoprostanes on primary immune responses in vivo. *The Journal of Allergy and Clinical Immunology*, *120*(2), 293–299. https://doi.org/10.1016/j.jaci.2007.03.017.

Guy, A., Flanagan, S., Durand, T., Oger, C., & Galano, J. M. (2015). Facile synthesis of cyclopentenone B1- and L1-type phytoprostanes. *Frontiers in Chemistry*, *3*, 41. https://doi.org/10.3389/fchem.2015.00041.

Imbusch, R., & Mueller, M. J. (2000a). Analysis of oxidative stress and wound-inducible dinor isoprostanes F(1) (phytoprostanes F(1)) in plants. *Plant Physiology*, *124*(3), 1293–1304. https://doi.org/10.1104/pp.124.3.1293.

Imbusch, R., & Mueller, M. J. (2000b). Formation of isoprostane F(2)-like compounds (phytoprostanes F(1)) from alpha-linolenic acid in plants. *Free Radical Biology & Medicine*, *28*(5), 720–726. https://doi.org/10.1016/s0891-5849(00)00154-4.

Iqbal, M., Evans, P., Lledo, A., Verdaguer, X., Pericas, M. A., Riera, A., et al. (2005). Total synthesis and biological activity of 13,14-dehydro-12-oxo-phytodienoic acids (deoxy-J1-phytoprostanes). *Chembiochem*, *6*(2), 276–280. https://doi.org/10.1002/cbic.200400259.

Jahn, U., Galano, J. M., & Durand, T. (2008). Beyond prostaglandins- -chemistry and biology of cyclic oxygenated metabolites formed by free-radical pathways from polyunsaturated fatty acids. *Angewandte Chemie (International Ed. in English)*, *47*(32), 5894–5955. https://doi.org/10.1002/anie.200705122.

Jahn, U., Galano, J. M., & Durand, T. (2010). A cautionary note on the correct structure assignment of phytoprostanes and the emergence of a new prostane ring system. *Prostaglandins, Leukotrienes, and Essential Fatty Acids*, *82*(2–3), 83–86. https://doi.org/10.1016/j.plefa.2009.10.005.

Karg, K., Dirsch, V. M., Vollmar, A. M., Cracowski, J. L., Laporte, F., & Mueller, M. J. (2007). Biologically active oxidized lipids (phytoprostanes) in the plant diet and parenteral lipid nutrition. *Free Radical Research*, *41*(1), 25–37. https://doi.org/10.1080/10715760600939734.

Lee, J. C., AlGhawas, D. S., Poutanen, K., Leung, K. S., Oger, C., Galano, J. M., et al. (2018). Dietary oat bran increases some proinflammatory polyunsaturated fatty-acid oxidation products and reduces anti-inflammatory products in apolipoprotein E(−/−) mice. *Lipids*, *53*(8), 785–796. https://doi.org/10.1002/lipd.12090.

León-Perez, D., Domínguez-Perles, R., Collado-González, J., Cano-Lamadrid, M., Durand, T., Guy, A., et al. (2020). Bioactive plant oxylipins-based lipidomics in eighty worldwide commercial dark chocolates: Effect of cocoa and fatty acid composition on their dietary burden. *Microchemical Journal, 157*, 105083. https://doi.org/10.1016/j.microc.2020.105083.

Leon-Perez, D., Medina, S., Londono-Londono, J., Cano-Lamadrid, M., Carbonell-Barrachina, A., Durand, T., et al. (2019). Comparative study of different cocoa (Theobroma cacao L.) clones in terms of their phytoprostanes and phytofurans contents. *Food Chemistry, 280*, 231–239. https://doi.org/10.1016/j.foodchem.2018.12.072.

Leung, K. S., Galano, J. M., Oger, C., Durand, T., & Lee, J. C. (2021). Enrichment of alpha-linolenic acid in rodent diet reduced oxidative stress and inflammation during myocardial infarction. *Free Radical Biology & Medicine, 162*, 53–64. https://doi.org/10.1016/j.freeradbiomed.2020.11.025.

Leung, K. S., Galano, J. M., Yau, Y. F., Oger, C., Durand, T., & Lee, J. C. (2020). Walnut-enriched diet elevated alpha-linolenic acid, phytoprostanes, and phytofurans in rat liver and heart tissues and modulated anti-inflammatory lipid mediators in the liver. *Journal of Agricultural and Food Chemistry*. https://doi.org/10.1021/acs.jafc.0c06690.

Lipan, L., Collado-Gonzalez, J., Dominguez-Perles, R., Corell, M., Bultel-Ponce, V., Galano, J. M., et al. (2020). Phytoprostanes and phytofurans-oxidative stress and bioactive compounds-in almonds are affected by deficit irrigation in almond trees. *Journal of Agricultural and Food Chemistry, 68*(27), 7214–7225. https://doi.org/10.1021/acs.jafc.0c02268.

Loeffler, C., Berger, S., Guy, A., Durand, T., Bringmann, G., Dreyer, M., et al. (2005). B1-phytoprostanes trigger plant defense and detoxification responses. *Plant Physiology, 137*(1), 328–340. https://doi.org/10.1104/pp.104.051714.

Lupette, J., Jaussaud, A., Vigor, C., Oger, C., Galano, J. M., Reversat, G., et al. (2018). Non-enzymatic synthesis of bioactive Isoprostanoids in the diatom Phaeodactylum following oxidative stress. *Plant Physiology, 178*(3), 1344–1357. https://doi.org/10.1104/pp.18.00925.

Marhuenda, J., Medina, S., Diaz-Castro, A., Martinez-Hernandez, P., Arina, S., Zafrilla, P., et al. (2015). Dependency of Phytoprostane fingerprints of must and wine on viticulture and enological processes. *Journal of Agricultural and Food Chemistry, 63*(41), 9022–9028. https://doi.org/10.1021/acs.jafc.5b03365.

Mariani, V., Gilles, S., Jakob, T., Thiel, M., Mueller, M. J., Ring, J., et al. (2007). Immunomodulatory mediators from pollen enhance the migratory capacity of dendritic cells and license them for Th2 attraction. *Journal of Immunology, 178*(12), 7623–7631. https://doi.org/10.4049/jimmunol.178.12.7623.

Martinez Sanchez, S., Dominguez-Perles, R., Montoro-Garcia, S., Gabaldon, J. A., Guy, A., Durand, T., et al. (2020). Bioavailable phytoprostanes and phytofurans from Gracilaria longissima have anti-inflammatory effects in endothelial cells. *Food & Function, 11*(6), 5166–5178. https://doi.org/10.1039/d0fo00976h.

Medina, S., Collado-Gonzalez, J., Ferreres, F., Londono-Londono, J., Jimenez-Cartagena, C., Guy, A., et al. (2017a). Valorization strategy of Banana passion fruit Shell wastes: An innovative source of Phytoprostanes and phenolic compounds and their potential use in pharmaceutical and cosmetic industries. *Journal of Food and Nutrition Research, 5*(11), 801–808.

Medina, S., Collado-Gonzalez, J., Ferreres, F., Londono-Londono, J., Jimenez-Cartagena, C., Guy, A., et al. (2017b). Quantification of phytoprostanes—bioactive oxylipins—and phenolic compounds of Passiflora edulis Sims shell using UHPLC-QqQ-MS/MS and LC-IT-DAD-MS/MS. *Food Chemistry, 229*, 1–8. https://doi.org/10.1016/j.foodchem.2017.02.049.

Medina, S., Collado-Gonzalez, J., Ferreres, F., Londono-Londono, J., Jimenez-Cartagena, C., Guy, A., et al. (2019). Potential of Physalis peruviana calyces as a low-cost valuable resource of phytoprostanes and phenolic compounds. *Journal of the Science of Food and Agriculture, 99*(5), 2194–2204. https://doi.org/10.1002/jsfa.9413.

Medina, S., Gil-Izquierdo, A., Abu-Reidah, I. M., Durand, T., Bultel-Ponce, V., Galano, J. M., et al. (2020). Evaluation of Phoenix dactylifera edible parts and byproducts as sources of Phytoprostanes and Phytofurans. *Journal of Agricultural and Food Chemistry, 68*(33), 8942–8950. https://doi.org/10.1021/acs.jafc.0c03364.

Medina, S., Gil-Izquierdo, A., Durand, T., Ferreres, F., & Dominguez-Perles, R. (2018). Structural/functional matches and divergences of Phytoprostanes and Phytofurans with bioactive human oxylipins. *Antioxidants (Basel)*, 7(11). https://doi.org/10.3390/antiox7110165.

Minghetti, L., Salvi, R., Lavinia Salvatori, M., Ajmone-Cat, M. A., De Nuccio, C., Visentin, S., et al. (2014). Nonenzymatic oxygenated metabolites of alpha-linolenic acid B1- and L1-phytoprostanes protect immature neurons from oxidant injury and promote differentiation of oligodendrocyte progenitors through PPAR-gamma activation. *Free Radical Biology & Medicine, 73*, 41–50. https://doi.org/10.1016/j.freeradbiomed.2014.04.025.

Mueller, S., Hilbert, B., Dueckershoff, K., Roitsch, T., Krischke, M., Mueller, M. J., et al. (2008). General detoxification and stress responses are mediated by oxidized lipids through TGA transcription factors in Arabidopsis. *Plant Cell, 20*(3), 768–785. https://doi.org/10.1105/tpc.107.054809.

Muench, M., Hsin, C. H., Ferber, E., Berger, S., & Mueller, M. J. (2016). Reactive electrophilic oxylipins trigger a heat stress-like response through HSFA1 transcription factors. *Journal of Experimental Botany, 67*(21), 6139–6148. https://doi.org/10.1093/jxb/erw376.

Parchmann, S., & Mueller, M. J. (1998). Evidence for the formation of dinor isoprostanes E1 from alpha-linolenic acid in plants. *The Journal of Biological Chemistry, 273*(49), 32650–32655. https://doi.org/10.1074/jbc.273.49.32650.

Phung, A. S., Bannenberg, G., Vigor, C., Reversat, G., Oger, C., Roumain, M., et al. (2020). Chemical compositional changes in over-oxidized fish oils. *Food, 9*(10). https://doi.org/10.3390/foods9101501.

Pinciroli, M., Dominguez-Perles, R., Abellan, A., Bultel-Ponce, V., Durand, T., Galano, J. M., et al. (2019). Statement of foliar fertilization impact on yield, composition, and oxidative biomarkers in Rice. *Journal of Agricultural and Food Chemistry, 67*(2), 597–605. https://doi.org/10.1021/acs.jafc.8b05808.

Pinciroli, M., Dominguez-Perles, R., Abellan, A., Guy, A., Durand, T., Oger, C., et al. (2017). Comparative study of the phytoprostane and phytofuran content of indica and japonica rice (Oryza sativa L.) flours. *Journal of Agricultural and Food Chemistry, 65*(40), 8938–8947. https://doi.org/10.1021/acs.jafc.7b03482.

Pinciroli, M., Dominguez-Perles, R., Garbi, M., Abellan, A., Oger, C., Durand, T., et al. (2018). Impact of salicylic acid content and growing environment on Phytoprostane and Phytofuran (stress biomarkers) in Oryza sativa L. *Journal of Agricultural and Food Chemistry, 66*(47), 12561–12570. https://doi.org/10.1021/acs.jafc.8b04975.

Pino Ramos, L. L., Jimenez-Aspee, F., Theoduloz, C., Burgos-Edwards, A., Dominguez-Perles, R., Oger, C., et al. (2019). Phenolic, oxylipin and fatty acid profiles of the Chilean hazelnut (Gevuina avellana): Antioxidant activity and inhibition of pro-inflammatory and metabolic syndrome-associated enzymes. *Food Chemistry, 298*, 125026. https://doi.org/10.1016/j.foodchem.2019.125026.

Porta, A., Chiesa, F., Quaroni, M., Persico, M., Moratti, R., Zanoni, G., et al. (2014). A divergent enantioselective synthesis of 9-J_1-phytoprostane and 9-A_1-phytoprostane methyl ester. *European Journal of Organic Chemistry, 10*, 2111–2119. https://doi.org/10.1002/ejoc.201301703.

Rac, M., Shumbe, L., Oger, C., Guy, A., Vigor, C., Ksas, B., et al. (2021). Luminescence imaging of leaf damage induced by lipid peroxidation products and its modulation by beta-cyclocitral. *Physiologia Plantarum, 171*(2), 246–259. https://doi.org/10.1111/ppl.13279.

Ruesgas-Ramon, M., Figueroa-Espinoza, M. C., Durand, E., Suarez-Quiroz, M. L., Gonzalez-Rios, O., Rocher, A., et al. (2019). Identification and quantification of phytoprostanes and phytofurans of coffee and cocoa by- and co-products. *Food & Function, 10*(10), 6882–6891. https://doi.org/10.1039/c9fo01528k.

Schmidt, A., & Boland, W. (2007). General strategy for the synthesis of B1 phytoprostanes, dinor isoprostanes, and analogs. *The Journal of Organic Chemistry, 72*(5), 1699–1706. https://doi.org/10.1021/jo062359x.

Sinclair, A. J., Attar-Bashi, N. M., & Li, D. (2002). What is the role of alpha-linolenic acid for mammals? *Lipids, 37*(12), 1113–1123. https://doi.org/10.1007/s11745-002-1008-x.

Smrček, J., Hajek, M., Hodek, O., Cizek, K., Pohl, R., Jahn, E., et al. (2021). First Total synthesis of Phytoprostanes with prostaglandin-like configuration, evidence for their formation in edible vegetable oils and orienting study of their biological activity. *Chemistry, 27*(37), 9556–9562. https://doi.org/10.1002/chem.202100872.

Stotz, H. U., Mueller, S., Zoeller, M., Mueller, M. J., & Berger, S. (2013). TGA transcription factors and jasmonate-independent COI1 signalling regulate specific plant responses to reactive oxylipins. *Journal of Experimental Botany, 64*(4), 963–975. https://doi.org/10.1093/jxb/ers389.

Thoma, I., Loeffler, C., Sinha, A. K., Gupta, M., Krischke, M., Steffan, B., et al. (2003). Cyclopentenone isoprostanes induced by reactive oxygen species trigger defense gene activation and phytoalexin accumulation in plants. *The Plant Journal, 34*(3), 363–375. https://doi.org/10.1046/j.1365-313x.2003.01730.x.

Tkacz, K., Gil-Izquierdo, Á., Medina, S., Turkiewicz, I. P., Domínguez-Perles, R., Nowicka, P., et al. (2021). Phytoprostanes, phytofurans, tocopherols, tocotrienols, carotenoids and free amino acids and biological potential of sea buckthorn juices. *Journal of the Science of Food and Agriculture*. https://doi.org/10.1002/jsfa.11345.

Traidl-Hoffmann, C., Mariani, V., Hochrein, H., Karg, K., Wagner, H., Ring, J., et al. (2005). Pollen-associated phytoprostanes inhibit dendritic cell interleukin-12 production and augment T helper type 2 cell polarization. *The Journal of Experimental Medicine, 201*(4), 627–636. https://doi.org/10.1084/jem.20041065.

Vázquez-Romero, A., Verdaguer, X., & Riera, A. (2013). General approach to Prostanes B1 by intermolecular Pauson–Khand reaction: Syntheses of methyl esters of prostaglandin B1 and Phytoprostanes 16-B1-PhytoP and 9-L1-PhytoP. *European Journal of Organic Chemistry, 2013*(9), 1716–1725. https://doi.org/10.1002/ejoc.201201442.

Vigor, C., Oger, C., Reversat, G., Rocher, A., Zhou, B., Linares-Maurizi, A., et al. (2020). Isoprostanoid profiling of marine microalgae. *Biomolecules, 10*(7). https://doi.org/10.3390/biom10071073.

Vigor, C., Reversat, G., Rocher, A., Oger, C., Galano, J. M., Vercauteren, J., et al. (2018). Isoprostanoids quantitative profiling of marine red and brown macroalgae. *Food Chemistry, 268*, 452–462. https://doi.org/10.1016/j.foodchem.2018.06.111.

Yonny, M. E., Rodriguez Torresi, A., Cuyamendous, C., Reversat, G., Oger, C., Galano, J. M., et al. (2016). Thermal stress in melon plants: Phytoprostanes and phytofurans as oxidative stress biomarkers and the effect of antioxidant supplementation. *Journal of Agricultural and Food Chemistry, 64*(44), 8296–8304. https://doi.org/10.1021/acs.jafc.6b03011.

FSC
www.fsc.org
MIX
Papier | Fördert
gute Waldnutzung
FSC® C083411

Kolibri 360 GmbH
Ferdinand-Jühlke-Straße 7
99095 Erfurt, Deutschland
produktsicherheit@kolibri360.de